International
REVIEW OF
Neurobiology
Volume 82

International
REVIEW OF
Neurobiology
Volume 82

The Neuroinflammation in Neuronal Death and Repair

EDITED BY

GIACINTO BAGETTA
Department of Pharmacobiology
University of Calabria
Arcavacata di Rende, CS, Italy

M. TIZIANA CORASANITI
Department of Pharmacobiological Sciences
University "Magna Graecia" of Catanzaro
Catanzaro, Italy

STUART A. LIPTON
Center for Neuroscience and Aging
The Burnham Institute
La Jolla, California, USA

ELSEVIER

AMSTERDAM • BOSTON • HEIDELBERG • LONDON
NEW YORK • OXFORD • PARIS • SAN DIEGO
SAN FRANCISCO • SINGAPORE • SYDNEY • TOKYO
Academic Press is an imprint of Elsevier

Academic Press is an imprint of Elsevier
525 B Street, Suite 1900, San Diego, California 92101-4495, USA
84 Theobald's Road, London WC1X 8RR, UK

This book is printed on acid-free paper. ♾

For information on all Academic Press publications
visit our Web site at www.books.elsevier.com

ISBN-13: 978-0-12-373989-6
ISBN-10: 0-12-373989-6

Printed and bound in the United Kingdom

Transferred to Digital Printing 2011

CONTENTS

The Role of Astrocytes and Complement System in Neural Plasticity

MILOS PEKNY, ULRIKA WILHELMSSON, YALDA RAHPEYMAI BOGESTÅL, AND MARCELA PEKNA

New Insights into the Roles of Metalloproteinases in Neurodegeneration and Neuroprotection

A. J. TURNER AND N. N. NALIVAEVA

Relevance of High-Mobility Group Protein Box 1 to Neurodegeneration

SILVIA FOSSATI AND ALBERTO CHIARUGI

Early Upregulation of Matrix Metalloproteinases Following Reperfusion Triggers Neuroinflammatory Mediators in Brain Ischemia in Rat

DIANA AMANTEA, ROSSELLA RUSSO, MICAELA GLIOZZI, VINCENZA FRATTO, LAURA BERLIOCCHI, G. BAGETTA, G. BERNARDI, AND M. TIZIANA CORASANITI

The (Endo)Cannabinoid System in Multiple Sclerosis and Amyotrophic Lateral Sclerosis

DIEGO CENTONZE, SILVIA ROSSI, ALESSANDRO FINAZZI-AGRÒ, GIORGIO BERNARDI, AND MAURO MACCARRONE

Chemokines and Chemokine Receptors: Multipurpose Players in Neuroinflammation

RICHARD M. RANSOHOFF, LIPING LIU, AND ASTRID E. CARDONA

Systemic and Acquired Immune Responses in Alzheimer's Disease

MARKUS BRITSCHGI AND TONY WYSS-CORAY

Neuroinflammation in Alzheimer's Disease and Parkinson's Disease: Are Microglia Pathogenic in Either Disorder?

JOSEPH ROGERS, DIEGO MASTROENI, BRIAN LEONARD, JEFFREY JOYCE, AND ANDREW GROVER

Cytokines and Neuronal Ion Channels in Health and Disease

BARBARA VIVIANI, FABRIZIO GARDONI, AND MARINA MARINOVICH

Cyclooxygenase-2, Prostaglandin E_2, and Microglial Activation in Prion Diseases

LUISA MINGHETTI AND MAURIZIO POCCHIARI

Glia Proinflammatory Cytokine Upregulation as a Therapeutic Target for Neurodegenerative Diseases: Function-Based and Target-Based Discovery Approaches

LINDA J. VAN ELDIK, WENDY L. THOMPSON, HANTAMALALA RALAY RANAIVO, HEATHER A. BEHANNA, AND D. MARTIN WATTERSON

Oxidative Stress and the Pathogenesis of Neurodegenerative Disorders

ASHLEY REYNOLDS, CHAD LAURIE, R. LEE MOSLEY, AND HOWARD E. GENDELMAN

Differential Modulation of Type 1 and Type 2 Cannabinoid Receptors Along the Neuroimmune Axis

SERGIO ODDI, PAOLA SPAGNUOLO, MONICA BARI, ANTONELLA D'AGOSTINO, AND MAURO MACCARRONE

Effects of the HIV-1 Viral Protein TAT on Central Neurotransmission: Role of Group I Metabotropic Glutamate Receptors

ELISA NERI, VERONICA MUSANTE, AND ANNA PITTALUGA

Evidence to Implicate Early Modulation of Interleukin-1β Expression in the Neuroprotection Afforded by 17β-Estradiol in Male Rats Undergone Transient Middle Cerebral Artery Occlusion

OLGA CHIAPPETTA, MICAELA GLIOZZI, ELISA SIVIGLIA, DIANA AMANTEA, LUIGI A. MORRONE, LAURA BERLIOCCHI, G. BAGETTA, AND M. TIZIANA CORASANITI

A Role for Brain Cyclooxygenase-2 and Prostaglandin-E_2 in Migraine: Effects of Nitroglycerin

CRISTINA TASSORELLI, ROSARIA GRECO, MARIE THERÈSE ARMENTERO, FABIO BLANDINI, GIORGIO SANDRINI, AND GIUSEPPE NAPPI

The Blockade of K^+-ATP Channels Has Neuroprotective Effects in an *In Vitro* Model of Brain Ischemia

ROBERT NISTICÒ, SILVIA PICCIRILLI, L. SEBASTIANELLI, GIUSEPPE NISTICÒ, G. BERNARDI, AND N. B. MERCURI

Retinal Damage Caused by High Intraocular Pressure–Induced Transient Ischemia Is Prevented by Coenzyme Q10 in Rat

CARLO NUCCI, ROSANNA TARTAGLIONE, ANGELICA CERULLI, R. MANCINO, A. SPANÒ, FEDERICA CAVALIERE, LAURA ROMBOLÀ, G. BAGETTA, M. TIZIANA CORASANITI, AND LUIGI A. MORRONE

Evidence Implicating Matrix Metalloproteinases in the Mechanism Underlying Accumulation of IL-1β and Neuronal Apoptosis in the Neocortex of HIV/gp120-Exposed Rats

Rossella Russo, Elisa Siviglia, Micaela Gliozzi, Diana Amantea, Annamaria Paoletti, Laura Berliocchi, G. Bagetta, and M. Tiziana Corasaniti

Neuroprotective Effect of Nitroglycerin in a Rodent Model of Ischemic Stroke: Evaluation of Bcl-2 Expression

Rosaria Greco, Diana Amantea, Fabio Blandini, Giuseppe Nappi, Giacinto Bagetta, M. Tiziana Corasaniti, and Cristina Tassorelli

CONTRIBUTORS

Numbers in parentheses indicate the pages on which the authors' contributions begin.

Antonella D'Agostino (327), Department of Biomedical Sciences, University of Teramo, Teramo 64100, Italy

Diana Amantea (149, 357, 407, 423), Department of Pharmacobiology, University of Calabria, Via P. Bucci, 87036 Arcavacata di Rende (Cosenza), Italy; University Centre for Adaptive Disorders and Headache, Section of Neuropharmacology for Normal and Pathological Neuronal Plasticity, University of Calabria, Via P. Bucci, 87036 Arcavacata di Rende (Cosenza), Italy

Marie Therèse Armentero (373), IRCCS "C. Mondino Institute of Neurology" Foundation, Pavia, Italy

Giacinto Bagetta (149, 357, 397, 407, 423), Department of Pharmacobiology, University of Calabria, Via P. Bucci, 87036 Arcavacata di Rende (CS), Italy; University Center for Adaptive Disorders and Headache (UCHAD), Center of Neuropharmacology for Normal and Pathological Neuronal Plasticity, University of Calabria, 87036 Arcavacata di Rende, Italy

Monica Bari (327), Department of Experimental Medicine and Biochemical Sciences, University of Rome Tor Vergata, Rome 00133, Italy

Heather A. Behanna (279), Center for Drug Discovery and Chemical Biology, Northwestern University, Chicago, Illinois 60611, USA

Laura Berliocchi (149, 357, 407), Center for Experimental Neuropharmacology, Mondino-Tor Vergata, University of Rome Tor Vergata, 00133 Rome, Italy

Giorgio Bernardi (149, 171, 383), Neurological Clinics, Department of Neurosciences, University of Rome Tor Vergata, Rome, Italy; European Center for Brain Research, (CERC)/IRCCS S. Lucia Foundation, Rome, Italy; CERC-Fondazione S. Lucia IRCCS, University of Rome Tor Vergata, 00133 Rome, Italy; Department of Experimental Neurology, S. Lucia Foundation IRCCS, Rome, Italy; Clinica Neurologica, University of Rome 'Tor Vergata,' Rome, Italy

Fabio Blandini (373, 423), IRCCS "C. Mondino Institute of Neurology" Foundation, 27100 Pavia, Italy

Yalda Rahpeymai Bogestål (95), Department of Medical Chemistry and Cell Biology, Institute of Biomedicine at Sahlgrenska Academy, Göteborg University, 405 30 Göteborg, Sweden

Giambattista Bonanno (73), Department of Experimental Medicine, Section of Pharmacology and Toxicology and Center of Excellence for Biomedical Research, University of Genoa, Italy

Liliana Brambilla (57), Department of Pharmacological Sciences, Center of Excellence on Neurodegenerative Diseases, University of Milan, Via Balzaretti 9, 20133 Milan, Italy

Markus Britschgi (205), Department of Neurology and Neurological Sciences, Stanford University School of Medicine, Stanford, California 94305, USA

Astrid E. Cardona (187), Neuroinflammation Research Center, Department of Neurosciences, Lerner Research Institute, Cleveland Clinic, Cleveland, Ohio 44195, USA

Federica Cavaliere (397), Department of Pharmacobiology, University of Calabria, 87036 Arcavacata di Rende, Italy

Diego Centonze (171), Neurological Clinics, Department of Neurosciences, University of Rome Tor Vergata, Rome, Italy; European Center for Brain Research, (CERC)/IRCCS S. Lucia Foundation, Rome, Italy

Angelica Cerulli (397), Physiopathological Optics, Department of Biopathology, University of Rome "Tor Vergata," 00133 Rome, Italy

Olga Chiappetta (357), Department of Pharmacobiology, University of Calabria, Via P. Bucci, 87036 Arcavacata di Rende (CS), Italy

Alberto Chiarugi (137), Department of Preclinical and Clinical Pharmacology, University of Florence, Florence, Italy

M. Tiziana Corasaniti (149, 357, 397, 407, 423), Department of Pharmacobiological Sciences, University Magna Graecia of Catanzaro, Italy; Center for Experimental Neuropharmacology, "Mondino-Tor Vergata," University of Rome Tor Vergata, Rome, Italy

Alessandro Finazzi-Agrò (171), Department of Experimental Medicine and Biochemical Sciences, University of Rome Tor Vergata, Rome, Italy

Silvia Fossati (137), Department of Preclinical and Clinical Pharmacology, University of Florence, Florence, Italy

Vincenza Fratto (149), Department of Pharmacobiological Sciences, Faculty of Pharmacy, University "Magna Graecia" of Catanzaro, 88100 Catanzaro, Italy

Fabrizio Gardoni (247), Department of Pharmacological Sciences, University of Milan, Italy

P. Gasque (29), Brain Inflammation and Immunity Group (BIIG), Department of Medical Biochemistry and Department of Histopathology, School of Medicine, Cardiff University, CF144XN Cardiff, United Kingdom; Groupe de recherche sur les maladies infectieuses et inflammatoires, GRII, LBGM, Faculty of Sciences and Technology, University of la Reunion, Saint Denis, Ile de la Reunion

Howard E. Gendelman (297), Department of Pharmacology and Experimental Neuroscience and Center for Neurovirology and Neurodegenerative Disorders, University of Nebraska Medical Center, Omaha, Nebraska 68198, USA

Micaela Gliozzi (149, 357, 407), Department of Pharmacobiological Sciences, Faculty of Pharmacy, University "Magna Graecia" of Catanzaro, 88100 Catanzaro, Italy

Rosaria Greco (373, 423), IRCCS "C. Mondino Institute of Neurology" Foundation, 27100 Pavia, Italy

M. Griffiths (29), Brain Inflammation and Immunity Group (BIIG), Department of Medical Biochemistry, School of Medicine, Cardiff University, CF144XN Cardiff, United Kingdom

Andrew Grover (235), The L. J. Roberts Center for Alzheimer's Research, Sun Health Research Institute, Sun City, Arizona 85351, USA; and The Christopher Center for Parkinson's Research, Sun Health Research Institute, Sun City, Arizona 85351, USA

Zezong Gu (1), Neuroscience and Aging Center, Burnham Institute for Medical Research, La Jolla, California 92037, USA

Jeffrey Joyce (235), The L. J. Roberts Center for Alzheimer's Research, Sun Health Research Institute, Sun City, Arizona 85351, USA; and The Christopher Center for Parkinson's Research, Sun Health Research Institute, Sun City, Arizona 85351, USA

Chad Laurie (297), Department of Pharmacology and Experimental Neuroscience and Center for Neurovirology and Neurodegenerative Disorders, University of Nebraska Medical Center, Omaha, Nebraska 68198, USA

Brian Leonard (235), The L. J. Roberts Center for Alzheimer's Research, Sun Health Research Institute Sun City, Arizona 85351, USA; and The Christopher Center for Parkinson's Research, Sun Health Research Institute Sun City, Arizona 85351, USA

Stuart A. Lipton (1), Neuroscience and Aging Center, Burnham Institute for Medical Research, La Jolla, California 92037, USA; Department of Neurosciences, University of California at San Diego, La Jolla, California 92039, USA; The Salk Institute for Biological Studies, La Jolla, California 92037, USA; The Scripps Research Institute, La Jolla, California 92037, USA

LiPing Liu (187), Neuroinflammation Research Center, Department of Neurosciences, Lerner Research Institute, Cleveland Clinic, Cleveland, Ohio 44195, USA

Mauro Maccarrone (171, 327), European Center for Brain Research, (CERC)/IRCCS S. Lucia Foundation, Rome 00143, Italy; Department of Biomedical Sciences, University of Teramo, Teramo 64100, Italy

R. Mancino (397), Physiopathological Optics, Department of Biopathology, University of Rome "Tor Vergata", 00133 Rome, Italy

Marina Marinovich (247), Laboratory of Toxicology, Department of Pharmacological Sciences, University of Milan, Italy

Diego Mastroeni (235), The L. J. Roberts Center for Alzheimer's Research, Sun Health Research Institute, Sun City, Arizona 85351, USA; and The Christopher Center for Parkinson's Research, Sun Health Research Institute, Sun City, Arizona 85351, USA

Edon Melloni (73), Department of Experimental Medicine, Center of Excellence for Biomedical Research and Section of Biochemistry, University of Genoa, Italy

N. B. Mercuri (383), Department of Experimental Neurology, S. Lucia Foundation IRCCS, Rome, Italy; Clinica Neurologica, University of Rome 'Tor Vergata', Rome, Italy

Marco Milanese (73), Department of Experimental Medicine, Section of Pharmacology and Toxicology, University of Genoa, Italy

Luisa Minghetti (265), Department of Cell Biology and Neurosciences, Degenerative and Inflammatory Neurological Diseases Unit, Istituto Superiore di Sanità, Viale Regina Elena 299, 00161 Rome, Italy

Luigi A. Morrone (357, 397), Department of Pharmacobiology, University of Calabria, Via P. Bucci, 87036 Arcavacata di Rende (CS), Italy; UCHAD University Center for Adaptive Disorders and Headache (UCHAD), Centre of Neuropharmacology for Normal and Pathological Neuronal Plasticity, University of Calabria, 87036 Arcavacata di Rende, Italy

R. Lee Mosley (297), Department of Pharmacology and Experimental Neuroscience and Center for Neurovirology and Neurodegenerative Disorders, University of Nebraska Medical Center, Omaha, Nebraska 68198, USA

Veronica Musante (339), Department of Experimental Medicine, Pharmacology and Toxicology Section, University of Genova, Genova, Italy

Tomohiro Nakamura (1), Neuroscience and Aging Center, Burnham Institute for Medical Research, La Jolla, California 92037, USA

N. N. Nalivaeva (113), Institute of Molecular and Cellular Biology, Faculty of Biological Sciences, University of Leeds, Leeds LS2 9JT, United Kingdom; I. M. Sechenov Institute of Evolutionary Physiology and Biochemistry, Russian Academy of Sciences, 194223 St. Petersburg, Russia

Giuseppe Nappi (373, 423), IRCCS "C. Mondino Institute of Neurology" Foundation, Pavia, Italy; University Centre for the Study of Adaptive Disorder and Headache (UCADH), Pavia, Italy; Department of Neurology and Otorhinolaryngology, University of Rome "La Sapienza," Rome, Italy

J. W. Neal (29), Department of Histopathology, School of Medicine, Cardiff University, CF144XN Cardiff, United Kingdom

Elisa Neri (339), Department of Experimental Medicine, Pharmacology and Toxicology Section, University of Genova, Genova, Italy

Giuseppe Nisticò (383), Centre of Pharmaceutical Biotechnology, University of Rome 'Tor Vergata,' Rome, Italy

Robert Nisticò (383), Department of Experimental Neurology, S. Lucia Foundation IRCCS, Rome, Italy; Department of Pharmacobiology and University Centre for Adaptive Disorders and Headache (UCHAD), Section of Neuropharmacology of Normal and Pathological Neuronal Plasticity, University of Calabria, Arcavacata di Rende, Italy

Carlo Nucci (397), Physiopathological Optics, Department of Biopathology, University of Rome "Tor Vergata" 00133 Rome, Italy; IRCCS (Istituti di Ricovero e Cura a Carattere Scientifico) Neurological Institute, C. Mondino Foundation, "Mondino-Tor Vergata", Center for Experimental Neuropharmacology, Laboratori of Neurochemistry, 00133 Rome, Italy

Sergio Oddi (327), Department of Biomedical Sciences, University of Teramo, Teramo 64100, Italy; European Center for Brain Research (CERC)/IRCCS S. Lucia Foundation, Rome 00143, Italy

Annamaria Paoletti (407), CNR Institute of Neurological Science, Section of Pharmacology, Roccelletta di Borgia, 88100 Catanzaro, Italy

Mario Passalacqua (73), Department of Experimental Medicine, Center of Excellence for Biomedical Research and Section of Biochemistry, University of Genoa, Italy

Marco Pedrazzi (73), Department of Experimental Medicine, Section of Biochemistry, University of Genoa, Italy

Marcela Pekna (95), Department of Medical Chemistry and Cell Biology, Institute of Biomedicine at Sahlgrenska Academy, Göteborg University, 405 30 Göteborg, Sweden

Milos Pekny (95), Center for Brain Repair and Rehabilitation, Department of Clinical Neuroscience and Rehabilitation, Institute for Neuroscience and Physiology at Sahlgrenska Academy, Göteborg University, 405 30 Göteborg, Sweden

Silvia Piccirilli (383), Department of Experimental Neurology, S. Lucia Foundation IRCCS, Rome, Italy

Anna Pittaluga (339), Department of Experimental Medicine, Pharmacology and Toxicology Section, University of Genova, Genova, Italy

Maurizio Pocchiari (265), Department of Cell Biology and Neurosciences, Degenerative and Inflammatory Neurological Diseases Unit, Istituto Superiore di Sanità, Viale Regina Elena 299, 00161 Rome, Italy

Luca Raiteri (73), Department of Experimental Medicine, Section of Pharmacology and Toxicology, University of Genoa, Italy

Hantamalala Ralay Ranaivo (277), Center for Drug Discovery and Chemical Biology, Northwestern University, Chicago, Illinois 60611, USA

Richard M. Ransohoff (187), Neuroinflammation Research Center, Department of Neurosciences, Lerner Research Institute, Cleveland Clinic, Cleveland, Ohio 44195, USA; Mellen Center for MS Treatment and Research, Cleveland Clinic, Cleveland Ohio 44195, USA

Ashley Reynolds (297), Department of Pharmacology and Experimental Neuroscience, University of Nebraska Medical Center, Omaha, Nebraska 68198, USA; and Center for Neurovirology and Neurodegenerative Disorders, University of Nebraska Medical Center, Omaha, Nebraska 68198, USA

Joseph Rogers (235), The L. J. Roberts Center for Alzheimer's Research, Sun Health Research Institute, Sun City, Arizona 85351, USA; and The Christopher Center for Parkinson's Research, Sun Health Research Institute, Sun City, Arizona 85351, USA

Laura Rombolà (397), Department of Pharmacobiology, University of Calabria, Via P. Bucci, 87036 Arcavacata di Rende (CS), Italy; 87036 Arcavacata di Rende (CS), Italy; UCHAD University Center for Adaptive Disorders and Headache (UCHAD), Center of Neuropharmacology for Normal and Pathological Neuronal Plasticity, University of Calabria, 87036 Arcavacata di Rende, Italy

Daniela Rossi (57), Department of Pharmacological Sciences, Center of Excellence on Neurodegenerative Diseases, University of Milan, Via Balzaretti 9, 20133 Milan, Italy

Silvia Rossi (171), Neurological Clinics, Department of Neurosciences, University of Rome Tor Vergata, Rome, Italy; European Center for Brain Research, (CERC)/IRCCS S. Lucia Foundation, Rome, Italy

Rossella Russo (149, 407), Department of Pharmacobiology, University of Calabria, Via P. Bucci, 87036 Arcavacata di Rende (CS), Italy; University Centre for the Study of Adaptive Disorder and Headache Center of Neuropharmacology for Normal and Pathological Neuronal Plasticity, University of Calabria, Via P. Bucci, 87036 Arcavacata di Rende (CS), Italy; Burnham Institute for Medical Research, Del E. Webb Center for Neurosciences and Aging, La Jolla, California 92037, USA

Giorgio Sandrini (373), IRCCS "C. Mondino Institute of Neurology" Foundation, Pavia, Italy; University Centre for the Study of Adaptive Disorder and Headache (UCADH), Pavia, Italy

L. Sebastianelli (383), Department of Experimental Neurology, S. Lucia Foundation IRCCS, Rome, Italy

Elisa Siviglia (357, 407), Department of Pharmacobiological Sciences, University "Magna Graecia" of Catanzaro, 88100 Catanzaro, Italy

Paola Spagnuolo (327), Department of Pharmacobiology and University Centre for the Study of Adaptive Disorder and Headache (UCADH), Section of Neuropharmacology of Normal and Pathological Neuronal Plasticity, University of Calabria, Rende (CS) 87036, Italy; Department of Experimental Medicine and Biochemical Sciences, University of Rome Tor Vergata, Rome 00133, Italy

A. Spanò (397), Physiopathological Optics, Department of Biopathology, University of Rome "Tor Vergata", 00133 Rome, Italy

Bianca Sparatore (73), Department of Experimental Medicine, Center of Excellence for Biomedical Research and Section of Biochemistry, University of Genoa, Italy

Carlo Tacchetti (73), Department of Experimental Medicine, Section of Human Anatomy, University of Genoa, Italy; FIRC Institute of Molecular Oncology (IFOM), Milan, Italy

Rosanna Tartaglione (397), Department of Pharmacobiology, University of Calabria, 87036 Arcavacata di Rende, Italy

Cristina Tassorelli (373, 423), IRCCS "C. Mondino Institute of Neurology" Foundation, Pavia, Italy; University Centre for the Study of Adaptive Disorder and Headache (UCADH), Pavia, Italy

Wendy L. Thompson (277), Center for Drug Discovery and Chemical Biology, Northwestern University, Chicago, Illinois 60611, USA

A. J. Turner (113), Institute of Molecular and Cellular Biology, Faculty of Biological Sciences, University of Leeds, Leeds LS2 9JT, United Kingdom

Cesare Usai (73), Institute of Biophysics, National Research Council, Genoa, Italy

Linda J. Van Eldik (277), Center for Drug Discovery and Chemical Biology, Northwestern University, Chicago, Illinois 60611, USA

Sabino Vesce (57), Department of Cell Biology and Morphology, University of Lausanne, Rue du Bugnon 9, 1005 Lausanne, Switzerland

Barbara Viviani (247), Laboratory of Toxicology, Department of Pharmacological Sciences, University of Milan, Italy

Andrea Volterra (57), Department of Cell Biology and Morphology, University of Lausanne, Rue du Bugnon 9, 1005 Lausanne, Switzerland; Department of Pharmacological Sciences, Center of Excellence on Neurodegenerative Diseases, University of Milan, Via Balzaretti 9, 20133 Milan, Italy

D. Martin Watterson (277), Center for Drug Discovery and Chemical Biology, Northwestern University, Chicago, Illinois 60611, USA

Ulrika Wilhelmsson (95), Center for Brain Repair and Rehabilitation, Department of Clinical Neuroscience and Rehabilitation, Institute for Neuroscience and

Physiology at Sahlgrenska Academy, Göteborg University, 405 30 Göteborg, Sweden

Tony Wyss-Coray (205), Department of Neurology and Neurological Sciences, Stanford University School of Medicine, Stanford, California 94305, USA; Veterans Affairs Palo Alto Health Care System, Palo Alto, California 94304, USA

Simona Zappettini (73), Department of Experimental Medicine, Section of Pharmacology and Toxicology, University of Genoa, Italy

PREFACE

This issue of *International Review on Neurobiology* features peer reviewed papers from the IX Workshop on Apoptosis in Biology and Medicine on *Neuroinflammation in Neuronal Death and Repair* held in Calabria at the Porto Pirgos Hotel in Parghelia (Vibo Valentia, Italy) from September 13 to September 16, 2006. The meeting has been organized in the frame of the Ph.D. course on Pharmacology and Biochemistry of Cell Death established by the Ministry for University Research since the XIV cycle at the University of Calabria in Consortium with the University Magna Graecia of Catanzaro and of Rome Tor Vergata.

The scope of this workshop was to bring together scientists whose research work has an international reputation to discuss current knowledge and views on the mechanisms of neuroinflammation, on its role in the context of acute and chronic neurodegenerative diseases, and on the potential approaches for future therapeutic strategies. This provided the students with the opportunity to meet with high-caliber scientists of worldwide reputation and to discuss with them their research projects. The articles published in this issue have been edited to reflect the topics discussed at the meeting by the invited speakers and by selected young scientists, including Ph.D. students, and these range from basic immune mechanisms and mediators, the interaction between the immune and nervous systems under normal and pathological conditions, to the development of novel therapeutic approaches to reduce neuronal death and to support neuronal repair.

The organization of the IX Workshop has been made possible by the highly professional and intelligent contribution of Mrs. Hilary Rowe and Mr. Nicola Fico to whom we address our most sincere thanks.

The latter event has been organized with the important financial support of the National Institute of Health (ISS, Rome). The financial support of the University Magna Graecia at Catanzaro (Calabria, Italy), the National Council for Research (CNR) Institute of Neurological Sciences, Section of Catanzaro, the Department of Pharmacobiology, University of Calabria (Calabria, Italy) and Elsevier (United States) is gratefully acknowledged. The participation of Ph.D. students and postdocs has been made possible by the generous financial support of the International Society for Neurochemistry (ISN) and the Italian Society of Pharmacology.

The editors would like to thank Prof. Luigi A. Morrone, Dr. Laura Rombolá, and Dr. Alessandra Levato for their precious contribution during the preparation of this volume.

Giacinto Bagetta
M. Tiziana Corasaniti
Stuart A. Lipton

FOREWORD

Over the last decade, increasing evidences have been accumulated in favor of an important role of neuroinflammation in the pathophysiology of neurodegenerative diseases.

Until recently, the central nervous system (CNS) was considered a relatively immunologically quiescent organ, thus determining the slow progress of our knowledge in this area of neuroscience research. The traditional view of glial cells as supportive partners of neurons has now undergone considerable reevaluation. Several studies, including those related to Alzheimer's disease and Parkinson's disease, have demonstrated that microglial cells can be activated by a wide range of stimuli, including infections of peripheral origin. It is also well established that astrocytes possess functional receptors for neurotransmitters and other signaling molecules and respond to their stimulation via release of gliotransmitters such as glutamate, the main excitatory neurotransmitter in the mammalian brain. Indeed, astrocytes react to synaptically released neurotransmitters via regulated exocytosis, and possibly other mechanisms, leading to a new concept of neuron-glia intercommunication where astrocytes play an unsuspected dynamic role by integrating neuronal inputs and modulating synaptic activity. On the other hand, glia-to-neuron signaling has been proved to be sensitive to the control operated by inflammatory mediators such as cytokines and autacoids.

When microglial cells and astrocytes are activated by pathological accumulation of these mediators, the transition to a reactive state of these cells is accompanied by the disturbance of the cross talk normally occurring with neurons, contributing to the disease development.

In the mammalian CNS, chemokines support developmental and neurophysiological functions and regulate the activation of glial cells. Therefore, chemokines and their receptors are among the most attractive therapeutic targets for the regulation of inflammation and immunity.

The fast progressing knowledge in the cross talk between peripheral and central immune cells and mediators is opening new therapeutical strategies to slow the progression of Parkinson's disease, based on active immunization. In fact, in animal model of neurodegeneration in the substantia nigra, immunization has been proven to induce T cells to enter inflamed nigrostriatal tissue, attenuate innate glial immunity, and increase local neurotrophic factor production, with a resulting neuroprotective effect on nigrostriatal neurons.

Inflammatory mediators such as nitric oxide (NO) and reactive oxygen species (ROS) have a well established role in neurodegenerative disorders.

Accumulation of radical species in the brain is, at least in part, due to an overactivation of the *N*-methyl-D-aspartate (NMDA) subtype of glutamate receptors, which produces excessive Ca^{2+} influx through the receptor-associated ion channels and the consequent free-radical formation. Memantine represents a possibility of therapeutic intervention by blocking the abnormal activation of NMDA receptors.

A recently discovered mechanism contributing to the damage and loss of neurons occurring in Alzheimer's disease and Parkinson's disease is *S*-nitrosylation. In fact, *S*-nitrosylation causes posttranslational modifications which can contribute to protein misfolding and thus to the progression of neurodegenerative diseases. On the contrary, *S*-nitrosylation of the NMDA receptors decreases its pathological activity and results in neuroprotection.

The expected lifetime's increment will likely lead in the near future to the raise of the incidence of neurodegenerative diseases in western countries. The *Neuroinflammation in Neuronal Death and Repair* book provides researchers and students with a critical view of the current knowledge in the field and stimulates future research.

INFLAMMATORY MEDIATORS LEADING TO PROTEIN MISFOLDING AND UNCOMPETITIVE/FAST OFF-RATE DRUG THERAPY FOR NEURODEGENERATIVE DISORDERS

Stuart A. Lipton,[*,†,‡,§] Zezong Gu,[*] and Tomohiro Nakamura[*]

[*]Neuroscience and Aging Center, Burnham Institute for Medical Research
La Jolla, California 92037, USA
[†]Department of Neurosciences, University of California at San Diego
La Jolla, California 92039, USA
[‡]The Salk Institute for Biological Studies, La Jolla, California 92037, USA
[§]The Scripps Research Institute, La Jolla, California 92037, USA

Inflammatory mediators, including free radicals such as nitric oxide (NO) and reactive oxygen species (ROS), can contribute to neurodegenerative diseases in part by triggering protein misfolding. In this chapter, we will discuss a newly discovered pathway for this phenomenon and possible novel treatments. Excitotoxicity, defined as overstimulation of glutamate receptors, has been implicated in a final common pathway contributing to neuronal injury and death in a wide range of acute and chronic neurological disorders, ranging from Parkinson's disease (PD), amyotrophic lateral sclerosis (ALS), multiple sclerosis, and Alzheimer's disease (AD) to stroke and trauma. Excitotoxic cell death is due, at least in part, to excessive activation of *N*-methyl-D-aspartate (NMDA)-type glutamate receptors, leading to excessive Ca^{2+} influx through the receptor's associated ion channel

INTERNATIONAL REVIEW OF
NEUROBIOLOGY, VOL. 82
DOI: 10.1016/S0074-7742(07)82001-0

0074-7742/07 $35.00

and subsequent free radical production, including NO and ROS. These free radicals can trigger a variety of injurious pathways, but newly discovered evidence suggests that some proteins are *S*-nitrosylated (transfer of NO to a critical thiol group), and this reaction can mimic the effect of rare genetic mutations. This posttranslational modification can contribute to protein misfolding, triggering neurodegenerative diseases. One such molecule affected is protein disulfide isomerase (PDI), an enzyme responsible for normal protein folding in the endoplasmic greticulum (ER). We found that when PDI is *S*-nitrosylation (forming SNO-PDI), the function of the enzyme is compromised, leading to misfolded proteins and contributing to neuronal cell injury and loss. Moreover, SNO-PDI occurs at pathological levels in several human diseases, including AD and PD. This discovery thus links protein misfolding to excitotoxicity and free radical formation in a number of neurodegenerative disorders. Another molecule whose S-nitrosylation can lead to abnormal protein accumulation is the E3 ubiquitin ligase, parkin, which contributes to the pathogenesis of PD. One way to ameliorate excessive NO production and hence abnormal *S*-nitrosylations would be to inhibit NMDA receptors. In fact, blockade of excessive NMDA receptor activity can in large measure protect neurons from this type of injury and death. However, inhibition of the NMDA receptor by high-affinity antagonists also blocks the receptor's normal function in synaptic transmission and leads to unacceptable side effects. For this reason, many NMDA receptor antagonists have disappointingly failed in advanced clinical trials. Our group was the first to demonstrate that gentle blockade of NMDA receptors by memantine, via a mechanism of uncompetitive open-channel block with a rapid "off-rate," can prevent this type of damage in a clinically efficacious manner without substantial side effects. For these *U*ncompetitive/*F*ast *O*ff-rate therapeutics, we use the term "UFO drugs" because like *U*nidentified *F*lying *O*bjects, they leave very quickly as soon as their job is finished. As a result, memantine blocks excessive NMDA receptor activity without disrupting normal activity. Memantine does this by preferentially entering the receptor-associated ion channel when it is excessively open, and, most importantly, when its off-rate from the channel is relatively fast so that it does not accumulate to interfere with normal synaptic transmission. Hence, memantine is clinically well tolerated, has been used in Europe for PD for many years, and recently passed multiple phase III trials for dementia, leading to its approval by the FDA and European Union for moderate-to-severe AD. Clinical studies of memantine for additional neurological disorders, including other dementias, neuropathic pain, and glaucoma, are underway. We have also developed a series of second-generation drugs that display greater neuroprotective properties than memantine. These second-generation drugs take advantage of the fact that the NMDA receptor has other modulatory sites, including critical thiol groups that are *S*-nitrosylated. In this case, in contrast to PDI or parkin, *S*-nitrosylation proves to be neuroprotective by decreasing excessive NMDA receptor activity. Targeted *S*-nitrosylation of the NMDA receptor can be achieved by coupling NO to memantine, yielding second-generation "UFO drugs" known as NitroMemantines.

I. Introduction

Excessive generation of reactive nitrogen species (RNS) and reactive oxygen species (ROS), which lead to neuronal cell injury and death, is a potential mediator of neurodegenerative disorders including: Parkinson's disease (PD), Alzheimer's disease (AD), amyotrophic lateral sclerosis (ALS), polyglutamine (polyQ) diseases such as Huntington's disease, glaucoma, *Human immunodeficiency virus*-associated dementia (HAD), multiple sclerosis, and ischemic brain injury, to name but a few (Barnham *et al.*, 2004; Beal, 2001; Emerit *et al.*, 2004; Lin and Beal, 2006; Muchowski and Wacker, 2005). While many intra- and extracellular molecules may participate in neuronal injury, accumulation of nitrosative stress due to excessive generation of nitric oxide (NO) appears to be a potential factor contributing to neuronal cell damage and death (Lipton, 2006a; Lipton and Rosenberg, 1994). A well-established model for NO production entails a central role of the *N*-methyl-D-aspartate (NMDA)-type glutamate receptors in nervous system. Excessive activation of NMDA receptors drives Ca^{2+} influx, which in turn activates neuronal NO synthase (nNOS) as well as the generation of ROS (Bredt *et al.*, 1991; Garthwaite *et al.*, 1988; Fig. 1). Importantly, normal mitochondrial respiration also generates free radicals, principally ROS, and one such molecule, superoxide anion (O_2^-), reacts rapidly with free radical NO^- to form the very toxic product peroxynitrite ($ONOO^-$) (Beckman *et al.*, 1990; Lipton *et al.*, 1993).

An additional feature of most neurodegenerative diseases is accumulation of misfolded and/or aggregated proteins (Bence *et al.*, 2001; Chaudhuri and Paul, 2006; Ciechanover and Brundin, 2003; Muchowski and Wacker, 2005). These protein aggregates can be cytosolic, nuclear, or extracellular. Importantly, protein aggregation can result from either (1) a mutation in the disease-related gene encoding the protein or (2) posttranslational changes to the protein engendered by nitrosative/oxidative stress (Zhang and Kaufman, 2006). A key theme of this chapter, therefore, is the hypothesis that nitrosative or oxidative stress contributes to protein misfolding in the brains of neurodegenerative

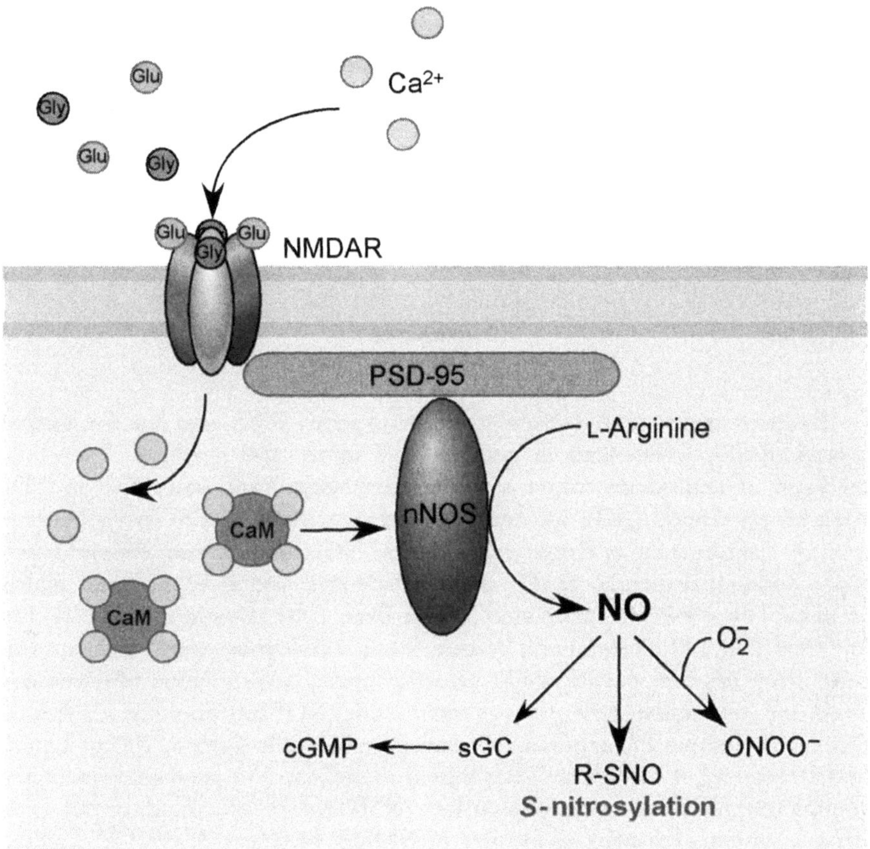

FIG. 1. Activation of the NMDA receptor by glutamate (Glu) and glycine (Gly) induces Ca^{2+} influx and consequent NO production via activation of nNOS. nNOS localizes to a complex attached to the NR1 subunit of the NMDA receptor via PDZ-domain binding to postsynaptic density protein (PSD-95). Subsequent effects of NO are mediated by chemical, enzymatic, and redox reactions within neurons. Specific interaction of NO with soluble guanylate cyclase (sGC) results in the production of cGMP, and cGMP could activate cGMP-dependent protein kinase to mediate the NO signaling. Excessive NMDA receptor activity, leading to the overproduction of NO can be neurotoxic. For example, *S*-nitrosylation or proteins such as PDI and parkin can contribute to neuronal cell damage and death. Neurotoxic effects of NO are also mediated by peroxynitrite ($ONOO^-$), a reaction product of NO and superoxide anion (O_2^-). Adapted from Nakamura *et al.* (2007), *Aging Cell*, Lipton laboratory Web site at www.burnham.org. (See Color Insert.)

patients. In this chapter, we discuss specific examples showing that *S*-nitrosylation of (1) ubiquitin E3 ligases such as parkin or (2) endoplasmic reticulum (ER) chaperones such as protein disulfide isomerase (PDI) is critical for the accumulation of misfolded proteins in neurodegenerative diseases such as PD, AD, and other conditions (Chung *et al.*, 2004; Lipton *et al.*, 2005; Uehara *et al.*, 2006; Yao *et al.*, 2004).

II. Protein Misfolding in Neurodegenerative Diseases

A feature in common to many neurodegenerative diseases is the accumulation of misfolded proteins that adversely affect neuronal connectivity and plasticity, and trigger cell death signaling pathways (Bence *et al.*, 2001; Muchowski and Wacker, 2005). For example, degenerating brain contains aberrant accumulations of misfolded, aggregated proteins such as α-synuclein and synphilin-1 in PD, and amyloid-β (Aβ) and tau in AD. The inclusions observed in PD are called Lewy bodies and are mostly found in the cytoplasm. AD brains show intracellular neurofibrillary tangles, which contain tau, and extracellular plaques, which contain Aβ. Other diseases with inclusions include Huntington's (polyQ), ALS, and prion disease (Ciechanover and Brundin, 2003). The above-mentioned aggregates consist of oligomeric complexes of nonnative secondary structures and demonstrate poor solubility in aqueous or detergent solvent. It has been suggested that either genetic mutations or an increase in nitrosative/oxidative stress can facilitate protein aggregation.

In general, protein aggregates do not accumulate in unstressed, healthy neurons due in part to the existence of cellular "quality control machineries." For example, molecular chaperones are believed to provide a defense mechanism against the toxicity of misfolded proteins because chaperones can prevent inappropriate interactions within and between polypeptides, and can promote refolding of proteins that have been misfolded because of cell stress. In addition to the quality control of proteins provided by molecular chaperones, the ubiquitin–proteasome system (UPS) is involved in the clearance of abnormal or aberrant proteins. When chaperones cannot repair misfolded proteins, they may be tagged via addition of polyubiquitin chains for degradation by the proteasome. In neurodegenerative conditions, intra- or extracellular protein aggregates are thought to accumulate in the brain as a result of a decrease in molecular chaperone or proteasome activities. In fact, several mutations that disturb the activity of molecular chaperones or UPS-associated enzymes can cause neurodegeneration (Cookson, 2005; Muchowski and Wacker, 2005; Zhao *et al.*, 2005).

Historically, lesions that contain aggregated proteins were considered to be pathogenic. Several lines of evidence have suggested that aggregates are formed through a complex multistep process by which misfolded proteins assemble into inclusion bodies; soluble oligomers of these aberrant proteins are thought to be the most toxic forms via interference with normal cell activities, while frank aggregates may be an attempt by the cell to wall off potentially toxic material (Arrasate *et al.*, 2004; Bredt *et al.*, 1991).

III. Generation of RNS/ROS

Glutamate is the major excitatory neurotransmitter in the brain and is important for normal functioning of the nervous system; however, excessive activation of glutamate receptors is implicated in neuronal damage in many neurological disorders ranging from acute hypoxic-ischemic brain injury to chronic neurodegenerative diseases. John Olney coined the term "excitotoxicity" to describe this phenomenon (Olney, 1969; Olney *et al.*, 1997). This form of toxicity is mediated at least in part by excessive activation of NMDA-type receptors (Chen and Lipton, 2006; Lipton, 2006a,b; Lipton and Rosenberg, 1994), resulting in excessive Ca^{2+} influx through a receptor's associated ion channel. Excessive Ca^{2+} leads to the production of damaging free radicals (e.g., NO and ROS) and other enzymatic processes, contributing to cell death (Bonfoco *et al.*, 1995; Budd *et al.*, 2000; Dawson *et al.*, 1991; Lafon-Cazal *et al.*, 1993; Lipton and Rosenberg, 1994; Lipton *et al.*, 1993). Intracellular Ca^{2+} triggers the generation of NO by activating nNOS in a Ca^{2+}/calmodulin (CaM)-dependent manner (Bredt *et al.*, 1991; Garthwaite *et al.*, 1988). It is thought that overstimulation of extrasynaptic NMDA receptors mediates this neuronal damage, while, in contrast, synaptic activity may activate survival pathways (Hardingham *et al.*, 2002; Papadia *et al.*, 2005). Intense hyperstimulation of excitatory receptors leads to necrotic cell death, but more mild or chronic overstimulation can result in apoptotic or other forms of cell death (Ankarcrona *et al.*, 1995; Bonfoco *et al.*, 1995; Budd *et al.*, 2000).

Increased levels of neuronal Ca^{2+}, in conjunction with the Ca^{2+}-binding protein CaM, trigger the activation of nNOS and subsequent generation of NO from the amino acid L-arginine (Abu-Soud and Stuehr, 1993; Bredt *et al.*, 1991). NO is a gaseous free radical (thus highly diffusible) and a key molecule that plays a vital role in normal signal transduction but in excess can lead to neuronal cell damage and death. Three subtypes of NOS have been identified; two constitutive forms of NOS—nNOS and endothelial NOS (eNOS)—take their names from the cell type in which they were first found. The name of the third subtype—inducible NOS (iNOS)—indicates that expression of the enzyme is induced by acute inflammatory stimuli. All three isoforms are widely distributed in the brain.

Recently, a novel cellular mechanism for Ca^{2+}-dependent release of NO was discovered in dorsal root ganglion neurons and pancreatic acinar cells. This Ca^{2+}-dependent NO release occurs not as a result of *de novo* synthesis by NO but instead via liberation of NO from an *S*-nitrosothiol (SNO) pool, whereby NO is reversibly bound to specific cysteine residues (see below for additional chemical information regarding this reaction). Interestingly, NOS-independent release of NO was mediated by calpain (a Ca^{2+}-dependent thiol protease) but not by CaM or protein kinase C (PKC) (Chvanov *et al.*, 2006). This finding leads to the question of whether glutamatergic signaling in neuronal cells can also trigger NO release from SNO pools.

Recent studies further pointed out the potential connection between ROS/RNS and mitochondrial dysfunction in neurodegenerative diseases, especially in PD (Beal, 2001; Betarbet *et al.*, 2000). Pesticide and other environmental toxins that inhibit mitochondrial complex I result in oxidative and nitrosative stress and consequent aberrant protein accumulation (Abou-Sleiman *et al.*, 2006; Chung *et al.*, 2004; He *et al.*, 2003; Uehara *et al.*, 2006; Yao *et al.*, 2004). Administration to animal models of complex I inhibitors, such as MPTP, 6-hydroxydopamine, rotenone, and paraquat, which result in overproduction of ROS/RNS, reproduces many of the features of sporadic PD such as dopaminergic neuron degeneration, upregulation and aggregation of α-synuclein, Lewy body–like intraneuronal inclusions, and behavioral impairment (Beal, 2001; Betarbet *et al.*, 2000). In addition, it has recently been proposed that mitochondrial cytochrome oxidase can produce NO in a nitrite (NO_2^-)- and pH-dependent but non-Ca^{2+}-dependent manner (Castello *et al.*, 2006).

Increased nitrosative and oxidative stress are associated with chaperone and proteasomal dysfunction, resulting in accumulation of misfolded aggregates (Isaacs *et al.*, 2006; Zhang and Kaufman, 2006). However, until recently little was known regarding the molecular and pathogenic mechanisms underlying contribution of NO to the formation of inclusion bodies such as amyloid plaques in AD or Lewy bodies in PD.

IV. Protein *S*-Nitrosylation and Neuronal Cell Death

Early investigations indicated that the NO group mediates cellular signaling pathways, which regulate broad aspects of brain function, including synaptic plasticity, normal development, and neuronal cell death (Bredt and Snyder, 1994; Dawson *et al.*, 1991; O'Dell *et al.*, 1991; Schuman and Madison, 1994). In general, NO exerts physiological and some pathophysiological effects via stimulation of guanylate cyclase to form cyclic guanosine-3′,5′-monophosphate (cGMP) or through *S*-nitros(yl)ation of regulatory protein thiol groups (Garthwaite *et al.*, 1988; Isaacs *et al.*, 2006; Kandel and O'Dell, 1992; Lei *et al.*, 1992; Lipton *et al.*, 1993; Stamler *et al.*, 1992a). *S*-nitrosylation is the covalent addition of an NO group to a critical cysteine thiol/sulfhydryl (RSH or, more properly, thiolate anion, RS^-) to form an *S*-nitrosothiol derivative (R-SNO). Such modification modulates the function of a broad spectrum of mammalian, plant, and microbial proteins. In general, a consensus motif of amino acids composed of nucleophilic residues (generally an acid and a base) surround a critical cysteine, which increases the cysteine sulfhydryl's susceptibility to *S*-nitrosylation (Hess *et al.*, 2005;

Stamler *et al.*, 1997). Our group first identified the physiological relevance of *S*-nitrosylation by showing that NO and related RNS exert paradoxical effects via redox-based mechanisms—NO is neuroprotective via *S*-nitrosylation of NMDA receptors (as well as other subsequently discovered targets, including caspases), and yet can also be neurodestructive by formation of peroxynitrite (or, as later discovered, reaction with additional molecules such as MMP-9 and GAPDH) (Choi *et al.*, 2000; Dimmeler *et al.*, 1997; Gu *et al.*, 2002; Hara *et al.*, 2005; Kim *et al.*, 1999; Lipton *et al.*, 1993; Mannick *et al.*, 1999; Melino *et al.*, 1997; Tenneti *et al.*, 1997). Over the past decade, accumulating evidence has suggested that *S*-nitrosylation can regulate the biological activity of a great variety of proteins, in some ways akin to phosphorylation (Chung *et al.*, 2004; Gu *et al.*, 2002; Haendeler *et al.*, 2002; Hara *et al.*, 2005; Hess *et al.*, 2005; Jaffrey *et al.*, 2001; Lipton *et al.*, 1993, 2002; Sliskovic *et al.*, 2005; Stamler, 1994; Stamler *et al.*, 1992b, 2001; Yao *et al.*, 2004). Chemically, NO is often a good "leaving group," facilitating further oxidation of critical thiol to disulfide bonds among neighboring (vicinal) cysteine residues or, via reaction with ROS, to sulfenic ($-SO^-$), sulfinic ($-SO_2^-$), or sulfonic ($-SO_3^-$) acid derivatization of the protein (Gu *et al.*, 2002; Stamler and Hausladen, 1998; Uehara *et al.*, 2006; Yao *et al.*, 2004). Alternatively, *S*-nitrosylation may possibly produce a nitroxyl disulfide in which the NO group is shared by close cysteine thiols (Houk *et al.*, 2003).

Analyses of mice deficient in either nNOS or iNOS confirmed that NO is an important mediator of cell injury and death after excitotoxic stimulation; NO generated from nNOS or iNOS is detrimental to neuronal survival (Huang *et al.*, 1994; Iadecola *et al.*, 1997). In addition, inhibition of NOS activity ameliorates the progression of disease pathology in animal models of PD, AD, and ALS, suggesting that excess generation of NO plays a pivotal role in the pathogenesis of several neurodegenerative diseases (Chabrier *et al.*, 1999; Hantraye *et al.*, 1996; Liberatore *et al.*, 1999; Przedborski *et al.*, 1996). Although the involvement of NO in neurodegeneration has been widely accepted, the chemical relationship between nitrosative stress and accumulation of misfolded proteins has remained obscure. Recent findings, however, have shed light on molecular events underlying this relationship. Specifically, we mounted physiological and chemical evidence that *S*-nitrosylation modulates the (1) ubiquitin E3 ligase activity of parkin (Chung *et al.*, 2004; Lipton *et al.*, 2005; Yao *et al.*, 2004) and (2) chaperone and isomerase activities of PDI (Uehara *et al.*, 2006), contributing to protein misfolding and neurotoxicity in models of neurodegenerative disorders. Additionally, Cohen *et al.* (2006) demonstrated that insulin/insulin-like growth factor-I (IGF-I) signaling, which influences longevity and life span in many species in part via downregulation of ROS/RNS generation, can affect aggregation of toxic proteins such as Aβ. This finding potentially provides an additional link between ROS/RNS production during the normal aging process and protein aggregation in neurodegenerative conditions.

V. Parkin and the UPS

The UPS is one of the principal mechanisms for proteolysis in mammalian cells. Formation of polyubiquitin chains constitutes the signal for proteasomal attack and degradation. The first ubiquitin in the polyubiquitin chain is covalently attached to the substrate through an isopeptide bond between the C-terminus of ubiquitin and a lysine residue of the target protein. The cascade of activating (E1), conjugating (E2), and ubiquitin-ligating (E3) type enzymes catalyzes the conjugation of the ubiquitin chain to proteins. In addition, individual E3 ubiquitin ligases play a key role in the recognition of specific substrates (Ross and Pickart, 2004).

One piece of direct evidence for UPS involvement in PD arises from the discovery that mutations in the parkin gene cause autosomal recessive juvenile parkinsonism (ARJP). Parkin is a member of a large family of E3 ubiquitin ligases that are related to one another by the presence of RING finger domains. RING fingers have varying numbers of cysteine and histidine residues that coordinate a structurally important zinc atom that is often involved in catalysis (Marin and Ferrus, 2002). Parkin has two RING finger domains separated by an "in between RING" (IBR) domain. This motif allows parkin to recruit substrate proteins as well as an E2 enzyme (e.g., UbcH7, UbcH8, or UbcH13). Point mutations, stop mutations, truncations, and deletions in both alleles of the parkin gene will eventually cause dysfunction in its activity (either an increase or a decrease) and are responsible for many cases of ARJP as well as rare adult forms of PD. Several putative target substrates have been identified for parkin E3 ligase activity. One group has reported that mutant parkin failed to bind glycosylated α-synuclein for ubiquitination, leading to α-synuclein accumulation (Shimura *et al.*, 2001), but most authorities do not feel that α-synuclein is a direct substrate of parkin. Synphilin-1 (α-synuclein interacting protein), on the other hand, is considered to be a substrate for parkin ubiquitination, and it is included in Lewy body–like inclusions in cultured cells when coexpressed with α-synuclein (Chung *et al.*, 2001). Other substrates for parkin include parkin-associated endothelin receptor-like receptor (Pael-R) (Imai *et al.*, 2001), cell division control-related protein (CDCrel-1) (Zhang *et al.*, 2000), cyclin E (Staropoli *et al.*, 2003), p38 tRNA synthase (Corti *et al.*, 2003), and synaptotagmin XI (Huynh *et al.*, 2003), as well as possibly parkin itself (autoubiquitination). It is generally accepted that accumulation of these substrates can lead to disastrous consequences for the survival of dopaminergic neurons in familial PD and possibly also in sporadic PD. Therefore, characterization of potential regulators that affect parkin E3 ligase activity may reveal important molecular mechanisms for the pathogenesis of PD. Heretofore, two cellular components have been shown to regulate the substrate specificity and ubiquitin E3 ligase activity of parkin. The first represents post-translational modification of parkin through *S*-nitrosylation (see below for details) or phosphorylation (Yamamoto *et al.*, 2005), and the second represents binding partners of parkin such as CHIP (Imai *et al.*, 2002) and BAG5 (Kalia *et al.*, 2004). CHIP enhances the ability of parkin to inhibit cell death through upregulation of parkin-mediated ubiquitination, while BAG5-mediated inhibition of parkin E3 ligase activity facilitates neuronal cell death. In addition, Fallon *et al.* (2006) reported another mechanism for parkin-mediated neuronal survival via a proteasome-independent pathway. In this model, parkin monoubiquitinates the epidermal growth factor receptor (EGFR)-associated protein, Eps15, leading to inhibition of EGFR endocytosis. The resulting prolongation of EGFR signaling via the phosphoinositide-3 kinase (PI3K)/Akt (PKB) signaling pathway is postulated to enhance neuronal survival.

Another important molecule that links aberrant UPS activity and PD is the ubiquitin hydrolase Uch-L1, a deubiquitinating enzyme that recycles ubiquitin. Autosomal dominant mutations of Uch-L1 have been identified in two siblings with PD (Leroy *et al.*, 1998). Interestingly, a study suggested that a novel ubiquitin-ubiquitin ligase activity of Uch-L1 might also be important in the pathogenesis of PD (Liu *et al.*, 2002). Additional mutations in α-synuclein, DJ-1, PINK1, and LRRK2 may contribute to UPS dysfunction and subsequently lead to PD.

VI. *S*-Nitrosylation and Parkin

PD is the second most prevalent neurodegenerative disease and is characterized by the progressive loss of (dopaminergic) neurons in the substantia nigra pars compacta. Appearance of Lewy bodies that contain misfolded and ubiquitinated proteins generally accompanies the loss of dopaminergic neurons in the PD brain. Such ubiquitinated inclusion bodies are the hallmark of many neurodegenerative disorders. Age-associated defects in intracellular proteolysis of misfolded or aberrant proteins might lead to accumulation and ultimately deposition of aggregates within neurons or glial cells. Although such aberrant protein accumulation had been observed in patients with genetically encoded mutant proteins, recent evidence from our laboratory suggests that nitrosative and oxidative stress are potential causal factors for protein accumulation in the much more common sporadic form of PD. As illustrated below, nitrosative/oxidative stress, commonly found during normal aging, can mimic rare genetic causes of disorders, such as PD, by promoting protein misfolding in the absence of a genetic mutation (Chung *et al.*, 2004; Lipton *et al.*, 2005; Yao *et al.*, 2004). For example, *S*-nitrosylation and further oxidation of parkin or Uch-L1 result in dysfunction of these enzymes and thus of the UPS (Choi *et al.*, 2004; Chung *et al.*, 2004, 2005; Gu *et al.*, 2005; Nishikawa *et al.*, 2003; Yao *et al.*, 2004). We and others recently discovered that nitrosative stress triggers *S*-nitrosylation of parkin (forming SNO-parkin) not only in rodent models of PD but also in the brains of human patients with PD and the related α-synucleinopathy, diffuse Lewy body disease (DLBD). SNO-parkin initially stimulates ubiquitin E3 ligase activity, resulting in enhanced ubiquitination as observed in Lewy bodies, followed by a decrease in enzyme activity, producing a futile cycle of dysfunctional UPS (Lim *et al.*, 2005; Lipton *et al.*, 2005; Yao *et al.*, 2004; Fig. 2; Chung *et al.*, 2004; Yao *et al.*, 2004). We also found that rotenone led to the generation of SNO-parkin and thus dysfunctional ubiquitin E3 ligase activity. Moreover, *S*-nitrosylation appears to compromise the neuroprotective effect of parkin (Chung *et al.*, 2004). These mechanisms involve *S*-nitrosylation of critical cysteine residues in the first RING domain of parkin (Yao *et al.*, 2004). Nitrosative stress has also been suggested to affect parkin misfolding and concomitantly compromise its protective function (Wang *et al.*, 2005). Additionally, it is likely that other ubiquitin E3 ligases with RING finger thiol motifs are *S*-nitrosylated in a similar manner to parkin to affect their enzymatic function; hence, *S*-nitrosylation of E3 ligases may be involved in a number of degenerative conditions.

The neurotransmitter dopamine (DA) may also impair parkin and contribute to neuronal demise via the modification of cysteine residue(s) (LaVoie *et al.*, 2005). DA can be oxidized to DA quinone, which can react with and inactivate proteins through covalent modification of cysteine sulfhydryl groups; peroxynitrite has been reported to promote oxidation of DA to form DA quinone (LaVoie and Hastings, 1999). LaVoie *et al.* (2005) showed that DA quinone can attack one or more cysteine residues in the RING domain(s) of parkin, forming a covalent adduct that abrogates its E3 uibiquitin ligase activity. DA quinone also reduced the solubility of parkin, possibly inducing parkin misfolding after disruption of the RING domain(s). Therefore, oxidative/nitrosative species may either directly or indirectly contribute to altered parkin activity within the brain, and subsequent loss of parkin-dependent neuroprotection results in increased cell death.

VII. The Unfolded Protein Response and PDI

The ER normally participates in protein processing and folding but undergoes a stress response when immature or misfolded proteins accumulate (Andrews and Johnson, 1996; Ellgaard *et al.*, 1999; Sidrauski *et al.*, 1998; Szegezdi *et al.*, 2006). ER stress stimulates two critical intracellular responses (Fig. 3). The first represents expression of chaperones that prevent protein aggregation via the unfolded protein response (UPR), and is implicated in protein refolding, posttranslational assembly of protein complexes, and protein degradation. This response is believed to contribute to adaptation during altered environmental conditions, promoting maintenance of cellular homeostasis. At least three ER transmembrane sensor proteins are involved in the UPR: pancreatic ER kinase (PKR)-like ER kinase

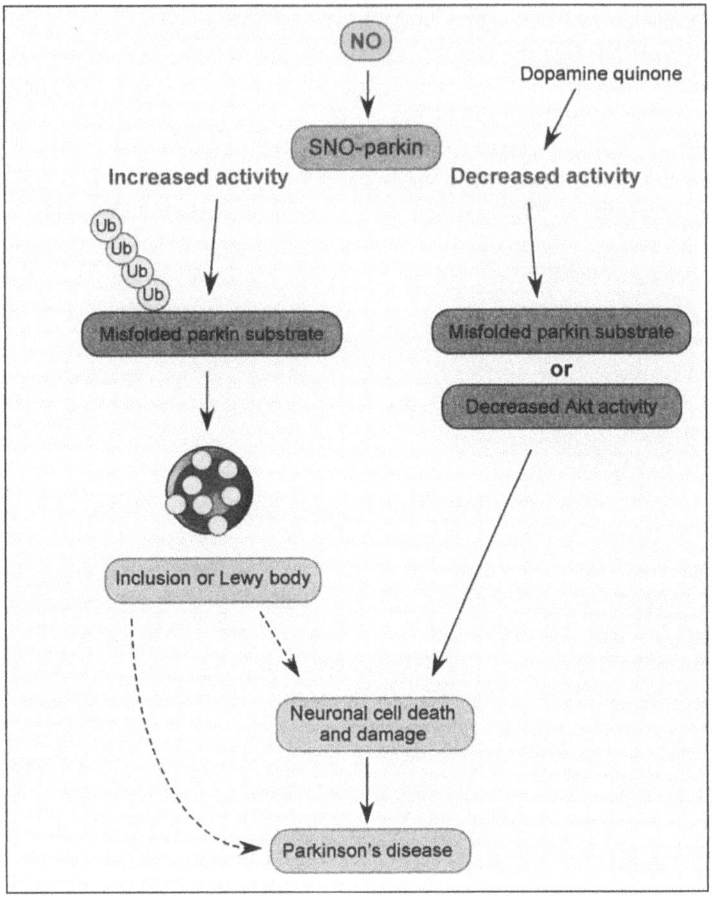

FIG. 2. Possible mechanism of *S*-nitrosylated parkin (SNO-parkin) contributing to the accumulation of aberrant proteins and damage or death of dopaminergic neurons. Nitrosative stress leads to *S*-nitrosylation of parkin, and, initially, to a dramatic increase followed by a decrease in its E3 ubiquitin ligase activity (Chung *et al.*, 2004; Lipton *et al.*, 2005; Yao *et al.*, 2004). The initial increase in this E3 ubiquitin ligase activity leads to enhanced ubiquitination of parkin substrates (e.g., synphilin-1, Pael-R, and parkin itself). Increased parkin E3 ubiquitin ligase activity may contribute to Lewy body formation and impair parkin function, as also suggested by Sriram *et al.* (2005). The subsequent decrease in parkin activity may allow misfolded proteins to accumulate. Downregulation of parkin may also result in decreased Akt neuroprotective activity because of enhanced EGFR internalization (Fallon *et al.*, 2006). Dopamine quinone can also modify cysteine thiols of parkin and reduce its activity (LaVoie *et al.*, 2005). Ub, ubiquitin. Adapted from Yao *et al.* (2004), Nakamura and Lipton (2007). (See Color Insert.)

(PERK), activating transcription factor 6 (ATF6), and inositol-requiring enzyme 1 (IRE1). The activation of all three proximal sensors results in the attenuation of protein synthesis via eukaryotic initiation factor-2 (eIF2) kinase and increased

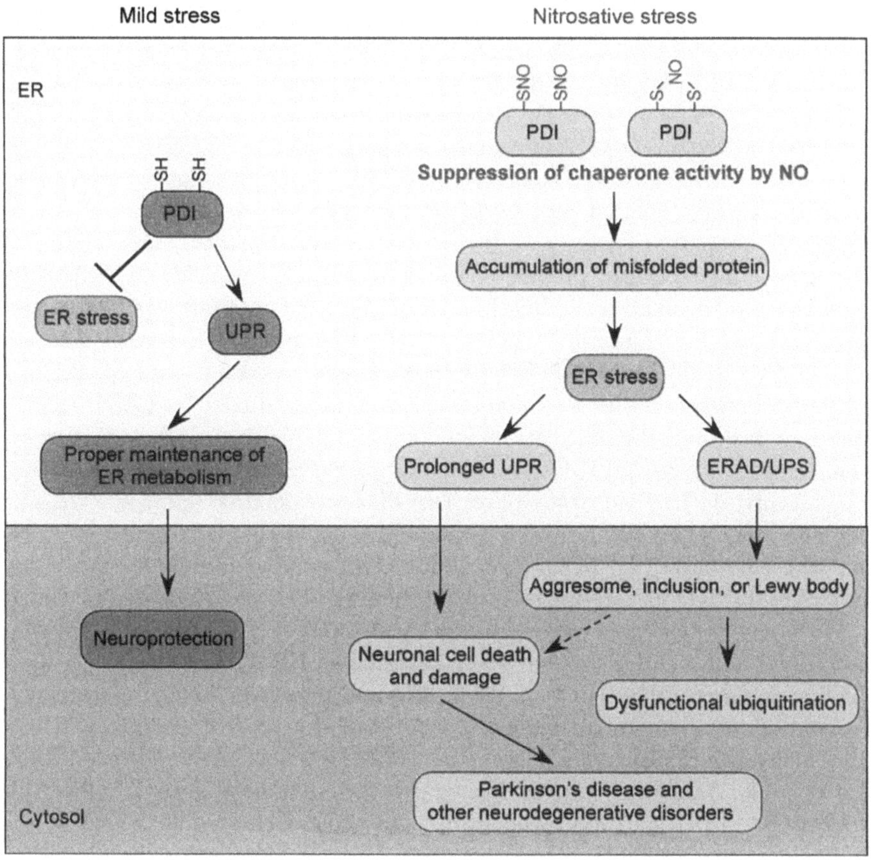

FIG. 3. Possible mechanism of *S*-nitrosylated PDI (SNO-PDI) contributing to the accumulation of aberrant proteins and neuronal cell damage or death. ER stress is triggered when misfolded proteins accumulate within the ER lumen, inducing the unfolded protein response (UPR). The UPR is usually a transient homeostatic mechanism for cell survival, while prolonged UPR elicits neuronal cell death. PDI modulates the activity of UPR sensors by mediating proper protein folding in the ER. Proteins that fail to attain their native folded state are eventually retrotranslocated across the ER membrane to be disposed of by cytosolic proteasomes. This process, known as ERAD, is essential in preventing protein accumulation and aggregation in the ER (Andrews and Johnson, 1996; Ellgaard *et al.*, 1999; Sidrauski *et al.*, 1998; Szegezdi *et al.*, 2006). Under conditions of severe nitrosative stress, *S*-nitrosylation of neuronal PDI inhibits normal protein folding in the ER, activates ER stress, and induces a prolonged UPR, thus contributing to protein aggregation and cell damage or death. For simplicity, *S*-nitrosylation of only one (of two) thioredoxin domains of PDI is shown, resulting in formation of SNO-PDI or possibly nitroxyl-PDI, as described in Uehara *et al.* (2006) and Forrester *et al.* (2006). Adapted from Uehara *et al.* (2006), Nakamura and Lipton (2007). (See Color Insert.)

protein folding capacity of the ER (Kaufman, 1999; Mori, 2000; Patil and Walter, 2001; Yoshida *et al.*, 2001). The second ER stress response, termed ER-associated degradation (ERAD), specifically recognizes terminally misfolded proteins and retrotranslocates them across the ER membrane into the cytosol, where they can be degraded by the UPS. Additionally, although severe ER stress can induce apoptosis, the ER withstands relatively mild insults via expression of stress proteins such as glucose-regulated protein (GRP) and PDI. These proteins behave as molecular chaperones that assist in the maturation, transport, and folding of secretory proteins. During protein folding in the ER, PDI catalyzes thiol/disulfide exchange, thus facilitating disulfide bond formation, rearrangement reactions, and structural stability (Lyles and Gilbert, 1991). PDI has four domains that are homologous to thioredoxin (TRX) (termed a, b, b′, and a′). Two of the four TRX-like domains (a and a′) contain a characteristic redox-active CXXC motif, and these two thiol/disulfide centers function as independent active sites (Edman *et al.*, 1985; Ellgaard and Ruddock, 2005; Gruber *et al.*, 2006; Vuori *et al.*, 1992). The recently determined structure of yeast PDI revealed that the four TRX-like domains form a twisted "U" shape with the two active sites facing each other on opposite sides of the "U" (Tian *et al.*, 2006). Hydrophobic residues line the inside surface of the "U," facilitating interactions between PDI and misfolded proteins.

In many neurodegenerative disorders and cerebral ischemia, the accumulation of immature and denatured proteins results in ER dysfunction (Atkin *et al.*, 2006; Conn *et al.*, 2004; Hu *et al.*, 2000; Rao and Bredesen, 2004), but upregulation of PDI represents an adaptive response promoting protein refolding and may offer neuronal cell protection (Conn *et al.*, 2004; Hetz *et al.*, 2005; Ko *et al.*, 2002; Tanaka *et al.*, 2000). In a recent study, we reported that the *S*-nitrosylation of PDI (to form SNO-PDI) disrupts its neuroprotective role (Uehara *et al.*, 2006).

VIII. *S*-Nitrosylation of PDI Mediates Protein Misfolding and Neurotoxicity in Cell Models of PD or AD

In contrast to the highly reducing environment of the cytosol and mitochondria, the ER manifests a relatively positive redox potential (e.g., the ratio of reduced:oxidized glutathione is ~3:1). This redox environment can contribute to the stability of protein *S*-nitrosylation or oxidation (Forrester *et al.*, 2006). We reported that excessive NO can lead to *S*-nitrosylation of the active site thiol groups of PDI, and this reaction inhibits both its isomerase and chaperone activities (Uehara *et al.*, 2006). Mitochondrial complex I insult by rotenone can also result in *S*-nitrosylation of PDI in cell culture models. Moreover, we found that PDI is *S*-nitrosylated in the brains of virtually all cases examined of sporadic AD and PD. Additionally, it is possible that vicinal (nearby) cysteine thiols reacting with NO can form nitroxyl disulfide (Houk *et al.*, 2003), and such reaction may potentially occur in the catalytic side of PDI to inhibit enzymatic activity. In order to determine the consequences of *S*-nitrosylated PDI (SNO-PDI) formation in neurons, we exposed cultured cerebrocortical neurons to neurotoxic concentrations of NMDA, thus inducing excessive Ca^{2+} influx and consequent NO production from nNOS. Under these conditions, we found that PDI was *S*-nitrosylated in a NOS-dependent manner. SNO-PDI formation led to the accumulation of polyubiquitinated/misfolded proteins and activation of the UPR. Moreover, *S*-nitrosylation abrogated the inhibitory effect of PDI on aggregation of proteins observed in Lewy body inclusions (Chung *et al.*, 2001; Uehara *et al.*, 2006). *S*-Nitrosylation of PDI also prevented its attenuation of neuronal cell death triggered by ER stress, misfolded proteins, or proteasome inhibition (Fig. 3). Further evidence suggested that SNO-PDI may in effect transport NO to the extracellular space, where it could conceivably exert additional adverse effects (Sliskovic *et al.*, 2005).

In addition to PDI, *S*-nitrosylation is likely to affect critical thiol groups on other chaperones such as HSP90 in the cytoplasm (Martinez-Ruiz *et al.*, 2005) and possibly GRP in the ER. Normally, HSP90 stabilizes misfolded proteins and modulates the activity of cell signaling proteins including NOS and calreticulin (Muchowski and Wacker, 2005). In AD brains, levels of HSP90 are increased in both the cytosolic and membranous fractions, where HSP90 is thought to maintain tau and Aβ in a soluble conformation, thereby averting their aggregation (Dou *et al.*, 2003; Kakimura *et al.*, 2002). Martinez-Ruiz *et al.* (2005) demonstrated that *S*-nitrosylation of HSP90 can occur in endothelial cells, and this modification abolishes its ATPase activity, which is required for its function as a molecular chaperone. These studies imply that *S*-nitrosylation of HSP90 in neurons of AD brains may contribute to the accumulation of tau and Aβ aggregates.

The UPS is apparently impaired in the aging brain. Additionally, inclusion bodies similar to those found in neurodegenerative disorders can appear in brains of normal aged individuals or those with subclinical manifestations of disease (Gray *et al.*, 2003). These findings suggest that the activity of the UPS and molecular chaperones may decline in an age-dependent manner (Paz Gavilan *et al.*, 2006). Given that we and others have not found detectable quantities of SNO-parkin and SNO-PDI in normal aged brain (Chung *et al.*, 2004; Uehara *et al.*, 2006; Yao *et al.*, 2004), we speculate that *S*-nitrosylation of these and similar proteins may represent a key event that contributes to susceptibility of the aging brain to neurodegenerative conditions.

IX. PDI Activity in ALS and Prion Disease

PDI has been implicated in the pathophysiology of familial ALS (Atkin *et al.*, 2006). Mutations in Cu/Zn superoxide dismutase (SOD1) are known to be involved in motor neuron death in some forms of familial ALS. SOD1 is an intracellular homodimeric metalloprotein that forms a stable intra-subunit disulfide bond. Biochemical evidence suggests that the disulfide-reduced monomer of mutant SOD1 (mtSOD1) forms inclusion bodies (Arnesano *et al.*, 2004; Doucette *et al.*, 2004; Furukawa and O'Halloran, 2005; Rakhit *et al.*, 2004; Tiwari and Hayward, 2003), and aggregates of misfolded mtSOD1 are commonly associated with the disease, as seen at postmortem examination. In addition, although wild-type (wt) SOD1 is found predominantly in the cytoplasm, mtSOD1 forms monomers or insoluble high-molecular-weight multimers within the ER (Kikuchi *et al.*, 2006). Atkin *et al.* (2006) showed that inhibition of PDI activity with bacitracin can increase aggregation of mtSOD1 in neuronal cells. Moreover, PDI colocalized and bound to intracellular aggregates of mtSOD1. Upregulation of the UPR was also observed in mtSOD1 mice. These findings suggest that ER stress may contribute to the pathophysiology of familial ALS, and PDI could potentially reduce mtSOD1 aggregation and affect neuronal survival. Interestingly, SNO levels have also been found to be abnormal in the spinal cords of mtSOD1 transgenic mice (Schonhoff *et al.*, 2006). Whether SNO-PDI is involved in SOD1 aggregation and motor neuron injury in ALS remains to be studied.

Additionally, transmissible spongiform encephalopathies (TSE), also known as prion diseases, are transmissible neurodegenerative disorders and include Creutzfeldt-Jakob disease, bovine spongiform encephalopathy, and scrapie. Cerebral accumulation of misfolded prion protein (PrP) and extensive neuronal apoptosis represent pathological hallmarks of these prion diseases. Reports have suggested that a prolonged UPR due to PrP misfolding in the ER may contribute to neuronal dysfunction (Hetz *et al.*, 2003, 2005; Yoo *et al.*, 2002). This ER stress response is mainly associated with upregulation of Grp58, an ER chaperone with PDI-like activity, suggesting that this chaperone may play an important role in the cellular response to prion infection (Hetz *et al.*, 2005). In fact, *in vitro* studies on Grp58, either overexpressing (via transfection) or downregulating (via RNAi), demonstrated that this ER chaperone protects cells against PrP misfolding and toxicity. Collectively, these studies raise the possibility that SNO-PDI and *S*-nitrosylation of other chaperone molecules may represent potential therapeutic targets to prevent protein aggregation in several neurodegenerative diseases.

X. Potential Treatment of Excessive NMDA-Induced Ca^{2+} Influx and *S*-Nitrosylation

One mechanism that could potentially curtail excessive Ca^{2+} influx and resultant overstimulation of nNOS activity would be inhibition of NMDA receptors. Until recently, however, drugs in this class blocked virtually all NMDA receptor activity and therefore manifest unacceptable side effects by inhibiting normal functions of the receptor. For this reason, many previous NMDA receptor antagonists have disappointingly failed in advanced clinical trials conducted for a number of neurodegenerative disorders. In contrast, studies in our laboratory first showed that the adamantane derivative, memantine, preferentially blocks excessive (pathological) NMDA receptor activity while relatively sparing normal (physiological) activity. Memantine does this in a surprising fashion because of its low (micromolar) affinity, even though its actions are quite selective for the NMDA receptor at that concentration. "Apparent" affinity of a drug is determined by the ratio of its "on-rate" to its "off-rate" for the target. The on-rate is not only a property of diffusion and availability of the target, but also the drug's concentration. In contrast, the off-rate is an intrinsic property of the drug–receptor complex, unaffected by drug concentration. A relatively fast off-rate is a major contributor to memantine's low affinity for the NMDA receptor. The inhibitory activity of memantine involves blockade of the NMDA receptor-associated ion channel when it is excessively open (termed open-channel block). The unique and subtle difference of the memantine blocking sites in the channel pore may explain the advantageous properties of memantine action.

Also critical for the clinical tolerability of memantine is its uncompetitive mechanism of action. An uncompetitive antagonist can be distinguished from a noncompetitive antagonist, which acts allosterically at a noncompetitive site, that is, at a site other than the agonist-binding site. An uncompetitive antagonist is defined as an inhibitor whose action is contingent on prior activation of the receptor by the agonist. Hence, the same amount of antagonist blocks higher concentrations of agonist relatively better than lower concentrations of agonist. Some open-channel blockers function as pure uncompetitive antagonists, depending on their exact properties of interaction with the ion channel. This uncompetitive mechanism of action coupled with a relatively fast off-rate from the channel yields a drug that preferentially blocks NMDA receptor-operated channels when they are excessively open while relatively sparing normal neurotransmission. In fact, the relatively fast off-rate is a major contributor to a drug like memantine's low affinity for the channel pore. While many factors determine a drug's clinical efficacy and tolerability, it appears that the relatively rapid off-rate is a predominant factor in memantine's tolerability in contrast to other NMDA-type receptor antagonists (reviewed in Chen and Lipton, 2006; Lipton, 2006a). Thus, the critical features of memantine's mode of action are its *U*ncompetitive mechanism and *F*ast *O*ff-rate, or what we call a UFO drug—which is there only when you need it and then quickly disappears. Memantine has been used for many years in Europe to treat PD, and regulatory groups in both Europe and the United States recently voted its approval as the first treatment for moderate-to-severe AD. It is under study for a number of other neurodegenerative disorders.

As promising as the results with memantine are, we are continuing to pursue ways to use additional modulatory sites on the NMDA receptor to block excitotoxicity even more effectively and safely than memantine alone. New approaches in this regard are explored below.

XI. Looking to the Future: NitroMemantines

NitroMemantines are second-generation memantine derivatives that were designed to have enhanced neuroprotective efficacy without sacrificing safety. As mentioned earlier, a nitrosylation site(s) is located on the *N*-terminus or extracellular domain of the NMDA receptor, and *S*-nitrosylation of this site, that is, NO reaction with the sulfhydryl group of a critical cysteine residue, downregulates (but does not completely shutoff) receptor activity (Fig. 4). The drug nitroglycerin, which generates NO-related species, can act at this site to limit excessive NMDA receptor activity. In fact, in rodent models, nitroglycerin can limit ischemic damage (Lipton and Wang, 1996), and there is some evidence that patients taking nitroglycerin for other medical reasons may be resistant to glaucomatous visual field loss as well (Zurakowski *et al.*, 1998). Consequently, we carefully characterized *S*-nitrosylation sites on the NMDA receptor in order to determine if we could design a nitroglycerin-like drug that could be more specifically targeted to the receptor. In brief, we found that five different cysteine residues on the NMDA

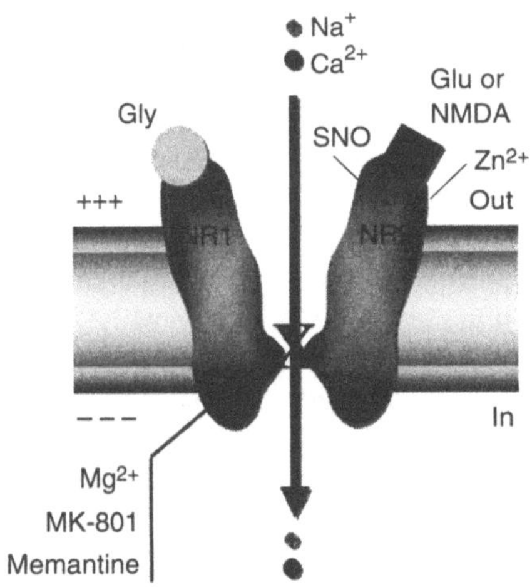

FIG. 4. NMDA receptor model illustrating important binding and modulatory sites. Glu or NMDA, glutamate or NMDA binding site; Gly, glycine binding site; Zn^{2+}, zinc binding site; NR1, NMDA receptor subunit 1; NR2, NMDA receptor subunit 2A; SNO, cysteine sulfhydryl group (-SH) reacting with nitric oxide species (NO); X, Mg^{2+}, MK-801, and memantine binding sites within the ion channel pore region. Adapted from Chen and Lipton (2006). (See Color Insert.)

receptor could interact with NO. One of these, located at cysteine residue 399 (Cys399) on the NR2A subunit of the NMDA receptor, mediates $\geq$90% of the effect of NO under our experimental conditions (Choi *et al.*, 2000). From crystal structure models and electrophysiological experiments, we further found that NO binding to the NMDA receptor at Cys 399 may induce a conformational change in the receptor protein that makes glutamate and Zn^{2+} bind more tightly to the receptor. The enhanced binding of glutamate and Zn^{2+} in turn causes the receptor to desensitize and, consequently, the ion channel to close (Lipton *et al.*, 2002). Electrophysiological studies have demonstrated this effect of NO on the NMDA channel (Choi *et al.*, 2000; Lei *et al.*, 1992; Lipton *et al.*, 1993).

Unfortunately, nitroglycerin is not very attractive as a neuroprotective agent. The same cardiovascular vasodilator effect that makes it useful in the treatment of angina could cause dangerously large drops in blood pressure in patients with dementia, stroke, traumatic injury, or glaucoma. However, the open-channel block mechanism of memantine not only leads to a higher degree of channel blockade in the presence of excessive levels of glutamate but also can be used as a homing signal for targeting drugs, for example, the NO group, to hyperactivated, open NMDA-gated channels. We have therefore been developing combinatorial drugs (NitroMemantines) that theoretically should be able to use memantine to target NO to the nitrosylation sites of the NMDA receptor in order to avoid the systemic side effects of NO. Two sites of modulation would be analogous to having two volume controls on your television set for fine-tuning the audio signal.

Preliminary studies have shown NitroMemantines to be highly neuroprotective in both *in vitro* and *in vivo* animal models (Lipton, 2006a). In fact, it appears to be more effective than memantine at lower dosage. Moreover, because of the targeting effect of the memantine moiety, NitroMemantines appear to lack the blood pressure lowering effects typical of nitroglycerin. More research still needs to be performed on NitroMemantine drugs, but, by combining two clinically tolerated drugs (memantine and nitroglycerin), we have created a new, improved class of UFO drugs that should be both clinically tolerated and neuroprotective.

XII. Conclusions

Excessive nitrosative and oxidative stress triggered by excessive NMDA receptor activation and/or mitochondrial dysfunction may result in malfunction of the UPS or molecular chaperones, thus contributing to abnormal protein accumulation and neuronal damage in sporadic forms of neurodegenerative diseases. Our elucidation of an NO-mediated pathway to dysfunction of parkin and PDI by *S*-nitrosylation provides a mechanistic link between free radical production, abnormal protein accumulation, and neuronal cell injury in neurodegenerative disorders such as PD. Elucidation of this new pathway may lead to the development of additional new therapeutic approaches to prevent aberrant protein misfolding by targeted disruption or prevention of nitrosylation of specific proteins such as parkin and PDI. The clinically tolerated NMDA receptor antagonist, memantine, is approved for moderate-to-severe Alzheimer's disease and is being tested for other neurological diseases. A second generation of memantine bifunctional drugs, termed NitroMemantines, offer promise for improved efficacy.

Acknowledgments

This work was supported in part by NIH grants P01 HD29587, R01 EY05477, R01 EY09024, and R01 NS046994, the American Parkinson's Disease Association, San Diego Chapter, and an Ellison Senior Scholars Award in Aging (to S.A.L.). The authors would like to acknowledge the seminal work of Dr. H.-S.V. Chen in his contributions to the mechanism of action of the drugs described here, namely memantine and the NitroMemantines.

References

Abou-Sleiman, P. M., Muqit, M. M., and Wood, N. W. (2006). Expanding insights of mitochondrial dysfunction in Parkinson's disease. *Nat. Rev. Neurosci.* **7,** 207–219.

Abu-Soud, H. M., and Stuehr, D. J. (1993). Nitric oxide synthases reveal a role for calmodulin in controlling electron transfer. *Proc. Natl. Acad. Sci. USA* **90,** 10769–10772.

Andrews, D. W., and Johnson, A. E. (1996). The translocon: More than a hole in the ER membrane? *Trends Biochem. Sci.* **21,** 365–369.

Ankarcrona, M., Dypbukt, J. M., Bonfoco, E., Zhivotovsky, B., Orrenius, S., Lipton, S. A., and Nicotera, P. (1995). Glutamate-induced neuronal death: A succession of necrosis or apoptosis depending on mitochondrial function. *Neuron* **15,** 961–973.

Arnesano, F., Banci, L., Bertini, I., Martinelli, M., Furukawa, Y., and O'Halloran, T. V. (2004). The unusually stable quaternary structure of human Cu,Zn-superoxide dismutase 1 is controlled by both metal occupancy and disulfide status. *J. Biol. Chem.* **279,** 47998–48003.

Arrasate, M., Mitra, S., Schweitzer, E. S., Segal, M. R., and Finkbeiner, S. (2004). Inclusion body formation reduces levels of mutant huntingtin and the risk of neuronal death. *Nature* **431,** 805–810.

Atkin, J. D., Farg, M. A., Turner, B.J., Tomas, D., Lysaght, J. A., Nunan, J., Rembach, A., Nagley, P., Beart, P. M., Cheema, S. S., and Horne, M. K. (2006). Induction of the unfolded protein response in familial amyotrophic lateral sclerosis and association of protein-disulfide isomerase with superoxide dismutase 1. *J. Biol. Chem.* **281,** 30152–30165.

Barnham, K. J., Masters, C. L., and Bush, A. I. (2004). Neurodegenerative diseases and oxidative stress. *Nat. Rev. Drug Discov.* **3,** 205–214.

Beal, M. F. (2001). Experimental models of Parkinson's disease. *Nat. Rev. Neurosci.* **2,** 325–334.

Beckman, J. S., Beckman, T. W., Chen, J., Marshall, P. A., and Freeman, B. A. (1990). Apparent hydroxyl radical production by peroxynitrite: Implications for endothelial injury from nitric oxide and superoxide. *Proc. Natl. Acad. Sci. USA* **87,** 1620–1624.

Bence, N. F., Sampat, R. M., and Kopito, R. R. (2001). Impairment of the ubiquitin-proteasome system by protein aggregation. *Science* **292,** 1552–1555.

Betarbet, R., Sherer, T. B., MacKenzie, G., Garcia-Osuna, M., Panov, A. V., and Greenamyre, J. T. (2000). Chronic systemic pesticide exposure reproduces features of Parkinson's disease. *Nat. Neurosci.* **3,** 1301–1306.

Bonfoco, E., Krainc, D., Ankarcrona, M., Nicotera, P., and Lipton, S. A. (1995). Apoptosis and necrosis: Two distinct events induced, respectively, by mild and intense insults with N-methyl-D-aspartate or nitric oxide/superoxide in cortical cell cultures. *Proc. Natl. Acad. Sci. USA* **92,** 7162–7166.

Bredt, D. S., and Snyder, S. H. (1994). Nitric oxide: A physiologic messenger molecule. *Annu. Rev. Biochem.* **63,** 175–195.

Bredt, D. S., Hwang, P. M., Glatt, C. E., Lowenstein, C., Reed, R. R., and Snyder, S. H. (1991). Cloned and expressed nitric oxide synthase structurally resembles cytochrome P-450 reductase. *Nature* **351,** 714–718.

Budd, S. L., Tenneti, L., Lishnak, T., and Lipton, S. A. (2000). Mitochondrial and extramitochondrial apoptotic signaling pathways in cerebrocortical neurons. *Proc. Natl. Acad. Sci. USA* **97,** 6161–6166.

Castello, P. R., David, P. S., McClure, T., Crook, Z., and Poyton, R. O. (2006). Mitochondrial cytochrome oxidase produces nitric oxide under hypoxic conditions: Implications for oxygen sensing and hypoxic signaling in eukaryotes. *Cell Metab.* **3,** 277–287.

Chabrier, P. E., Demerle-Pallardy, C., and Auguet, M. (1999). Nitric oxide synthases: Targets for therapeutic strategies in neurological diseases. *Cell Mol. Life Sci.* **55,** 1029–1035.

Chaudhuri, T. K., and Paul, S. (2006). Protein-misfolding diseases and chaperone-based therapeutic approaches. *FEBS J.* **273,** 1331–1349.

Chen, H. S., and Lipton, S. A. (2006). The chemical biology of clinically tolerated NMDA receptor antagonists. *J. Neurochem.* **97,** 1611–1626.

Choi, J., Levey, A. I., Weintraub, S. T., Rees, H. D., Gearing, M., Chin, L. S., and Li, L. (2004). Oxidative modifications and down-regulation of ubiquitin carboxyl-terminal hydrolase L1 associated with idiopathic Parkinson's and Alzheimer's diseases. *J. Biol. Chem.* **279,** 13256–13264.

Choi, Y. B., Tenneti, L., Le, D. A., Ortiz, J., Bai, G., Chen, H. S., and Lipton, S. A. (2000). Molecular basis of NMDA receptor-coupled ion channel modulation by *S*-nitrosylation. *Nat. Neurosci.* **3,** 15–21.

Chung, K. K., Zhang, Y., Lim, K. L., Tanaka, Y., Huang, H., Gao, J., Ross, C. A., Dawson, V. L., and Dawson, T. M. (2001). Parkin ubiquitinates the α-synuclein-interacting protein, synphilin-1: Implications for Lewy-body formation in Parkinson disease. *Nat. Med.* **7,** 1144–1150.

Chung, K. K., Thomas, B., Li, X., Pletnikova, O., Troncoso, J. C., Marsh, L., Dawson, V. L., and Dawson, T. M. (2004). *S*-nitrosylation of parkin regulates ubiquitination and compromises parkin's protective function. *Science* **304,** 1328–1331.

Chung, K. K., Dawson, T. M., and Dawson, V. L. (2005). Nitric oxide, *S*-nitrosylation and neurodegeneration. *Cell. Mol. Biol. (Noisy-le-grand)* **51,** 247–254.

Chvanov, M., Gerasimenko, O. V., Petersen, O. H., and Tepikin, A. V. (2006). Calcium-dependent release of NO from intracellular *S*-nitrosothiols. *EMBO J.* **25,** 3024–3032.

Ciechanover, A., and Brundin, P. (2003). The ubiquitin proteasome system in neurodegenerative diseases: Sometimes the chicken, sometimes the egg. *Neuron* **40,** 427–446.

Cohen, E., Bieschke, J., Perciavalle, R. M., Kelly, J. W., and Dillin, A. (2006). Opposing activities protect against age-onset proteotoxicity. *Science* **313,** 1604–1610.

Conn, K. J., Gao, W., McKee, A., Lan, M. S., Ullman, M. D., Eisenhauer, P. B., Fine, R. E., and Wells, J. M. (2004). Identification of the protein disulfide isomerase family member PDIp in experimental Parkinson's disease and Lewy body pathology. *Brain Res.* **1022,** 164–172.

Cookson, M. R. (2005). The biochemistry of Parkinson's disease. *Annu. Rev. Biochem.* **74,** 29–52.

Corti, O., Hampe, C., Koutnikova, H., Darios, F., Jacquier, S., Prigent, A., Robinson, J. C., Pradier, L., Ruberg, M., Mirande, M., Hirsch, E., Rooney, T., *et al.* (2003). The p38 subunit of the aminoacyl-tRNA synthetase complex is a Parkin substrate: Linking protein biosynthesis and neurodegeneration. *Hum. Mol. Genet.* **12,** 1427–1437.

Dawson, V. L., Dawson, T. M., London, E. D., Bredt, D. S., and Snyder, S. H. (1991). Nitric oxide mediates glutamate neurotoxicity in primary cortical cultures. *Proc. Natl. Acad. Sci. USA* **88,** 6368–63671.

Dimmeler, S., Haendeler, J., Nehls, M., and Zeiher, A. M. (1997). Suppression of apoptosis by nitric oxide via inhibition of interleukin-1β-converting enzyme (ICE)-like and cysteine protease protein (CPP)-32-like proteases. *J. Exp. Med.* **185,** 601–607.

Dou, F., Netzer, W. J., Tanemura, K., Li, F., Hartl, F. U., Takashima, A., Gouras, G. K., Greengard, P., and Xu, H. (2003). Chaperones increase association of tau protein with microtubules. *Proc. Natl. Acad. Sci. USA* **100,** 721–726.

Doucette, P. A., Whitson, L. J., Cao, X., Schirf, V., Demeler, B., Valentine, J. S., Hansen, J. C., and Hart, P. J. (2004). Dissociation of human copper-zinc superoxide dismutase dimers using chaotrope and reductant. Insights into the molecular basis for dimer stability. *J. Biol. Chem.* **279,** 54558–54566.

Edman, J. C., Ellis, L., Blacher, R. W., Roth, R. A., and Rutter, W. J. (1985). Sequence of protein disulphide isomerase and implications of its relationship to thioredoxin. *Nature* **317,** 267–270.

Ellgaard, L., and Ruddock, L. W. (2005). The human protein disulphide isomerase family: Substrate interactions and functional properties. *EMBO Rep.* **6,** 28–32.

Ellgaard, L., Molinari, M., and Helenius, A. (1999). Setting the standards: Quality control in the secretory pathway. *Science* **286,** 1882–1888.

Emerit, J., Edeas, M., and Bricaire, F. (2004). Neurodegenerative diseases and oxidative stress. *Biomed. Pharmacother.* **58,** 39–46.

Fallon, L., Belanger, C. M., Corera, A. T., Kontogiannea, M., Regan-Klapisz, E., Moreau, F., Voortman, J., Haber, M., Rouleau, G., Thorarinsdottir, T., Brice, A., van Bergen En Henegouwen, P. M., *et al.* (2006). A regulated interaction with the UIM protein Eps15 implicates parkin in EGF receptor trafficking and PI(3)K-Akt signalling. *Nat. Cell Biol.* **8,** 834–842.

Forrester, M. T., Benhar, M., and Stamler, J. S. (2006). Nitrosative stress in the ER: A new role for *S*-nitrosylation in neurodegenerative diseases. *ACS Chem. Biol.* **1,** 355–358.

Furukawa, Y., and O'Halloran, T. V. (2005). Amyotrophic lateral sclerosis mutations have the greatest destabilizing effect on the apo- and reduced form of SOD1, leading to unfolding and oxidative aggregation. *J. Biol. Chem.* **280,** 17266–17274.

Garthwaite, J., Charles, S. L., and Chess-Williams, R. (1988). Endothelium-derived relaxing factor release on activation of NMDA receptors suggests role as intercellular messenger in the brain. *Nature* **336,** 385–388.

Gray, D. A., Tsirigotis, M., and Woulfe, J. (2003). Ubiquitin, proteasomes, and the aging brain. *Sci. Aging Knowledge Environ.* **2003,** RE6.

Gruber, C. W., Cemazar, M., Heras, B., Martin, J. L., and Craik, D. J. (2006). Protein disulfide isomerase: The structure of oxidative folding. *Trends Biochem. Sci.* **31,** 455–464.

Gu, Z., Kaul, M., Yan, B., Kridel, S. J., Cui, J., Strongin, A., Smith, J. W., Liddington, R. C., and Lipton, S. A. (2002). *S*-nitrosylation of matrix metalloproteinases: Signaling pathway to neuronal cell death. *Science* **297,** 1186–1190.

Gu, Z., Nakamura, T., Yao, D., Shi, Z. Q., and Lipton, S. A. (2005). Nitrosative and oxidative stress links dysfunctional ubiquitination to Parkinson's disease. *Cell Death Differ.* **12,** 1202–1204.

Haendeler, J., Hoffmann, J., Tischler, V., Berk, B. C., Zeiher, A. M., and Dimmeler, S. (2002). Redox regulatory and anti-apoptotic functions of thioredoxin depend on *S*-nitrosylation at cysteine 69. *Nat. Cell Biol.* **4,** 743–749.

Hantraye, P., Brouillet, E., Ferrante, R., Palfi, S., Dolan, R., Matthews, R. T., and Beal, M. F. (1996). Inhibition of neuronal nitric oxide synthase prevents MPTP-induced parkinsonism in baboons. *Nat. Med.* **2,** 1017–1021.

Hara, M. R., Agrawal, N., Kim, S. F., Cascio, M. B., Fujimuro, M., Ozeki, Y., Takahashi, M., Cheah, J. H., Tankou, S. K., Hester, L. D., Ferris, C. D., Hayward, S. D., *et al.* (2005). *S*-nitrosylated GAPDH initiates apoptotic cell death by nuclear translocation following Siah1 binding. *Nat. Cell Biol.* **7,** 665–674.

Hardingham, G. E., Fukunaga, Y., and Bading, H. (2002). Extrasynaptic NMDARs oppose synaptic NMDARs by triggering CREB shut-off and cell death pathways. *Nat. Neurosci.* **5,** 405–414.

He, Y., Imam, S. Z., Dong, Z., Jankovic, J., Ali, S. F., Appel, S. H., and Le, W. (2003). Role of nitric oxide in rotenone-induced nigro-striatal injury. *J. Neurochem.* **86,** 1338–1345.

Hess, D. T., Matsumoto, A., Kim, S. O., Marshall, H. E., and Stamler, J. S. (2005). Protein *S*-nitrosylation: Purview and parameters. *Nat. Rev. Mol. Cell Biol.* **6,** 150–166.

Hetz, C., Russelakis-Carneiro, M., Maundrell, K., Castilla, J., and Soto, C. (2003). Caspase-12 and endoplasmic reticulum stress mediate neurotoxicity of pathological prion protein. *EMBO J.* **22,** 5435–5445.

Hetz, C., Russelakis-Carneiro, M., Walchli, S., Carboni, S., Vial-Knecht, E., Maundrell, K., Castilla, J., and Soto, C. (2005). The disulfide isomerase Grp58 is a protective factor against prion neurotoxicity. *J. Neurosci.* **25,** 2793–2802.

Houk, K. N., Hietbrink, B. N., Bartberger, M. D., McCarren, P. R., Choi, B. Y., Voyksner, R. D., Stamler, J. S., and Toone, E. J. (2003). Nitroxyl disulfides, novel intermediates in transnitrosation reactions. *J. Am. Chem. Soc.* **125,** 6972–6976.

Hu, B. R., Martone, M. E., Jones, Y. Z., and Liu, C. L. (2000). Protein aggregation after transient cerebral ischemia. *J. Neurosci.* **20,** 3191–3199.

Huang, Z., Huang, P. L., Panahian, N., Dalkara, T., Fishman, M. C., and Moskowitz, M. A. (1994). Effects of cerebral ischemia in mice deficient in neuronal nitric oxide synthase. *Science* **265,** 1883–1885.

Huynh, D. P., Scoles, D. R., Nguyen, D., and Pulst, S. M. (2003). The autosomal recessive juvenile Parkinson disease gene product, parkin, interacts with and ubiquitinates synaptotagmin XI. *Hum. Mol. Genet.* **12,** 2587–2597.

Iadecola, C., Zhang, F., Casey, R., Nagayama, M., and Ross, M. E. (1997). Delayed reduction of ischemic brain injury and neurological deficits in mice lacking the inducible nitric oxide synthase gene. *J. Neurosci.* **17,** 9157–9164.

Imai, Y., Soda, M., Inoue, H., Hattori, N., Mizuno, Y., and Takahashi, R. (2001). An unfolded putative transmembrane polypeptide, which can lead to endoplasmic reticulum stress, is a substrate of Parkin. *Cell* **105,** 891–902.

Imai, Y., Soda, M., Hatakeyama, S., Akagi, T., Hashikawa, T., Nakayama, K. I., and Takahashi, R. (2002). CHIP is associated with Parkin, a gene responsible for familial Parkinson's disease, and enhances its ubiquitin ligase activity. *Mol. Cell* **10,** 55–67.

Isaacs, A. M., Senn, D. B., Yuan, M., Shine, J. P., and Yankner, B. A. (2006). Acceleration of amyloid β-peptide aggregation by physiological concentrations of calcium. *J. Biol. Chem.* **281,** 27916–27923.

Jaffrey, S. R., Erdjument-Bromage, H., Ferris, C. D., Tempst, P., and Snyder, S. H. (2001). Protein *S*-nitrosylation: A physiological signal for neuronal nitric oxide. *Nat. Cell Biol.* **3,** 193–197.

Kakimura, J., Kitamura, Y., Takata, K., Umeki, M., Suzuki, S., Shibagaki, K., Taniguchi, T., Nomura, Y., Gebicke-Haerter, P. J., Smith, M. A., Perry, G., and Shimohama, S. (2002). Microglial activation and amyloid-β clearance induced by exogenous heat-shock proteins. *FASEB J.* **16,** 601–603.

Kalia, S. K., Lee, S., Smith, P. D., Liu, L., Crocker, S. J., Thorarinsdottir, T. E., Glover, J. R., Fon, E. A., Park, D. S., and Lozano, A. M. (2004). BAG5 inhibits parkin and enhances dopaminergic neuron degeneration. *Neuron* **44,** 931–945.

Kandel, E. R., and O'Dell, T. J. (1992). Are adult learning mechanisms also used for development? *Science* **258,** 243–245.

Kaufman, R. J. (1999). Stress signaling from the lumen of the endoplasmic reticulum: Coordination of gene transcriptional and translational controls. *Genes Dev.* **13,** 1211–1233.

Kikuchi, H., Almer, G., Yamashita, S., Guegan, C., Nagai, M., Xu, Z., Sosunov, A. A., McKhann, G. M., II, and Przedborski, S. (2006). Spinal cord endoplasmic reticulum stress associated with a microsomal accumulation of mutant superoxide dismutase-1 in an ALS model. *Proc. Natl. Acad. Sci. USA* **103,** 6025–6030.

Kim, W. K., Choi, Y. B., Rayudu, P. V., Das, P., Asaad, W., Arnelle, D. R., Stamler, J. S., and Lipton, S. A. (1999). Attenuation of NMDA receptor activity and neurotoxicity by nitroxyl anion, NO. *Neuron* **24,** 461–469.

Ko, H. S., Uehara, T., and Nomura, Y. (2002). Role of ubiquilin associated with protein-disulfide isomerase in the endoplasmic reticulum in stress-induced apoptotic cell death. *J. Biol. Chem.* **277,** 35386–35392.

Lafon-Cazal, M., Pietri, S., Culcasi, M., and Bockaert, J. (1993). NMDA-dependent superoxide production and neurotoxicity. *Nature* **364,** 535–537.

LaVoie, M. J., and Hastings, T. G. (1999). Peroxynitrite- and nitrite-induced oxidation of dopamine: Implications for nitric oxide in dopaminergic cell loss. *J. Neurochem.* **73,** 2546–2554.

LaVoie, M. J., Ostaszewski, B. L., Weihofen, A., Schlossmacher, M. G., and Selkoe, D. J. (2005). Dopamine covalently modifies and functionally inactivates parkin. *Nat. Med.* **11,** 1214–1221.

Lei, S. Z., Pan, Z. H., Aggarwal, S. K., Chen, H. S., Hartman, J., Sucher, N. J., and Lipton, S. A. (1992). Effect of nitric oxide production on the redox modulatory site of the NMDA receptor-channel complex. *Neuron* **8,** 1087–1099.

Leroy, E., Boyer, R., Auburger, G., Leube, B., Ulm, G., Mezey, E., Harta, G., Brownstein, M. J., Jonnalagada, S., Chernova, T., Dehejia, A., Lavedan, C., *et al.* (1998). The ubiquitin pathway in Parkinson's disease. *Nature* **395,** 451–452.

Liberatore, G. T., Jackson-Lewis, V., Vukosavic, S., Mandir, A. S., Vila, M., McAuliffe, W. G., Dawson, V. L., Dawson, T. M., and Przedborski, S. (1999). Inducible nitric oxide synthase stimulates dopaminergic neurodegeneration in the MPTP model of Parkinson disease. *Nat. Med.* **5,** 1403–1409.

Lim, K. L., Chew, K. C., Tan, J. M., Wang, C., Chung, K. K., Zhang, Y., Tanaka, Y., Smith, W., Engelender, S., Ross, C. A., Dawson, V. L., and Dawson, T. M. (2005). Parkin mediates nonclassical, proteasomal-independent ubiquitination of synphilin-1: Implications for Lewy body formation. *J. Neurosci.* **25,** 2002–2009.

Lin, M. T., and Beal, M. F. (2006). Mitochondrial dysfunction and oxidative stress in neurodegenerative diseases. *Nature* **443,** 787–795.

Lipton, S. A. (2006a). Paradigm shift in neuroprotection by NMDA receptor blockade: Memantine and beyond. *Nat. Rev. Drug Discov.* **5,** 160–170.

Lipton, S. A. (2006b). NMDA receptors, glial cells, and clinical medicine. *Neuron* **6,** 9–11.

Lipton, S. A., and Rosenberg, P. A. (1994). Excitatory amino acids as a final common pathway for neurologic disorders. *N. Engl. J. Med.* **330,** 613–622.

Lipton, S. A., and Wang, Y. F. (1996). NO-related species can protect from focal cerebral ischemia/reperfusion. *In* "Pharmacology of Cerebral Ischemia" (J. Krieglstein and H. Oberpichler-Schwenk, Eds.), pp. 183–191. Wissenschaftliche Verlagsgesellschaft, Stuttgart.

Lipton, S. A., Choi, Y. B., Pan, Z. H., Lei, S. Z., Chen, H. S., Sucher, N. J., Loscalzo, J., Singel, D. J., and Stamler, J. S. (1993). A redox-based mechanism for the neuroprotective and neurodestructive effects of nitric oxide and related nitroso-compounds. *Nature* **364,** 626–632.

Lipton, S. A., Choi, Y. B., Takahashi, H., Zhang, D., Li, W., Godzik, A., and Bankston, L. A. (2002). Cysteine regulation of protein function—as exemplified by NMDA-receptor modulation. *Trends Neurosci.* **25,** 474–480.

Lipton, S. A., Nakamura, T., Yao, D., Shi, Z. Q., Uehara, T., and Gu, Z. (2005). Comment on "S-nitrosylation of parkin regulates ubiquitination and compromises parkin's protective function." *Science* **308,** 1870.

Liu, Y., Fallon, L., Lashuel, H. A., Liu, Z., and Lansbury, P. T., Jr. (2002). The UCH-L1 gene encodes two opposing enzymatic activities that affect α-synuclein degradation and Parkinson's disease susceptibility. *Cell* **111,** 209–218.

Lyles, M. M., and Gilbert, H. F. (1991). Catalysis of the oxidative folding of ribonuclease A by protein disulfide isomerase: Dependence of the rate on the composition of the redox buffer. *Biochemistry* **30,** 613–619.

Mannick, J. B., Hausladen, A., Liu, L., Hess, D. T., Zeng, M., Miao, Q. X., Kane, L. S., Gow, A. J., and Stamler, J. S. (1999). Fas-induced caspase denitrosylation. *Science* **284,** 651–654.

Marin, I., and Ferrus, A. (2002). Comparative genomics of the RBR family, including the Parkinson's disease-related gene parkin and the genes of the ariadne subfamily. *Mol. Biol. Evol.* **19,** 2039–2050.

Martinez-Ruiz, A., Villanueva, L., Gonzalez de Orduna, C., Lopez-Ferrer, D., Higueras, M. A., Tarin, C., Rodriguez-Crespo, I., Vazquez, J., and Lamas, S. (2005). *S*-nitrosylation of Hsp90 promotes the inhibition of its ATPase and endothelial nitric oxide synthase regulatory activities. *Proc. Natl. Acad. Sci. USA* **102,** 8525–8530.

Melino, G., Bernassola, F., Knight, R. A., Corasaniti, M. T., Nistico, G., and Finazzi-Agro, A. (1997). *S*-nitrosylation regulates apoptosis. *Nature* **388,** 432–433.

Mori, K. (2000). Tripartite management of unfolded proteins in the endoplasmic reticulum. *Cell* **101,** 451–454.

Muchowski, P. J., and Wacker, J. L. (2005). Modulation of neurodegeneration by molecular chaperones. *Nat. Rev. Neurosci.* **6,** 11–22.

Nakamura, T., Gu, Z., and Lipton, S. A. (2007). Contribution of glutamatergic signaling to nitrosative stress-induced protein misfolding in normal brain aging and neurodegenerative diseases. [Epub ahead of print: March 26, 2007].

Nakamura, T., and Lipton, S. A. (2007). Molecular mechanisms of nitrosative stress-mediated protein misfolding in neurodegenerative diseases. *Cell. Mol. Life Sci.* [Epub ahead of print: Aprint 24, 2007].

Nishikawa, K., Li, H., Kawamura, R., Osaka, H., Wang, Y. L., Hara, Y., Hirokawa, T., Manago, Y., Amano, T., Noda, M., Aoki, S., and Wada, K. (2003). Alterations of structure and hydrolase activity of parkinsonism-associated human ubiquitin carboxyl-terminal hydrolase L1 variants. *Biochem. Biophys. Res. Commun.* **304,** 176–183.

O'Dell, T. J., Hawkins, R. D., Kandel, E. R., and Arancio, O. (1991). Tests of the roles of two diffusible substances in long-term potentiation: Evidence for nitric oxide as a possible early retrograde messenger. *Proc. Natl. Acad. Sci. USA* **88,** 11285–11289.

Olney, J. W. (1969). Brain lesions, obesity, and other disturbances in mice treated with monosodium glutamate. *Science* **164,** 719–721.

Olney, J. W., Wozniak, D. F., and Farber, N. B. (1997). Excitotoxic neurodegeneration in Alzheimer disease. New hypothesis and new therapeutic strategies. *Arch. Neurol.* **54,** 1234–1240.

Papadia, S., Stevenson, P., Hardingham, N. R., Bading, H., and Hardingham, G. E. (2005). Nuclear Ca^{2+} and the cAMP response element-binding protein family mediate a late phase of activity-dependent neuroprotection. *J. Neurosci.* **25,** 4279–4287.

Patil, C., and Walter, P. (2001). Intracellular signaling from the endoplasmic reticulum to the nucleus: The unfolded protein response in yeast and mammals. *Curr. Opin. Cell Biol.* **13,** 349–355.

Paz Gavilan, M., Vela, J., Castano, A., Ramos, B., del Rio, J. C., Vitorica, J., and Ruano, D. (2006). Cellular environment facilitates protein accumulation in aged rat hippocampus. *Neurobiol. Aging* **27,** 973–982.

Przedborski, S., Jackson-Lewis, V., Yokoyama, R., Shibata, T., Dawson, V. L., and Dawson, T. M. (1996). Role of neuronal nitric oxide in 1-methyl-4-phenyl-1,2,3,6-tetrahydropyridine (MPTP)-induced dopaminergic neurotoxicity. *Proc. Natl. Acad. Sci. USA* **93,** 4565–4571.

Rakhit, R., Crow, J. P., Lepock, J. R., Kondejewski, L. H., Cashman, N. R., and Chakrabartty, A. (2004). Monomeric Cu,Zn-superoxide dismutase is a common misfolding intermediate in the oxidation models of sporadic and familial amyotrophic lateral sclerosis. *J. Biol. Chem.* **279,** 15499–15504.

Rao, R. V., and Bredesen, D. E. (2004). Misfolded proteins, endoplasmic reticulum stress and neurodegeneration. *Curr. Opin. Cell Biol.* **16,** 653–662.

Ross, C. A., and Pickart, C. M. (2004). The ubiquitin-proteasome pathway in Parkinson's disease and other neurodegenerative diseases. *Trends Cell Biol.* **14,** 703–711.

Schonhoff, C. M., Matsuoka, M., Tummala, H., Johnson, M. A., Estevez, A. G., Wu, R., Kamaid, A., Ricart, K. C., Hashimoto, Y., Gaston, B., Macdonald, T. L., Xu, Z., *et al.* (2006). *S*-nitrosothiol depletion in amyotrophic lateral sclerosis. *Proc. Natl. Acad. Sci. USA* **103,** 2404–2409.

Schuman, E. M., and Madison, D. V. (1994). Locally distributed synaptic potentiation in the hippocampus. *Science* **263,** 532–536.

Shimura, H., Schlossmacher, M. G., Hattori, N., Frosch, M. P., Trockenbacher, A., Schneider, R., Mizuno, Y., Kosik, K. S., and Selkoe, D. J. (2001). Ubiquitination of a new form of α-synuclein by parkin from human brain: Implications for Parkinson's disease. *Science* **293,** 263–269.

Sidrauski, C., Chapman, R., and Walter, P. (1998). The unfolded protein response: An intracellular signalling pathway with many surprising features. *Trends Cell Biol.* **8,** 245–249.

Sliskovic, I., Raturi, A., and Mutus, B. (2005). Characterization of the *S*-denitrosation activity of protein disulfide isomerase. *J. Biol. Chem.* **280,** 8733–8741.

Sriram, S. R., Li, X., Ko, H. S., Chung, K. K., Wong, E., Lim, K. L., Dawson, V. L., and Dawson, T. M. (2005). Familial-associated mutations differentially disrupt the solubility, localization, binding and ubiquitination properties of parkin. *Hum. Mol. Genet.* **14,** 2571–2586.

Stamler, J. S. (1994). Redox signaling: Nitrosylation and related target interactions of nitric oxide. *Cell* **78,** 931–936.

Stamler, J. S., and Hausladen, A. (1998). Oxidative modifications in nitrosative stress. *Nat. Struct. Biol.* **5,** 247–249.

Stamler, J. S., Simon, D. I., Osborne, J. A., Mullins, M. E., Jaraki, O., Michel, T., Singel, D. J., and Loscalzo, J. (1992a). *S*-nitrosylation of proteins with nitric oxide: Synthesis and characterization of biologically active compounds. *Proc. Natl. Acad. Sci. USA* **89,** 444–448.

Stamler, J. S., Singel, D. J., and Loscalzo, J. (1992b). Biochemistry of nitric oxide and its redox-activated forms. *Science* **258,** 1898–1902.

Stamler, J. S., Jia, L., Eu, J. P., McMahon, T. J., Demchenko, I. T., Bonaventura, J., Gernert, K., and Piantadosi, C. A. (1997). Blood flow regulation by *S*-nitrosohemoglobin in the physiological oxygen gradient. *Science* **276,** 2034–2037.

Stamler, J. S., Lamas, S., and Fang, F. C. (2001). Nitrosylation. The prototypic redox-based signaling mechanism. *Cell* **106,** 675–683.

Staropoli, J. F., McDermott, C., Martinat, C., Schulman, B., Demireva, E., and Abeliovich, A. (2003). Parkin is a component of an SCF-like ubiquitin ligase complex and protects postmitotic neurons from kainate excitotoxicity. *Neuron* **37,** 735–749.

Szegezdi, E., Logue, S. E., Gorman, A. M., and Samali, A. (2006). Mediators of endoplasmic reticulum stress-induced apoptosis. *EMBO Rep.* **7,** 880–885.

Tanaka, S., Uehara, T., and Nomura, Y. (2000). Up-regulation of protein-disulfide isomerase in response to hypoxia/brain ischemia and its protective effect against apoptotic cell death. *J. Biol. Chem.* **275,** 10388–10393.

Tenneti, L., D'Emilia, D. M., and Lipton, S. A. (1997). Suppression of neuronal apoptosis by *S*-nitrosylation of caspases. *Neurosci. Lett.* **236,** 139–142.

Tian, G., Xiang, S., Noiva, R., Lennarz, W. J., and Schindelin, H. (2006). The crystal structure of yeast protein disulfide isomerase suggests cooperativity between its active sites. *Cell* **124,** 61–73.

Tiwari, A., and Hayward, L. J. (2003). Familial amyotrophic lateral sclerosis mutants of copper/zinc superoxide dismutase are susceptible to disulfide reduction. *J. Biol. Chem.* **278,** 5984–5992.

Uehara, T., Nakamura, T., Yao, D., Shi, Z. Q., Gu, Z., Ma, Y., Masliah, E., Nomura, Y., and Lipton, S. A. (2006). *S*-nitrosylated protein-disulphide isomerase links protein misfolding to neurodegeneration. *Nature* **441,** 513–517.

Vuori, K., Pihlajaniemi, T., Myllyla, R., and Kivirikko, K. I. (1992). Site-directed mutagenesis of human protein disulphide isomerase: Effect on the assembly, activity and endoplasmic reticulum retention of human prolyl 4-hydroxylase in Spodoptera frugiperda insect cells. *EMBO J.* **11,** 4213–4217.

Wang, C., Ko, H. S., Thomas, B., Tsang, F., Chew, K. C., Tay, S. P., Ho, M. W., Lim, T. M., Soong, T. W., Pletnikova, O., Troncoso, J., Dawson, V. L., *et al.* (2005). Stress-induced alterations in parkin solubility promote parkin aggregation and compromise parkin's protective function. *Hum. Mol. Genet.* **14,** 3885–3897.

Yamamoto, A., Friedlein, A., Imai, Y., Takahashi, R., Kahle, P. J., and Haass, C. (2005). Parkin phosphorylation and modulation of its E3 ubiquitin ligase activity. *J. Biol. Chem.* **280,** 3390–3399.

Yao, D., Gu, Z., Nakamura, T., Shi, Z. Q., Ma, Y., Gaston, B., Palmer, L. A., Rockenstein, E. M., Zhang, Z., Masliah, E., Uehara, T., and Lipton, S. A. (2004). Nitrosative stress linked to sporadic Parkinson's disease: *S*-nitrosylation of parkin regulates its E3 ubiquitin ligase activity. *Proc. Natl. Acad. Sci. USA* **101,** 10810–10814.

Yoo, B. C., Krapfenbauer, K., Cairns, N., Belay, G., Bajo, M., and Lubec, G. (2002). Overexpressed protein disulfide isomerase in brains of patients with sporadic Creutzfeldt-Jakob disease. *Neurosci. Lett.* **334,** 196–200.

Yoshida, H., Matsui, T., Yamamoto, A., Okada, T., and Mori, K. (2001). XBP1 mRNA is induced by ATF6 and spliced by IRE1 in response to ER stress to produce a highly active transcription factor. *Cell* **107,** 881–891.

Zhang, K., and Kaufman, R. J. (2006). The unfolded protein response: A stress signaling pathway critical for health and disease. *Neurology* **66,** S102–S109.

Zhang, Y., Gao, J., Chung, K. K., Huang, H., Dawson, V. L., and Dawson, T. M. (2000). Parkin functions as an E2-dependent ubiquitin-protein ligase and promotes the degradation of the synaptic vesicle-associated protein, CDCrel-1. *Proc. Natl. Acad. Sci. USA* **97,** 13354–13359.

Zhao, L., Longo-Guess, C., Harris, B. S., Lee, J. W., and Ackerman, S. L. (2005). Protein accumulation and neurodegeneration in the woozy mutant mouse is caused by disruption of SIL1, a cochaperone of BiP. *Nat. Genet.* **37,** 974–979.

Zurakowski, D., Vorwerk, C. K., Gorla, M., Kanellopoulos, A. J., Chaturvedi, N., Grosskreutz, C. L., Lipton, S. A., and Dreyer, E. B. (1998). Nitrate therapy may retard glaucomatous optic neuropathy, perhaps through modulation of glutamate receptors. *Vision Res.* **38,** 1489–1494.

INNATE IMMUNITY AND PROTECTIVE NEUROINFLAMMATION: NEW EMPHASIS ON THE ROLE OF NEUROIMMUNE REGULATORY PROTEINS

M. Griffiths,* J. W. Neal,† and P. Gasque*,†,‡

*Brain Inflammation and Immunity Group (BIIG), Department of Medical Biochemistry
School of Medicine, Cardiff University, CF144XN Cardiff, United Kingdom
†Department of Histopathology, School of Medicine, Cardiff University
CF144XN Cardiff, United Kingdom
‡Groupe de recherche sur les maladies infectieuses et inflammatoires
GRII, LBGM, Faculty of Sciences and Technology, University of la Reunion
Saint Denis, Ile de la Reunion

Brain inflammation due to infection, hemorrhage, and aging is associated with activation of the local innate immune system as expressed by infiltrating cells, resident glial cells, and neurons. The innate immune response relies on the detection of "nonself" and "danger-self" ligands behaving as "eat me signals" by a plethora of pattern recognition receptors (PRRs) expressed by professional and amateur phagocytes to promote the clearance of pathogens, toxic cell debris (amyloid fibrils, aggregated synucleins, prions), and apoptotic cells accumulating within the brain parenchyma and the cerebrospinal fluid (CSF). These PRRs (e.g., complement, TLR, CD14, scavenger receptors) are highly conserved between vertebrates and invertebrates and may represent the most ancestral innate scavenging system involved in tissue homeostasis. However, in some diseases, these protective mechanisms lead to neurodegeneration on the ground that

INTERNATIONAL REVIEW OF
NEUROBIOLOGY, VOL. 82
DOI: 10.1016/S0074-7742(07)82002-2

0074-7742/07 $35.00

several innate immune molecules have neurocytotoxic activities. The response is a "double-edged sword" representing a fine balance between protective and detrimental effects. Several key regulatory mechanisms have now been evidenced in the control of CNS innate immunity, and these could be harnessed to explore novel therapeutic avenues. We will herein provide new emphasis on the role of neuroimmune regulatory proteins (NIRegs), such as CD95L, TNF, CD200, CD47, sialic acids, CD55, CD46, fH, C3a, HMGB1, which are involved in silencing innate immunity at the cellular and molecular levels and suppression of inflammation. For instance, NIRegs may play an important role in controlling lymphocyte/macrophage/microglia hyperinflammatory responses, while sparing host defense and repair mechanisms. Moreover, NIRegs have direct beneficial effects on neurogenesis and contributing to brain tissue remodeling.

I. Characteristics of the Cellular and Molecular Innate Immune Responses in the Brain

Classically, innate immune cells are known as neutrophils, natural killer (NK) cells, dendritic cells, and macrophages involved in the selective recognition and the clearance of pathogens and toxic cell debris during infection or tissue injury (Gordon, 2002; Medzhitov and Janeway, 2002; Stuart and Ezekowitz, 2005). However, there is little evidence of an immunosurveillance of the brain by these peripheral cells, and it is now evident that resident cells, glial cells, ependymal cells, and neurons are capable of mounting a robust innate immune response on their own (Hauwel *et al.*, 2005; Nguyen *et al.*, 2002; Schwartz *et al.*, 1999). The local innate immune response is based on the recognition of "nonself" and "altered-self" patterns, also called "danger signals," by molecules and receptors expressed essentially by microglia but also found on astrocytes, oligodendrocytes, and neurons (Elward and Gasque, 2003). These molecules and receptors are called pattern recognition receptors (PRRs) (see Fig. 1) and are released in soluble forms or displayed on the cell membrane. These receptors act alone and in concert to bind, phagocytose, and transduce cellular signals derived from the molecular patterns. The outcome of these interactions is dependent on the nature of the ligands and on the nature and combination of the ligated receptors. Whereas much attention has been focused on the properties and activities of the Toll-like receptors (TLRs) in this process, many other CNS innate immune molecules have been described (e.g., complement, scavenger receptors, IFN-α) (Gasque *et al.*, 2000; Nguyen *et al.*, 2002). The decoding and sampling of the microenvironment for danger signals will contribute to the removal of the harmful intruders (Matzinger, 2007). Interestingly, we will describe how the recognition of

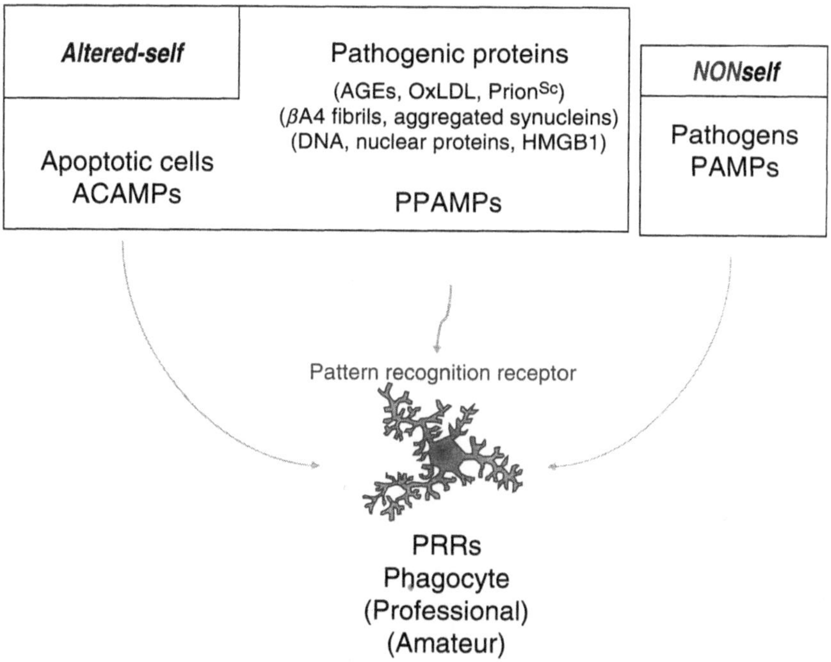

FIG. 1. The innate immune response relies on the detection of "nonself" and "altered-self" ligands behaving as "eat me signals" by a plethora of pattern recognition receptors (PRRs) expressed by professional and amateur phagocytes to promote the clearance of pathogens, toxic cell debris (amyloid fibrils, aggregated synucleins, prion), and apoptotic cells accumulating within the brain parenchyma and the CSF. Receptor complexes contribute to discriminating and initiating appropriate responses, that is, phlogistic or nonphlogistic responses in the case of pathogen or apoptotic cells, respectively. PAMPs, pathogen-associated molecular patterns; PPAMPs, pathogenic protein-associated molecular patterns; and ACAMPs, apoptotic cell-associated molecular patterns. (See Color Insert.)

apoptotic cells or altered proteins via common PRRs appears to result in responses that differ from those elicited by microbial ligands (Savill *et al.*, 2002). From a therapeutic standpoint, it will be important to manipulate this innate immune response to promote the clearance of pathogens or toxic proteins accumulating in the brain (Hauwel *et al.*, 2005; Lazarov-Spiegler *et al.*, 1996). In Alzheimer's disease (AD), it may be possible to reeducate and to boost the phagocytic activity of microglia to limit the accumulation of toxic cell debris (Baron *et al.*, 2007; Butovsky *et al.*, 2006; Simard *et al.*, 2006).

However and critically, several innate immune molecules can contribute to cytotoxic and cytolytic activities and must be controlled to avoid neuronal loss (neurodegeneration) and robust inflammation (Gendelman, 2002; McGeer and McGeer, 2002; Minghetti, 2005). Several TLRs and complement have recently been involved in neuroinflammation and demyelinating diseases. In this context, it is plausible that these innate immune molecules (e.g., TLR4, TLR9, TLR8, C1q, C3) are sensing and responding to endogenous ligands accumulating at the site of injury and eliciting a chronic autoimmune response (Elward *et al.*, 2005; Lehnardt *et al.*, 2003; Ma *et al.*, 2006; Mead *et al.*, 2002; Prinz *et al.*, 2006; Rus *et al.*, 2006).

Cellular and regulatory mechanisms of innate immunity have recently been described and thought to be at the root of neuroprotective mechanisms. Hence, there is new emphasis on the role of neuroimmune regulatory proteins (NIRegs) which in analogy to T regulatory cells (Tregs) are involved in silencing and reshaping an adverse innate immune response and polarizing macrophages and microglia toward a protective phenotype. Macrophages and microglia are programmed to kill and phagocytose target cells, and it is equally important to understand the mechanisms controlling their eager appetite (Marin-Teva *et al.*, 2004). NIRegs were originally termed self-associated molecular patterns (SAMPs) on the ground that they are expressed by self-viable cells to control and maintain the inflammatory reaction to the site of injury (Elward and Gasque, 2003; Grimsley and Ravichandran, 2003). Several SAMPs have now been described such as CD47 and behaving as "don't eat me" signals to prevent unwarranted phagocytosis (Elward and Gasque, 2003; Elward *et al.*, 2005; Gardai *et al.*, 2005). Moreover, there is mounting evidence that innate immune molecules (e.g., C3a) can contribute to tissue repair notably by stimulating the mobilization of neural stem cells and with the production of growth factors (Bsibsi *et al.*, 2006; Glezer *et al.*, 2006; Rahpeymai *et al.*, 2006). These different brain-derived innate immune properties are described below.

II. Innate Immune Response in Health: The Key Role of Physical Barriers

The brain is isolated from the systemic circulation by a protective blood–brain barrier (BBB) composed of endothelial cells linked by tight junctions and surrounded by the end-feet of astrocytes (Fig. 2; Pachter *et al.*, 2003). It has emerged that astrocytes and endothelial cells can produce innate immune molecules such as antimicrobial peptides (defensins, CRAMP), which actively control the infiltration of pathogens at the level of the BBB (Bergman *et al.*, 2006; Hao *et al.*, 2001). A further protective barrier composed of specialized ciliated glia, the ependyma, lines the ventricles preventing entry of pathogens from the cerebrospinal fluid (CSF) into the brain (Canova *et al.*, 2006; Martino *et al.*, 2001). Within the ventricles is the choroid plexus containing Kolmer phagocyte cells, similar to dendritic cells of the skin, which prevent the infiltration of pathogens. Kolmer cells express MHC II and are capable of presenting antigens to T lymphocyte cells and controlling T cell expansion by responding to various cytokines. Cells within the choroid plexus and the ependymal layer also express receptors that are capable of detecting pathogens in the CSF and regulating the inflammatory

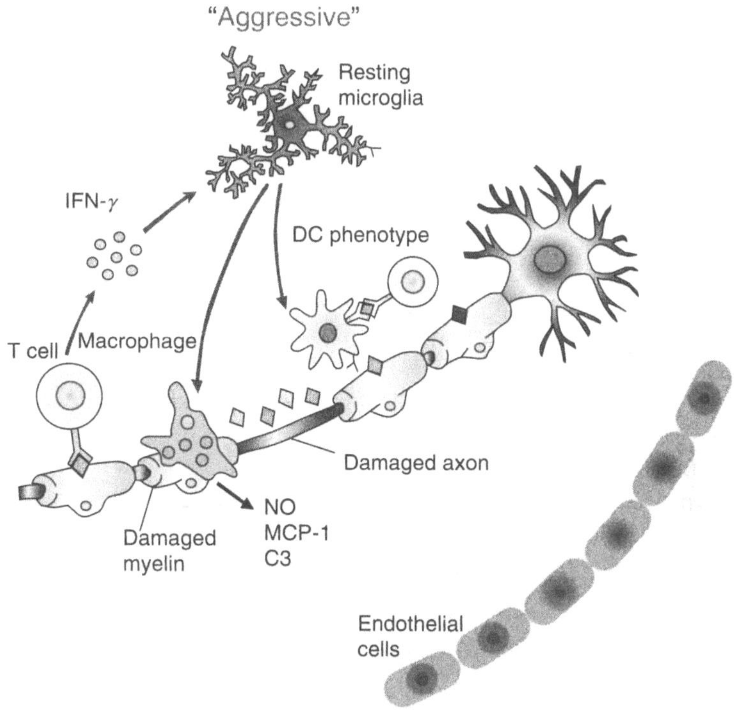

FIG. 2. The aggressive proinflammatory activity of microglia. Microglia, the resident macrophage in the brain, has long been considered as a key immune cell driving inflammation and hence contributing to disease processes. For instance, microglia was shown to activate dendritic cells (DC), autoimmune T cells and macrophages to release proinflammatory molecules such as TNF-α, IL1-β, cytotoxic complement, proteins, NO, chemoattractants, all in all contributing to demyelination axonal injuries, and neurodegeneration. (See Color Insert.)

response (McMenamin, 1999; Serafini *et al.*, 2000). These innate immune receptors are highly conserved and include for example TLRs, CD14, and complement receptors (CRs) CR3 and CR4 (Bsibsi *et al.*, 2002; Laflamme and Rivest, 2001; Singhrao *et al.*, 2000). The details regarding the recognition and interaction between glia and pathogens will be discussed below.

III. Immunoprivileged Status of the Brain by Preventing the Infiltration of Potentially Harmful Systemic Immune Cells: Roles of ACAMPs

The active destruction of infiltrating T lymphocytes through induction of apoptosis provides the brain with a degree of low immunosurveillance, preventing the entry of autoimmune lymphocytes and downregulating inflammation

(Pender *et al.*, 1991; Zipp *et al.*, 1999). Active apoptosis of infiltrating T lymphocytes is induced by neurons and glia utilizing the "death signaling pathways" based on CD95L (FasL)/CD95 (Fas), perforin, and the TNF lymphotoxin receptor–TNF receptor 1 (TNFR1) pathway (Gasque *et al.*, 1998; Magnus *et al.*, 2001; Medana *et al.*, 2001; Raine *et al.*, 1998; Zipp *et al.*, 1999). The initiator of apoptosis, CD95L, is expressed by both astrocytes and oligodendroglia and transmits an apoptotic signal to target T cells. In EAE (experimental autoimmune encephalomyelitis), it was found that T cells infiltrating preferentially the injured brain parenchyma were found in the immediate vicinity of CD95L expressing neurons and even inside their perikarya (Flugel *et al.*, 2000). This interaction at the cell surface induces the activation of caspases and subsequent apoptosis of the target cell, with engulfment by microglia and downregulation of their activation (Magnus *et al.*, 2001). As such, CD95L, astrocyte-derived perforin and TNF-α could be regarded as NIRegs controlling innate and acquired immune responses. Recovery from EAE is increased through induction of apoptosis in inflammatory T cells by the TNFR signaling pathway, and in TNFR knockout mice T cell apoptosis is reduced by 50% in the periphery of demyelinating plaques (Eugster *et al.*, 1999). In multiple sclerosis (MS), it is possible that this death signaling pathway is functional in reducing the severity of demyelination (Dowling *et al.*, 1996).

The induction of apoptosis in infiltrating T lymphocytes and virally infected cells will reduce host tissue destruction, but only if they are rapidly cleared by resident cells. Apoptotic cells contain large amounts of toxic enzymes that are able to activate proinflammatory cytokine release. Phagocytosis of apoptotic cells is termed nonphlogistic and is accompanied by a downregulation of proinflammatory cytokines contributing to the limitation of tissue damage and "self" defense (Savill *et al.*, 2002). Apoptotic cell removal may also play a role in tissue homeostasis by inducing growth factors which could serve to drive replacement of deleted cells. However, these beneficial responses may have been subverted by pathogens to gain entry in the CNS (Everett and McFadden, 1999).

Apoptotic cells are "altered-self" and express apoptotic cell–associated molecular patterns, ACAMPs (Elward and Gasque, 2003; Gregory and Devitt, 2004; see Fig. 1). The identity of ACAMPS has not been fully elucidated but includes nucleic acids, sugars, oxidized low density lipoproteins, and alteration of membrane electrical charge (Elward and Gasque, 2003). The best-characterized ACAMP to date is phosphotidylserine (PS) (Fadok *et al.*, 2000). Glia and macrophages express a range of PRRs that recognize ACAMPS, including the phosphotidylserine receptor (PSR), CD14, CD36, and milk fat globulin (MFG-EGF 8) (De Simone *et al.*, 2004; Fadok *et al.*, 1998, 2000; Gregory, 2000; Hanayama *et al.*, 2002; Leonardi-Essmann *et al.*, 2005). In PSR-deficient mice, dead cells accumulated in the lung and brain, causing abnormal development and leading to neonatal lethality. A fraction of PSR knockout mice manifested a hyperplasic brain phenotype resembling that of mice deficient in the cell death-associated genes encoding Apaf-1, caspase-3, and caspase-9, which suggests that phagocytes may also be involved in promoting apoptosis (Li *et al.*, 2003).

Activation of the classical innate immune complement pathway by virtue of C1q recognition of nucleic acids (DNA) initiates the generation of opsonins C3b and iC3b that bind to apoptotic cells and act as targets for the phagocytic CR3 and CR4 receptors expressed by macrophages and microglia (Elward *et al.*, 2005; Korb and Ahearn, 1997).

IV. Protective Innate Immune Response During Brain Infection and Inflammation to Promote the Clearance of Pathogens: Roles of PAMPs

After penetrating a damaged BBB, an infiltrating pathogen will encounter the innate immune response delivered by professional (microglia) and nonprofessional phagocytes (endothelial cells, astrocytes, ependyma, neurons, and oligodendroglia). These cells provide the second line of the brain's innate defense against infection, but must also be regulated to prevent destruction of host "self" cells. Astrocytes and microglia express a large array of membrane and soluble PRRs. These include lectins [e.g., phagocytic macrophage mannose receptor (MMR)], scavenger receptors (e.g., SRA), TLR molecules (Toll-like receptors, TLR2 and TLR4, CD14), NOD, and associated complement molecules (e.g., C1q, C3, CR1, CR3, CR4) (Bsibsi *et al.*, 2002; Burudi *et al.*, 1999; Dalpke *et al.*, 2002; Esen *et al.*, 2004; Gasque *et al.*, 2000; Husemann *et al.*, 2002; Kielian *et al.*, 2002; Laflamme *et al.*, 2003; Linehan *et al.*, 1999). Neurons and endothelial cells can also contribute to the innate immune response against pathogens. For example, the platelet-activating factor receptor (PAFr) expressed by cerebral endothelial cells and neurons have been involved in the innate immune response to gram-positive bacterial cell wall (Fillon *et al.*, 2006). Moreover, it has been shown that neurons are the main producers of IFN-α/β, a major innate immune cytokine, required to mount a robust response against virally infected cells (Delhaye *et al.*, 2006).

Innate immune molecules are capable of detecting unique arrangements of lipopolysaccharides (LPS) and peptidoglycan (PG) molecules termed pathogen-associated molecular patterns (PAMPs) within the cell walls of microorganisms. The complement system is uniquely armed to recognize and destroy viruses, prion, and other microbes infiltrating the CNS (Speth *et al.*, 2002). PAMPs are exclusively expressed by pathogens and therefore distinguish "self" (host) from "nonself" (pathogen). Removal of pathogens therefore reduces the severity of an inflammatory response before its destructive effects outweigh the benefits of pathogen clearance. The recognition of pathogens also relies on the defense collagens, a group of PRRs composed of a globular C-terminal sequence that recognizes PAMPS together with an N-terminal sequence that binds to specific phagocytic receptors on the surface of phagocytic cells (Lu *et al.*, 2002; Tenner, 1999). These include C1q, the first component of the complement classical pathway, mannan-binding lectin (MBL), and surfactant protein A (SPA). C1q and MBL have been shown to be expressed by astrocytes whereas the expression of SPA in the CNS is yet to be defined (Wagner *et al.*, 2003). The CRs CR1, CR3, and CR4 belong to the family of β-integrins, and they are important for identifying pathogens and apoptotic cells opsonized with complement component C3 fragments (C3b and iC3b) (Ehlers, 2000; Gasque, 2004). A new CR has been described, CRIg, of the immunoglobulin superfamily and which also binds C3b and iC3b (Helmy *et al.*, 2006). This receptor was best known as the VSIG4 or Z39Ig molecules and being weakly expressed in the brain. The neuroinflammatory phenotype of CR knockout animals has yet to be reported to firmly establish the roles of these receptors in pathogen handling and subsequent inflammatory processes. What is known, however, is that animals deficient for complement are highly susceptible to infectious challenges and corroborating human clinical settings (for review Walport, 2001).

V. Interactions of Innate Immune Molecules with Toxic Proteins: Roles of PPAMPs

In analogy to PAMPs, PPAMPs are pathogenic protein-associated molecular patterns which for some can be recognized by the same PRRs involved in the antipathogen response. Any disruption of the BBB invariably permits entry into the brain of neurotoxic systemic proteins, but it is in aging that the innate immune response probably plays a critical role in the elimination of toxic PPAMPs such as prion, amyloid fibrils, aggregated synucleins, and other toxic cell debris such as nuclear proteins (e.g., HMGB1, S100) released from damaged cells (Kovacs *et al.*, 2004; Rogers *et al.*, 2002; Tenner, 2001; Zhang *et al.*, 2005).

Thrombin is a serine protease vital for blood coagulation. High levels of thrombin persist at the site of brain injury for up to a week and have been shown to contribute to the formation of edema and cerebral ischemia (Gingrich and Traynelis, 2000; Lee *et al.*, 1996; Xi *et al.*, 2003). Thrombin is generated in the systemic circulation by cleavage of prothrombin (PT) by factor Xa, with subsequent generation of fibrin. Thrombin is also endogenously synthesized in the brain as confirmed by the presence of mRNAs for PT and factor Xa in both neurons and astrocytes (Citron *et al.*, 2000; Dihanich *et al.*, 1991; Shikamoto and Morita, 1999; Xi *et al.*, 2003). Functional receptors for thrombin and the corresponding mRNA PAR 1 and PAR 2 are expressed by neurons and astrocytes *in vivo* and *in vitro* (Gingrich and Traynelis, 2000; Ubl *et al.*, 2000; Wang and Reiser, 2003; Weinstein *et al.*, 1995). Thrombin, at low concentrations (50–100 pM) is neuroprotective, regulating NGF synthesis and synaptic outgrowth (Vivien and Buisson, 2000); it is protective against oxygen and glucose deprivation due to its modulation of intracellular calcium (Grand *et al.*, 1996; Masada *et al.*, 2000; Striggow *et al.*, 2000; Vaughan *et al.*, 1995). At high levels (500 nM), as found following intracerebral hemorrhage, thrombin is neurotoxic as it activates NMDA excitotoxic receptors and PAR-1 inhibiting neurite extension and neuronal repair (Gingrich and Traynelis, 2000; Xi *et al.*, 2003). To counter the neurotoxic effects of thrombin, both glia and neurons express a range of "self" defense proteins. These include the serine protease inhibitors (serpins) that prevent the synthesis of thrombin (Docagne *et al.*, 1999; Pindon *et al.*, 1997). The serpins include the antithrombin colligin (Hsp47) expressed by microglia and astrocytes, plasminogen activator inhibitor (PAI-1) and protease glial-derived nexin-1 (PN-1), both of which are expressed by astrocytes and neurons (Buisson *et al.*, 1998, 2003; Cavanaugh *et al.*, 1990). Pigment epithelium–derived factor (PEDF) is a serpin that is selectively trophic for motorneurons, protecting them *in vitro* against glutamate toxicity and also blocking microglial proliferation (Bilak *et al.*, 2002; Sugita *et al.*, 1997). Neuroserpin, an inhibitor of tissue plasminogen (tPA) expression, is restricted to neurons and astrocytes (Docagne *et al.*, 1999; Gingrich and Traynelis, 2000; Yepes *et al.*, 2000). In a rat model of stroke, neuroserpin and glial PN-1 expression was increased around the penumbra and administration of neuroserpin reduced infarct volume by inhibiting thrombin synthesis and promoting neuron survival (Yepes *et al.*, 2000). TGF-α and TGF-β are expressed by neurons *in vitro* and stimulate astrocyte expression of PAI-1, which inhibits thrombin synthesis and reduces its neurotoxic effects (Cavanaugh *et al.*, 1990; Docagne *et al.*, 1999; Gabriel *et al.*, 2003).

The innate immune system and in particular the complement system plays an important role in the selective recognition and phagocytosis of PPAMPs accumulating in the aging brain. Amyloid fibrils and aggregated forms of prion fibrils have been shown to be opsonized by complement components (C1q, C3) to promote clearance by macrophages and microglia bearing CRs (Kovacs *et al.*, 2004; Tenner, 2001). In AD, it has been suggested that CD14, TLR, and scavenger receptors contribute to the removal of β-amyloid fibrils (El Khoury *et al.*, 2003; Fassbender *et al.*, 2004; Tahara *et al.*, 2006). Many other neurodegenerative diseases, including Parkinson's disease, Huntington's disease (HD), and Pick's disease, are associated with the formation of intracellular aggregates of toxic proteins leading to neuronal apoptosis. The degradation pathways acting on such aggregate-prone cytosolic proteins include the ubiquitin–proteasome system and macroautophagy, but it is also possible that innate immune molecules recognize these altered neurons through apoptotic markers and prompting safe removal in a nonphlogistic manner.

Necrotic cells are known to release very powerful inflammatory molecules such as heat shock proteins, S100 and HMGB1, the high-mobility group Box 1 protein.

These molecules are PPAMPs and interact with several PRRs such as TLR to mediate inflammatory responses in the CNS (Foell *et al.*, 2007; Lotze and Tracey, 2005; Roth *et al.*, 2003). Very little is known about the regulatory mechanisms controlling the toxic activity of these factors although the role of CD141 (thrombomodulin), a natural anticoagulant through its capacity to bind thrombin, has been evidenced (Abeyama *et al.*, 2005). In the brain, CD141 is expressed by endothelial cells and microglia and could rapidly sequester HMGB1 before signaling to TLR and or RAGE (receptor for advanced glycated end products).

VI. Regulating the Innate Immune Response in the CNS While Promoting Tissue Repair: Roles of Neuroimmune Regulatory Molecules

Critically, innate immune cells face an important paradox, namely that seemingly identical armory of PRRs and integrating signaling pathways are involved to discriminate between a dangerous intruder (pathogen) by mounting a strong proinflammatory response, while the clearance of apoptotic cells or toxic cell debris seem to engage a rather nonphlogistic response. To counter the neurotoxic effects of innate immune molecules, both glia and neurons express a range of "self" defense proteins. Interestingly, there is a growing body of evidence that neurons may be capable of evading detection by activated microglia and macrophages by expressing the so-called "don't eat me" signals or SAMPs. SAMPs are markers of "self" preventing recognition of host cells and reducing the severity of any inflammatory response through inhibition of innate immune cells such as microglia and infiltrating macrophages (Elward and Gasque, 2003; Gardai *et al.*, 2005; Grimsley and Ravichandran, 2003; Neumann, 2001; Fig. 3). SAMPs may also polarize (educate) innate immune cells toward a protective phenotype while sparing antipathogen responses (Fig. 4). Indeed, pathogens lack SAMPs, but it is plausible that this strategy have been subverted by viruses, for example in their continuing evolutionary struggle with their hosts (Cameron *et al.*, 2005; Foster-Cuevas *et al.*, 2004).

SAMPs must have important features that make them instrumental in the control of the innate immune response: First, SAMPs is expressed by almost all host "self" cells, and not by pathogens. Second, SAMPs are membrane-bound or secreted molecules that control the level of humoral and/or cellular innate immune responses. Third, the interaction between SAMPs from the host cells and novel inhibitory PRRs expressed by macrophages engage immunoregulatory activities to maintain the nonphlogistic response and promoting tissue repair. The hypothesis that self-patterns must exist to control the humoral and cellular innate immune responses is in-line with the well-documented concept that host cells

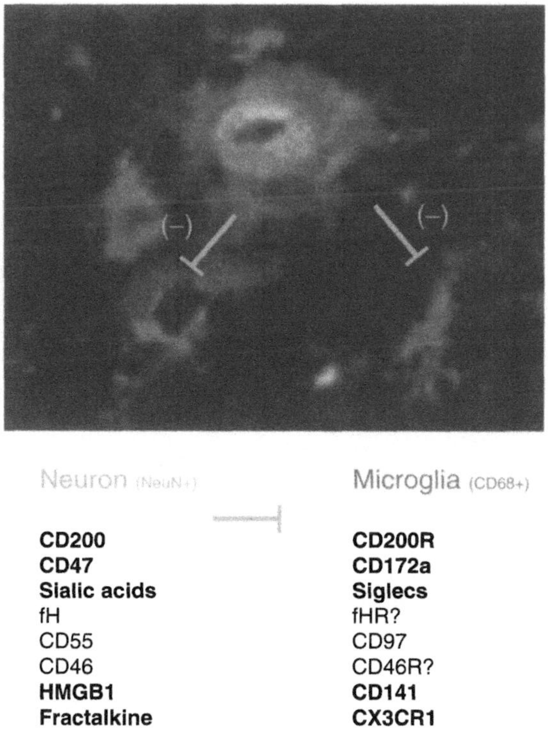

FIG. 3. Identification of novel neuroimmune regulatory pathways (NIRegs). NIRegs (also known as self-associated molecular patterns, SAMPs) are expressed by neurons to signal "don't eat me" to microglia and macrophages. Failure to express these protective signals (e.g., apoptosis) will induce robust phagocytosis and neurodegeneration. The identity of the fHR and CD46R (R, receptor) is presently unknown. (See Color Insert.)

express high levels of MHC class I molecules to avoid killing activities by NK cells (Cerwenka and Lanier, 2001).

Recognition of missing self is not unique to NK cell function, and another well-known example of the use of this strategy is the regulation of the innate humoral and cellular complement system. To avoid this self-destructive tendency, host cells use a wide armamentarium of regulatory molecules (inhibitors), which inhibit assembly of either the C3-cleaving enzymes or the formation of the membrane attack complex (MAC). As pathogens lack these inhibitors, activation of the complement cascade can proceed on their surfaces and results in lysis or phagocytosis of microbial intruders. Similarly, as self-cells progress to altered-self (apoptotic cells), downregulation of complement inhibitors at the cell surface (low sialic acid content lowering fH binding; loss of membrane inhibitors such as CD46) can lead to moderate and limited complement opsonization (C3b, iC3b)

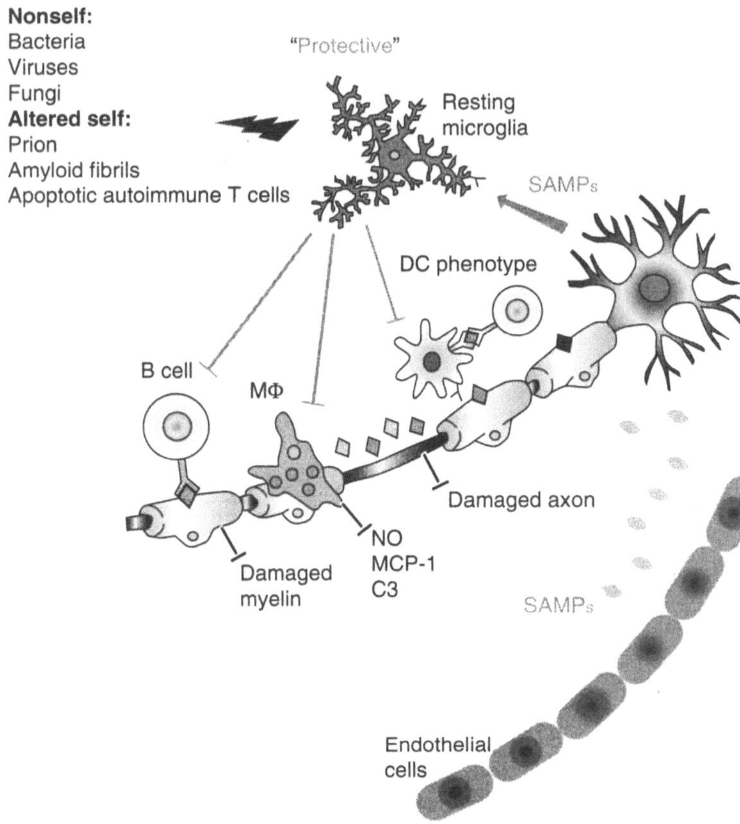

FIG. 4. Innate immune neuroprotective inflammation. Macrophages and microglia are programmed to kill and phagocytose target cells, and it is equally important to understand the mechanisms controlling their eager appetite. NIRegs (e.g., CD47, CD200 . . .) were originally termed self-associated molecular patterns (SAMPs) on the ground that they are expressed by neurons to control phagocyte activation which are polarized toward a neuroprotective phenotype. Importantly, this neurocontrol does not affect the phagocytic activity of microglia in the removal of pathogens and toxic cell debris. (See Color Insert.)

to promote phagocytosis. C1 inhibitor (C1-INH), C4b-binding protein (C4bp), factor H (fH), factor I (fI), S protein (Sp), and clusterin are all soluble C inhibitors secreted and released in the fluid phase. The other complement inhibitors are expressed on the cell membrane and include CR1, membrane cofactor protein (MCP, CD46), decay accelerating factor (DAF, CD55), and CD59 (see comprehensive review Morgan and Meri, 1994). From a SAMP standpoint, fH, C1inh, clusterin, CD46, and CD55 seem to fit the bill given that they are broadly expressed and extremely important in the control of complement activation on self-cells. Moreover, since CD55 is a ligand for CD97 on macrophages it is tantalizing to speculate that CD55–CD97 interactions could play an important role to control phagocytosis (Hamann *et al.*, 1996).

Another example of a molecular signal that functions as a marker of normal self is the carbohydrate structures that decorate glycoproteins and glycolipids on the cell surface. They usually terminate with sialic acids, which can be recognized by a variety of molecules (fH) or receptors involved in intracellular signaling (Meri and Pangburn, 1990). One group of receptors that can bind to sialylated glycoproteins and glycolipids comprise a family of proteins known as siglecs (Crocker, 2005). Similar to other inhibitory receptors, siglecs contain immunoreceptor tyrosine-based inhibitory motifs (ITIMs) in their cytoplasmic tail which may negatively regulate the function of phagocytes, including microglia (Angata *et al.*, 2002; Crocker, 2005; Jones *et al.*, 2003; Lock *et al.*, 2004). Lack of sialic acid expression on most microorganisms, and in some cases, on virally infected or transformed cells may be recognized as missing self-signals to promote phagocytosis. Reduced expression of sialic acids on apoptotic cells may also provide a missing self-signal for the safe disposal of the toxic cell debris.

Other possible SAMPs have been identified including CD200 (and its receptor CD200R), the integrin CD47 with its receptor SIRPα (CD172a), both regulating myeloid cell and lymphocyte activity (Barclay *et al.*, 2002; Brown and Frazier, 2001; Hoek *et al.*, 2000). CD200 is a 41- to 47-kDa surface molecule and a member of the immunoglobulin Ig supergene family characterized by two IgSF domains (Barclay *et al.*, 2002; Wright *et al.*, 2001). It is a highly conserved molecule found in invertebrates and vertebrates and many of the glycoproteins containing this arrangement are involved with regulation of the immune system. In the brain, OX2, now CD200, is expressed by cerebellar and retinal neurons, together with vascular endothelium. Intuitively, CD200 may also play an important immunoregulatory role on neural progenitors (Sergent-Tanguy *et al.*, 2006). Astrocytes do not express CD200 in contrast to microglia (Broderick *et al.*, 2002). The counter-receptor to CD200, CD200R, also contains two IgSF domains and is expressed by myeloid cells and brain microglia. In CD200 deficient mice the number of activated microglia and macrophages were more numerous after a lesion than the wild-type animals, providing evidence that the CD200–CD200R interaction is related to regulation of microglial activation and local inflammation (Hoek *et al.*, 2000). This interpretation is supported by experiments in mice inoculated with MOG peptide to induce EAE.

CD47 is constitutively expressed by endothelium, neurons, astrocytes, macrophages, and dendritic cells (de Vries *et al.*, 2002; Reinhold *et al.*, 1995). CD47 has five transmembrane regions with alternatively spliced isoforms of CD47 having a tissue-specific expression, form 2 is present in bone marrow, whereas form 4 is highly expressed in brain (Reinhold *et al.*, 1995). The counter-receptor for CD47 is signal regulatory protein SIRPα (CD172a), a plasma membrane protein with three

Ig domains in its extracellular component (Brown and Frazier, 2001). CD172a is expressed by myeloid cells, microglia, and neurons. The interaction between CD47 on a host cell and CD172a recruits tyrosine phosphotases SHP-1 and SHP-2 with downregulation of macrophage phagocytosis, complement activation, and cytokine synthesis including TGF-β all contributing to the reduction of any inflammatory response (Brown and Frazier, 2001; Oldenborg *et al.*, 2001; Seiffert *et al.*, 2001; Vernon-Wilson *et al.*, 2000). The protective activity of CD47 could also be extended to its beneficial role in supporting neural development and promoting clearance of amyloid fibrils albeit by mechanisms that remain ill-characterized (Bamberger *et al.*, 2003; Murata *et al.*, 2006).

The interaction between CD47 and CD172a has been shown to reduce neutrophil migration across endothelium and blocking CD47 reduced bacterial-induced expression of inflammatory cytokines by dendritic cells. Furthermore, CD47 is capable of inducing apoptosis in T cells, and cells deficient in CD47 are rapidly cleared from the systemic circulation by the spleen. Hence, CD47 represents an important "don't eat me signal" preventing inappropriate phagocytosis of host cells. Apoptotic cells rapidly loose CD47, reducing its ability to bind and phosphorylate CD172a to recruit inhibitory signals. Hence, it has been shown that the loss or inactivation of CD47 will prompt phagocytosis (Gardai *et al.*, 2005). The finding that viable cells are readily ingested if "don't eat me signals" are disrupted raises the intriguing and challenging possibility that recognition and removal by phagocytosis is a default process that is actively prevented by inhibitory ligands on viable cells. Whether or not the CD47–CD172a pathway is capable of regulating microglial activity in disease remains to be determined.

Other neural NIRegs are likely to play a critical role in neuroprotection and it will be important to precisely identify the emerging role of both fractalkine or HMGB1 signaling to CX3CR1 and CD141 expressed by microglia (Abeyama *et al.*, 2005; Murata *et al.*, 2006).

VII. Innate Immunity and Neurogenesis

Activation of the complement pathway produces C3a and C5a anaphylatoxins, chemoattractant to myeloid cells expressing the anaphylatoxin receptors C3aR and C5aR and contributing to proinflammatory activities. However, C3a also has regulatory properties based on its capacity to block LPS stimulation of macrophage TNF-α cytokine expression as well as IL-6- and IL-1-β expression by lymphocytes increasing synthesis of IL-10 and NGF (Fischer and Hugli, 1997). The immunoregulatory role of C3a has been evidenced in an animal model of MS (Boos *et al.*, 2005). This new "self-defense" role for C3a is further supported by the presence of C3aR on adrenal and pituitary gland cells, both glands having important roles in the synthesis of corticosteroids to control systemic and central inflammation and infection (Francis *et al.*, 2003). It has been shown that complement anaphylatoxins may be involved in the control of neurogenesis and with antiapoptotic activities by its capacity to reduce NMDA-induced neuronal death (Rahpeymai *et al.*, 2006; van Beek *et al.*, 2001). Signaling through other innate immune receptors such as TLR2 and TLR3 may also be protective as recently suggested (Bsibsi *et al.*, 2006; Setzu *et al.*, 2006).

VIII. The Canonical Innate Immune System in the CNS: The Complement System

The most effective innate immune system for the CNS will have to combine the activity of TLRs in rapidly sensing harmful intruders and signaling antipathogen activities together with the capacity of phagocytic receptors to engage engulfment. Furthermore, this ideal innate immune system will have intrinsic means of controlling adverse hyperinflammatory responses by the expression of key soluble and membrane regulatory proteins. To our knowledge, the innate immune complement system fulfills all these aspects as described below.

The complement system is a vital component of the CNS innate immune defense system as it recognizes and clears pathogens and apoptotic cells, but its activation must be closely regulated to prevent excessive tissues destruction (Gasque *et al.*, 2000; Van Beek *et al.*, 2000). The expression of complement components of both the classical and alternative pathways has been well described in astrocytes, microglia, and neurons *in vitro* and *in vivo* (Gasque *et al.*, 2000). The classical complement pathway is activated by C1q interacting with neurons, myelin basic protein (MBP), and myelin oligodendrocytic glycoprotein (MOG). The alternative pathway is activated independently of C1q and immune complex formation, but through a binding of C3 to activating surfaces that cleave C5 to initiate the terminal pathway. One component C3b is abundantly deposited on target cells and functions as an opsonin for microglia and Kolmer cells expressing the PRRs CR1 (CD35), CR3, and CR4 receptors to promote robust phagocytosis. The terminal pathway for both classical and alternative pathways provides the MAC composed of C5b-9. This complex is capable of producing cell lysis unless inhibited by complement regulator proteins (regulators of complement activation, RCA), see below. The inappropriate activation of the complement pathways with MAC formation will therefore indiscriminately damage host tissue and rapidly change the protective nature of the innate immune response into a detrimental one as evidenced in demyelinating diseases (Compston *et al.*, 1989; Linington *et al.*, 1989; Lucchinetti *et al.*, 2000; Mollnes and Harboe, 1987). Knockout mice for CD59, a natural MAC inhibitor, increased the severity of EAE whereas the failure to produce MAC in C6-deficient animals reduced the severity of axon damage in diseased animals compared with the control group (Mead *et al.*, 2002, 2004). In contrast, complement is activated in AD as the result of C1q binding to fibrillary β-amyloid, activating complement and increasing C1q, C3, and C5 as part of the protective response promoting clearance of the amyloid plaque (Eikelenboom *et al.*, 2002; Jiang *et al.*, 1994; McGeer *et al.*, 1989). This was supported by the observation that inhibition of the complement cascade increased amyloid plaque burden (Wyss-Coray and Mucke, 2002). However, in the context of acute inflammation associated with robust expression of complement proteins by microglia and astrocytes, it is plausible that complement activation on myelin/neuronal debris contributes to secondary brain injury. The formation of the MAC and nonspecific binding to surrounding cells would cause bystander damage. In HD, the expression of C1q, C3, iC3b, and C4 is increased in microglia and astrocytes adding to the inflammatory response and secondary brain injury. One further possible mechanism for complement activation in HD is the presence of the expanded glutamine sequence in the mutant huntingtin protein, initiating apoptosis and the activation of complement through C1q binding to apoptotic cells (Singhrao *et al.*, 1999). The formation of soluble MAC and nonspecific binding to surrounding cells would cause bystander damage. Complement activation is also present in the tauopathies (including Pick's disease) resulting in localization of complement products on ballooned neurons; the details regarding the identity of the C1q binding molecule in these diseases is not yet known (Singhrao *et al.*, 1996; Yasuhara *et al.*, 1994). Complement activation is also present in experimental models of ischemia (Barnum *et al.*, 2002; Van Beek *et al.*, 2000; van Beek *et al.*, 2001). Interestingly, administration of a C1 inhibitor C1-INH, resulted in neuroprotection after experimental ischemia, but its protective effect was interpreted as independent of C1q activating the complement pathway (De Simoni *et al.*, 2004).

Complement pathways are tightly regulated in order to prevent unrestrained activity resulting in cytolytic destruction of host cells, brain cells being particularly vulnerable to complement attack. The complement regulators can be divided into two groups, the membrane bound and those located within the extracellular fluid, the so-called fluid phase regulators. Together they inhibit the activation of the complement pathways (for detailed review see Morgan and Meri, 1994). In brief, the membrane-bound regulators, CR1 (CD35) and MCP (CD46) expressed on nucleated cell membranes bind to C4b or serve as cofactors to increase its cleavage inhibiting this step in the complement pathway. DAF (CD55) inhibits the C3/C5 convertase step and CD59 blocks MAC formation in the terminal pathway. CD35 and CD55 also regulate the alternative complement pathway by accelerating the decay of C3 and C5 convertases. The fluid phase regulators are composed of C1 inhibitor (C1-INH) an effective inhibitor of the C1 component of the classical pathway; factors H and FI (alternative pathway) accelerate C3b/C4b degeneration, whereas S protein and clusterin prevent C5b-7 assisting formation of the MAC pathway, inhibiting the final terminal pathway of the alternative and classical C pathways.

In primary culture, fetal neurons spontaneously activate C classical pathway with formation of MAC resulting in significant lysis because *in vitro* they express low levels of the membrane regulators (CD59, CD46, C1 INH, and fH) and no CD55 (Singhrao *et al.*, 2000). Neuroblastoma cell lines have been found to express CD59 and low levels of CD46, whereas IMR32 and Paju cell lines fail to express DAF (CD55) but did express the fluid phase regulators Sp, clusterin, fH, and C1-INH (Agoropoulou *et al.*, 1996; Gasque *et al.*, 1995). A study by Zhang *et al.* (1998), however, found Paju cell lines actively expressing DAF on growth cones and neural processes raising the possibility that DAF not only prevents complement-mediated lysis, but also represents a marker for regenerating neurons. A study by van Beek *et al.* (2005) has demonstrated neuron protection against complement activation by DAF expression. Under physiological conditions using tissue immunohistochemistry, ISH, and RT-PCR, neurons have been shown to express CD46, but not CD59, CD55, or CD35 (Gasque *et al.*, 1996; Pasinetti *et al.*, 1994; Walker *et al.*, 1995). By comparison, astrocytes and microglia express a wider range of inhibitors such as CD46, CD59, and CD55 together with the fluid phase regulators (Gasque *et al.*, 1995; Gordon *et al.*, 1992; Pasinetti *et al.*, 1994). Initial observations demonstrated rat oligodendrocytes in culture were susceptible to C attack and did not express CD59 (Scolding and Franklin, 1998). However in human, CD59, CD46, and CD55, together with C1 INH, fH, S protein, and clusterin are all expressed and do not spontaneously activate complement. In AD and Pick's disease, neurons did not express CD35, CD59, CD46, and CD55, whereas in HD, neurons expressed high levels of CD46. Overall, the combined data from *in vitro* and *in vivo* experiments indicates that astrocytes, microglia, and oligodendrocytes are well protected from the effects of direct or bystander complement lysis because they express high levels of CD59, CD46, and CD55. Neurons, particularly lacking CD55, are vulnerable to the detrimental effects of complement attack and highlighting the therapeutic applications of complement regulatory proteins to prevent neurodegeneration.

IX. Conclusion: Elements to Drive Innate Immune Neuroprotective Activities

The balance between the protective and harmful effects of the innate immune response mounted against pathogen invasion, accumulation of toxic proteins, and brain injury has been termed a "double-edged sword" (Wyss-Coray and Mucke, 2002). This balance must be critically regulated in order to promote conditions supportive of brain repair but without excessive destruction of "self" or host cells. The CNS innate immune response is regulated by a number of "self-defense" pathways. "Self" is distinguished from "nonself" by the detection of surface PAMP and ACAMP molecules and by the expression of SAMPs by host cells. The CNS has a range of defense strategies at its disposal, each capable of regulating the protective components of the innate immune response while at the same time limiting the extent of accompanying brain injury and possible autoimmune responses. In the presence of apoptotic cells due to tissue injury or as a consequence of viral infection, it is essential that the local innate immune response is rapidly engaged to remove potentially toxic and proautoimmune molecules. The therapeutic manipulation of the immunoregulatory and defense strategies designed to reduce brain injury and promote repair, as described in this chapter, is now becoming a clinical reality.

Glossary

Innate immunity: First line of defense against infection in which phagocytic cells use primitive nonspecific recognition systems to kill pathogens. Innate immune molecules can also recognize toxic cell debris such as apoptotic cells and misfolded or aggregated proteins, and this function may represent the most ancestral role of the innate immune system.

Complement: An important innate immune system composed of almost 30 proteins expressed by phagocytes, glial cells, neurons, and most other cell types. C3 is the canonical complement protein with the capacity to bind to pathogens and promoting clearance by phagocytes expressing C3 receptors. Small fragments of C3 called C3a and C5a anaphylatoxins have stimulatory activities through signaling to G-protein–coupled seven transmembrane receptors.

Danger signals: Pathogens and toxic cell debris (apoptotic/necrotic cells) express molecules that will alert the innate immune response, the first line of defense of the host, against infection and tissue injury.

Self, nonself, and altered-self: Host cells in physiological conditions are capable of controlling the innate immune response through the expression of self molecules such as HLA class I and "don't eat me" signals (CD46, CD200). When these cells are abnormal and present an altered membrane, they will be recognized and phagocytosed by macrophages to prevent the release of toxic potentially autoimmune cell debris (nucleic acids). Pathogens are nonself and will be recognized and eliminated by innate immune cells.

Opsonin: A terminology derived from the Greek and meaning, sauce or seasoning, in other words making the target cells such as pathogen more palatable to the phagocyte and more easily eaten. For example, C3b is an opsonin bound to target cells following complement activation and promoting phagocytosis by macrophages expressing C3 receptors.

References

Abeyama, K., Stern, D. M., Ito, Y., Kawahara, K., Yoshimoto, Y., Tanaka, M., Uchimura, T., Ida, N., Yamazaki, Y., Yamada, S., Yamamoto, Y., Yamamoto, H., *et al.* (2005). The N-terminal domain of thrombomodulin sequesters high-mobility group-B1 protein, a novel antiinflammatory mechanism. *J. Clin. Invest.* **115,** 1267–1274.

Agoropoulou, C., Wing, M. G., and Wood, A. (1996). CD59 expression and complement susceptibility of human neuronal cell line (NTera2). *Neuroreport* **7,** 997–1004.

Angata, T., Kerr, S. C., Greaves, D. R., Varki, N. M., Crocker, P. R., and Varki, A. (2002). Cloning and characterization of human Siglec-11. A recently evolved signaling that can interact with SHP-1 and SHP-2 and is expressed by tissue macrophages, including brain microglia. *J. Biol. Chem.* **277,** 24466–24474.

Bamberger, M. E., Harris, M. E., McDonald, D. R., Husemann, J., and Landreth, G. E. (2003). A cell surface receptor complex for fibrillar beta-amyloid mediates microglial activation. *J. Neurosci.* **23,** 2665–2674.

Barclay, A. N., Wright, G. J., Brooke, G., and Brown, M. H. (2002). CD200 and membrane protein E interactions in the control of myeloid cells. *Trends Immunol.* **23,** 285–290.

Barnum, S. R., Ames, R. S., Maycox, P. R., Hadingham, S. J., Meakin, J., Harrison, D., and Parsons, A. A. (2002). Expression of the complement C3a and C5a receptors after permanent focal ischemia: An alternative interpretation. *Glia* **38,** 169–173.

Baron, R., Harpaz, I., Nemirovsky, A., Cohen, H., and Monsonego, A. (2007). Immunity and neuronal repair in the progression of Alzheimer's disease: A brief overview. *Exp. Gerontol.* **42,** 64–69.

Bergman, P., Johansson, L., Wan, H., Jones, A., Gallo, R. L., Gudmundsson, G. H., Hokfelt, T., Jonsson, A. B., and Agerberth, B. (2006). Induction of the antimicrobial peptide CRAMP in the blood-brain barrier and meninges after meningococcal infection. *Infect. Immun.* **74,** 6982–6991.

Bilak, M. M., Becerra, S. P., Vincent, A. M., Moss, B. H., Aymerich, M. S., and Kuncl, R. W. (2002). Identification of the neuroprotective molecular region of pigment epithelium-derived factor and its binding sites on motor neurons. *J. Neurosci.* **22,** 9378–9386.

Boos, L., Szalai, A. J., and Barnum, S. R. (2005). C3a expressed in the central nervous system protects against LPS-induced shock. *Neurosci. Lett.* **387,** 68–71.

Broderick, C., Hoek, R. M., Forrester, J. V., Liversidge, J., Sedgwick, J. D., and Dick, A. D. (2002). Constitutive retinal CD200 expression regulates resident microglia and activation state of inflammatory cells during experimental autoimmune uveoretinitis. *Am. J. Pathol.* **161,** 1669–1677.

Brown, E. J., and Frazier, W. A. (2001). Integrin-associated protein (CD47) and its ligands. *Trends Cell Biol.* **11,** 130–135.

Bsibsi, M., Ravid, R., Gveric, D., and van Noort, J. M. (2002). Broad expression of Toll-like receptors in the human central nervous system. *J. Neuropathol. Exp. Neurol.* **61,** 1013–1021.

Bsibsi, M., Persoon-Deen, C., Verwer, R. W., Meeuwsen, S., Ravid, R., and Van Noort, J. M. (2006). Toll-like receptor 3 on adult human astrocytes triggers production of neuroprotective mediators. *Glia* **53,** 688–695.

Buisson, A., Nicole, O., Docagne, F., Sartelet, H., Mackenzie, E. T., and Vivien, D. (1998). Up-regulation of a serine protease inhibitor in astrocytes mediates the neuroprotective activity of transforming growth factor beta1. *FASEB J.* **12,** 1683–1691.

Buisson, A., Lesne, S., Docagne, F., Ali, C., Nicole, O., MacKenzie, E. T., and Vivien, D. (2003). Transforming growth factor-beta and ischemic brain injury. *Cell. Mol. Neurobiol.* **23,** 539–550.

Burudi, E. M., Riese, S., Stahl, P. D., and Regnier-Vigouroux, A. (1999). Identification and functional characterisation of the mannose receptor in astrocytes. *Glia* **25,** 44–55.

Butovsky, O., Koronyo-Hamaoui, M., Kunis, G., Ophir, E., Landa, G., Cohen, H., and Schwartz, M. (2006). Glatiramer acetate fights against Alzheimer's disease by inducing dendritic-like microglia expressing insulin-like growth factor 1. *Proc. Natl. Acad. Sci. USA* **103,** 11784–11789.

Cameron, C. M., Barrett, J. W., Mann, M., Lucas, A., and McFadden, G. (2005). Myxoma virus M128L is expressed as a cell surface CD47-like virulence factor that contributes to the down-regulation of macrophage activation *in vivo*. *Virology* **337,** 55–67.

Canova, C., Neal, J. W., and Gasque, P. (2006). Expression of innate immune complement regulators on brain epithelial cells during human bacterial meningitis. *J. Neuroinflammation* **3,** 22.

Cavanaugh, K. P., Gurwitz, D., Cunningham, D. D., and Bradshaw, R. A. (1990). Reciprocal modulation of astrocyte stellation by thrombin and protease nexin-1. *J. Neurochem.* **54,** 1735–1743.

Cerwenka, A., and Lanier, L. L. (2001). Natural killer cells, viruses and cancer. *Nat. Rev. Immunol.* **1,** 41–49.

Citron, B. A., Smirnova, I. V., Arnold, P. M., and Festoff, B. W. (2000). Upregulation of neurotoxic serine proteases, prothrombin, and protease-activated receptor 1 early after spinal cord injury. *J. Neurotrauma* **17,** 1191–1203.

Compston, D. A., Morgan, B. P., Campbell, A. K., Wilkins, P., Cole, G., Thomas, N. D., and Jasani, B. (1989). Immunocytochemical localization of the terminal complement complex in multiple sclerosis. *Neuropathol. Appl. Neurobiol.* **15,** 307–316.

Crocker, P. R. (2005). Siglecs in innate immunity. *Curr. Opin. Pharmacol.* **5**(4), 431–437.

Dalpke, A. H., Schafer, M. K., Frey, M., Zimmermann, S., Tebbe, J., Weihe, E., and Heeg, K. (2002). Immunostimulatory CpG-DNA activates murine microglia. *J. Immunol.* **168,** 4854–4863.

De Simone, R., Ajmone-Cat, M. A., and Minghetti, L. (2004). Atypical antiinflammatory activation of microglia induced by apoptotic neurons: Possible role of phosphatidylserine-phosphatidylserine receptor interaction. *Mol. Neurobiol.* **29,** 197–212.

De Simoni, M. G., Rossi, E., Storini, C., Pizzimenti, S., Echart, C., and Bergamaschini, L. (2004). The powerful neuroprotective action of C1-inhibitor on brain ischemia-reperfusion injury does not require C1q. *Am. J. Pathol.* **164,** 1857–1863.

de Vries, H. E., Hendriks, J. J., Honing, H., De Lavalette, C. R., van der Pol, S. M., Hooijberg, E., Dijkstra, C. D., and van den Berg, T. K. (2002). Signal-regulatory protein alpha-CD47 interactions are required for the transmigration of monocytes across cerebral endothelium. *J. Immunol.* **168,** 5832–5839.

Delhaye, S., Paul, S., Blakqori, G., Minet, M., Weber, F., Staeheli, P., and Michiels, T. (2006). Neurons produce type I interferon during viral encephalitis. *Proc. Natl. Acad. Sci. USA* **103,** 7835–7840.

Dihanich, M., Kaser, M., Reinhard, E., Cunningham, D., and Monard, D. (1991). Prothrombin mRNA is expressed by cells of the nervous system. *Neuron* **6,** 575–581.

Docagne, F., Nicole, O., Marti, H. H., MacKenzie, E. T., Buisson, A., and Vivien, D. (1999). Transforming growth factor-beta1 as a regulator of the serpins/t-PA axis in cerebral ischemia. *FASEB J.* **13,** 1315–1324.

Dowling, P., Shang, G., Raval, S., Menonna, J., Cook, S., and Husar, W. (1996). Involvement of the CD95 (APO-1/Fas) receptor/ligand system in multiple sclerosis brain. *J. Exp. Med.* **184,** 1513–1518.

Ehlers, M. R. (2000). CR3: A general purpose adhesion-recognition receptor essential for innate immunity. *Microbes Infect.* **2,** 289–294.

Eikelenboom, P., Bate, C., Van Gool, W. A., Hoozemans, J. J., Rozemuller, J. M., Veerhuis, R., and Williams, A. (2002). Neuroinflammation in Alzheimer's disease and prion disease. *Glia* **40,** 232–239.

El Khoury, J. B., Moore, K. J., Means, T. K., Leung, J., Terada, K., Toft, M., Freeman, M. W., and Luster, A. D. (2003). CD36 mediates the innate host response to beta-amyloid. *J. Exp. Med.* **197,** 1657–1666.

Elward, K., and Gasque, P. (2003). "Eat me" and "don't eat me" signals govern the innate immune response and tissue repair in the CNS: Emphasis on the critical role of the complement system. *Mol. Immunol.* **40,** 85–94.

Elward, K., Griffiths, M., Mizuno, M., Harris, C. L., Neal, J. W., Morgan, B. P., and Gasque, P. (2005). CD46 plays a key role in tailoring innate immune recognition of apoptotic and necrotic cells. *J. Biol. Chem.* **280,** 36342–36354.

Esen, N., Tanga, F. Y., DeLeo, J. A., and Kielian, T. (2004). Toll-like receptor 2 (TLR2) mediates astrocyte activation in response to the Gram-positive bacterium *Staphylococcus aureus. J. Neurochem.* **88,** 746–758.

Eugster, H. P., Frei, K., Bachmann, R., Bluethmann, H., Lassmann, H., and Fontana, A. (1999). Severity of symptoms and demyelination in MOG-induced EAE depends on TNFR1. *Eur. J. Immunol.* **29,** 626–632.

Everett, H., and McFadden, G. (1999). Apoptosis: An innate immune response to virus infection. *Trends Microbiol.* **7,** 160–165.

Fadok, V. A., Warner, M., Bratton, D. L., and Henson, P. M. (1998). CD36 is required for phagocytosis of apoptotic cells by human macrophages that use either a phosphatidylserine receptor or the vitronectin receptor (v beta 3). *J. Immunol.* **161,** 6350–6357.

Fadok, V. A., Bratton, D. L., Rose, D. M., Pearson, A., Ezekewitz, A. B., and Henson, P. M. (2000). A receptor for phosphatidylserine specific clearance of apoptotic cells. *Nature* **405,** 85–90.

Fassbender, K., Walter, S., Kuhl, S., Landmann, R., Ishii, K., Bertsch, T., Stalder, A. K., Muehlhauser, F., Liu, Y., Ulmer, A. J., Rivest, S., Lentschat, A., *et al.* (2004). The LPS receptor (CD14) links innate immunity with Alzheimer's disease. *FASEB J.* **18,** 203–205.

Fillon, S., Soulis, K., Rajasekaran, S., Benedict-Hamilton, H., Radin, J. N., Orihuela, C. J., El Kasmi, K. C., Murti, G., Kaushal, D., Gaber, M. W., Weber, J. R., Murray, P. J., *et al.* (2006). Platelet-activating factor receptor and innate immunity: Uptake of gram-positive bacterial cell wall into host cells and cell-specific pathophysiology. *J. Immunol.* **177,** 6182–6191.

Fischer, W. H., and Hugli, T. E. (1997). Regulation of B cell functions by C3a and C3a(desArg): Suppression of TNF-alpha, IL-6, and the polyclonal immune response. *J. Immunol.* **159,** 4279–4286.

Flugel, A., Schwaiger, F. W., Neumann, H., Medana, I., Willem, M., Wekerle, H., Kreutzberg, G. W., and Graeber, M. B. (2000). Neuronal FasL induces cell death of encephalitogenic T lymphocytes. *Brain Pathol.* **10,** 353–364.

Foell, D., Wittkowski, H., Vogl, T., and Roth, J. (2007). S100 proteins expressed in phagocytes: A novel group of damage-associated molecular pattern molecules. *J. Leukoc. Biol.* **81,** 28–37.

Foster-Cuevas, M., Wright, G. J., Puklavec, M. J., Brown, M. H., and Barclay, A. N. (2004). Human herpesvirus 8 K14 protein mimics CD200 in down-regulating macrophage activation through CD200 receptor. *J. Virol.* **78,** 7667–7676.

Francis, K., Lewis, B. M., Akatsu, H., Monk, P. N., Cain, S. A., Scanlon, M. F., Morgan, B. P., Ham, J., and Gasque, P. (2003). Complement C3a receptors in the pituitary gland: A novel pathway by which an innate immune molecule releases hormones involved in the control of inflammation. *FASEB J.* **17,** 2266–2268.

Gabriel, C., Ali, C., Lesne, S., Fernandez-Monreal, M., Docagne, F., Plawinski, L., MacKenzie, E. T., Buisson, A., and Vivien, D. (2003). Transforming growth factor alpha-induced expression of type 1 plasminogen activator inhibitor in astrocytes rescues neurons from excitotoxicity. *FASEB J.* **17,** 277–279.

Gardai, S. J., McPhillips, K. A., Frasch, S. C., Janssen, W. J., Starefeldt, A., Murphy-Ullrich, J. E., Bratton, D. L., Oldenborg, P. A., Michalak, M., and Henson, P. M. (2005). Cell-surface calreticulin initiates clearance of viable or apoptotic cells through trans-activation of LRP on the phagocyte. *Cell* **123,** 321–334.

Gasque, P. (2004). Complement: A unique innate immune sensor for danger signals. *Mol. Immunol.* **41,** 1089–1098.

Gasque, P., Fontaine, M., and Morgan, B. P. (1995). Complement expression in human brain. Biosynthesis of terminal pathway components and regulators in human glial cells and cell lines. *J. Immunol.* **154,** 4726–4733.

Gasque, P, Singhrao, S. K., Neal, J. W., Gotze, O., and Morgan, B. P. (1996). Expression of the receptor for complement C5a (CD88) is up-regulated on reactive astrocytes, microglia, and endothelial cells in the inflammed human central nervous system. *Am. J. Pathol.* **150,** 2–11.

Gasque, P., Jones, J., Singhrao, S. K., and Morgan, B. P. (1998). Identification of an astrocyte cell population from human brain that expresses perforin, a cytotoxic protein implicated in immune defense. *J. Exp. Med.* **187,** 451–460.

Gasque, P., Dean, Y. D., McGreal, E. P., VanBeek, J., and Morgan, B. P. (2000). Complement components of the innate immune system in health and disease in the CNS. *Immunopharmacology* **49,** 171–186.

Gendelman, H. E. (2002). Neural immunity: Friend or foe? *J. Neurovirol.* **8,** 474–479.

Gingrich, M. B., and Traynelis, S. F. (2000). Serine proteases and brain damage—is there a link? *Trends Neurosci.* **23,** 399–407.

Glezer, I., Lapointe, A., and Rivest, S. (2006). Innate immunity triggers oligodendrocyte progenitor reactivity and confines damages to brain injuries. *FASEB J.* **20,** 750–752.

Gordon, D. L., Sadlon, T. A., Wesselingh, S. L., Russell, S. M., Johnstone, R. W., and Purcell, D. F. (1992). Human astrocytes express membrane cofactor protein (CD46), a regulator of complement activation. *J. Neuroimmunol.* **36,** 199–208.

Gordon, S. (2002). Pattern recognition receptors: Doubling up for the innate immune response. *Cell* **111,** 927–930.

Grand, R. J., Turnell, A. S., and Grabham, P. W. (1996). Cellular consequences of thrombin-receptor activation. *Biochem. J.* **313**(Pt. 2), 353–368.

Gregory, C. D. (2000). CD14 dependent clearance of apoptotic cells: Relevance to the immune system. *Curr. Opin. Immunol.* **12,** 27–34.

Gregory, C. D., and Devitt, A. (2004). The macrophage and the apoptotic cell: An innate immune interaction viewed simplistically? *Immunology* **113,** 1–14.

Grimsley, C., and Ravichandran, K. S. (2003). Cues for apoptotic cell engulfment: Eat-me, don't eat-me and come-get-me signals. *Trends Cell Biol.* **13,** 648–656.

Hamann, J., Vogel, B., van Schijndel, G. M., and van Lier, R. A. (1996). The seven-span transmembrane receptor CD97 has a cellular ligand (CD55, DAF). *J. Exp. Med.* **184,** 1185–1189.

Hanayama, R., Tanaka, M., Miwa, K., Shinohara, A., Iwamatsu, A., and Nagata, S. (2002). Identification of a factor that links apoptotic cells to phagocytes. *Nature* **417,** 182–187.

Hao, H. N., Zhao, J., Lotoczky, G., Grever, W. E., and Lyman, W. D. (2001). Induction of human beta-defensin-2 expression in human astrocytes by lipopolysaccharide and cytokines. *J. Neurochem.* **77,** 1027–1035.

Hauwel, M., Furon, E., Canova, C., Griffiths, M., Neal, J., and Gasque, P. (2005). Innate (inherent) control of brain infection, brain inflammation and brain repair: The role of microglia, astrocytes, "protective" glial stem cells and stromal ependymal cells. *Brain Res. Brain Res. Rev.* **48,** 220–233.

Helmy, K. Y., Katschke, K. J., Jr., Gorgani, N. N., Kljavin, N. M., Elliott, J. M., Diehl, L., Scales, S. J., Ghilardi, N., and van Lookeren Campagne, M. (2006). CRIg: A macrophage complement receptor required for phagocytosis of circulating pathogens. *Cell* **124,** 915–927.

Hoek, R. M., Ruuls, S. R., Murphy, C. A., Wright, G. J., Goddard, R., Zurawski, S. M., Blom, B., Homola, M. E., Streit, W. J., Brown, M. H., Barclay, A. N., and Sedgwick, J. D. (2000). Down-regulation of the macrophage lineage through interaction with OX2 (CD200). *Science* **290,** 1768.

Husemann, J., Loike, J. D., Anankov, R., Febbraio, M., and Silverstein, S. C. (2002). Scavenger receptors in neurobiology and neuropathology: Their role on microglia and other cells of the nervous system. *Glia* **40,** 195–205.

Jiang, H., Burdick, D., Glabe, C. G., Cotman, C. W., and Tenner, A. J. (1994). beta-Amyloid activates complement by binding to a specific region of the collagen-like domain of the C1q A chain. *J. Immunol.* **152,** 5050–5059.

Jones, C., Virji, M., and Crocker, P. R. (2003). Recognition of sialylated meningococcal lipopolysaccharide by siglecs expressed on myeloid cells leads to enhanced bacterial uptake. *Mol. Microbiol.* **49,** 1213–1225.

Kielian, T., Mayes, P., and Kielian, M. (2002). Characterization of microglial responses to Staphylococcus aureus: Effects on cytokine, costimulatory molecule, and Toll-like receptor expression. *J. Neuroimmunol.* **130,** 86–99.

Korb, L. C., and Ahearn, J. M. (1997). C1q binds directly and specifically to surface blebs of apoptotic human keratinocytes: Complement deficiency and systemic lupus erythematosus revisited. *J. Immunol.* **158,** 4525–4528.

Kovacs, G. G., Gasque, P., Strobel, T., Lindeck-Pozza, E., Strohschneider, M., Ironside, J. W., Budka, H., and Guentchev, M. (2004). Complement activation in human prion disease. *Neurobiol. Dis.* **15,** 21–28.

Laflamme, N., and Rivest, S. (2001). Toll-like receptor 4: The missing link of the cerebral innate immune response triggered by circulating gram-negative bacterial cell wall components. *FASEB J.* **15,** 155–163.

Laflamme, N., Echchannaoui, H., Landmann, R., and Rivest, S. (2003). Cooperation between toll-like receptor 2 and 4 in the brain of mice challenged with cell wall components derived from gram-negative and gram-positive bacteria. *Eur. J. Immunol.* **33,** 1127–1138.

Lazarov-Spiegler, O., Solomon, A. S., Zeev-Brann, A. B., Hirschberg, D. L., Lavie, V., and Schwartz, M. (1996). Transplantation of activated macrophages overcomes central nervous system regrowth failure. *FASEB J.* **10,** 1296–1302.

Lee, K. R., Colon, G. P., Betz, A. L., Keep, R. F., Kim, S., and Hoff, J. T. (1996). Edema from intracerebral hemorrhage: The role of thrombin. *J. Neurosurg.* **84,** 91–96.

Lehnardt, S., Massillon, L., Follett, P., Jensen, F. E., Ratan, R., Rosenberg, P. A., Volpe, J. J., and Vartanian, T. (2003). Activation of innate immunity in the CNS triggers neurodegeneration through a Toll-like receptor 4-dependent pathway. *Proc. Natl. Acad. Sci. USA* **100,** 8514–8519.

Leonardi-Essmann, F., Emig, M., Kitamura, Y., Spanagel, R., and Gebicke-Haerter, P. J. (2005). Fractalkine-upregulated milk-fat globule EGF factor-8 protein in cultured rat microglia. *J. Neuroimmunol.* **160,** 92–101.

Li, M. O., Sarkisian, M. R., Mehal, W. Z., Rakic, P., and Flavell, R. A. (2003). Phosphatidylserine receptor is required for clearance of apoptotic cells. *Science* **302,** 1560–1563.

Linehan, S. A., Martinez-Pomares, L., Stahl, P. D., and Gordon, S. (1999). Mannose receptor and its putative ligands in normal murine lymphoid and nonlymphoid organs: *In situ* expression of mannose receptor by selected macrophages, endothelial cells, perivascular microglia, and mesangial cells, but not dendritic cells. *J. Exp. Med.* **189,** 1961–1972.

Linington, C., Lassmann, H., Morgan, B. P., and Compston, D. A. (1989). Immunohistochemical localisation of terminal complement component C9 in experimental allergic encephalomyelitis. *Acta Neuropathol. (Berl.)* **79,** 78–85.

Lock, K., Zhang, J., Lu, J., Lee, S. H., and Crocker, P. R. (2004). Expression of CD33-related siglecs on human mononuclear phagocytes, monocyte-derived dendritic cells and plasmacytoid dendritic cells. *Immunobiology* **209,** 199–207.

Lotze, M. T., and Tracey, K. J. (2005). High-mobility group box 1 protein (HMGB1): Nuclear weapon in the immune arsenal. *Nat. Rev. Immunol.* **5,** 331–342.

Lu, J., Teh, C., Kishore, U., and Reid, K. B. (2002). Collectins and ficolins: Sugar pattern recognition molecules of the mammalian innate immune system. *Biochim. Biophys. Acta* **1572,** 387–400.

Lucchinetti, C., Bruck, W., Parisi, J., Scheithauer, B., Rodriguez, M., and Lassmann, H. (2000). Heterogeneity of multiple sclerosis lesions: Implications for the pathogenesis of demyelination. *Ann. Neurol.* **47,** 707–717.

Ma, Y., Li, J., Chiu, I., Wang, Y., Sloane, J. A., Lu, J., Kosaras, B., Sidman, R. L., Volpe, J. J., and Vartanian, T. (2006). Toll-like receptor 8 functions as a negative regulator of neurite outgrowth and inducer of neuronal apoptosis. *J. Cell Biol.* **175,** 209–215.

Magnus, T., Chan, A., Grauer, O., Toyka, K. V., and Gold, R. (2001). Microglial phagocytosis of apoptotic inflammatory T cells leads to down regulation of microglial immune activation. *J. Immunol.* **167,** 5004–5010.

Marin-Teva, J. L., Dusart, I., Colin, C., Gervais, A., van Rooijen, N., and Mallat, M. (2004). Microglia promote the death of developing Purkinje cells. *Neuron* **41,** 535–547.

Martino, G., Furlan, R., Comi, G., and Adorini, L. (2001). The ependymal route to the CNS: An emerging gene-therapy approach for MS. *Trends Immunol.* **22,** 483–490.

Masada, T., Xi, G., Hua, Y., and Keep, R. F. (2000). The effects of thrombin preconditioning on focal cerebral ischemia in rats. *Brain Res.* **867,** 173–179.

Matzinger, P. (2007). Friendly and dangerous signals: Is the tissue in control? *Nat. Immunol.* **8,** 11–13.

McGeer, P. L., and McGeer, E. G. (2002). Innate immunity, local inflammation, and degenerative disease. *Sci. Aging Knowledge Environ.* **2002,** re3.

McGeer, P. L., Akiyama, H., Itagaki, S., and McGeer, E. G. (1989). Activation of the classical complement pathway in brain tissue of Alzheimer patients. *Neurosci. Lett.* **107,** 341–346.

McMenamin, P. G. (1999). Distribution and phenotype of dendritic cells and resident tissue macrophages in the dura mater, leptomeninges, and choroid plexus of the rat brain as demonstrated in wholemount preparations. *J. Comp. Neurol.* **405,** 553–562.

Mead, R. J., Singhrao, S. K., Neal, J. W., Lassmann, H., and Morgan, B. P. (2002). The membrane attack complex of complement causes severe demyelination associated with acute axonal injury. *J. Immunol.* **168,** 458–465.

Mead, R. J., Neal, J. W., Griffiths, M. R., Linington, C., Botto, M., Lassmann, H., and Morgan, B. P. (2004). Deficiency of the complement regulator CD59a enhances disease severity, demyelination and axonal injury in murine acute experimental allergic encephalomyelitis. *Lab. Invest.* **84,** 21–28.

Medana, I., Li, Z., Flugel, A., Tschopp, J., Wekerle, H., and Neumann, H. (2001). Fas ligand (CD95L) protects neurons against perforin-mediated T lymphocyte cytotoxicity. *J. Immunol.* **167,** 674–681.

Medzhitov, R., and Janeway, C. A., Jr. (2002). Decoding the patterns of self and nonself by the innate immune system. *Science* **296,** 298–300.

Meri, S., and Pangburn, M. K. (1990). Discrimination between activators and nonactivators of the alternative pathway of complement: Regulation via a sialic acid/polyanion binding site on factor H. *Proc. Natl. Acad. Sci. USA* **87,** 3982–3986.

Minghetti, L. (2005). Role of inflammation in neurodegenerative diseases. *Curr. Opin. Neurol.* **18,** 315–321.

Mollnes, T. E., and Harboe, M. (1987). Immunohistochemical detection of the membrane and fluid-phase terminal complement complexes C5b-9(m) and SC5b-9. Consequences for interpretation and terminology. *Scand. J. Immunol.* **26,** 381–386.

Morgan, B. P., and Meri, S. (1994). Membrane proteins that protect against complement lysis. *Springer Semin. Immunopathol.* **15,** 369–396.

Murata, T., Ohnishi, H., Okazawa, H., Murata, Y., Kusakari, S., Hayashi, Y., Miyashita, M., Itoh, H., Oldenborg, P. A., Furuya, N., and Matozaki, T. (2006). CD47 promotes neuronal development through Src- and FRG/Vav2-mediated activation of Rac and Cdc42. *J. Neurosci.* **26,** 12397–12407.

Neumann, H. (2001). Control of glial immune function by neurons. *Glia* **36,** 191–199.

Nguyen, M. D., Julien, J. P., and Rivest, S. (2002). Innate immunity: The missing link in neuroprotection and neurodegeneration? *Nat. Rev. Neurosci.* **3,** 216–227.

Oldenborg, P. A., Gresham, H. D., and Lindberg, F. P. (2001). CD47-signal regulatory protein alpha (SIRPalpha) regulates Fcgamma and complement receptor-mediated phagocytosis. *J. Exp. Med.* **193,** 855–862.

Pachter, J. S., de Vries, H. E., and Fabry, Z. (2003). The blood-brain barrier and its role in immune privilege in the central nervous system. *J. Neuropathol. Exp. Neurol.* **62,** 593–604.

Pasinetti, G. M., Johnson, S. A., Oda, T., Rozovsky, I., and Finch, C. E. (1994). Clusterin (SGP-2): A multifunctional glycoprotein with regional expression in astrocytes and neurons of the adult rat brain. *J. Comp. Neurol.* **339,** 387–400.

Pender, M. P., Nguyen, K. B., McCombe, P. A., and Kerr, J. F. (1991). Apoptosis in the nervous system in experimental allergic encephalomyelitis. *J. Neurol. Sci.* **104,** 81–87.

Pindon, A., Hantai, D., Jandrot-Perrus, M., and Festoff, B. W. (1997). Novel expression and localization of active thrombomodulin on the surface of mouse brain astrocytes. *Glia* **19,** 259–268.

Prinz, M., Garbe, F., Schmidt, H., Mildner, A., Gutcher, I., Wolter, K., Piesche, M., Schroers, R., Weiss, E., Kirschning, C. J., Rochford, C. D., Bruck, W., *et al.* (2006). Innate immunity mediated by TLR9 modulates pathogenicity in an animal model of multiple sclerosis. *J. Clin. Invest.* **116,** 456–464.

Rahpeymai, Y., Hietala, M. A., Wilhelmsson, U., Fotheringham, A., Davies, I., Nilsson, A. K., Zwirner, J., Wetsel, R. A., Gerard, C., Pekny, M., and Pekna, M. (2006). Complement: A novel factor in basal and ischemia-induced neurogenesis. *EMBO J.* **25,** 1364–1374.

Raine, C. S., Bonetti, B., and Cannella, B. (1998). Multiple sclerosis: Expression of molecules of the tumor necrosis factor ligand and receptor families in relationship to the demyelinated plaque. *Rev. Neurol. (Paris)* **154,** 577–585.

Reinhold, M. I., Lindberg, F. P., Plas, D., Reynolds, S., Peters, M. G., and Brown, E. J. (1995). *In vivo* expression of alternatively spliced forms of integrin-associated protein (CD47). *J. Cell Sci.* **108** (Pt. 11), 3419–3425.

Rogers, J., Strohmeyer, R., Kovelowski, C. J., and Li, R. (2002). Microglia and inflammatory mechanisms in the clearance of amyloid beta peptide. *Glia* **40,** 260–269.

Roth, J., Vogl, T., Sorg, C., and Sunderkotter, C. (2003). Phagocyte-specific S100 proteins: A novel group of proinflammatory molecules. *Trends Immunol.* **24,** 155–158.

Rus, H., Cudrici, C., David, S., and Niculescu, F. (2006). The complement system in central nervous system diseases. *Autoimmunity* **39,** 395–402.

Savill, J., Dransfield, I., Gregory, C., and Haslett, C. (2002). A blast from the past: Clearance of apoptotic cells regulates immune responses. *Nat. Rev. Immunol.* **2,** 965–975.

Schwartz, M., Moalem, G., Leibowitz-Amit, R., and Cohen, I. R. (1999). Innate and adaptive immune responses can be beneficial for CNS repair. *Trends Neurosci.* **22,** 295–299.

Scolding, N., and Franklin, R. (1998). Axon loss in multiple sclerosis. *Lancet* **352,** 340–341.

Seiffert, M., Brossart, P., Cant, C., Cella, M., Colonna, M., Brugger, W., Kanz, L., Ullrich, A., and Buhring, H. J. (2001). Signal-regulatory protein alpha (SIRPalpha) but not SIRPbeta is involved in T-cell activation, binds to CD47 with high affinity, and is expressed on immature CD34(+) CD38(−) hematopoietic cells. *Blood* **97,** 2741–2749.

Serafini, B., Columba-Cabezas, S., Di Rosa, F., and Aloisi, F. (2000). Intracerebral recruitment and maturation of dendritic cells in the onset and progression of experimental autoimmune encephalomyelitis. *Am. J. Pathol.* **157,** 1991–2002.

Sergent-Tanguy, S., Veziers, J., Bonnamain, V., Boudin, H., Neveu, I., and Naveilhan, P. (2006). Cell surface antigens on rat neural progenitors and characterization of the CD3(+)/CD3(−) cell populations. *Differentiation* **74,** 530–541.

Setzu, A., Lathia, J. D., Zhao, C., Wells, K., Rao, M. S., Ffrench-Constant, C., and Franklin, R. J. (2006). Inflammation stimulates myelination by transplanted oligodendrocyte precursor cells. *Glia* **54,** 297–303.

Shikamoto, Y., and Morita, T. (1999). Expression of factor X in both the rat brain and cells of the central nervous system. *FEBS Lett.* **463,** 387–389.

Simard, A. R., Soulet, D., Gowing, G., Julien, J. P., and Rivest, S. (2006). Bone marrow-derived microglia play a critical role in restricting senile plaque formation in Alzheimer's disease. *Neuron* **49,** 489–502.

Singhrao, S. K., Neal, J. W., Gasque, P., Morgan, B. P., and Newman, G. R. (1996). Role of complement in the aetiology of Pick's disease? *J. Neuropathol. Exp. Neurol.* **55**(5), 578–593.

Singhrao, S. K., Neal, J. W., Morgan, B. P., and Gasque, P. (1999). Increased complement biosynthesis by microglia and complement activation on neurons in Huntington's disease. *Exp. Neurol.* **159,** 362–376.

Singhrao, S. K., Neal, J. W., Rushmere, N. K., Morgan, B. P., and Gasque, P. (2000). Spontaneous classical pathway activation and deficiency of membrane regulators render human neurons susceptible to complement lysis. *Am. J. Pathol.* **157,** 905–918.

Speth, C., Dierich, M. P., and Gasque, P. (2002). Neuroinvasion by pathogens: A key role of the complement system. *Mol. Immunol.* **38,** 669–679.

Striggow, F., Riek, M., Breder, J., Henrich-Noack, P., Reymann, K. G., and Reiser, G. (2000). The protease thrombin is an endogenous mediator of hippocampal neuroprotection against ischemia at low concentrations but causes degeneration at high concentrations. *Proc. Natl. Acad. Sci. USA* **97,** 2264–2269.

Stuart, L. M., and Ezekowitz, R. A. (2005). Phagocytosis: Elegant complexity. *Immunity* **22,** 539–550.

Sugita, Y., Becerra, S. P., Chader, G. J., and Schwartz, J. P. (1997). Pigment epithelium-derived factor (PEDF) has direct effects on the metabolism and proliferation of microglia and indirect effects on astrocytes. *J. Neurosci. Res.* **49,** 710–718.

Tahara, K., Kim, H. D., Jin, J. J., Maxwell, J. A., Li, L., and Fukuchi, K. (2006). Role of toll-like receptor signalling in Abeta uptake and clearance. *Brain* **129,** 3006–3019.

Tenner, A. J. (1999). Membrane receptors for soluble defense collagens. *Curr. Opin. Immunol.* **11,** 34–41.

Tenner, A. J. (2001). Complement in Alzheimer's disease: Opportunities for modulating protective and pathogenic events. *Neurobiol. Aging* **22,** 849–861.

Ubl, J. J., Sergeeva, M., and Reiser, G. (2000). Desensitisation of protease-activated receptor-1 (PAR-1) in rat astrocytes: Evidence for a novel mechanism for terminating Ca2+ signalling evoked by the tethered ligand. *J. Physiol.* **525**(Pt. 2), 319–330.

Van Beek, J., Bernaudin, M., Petit, E., Gasque, P., Nouvelot, A., MacKenzie, E. T., and Fontaine, M. (2000). Expression of receptors for complement anaphylatoxins C3a and C5a following permanent focal cerebral ischemia in the mouse. *Exp. Neurol.* **161,** 373–382.

van Beek, J., Nicole, O., Ali, C., Ischenko, A., MacKenzie, E. T., Buisson, A., and Fontaine, M. (2001). Complement anaphylatoxin C3a is selectively protective against NMDA-induced neuronal cell death. *Neuroreport* **12,** 289–293.

van Beek, J., van Meurs, M., t Hart, B. A., Brok, H. P., Neal, J. W., Chatagner, A., Harris, C. L., Omidvar, N., Morgan, B. P., Laman, J. D., and Gasque, P. (2005). Decay-accelerating factor (CD55) is expressed by neurons in response to chronic but not acute autoimmune central nervous system inflammation associated with complement activation. *J. Immunol.* **174,** 2353–2365.

Vaughan, P. J., Pike, C. J., Cotman, C. W., and Cunningham, D. D. (1995). Thrombin receptor activation protects neurons and astrocytes from cell death produced by environmental insults. *J. Neurosci.* **15,** 5389–5401.

Vernon-Wilson, E. F., Kee, W. J., Willis, A. C., Barclay, A. N., Simmons, D. L., and Brown, M. H. (2000). CD47 is a ligand for rat macrophage membrane signal regulatory protein SIRP (OX41) and human SIRPalpha 1. *Eur. J. Immunol.* **30,** 2130–2137.

Vivien, D., and Buisson, A. (2000). Serine protease inhibitors: Novel therapeutic targets for stroke? *J. Cereb. Blood Flow Metab.* **20,** 755–764.

Wagner, S., Lynch, N. J., Walter, W., Schwaeble, W. J., and Loos, M. (2003). Differential expression of the murine mannose-binding lectins A and C in lymphoid and nonlymphoid organs and tissues. *J. Immunol.* **170,** 1462–1465.

Walker, D. G., Yasuhara, O., Patston, P. A., McGeer, E. G., and McGeer, P. L. (1995). Complement C1 inhibitor is produced by brain tissue and is cleaved in Alzheimer disease. *Brain Res.* **675,** 75–82.

Walport, M. J. (2001). Complement. First of two parts. *N. Engl. J. Med.* **344,** 1058–1066.

Wang, H., and Reiser, G. (2003). Thrombin signaling in the brain: The role of protease-activated receptors. *Biol. Chem.* **384,** 193–202.

Weinstein, J. R., Gold, S. J., Cunningham, D. D., and Gall, C. M. (1995). Cellular localization of thrombin receptor mRNA in rat brain: Expression by mesencephalic dopaminergic neurons and codistribution with prothrombin mRNA. *J. Neurosci.* **15,** 2906–2919.

Wright, G. J., Jones, M., Puklavec, M. J., Brown, M. H., and Barclay, A. N. (2001). The unusual distribution of the neuronal/lymphoid cell surface CD200 (OX2) glycoprotein is conserved in humans. *Immunology* **102,** 173–179.

Wyss-Coray, T., and Mucke, L. (2002). Inflammation in neurodegenerative disease—A double-edged sword. *Neuron* **35,** 419–432.

Xi, G., Reiser, G., and Keep, R. F. (2003). The role of thrombin and thrombin receptors in ischemic, hemorrhagic and traumatic brain injury: Deleterious or protective? *J. Neurochem.* **84,** 3–9.

Yasuhara, O., Aimi, Y., McGeer, E. G., and McGeer, P. L. (1994). Expression of the complement membrane attack complex and its inhibitors in Pick disease brain. *Brain Res.* **652,** 346–349.

Yepes, M., Sandkvist, M., Wong, M. K., Coleman, T. A., Smith, E., Cohan, S. L., and Lawrence, D. A. (2000). Neuroserpin reduces cerebral infarct volume and protects neurons from ischemia-induced apoptosis. *Blood* **96,** 569–576.

Zhang, K. Z., Junnikkala, S., Erlander, M. G., Guo, H., Westberg, J. A., Meri, S., and Andersson, L. C. (1998). Up-regulated expression of decay-accelerating factor (CD55) confers increased complement resistance to sprouting neural cells. *Eur. J. Immunol.* **28,** 1189–1196.

Zhang, W., Wang, T., Pei, Z., Miller, D. S., Wu, X., Block, M. L., Wilson, B., Zhang, W., Zhou, Y., Hong, J. S., and Zhang, J. (2005). Aggregated alpha-synuclein activates microglia: A process leading to disease progression in Parkinson's disease. *FASEB J.* **19,** 533–542.

Zipp, F., Krammer, P. H., and Weller, M. (1999). Immune (dys) regulation in multiple sclerosis: Role of the CD95-CD95 ligand system. *Immunol. Today* **20,** 550–554.

GLUTAMATE RELEASE FROM ASTROCYTES IN PHYSIOLOGICAL CONDITIONS AND IN NEURODEGENERATIVE DISORDERS CHARACTERIZED BY NEUROINFLAMMATION

Sabino Vesce,* Daniela Rossi,† Liliana Brambilla,† and Andrea Volterra*,†

*Department of Cell Biology and Morphology, University of Lausanne
Rue du Bugnon 9, 1005 Lausanne, Switzerland
†Department of Pharmacological Sciences
Center of Excellence on Neurodegenerative Diseases
University of Milan, Via Balzaretti 9, 20133 Milan, Italy

Although glial cells have been traditionally viewed as supportive partners of neurons, studies of the last 20 years demonstrate that astrocytes possess functional receptors for neurotransmitters and other signaling molecules and respond to their stimulation via release of chemical transmitters (called gliotransmitters) such as glutamate, ATP, and D-serine. Notably, astrocytes react to synaptically released neurotransmitters with intracellular calcium ($[Ca^{2+}]_i$) elevations, which result in the release of glutamate via regulated exocytosis and possibly other mechanisms. These findings have led to a new concept of neuron–glia intercommunication where astrocytes play an unsuspected dynamic role by integrating neuronal inputs and modulating synaptic activity. The additional discovery that glutamate release from astrocytes is controlled by molecules linked to inflammatory reactions, such as the cytokine tumor necrosis factor-α (TNF-α) and prostaglandins, suggests that glia-to-neuron signaling may be sensitive to changes in production of these mediators in pathological conditions. Indeed, a local, parenchymal brain inflammatory reaction (neuroinflammation) characterized by astrocytic and microglial activation has been reported in several neurodegenerative disorders, including Alzheimer's disease and AIDS dementia complex. This transition to a reactive state may be accompanied by a disruption of the cross talk normally occurring between astrocytes and neurons and so contribute to disease development.

INTERNATIONAL REVIEW OF
NEUROBIOLOGY, VOL. 82
DOI: 10.1016/S0074-7742(07)82003-4

0074-7742/07 $35.00

The findings reported in this chapter suggest that a better comprehension of the glutamatergic interplay between neurons and glia may provide information about normal brain function and also highlight possible molecular targets for therapeutic interventions in pathology.

I. Introduction

Glia are a group of cells largely represented in the central nervous system (CNS). Based on their diverse morphology and function they can be distinguished in various subclasses, including oligodendrocytes, responsible for axon myelination; microglia, immunocompetent, specialized brain macrophages; and astrocytes, which represent the major CNS population of glial cells and perform multiple tasks owing to their strategic anatomical location between neurons and blood vessels (Kettenmann and Ransom, 1995).

On the one hand, astrocytes are intimately associated with neurons: in the hippocampus, for example, their processes tightly enwrap 50% of the synapses (Pfrieger and Barres, 1996; Ventura and Harris, 1999). This close interaction allows astrocytes to provide nerve cells with structural, metabolic, and trophic support. On the other hand, glucose, the main source of energy for nerve cells, enters the brain parenchyma from the cerebral circulation via uptake by the astrocytes. In the astrocytes, glucose is either stored in the form of glycogen or enters glycolysis to produce lactate, a major metabolic substrate for neurons (Tsacopoulos and Magistretti, 1996). By means of permeable channels and active pumps located on the plasma membrane, astrocytes play also a critical role in maintaining ionic homeostasis. For instance, they maintain the extracellular potassium concentration within the physiological range (Karwoski *et al.*, 1989), which is critical for assuring normal neuronal excitability. Excess extracellular potassium may indeed result in neuronal depolarization and eventually cause action potential blockage. Moreover, astrocytes control the extracellular concentration of synaptically released neurotransmitters by means of specific plasma membrane transporter proteins (Coco *et al.*, 1997; Dehnes *et al.*, 1998). At glutamatergic synapses, glutamate uptake from astrocytes is the main mechanism for removing the neurotransmitter from the extracellular compartment.

Because of their established functions of energy suppliers for neurons and regulators of the extracellular brain fluid composition, for decades astrocytes have been regarded solely as passive supportive elements maintaining the optimal neuronal microenvironment but devoid of any direct role in brain communication. However, this vision has been challenged by studies of the last 20 years showing that astrocytic cells modulate synaptic activity and actively participate to

CNS function and formation. One initial puzzling discovery was that astrocytes express a large repertoire of receptors for neurotransmitters, often mirroring the ones seen in the neighboring synapses. Subsequently, it was found that such receptors can be activated by the spillover of neurotransmitters during synaptic activity and induce $[Ca^{2+}]_i$ elevations in the astrocytes. This event can in turn drive the release of gliotransmitters such as glutamate from the astrocytes and trigger intercellular communication, including astrocyte-to-neuron signaling (Araque *et al.*, 1999). Astrocyte-released gliotransmitters are able to activate neuronal receptors and thereby modify the neuronal electrical excitability and synaptic transmission (Schipke and Kettenmann, 2004; Volterra and Meldolesi, 2005).

The amino acid glutamate can be released from astrocytes through multiple pathways that are activated under different conditions, at different loci and/or with different modalities. Both Ca^{2+}-dependent and Ca^{2+}-independent processes have been described. Ca^{2+}-independent ones include reversed operation of reuptake carriers, notably under ischemic conditions (Rossi *et al.*, 2000), exchange with cystine, the essential substrate for astrocytic production of glutathione, mediated by specific cystine-glutamate antiporters (Warr *et al.*, 1999), and molecular permeation of large pore channels, including $P2X_7$ receptors (Duan *et al.*, 2003), gap-junction hemichannels (Ye *et al.*, 2002), and volume-sensitive organic anion channels (Kimelberg *et al.*, 1990), although for the latter a Ca^{2+}-dependent mechanism has also been described (Takano *et al.*, 2005). In this chapter, we will focus on the mechanism(s) underlying astrocytic Ca^{2+}-dependent glutamate release in physiological circumstances and discuss the implications that alterations of this mechanism may have in neuroinflammatory and degenerative processes, notably in AIDS dementia complex (ADC) and Alzheimer's disease (AD).

II. Ca^{2+}-Dependent Glutamate Release from Astrocytes

The initial demonstration that astrocytes signal to neurons in response to physiological stimuli by Ca^{2+}-dependent glutamate release came in 1994 by the group of Philip Haydon. The authors described that stimulation of mixed neuron–glia cultures with the peptide bradykinin triggered $[Ca^{2+}]_i$ rises and glutamate release from astrocytes. In addition, they observed $[Ca^{2+}]_i$ elevations in neurons following those in astrocytes and showed that neuronal $[Ca^{2+}]_i$ elevations were mediated by astrocyte-released glutamate (Parpura *et al.*, 1994).

Support to these first *in vitro* studies and to the hypothesis that a bidirectional neuron–glia signaling exists in intact brain tissue was subsequently provided by Pasti *et al.* (1997). These authors showed that stimulation of neuronal afferents in acute cortical and hippocampal slices induced $[Ca^{2+}]_i$ oscillations in surrounding astrocytes. Synaptically released glutamate seemed the most likely initiator of such events via stimulation of metabotropic receptors (mGluRs) located in the astrocytic plasma membrane. $[Ca^{2+}]_i$ oscillations in astrocytes were followed by $[Ca^{2+}]_i$ rises in the surrounding neurons, possibly as a result of glutamate release from astrocytes.

Indeed, a parallel study by our group (Bezzi *et al.*, 1998) provided the direct demonstration that activation of astrocyte mGluRs (and even more effectively the combined activation of AMPA/kainate and mGluRs) triggers glutamate release from astrocytes via a Ca^{2+}-dependent mechanism involving prostaglandins. In addition, such glutamate- and prostaglandin-dependent glutamate release from astrocytes in the hippocampal stratum radiatum induced GluR-dependent $[Ca^{2+}]_i$ responses in neighboring pyramidal neurons, definitively proving the existence of a glutamatergic neuron-astrocyte cross talk *in situ*. Interestingly, glutamate-induced glutamate release from astrocytes was occlusive with the release (Ca^{2+}-dependent) triggered by bradykinin, but additive to the release (Ca^{2+}-independent) mediated by glutamate transporters, suggesting that distinct astrocyte neuroligand receptors are coupled to a common Ca^{2+}-dependent transduction pathway ultimately leading to glutamate release; and that Ca^{2+}-dependent and Ca^{2+}-independent release mechanisms probably use different glial glutamate pools.

A wide range of more recent studies have confirmed these initial observations, reinforcing the view that the Ca^{2+}-dependent pathway is triggered by activation of G-protein–coupled receptors (GPCR) and mediated by inositol 1,4,5-triphosphate-induced Ca^{2+} release from stores of the endoplasmic reticulum (Bezzi *et al.*, 2001; Jeremic *et al.*, 2001; Kang *et al.*, 2005; Sanzgiri *et al.*, 1999; Takano *et al.*, 2005). However, the ultimate process by which glutamate is released, whether regulated exocytosis or permeation via volume-regulated anion channels, has been more debated (Nedergaard *et al.*, 2003).

The evidence for the existence of a regulated exocytosis pathway in astrocytes is now overwhelming. Notably, our group has provided the first ultrastructural demonstration that astrocytes in adult brain tissue contain a vesicular compartment whose vesicles strictly resemble microvesicles of glutamatergic nerve terminals. These synaptic-like microvesicles (SLMVs), found in perisynaptic astrocytic processes, express vesicular glutamate transporters (VGLUT1/2) and vesicular SNARE proteins such as cellubrevin, indicating that they are equipped for taking up, storing, and releasing glutamate (Bezzi *et al.*, 2004). By complementing the ultrastructural studies *in situ* with dynamic total internal reflection fluorescence (TIRF) real-time imaging studies in cultured astrocytes, we could directly document individual vesicle fusion events accompanied by glutamate release in response to mGluR stimulation (Bezzi *et al.*, 2004). Such exocytic fusions occurred within tens of milliseconds from GPCR stimulation and were abolished by a classical blocker of exocytosis, tetanus neurotoxin. The presence of regulated exocytosis in astrocytes has been now documented with a wide range of experimental approaches, including optical detection techniques different from TIRF

(Crippa *et al.*, 2006; Kang *et al.*, 2005), membrane capacitance measurements (Kreft *et al.*, 2004; Zhang *et al.*, 2004a), electrochemical amperometry (Chen *et al.*, 2005), as well as by selective interference with proteins of the exocytotic machinery (Montana *et al.*, 2004; Zhang *et al.*, 2004b).

Although astrocytes express the machinery for regulated exocytosis, several relevant differences have emerged in terms of protein composition, organization, and kinetics with respect to synaptic exocytosis. Molecularly, four vesicular proteins (SNAP23, complexin 2, Munc18a, and synaptotagmin IV) are certainly expressed in astrocytes, while other exocytotic proteins typically expressed in synapses have not been detected (for a review, see Volterra and Meldolesi, 2005). Although similar in size to their neuronal counterparts, astrocytic vesicles are less densely packed and not as orderly arranged. However, they are frequently observed at sites adjacent to neuronal structures that bear glutamate receptors. The exocytotic event itself appears to be relatively slow in astrocytes, about 100 times slower than at synapses (Kreft *et al.*, 2004). Such property, however, is justified by the slower stimulus-secretion coupling mechanism via GPCR signaling and Ca^{2+} release from the internal stores, and may well fit the modulatory role that astrocytes are supposed to play in synaptic function.

In addition to mGluRs, other GPCRs trigger Ca^{2+}-dependent glutamate exocytosis from astrocytes. In a study, Domercq *et al.* (2006) documented an exocytic response to stimulation of purinergic P2Y1 receptors (P2Y1R, activated by ATP in physiological conditions). In complement to TIRF imaging evidence in cultured astrocytes, the authors showed that P2Y1R stimulation induces Ca^{2+}-dependent glutamate release in hippocampal slices (but not in hippocampal synaptosomes), and that the release *in situ* shares the same pharmacological properties with the P2Y1R-evoked release in cultured astrocytes. Notably, the process *in situ* is sensitive to the exocytosis blocker, bafilomycin A1 (Baf A1). The rate by which Baf A1 inhibits P2Y1R-dependent glutamate release is slower than that by which the drug acts on high K^+-evoked neuronal exocytosis, suggesting that the P2Y1R-evoked process is distinct from the one in nerve terminals.

Also, Takano *et al.* (2005) found that stimulation of metabotropic P2Y receptors evokes Ca^{2+}-dependent glutamate release in cultured astrocytes. These authors, however, reported the release to be insensitive to blockers of exocytosis and sensitive to inhibitors of volume-regulated anion channels. According to the purinergic receptor pharmacology for this release, ATP activates P2Y2 or P2Y4 receptors, not P2Y1R. Why the same endogenous ligand activates distinct P2Y receptor subtypes in the two studies remains to be established. One possibility is that astrocytes express different P2YR subtypes in different culture conditions. In this context, it remains to be confirmed that the mechanism described by Takano *et al.* takes place *in situ*. If so, one can hypothesize that ATP leads to glutamate release from astrocytes via activation of multiple P2YR subtypes coupled to distinct Ca^{2+}-dependent intracellular pathways.

Shedding of the cytokine tumor necrosis factor-α (TNF-α) as well as formation of prostaglandin E_2 (PGE_2) (see next paragraph) are apparently primordial signaling events accompanying ATP-induced glutamate exocytosis. Thus, this release is strongly diminished in the presence of pharmacological inhibitors of TNF-α and prostaglandins, and the same occurs in cultures of astrocytes from TNF-α knockout mice (Domercq *et al.*, 2006). The relatively fast kinetics of glutamate secretion from astrocytes indicates that the production of TNF-α and PGE_2 should occur in the timescale of milliseconds following P2Y1R activation. This is perhaps possible according to observations (Goddard *et al.*, 2001; Zonta *et al.*, 2003), but remains to be directly demonstrated. Alternatively, a tonic basal release of TNF-α and PGE_2 from astrocytes might appropriately tune the secretory apparatus for efficient glutamate liberation. Moreover, both TNF-α and PGE_2 are connected to $[Ca^{2+}]_i$ elevations in astrocytes (Bezzi *et al.*, 1998; Domercq *et al.*, 2006) and could boost the intracellular Ca^{2+} response initiated by P2Y1R activation.

III. Excitotoxicity Involving Ca^{2+}-Dependent Glutamate Release from Astrocytes in Pathological Conditions: The Case of ADC

Several lines of evidence indicate the presence of a local, parenchymal inflammatory response in a number of chronic neurodegenerative disorders. Such condition is characterized by specific morphological and functional changes of astrocytes and microglia, broadly defined as "reactive gliosis." The signals exchanged between the two glial cell types during these events are largely unknown, yet their transition from the resting to the activated state appears to be associated with a marked upregulation of several genes and the secretion of factors like cytokines, eicosanoids, reactive oxygen species, nitric oxide, and excitatory amino acids (Perry *et al.*, 1995). Although the inflammatory glial reaction was originally thought to be mostly beneficial for tissue-repairing processes, in several cases it may actually contribute to the exacerbation of neurodamaging processes. In view of the control exerted by prostaglandins and TNF-α on Ca^{2+}-dependent glutamate release from astrocytes, overproduction of these mediators during neuroinflammation might favor an increased and deleterious glutamatergic input from astrocytes to neurons. This hypothesis is substantiated by direct experimental evidence.

Thus, our group discovered that, in addition to classical transmitters, such as glutamate or ATP, the chemotactic cytokine (chemokine) stromal-derived factor-1α (SDF-1α) also induces Ca^{2+}-dependent glutamate release from astrocytes (Bezzi *et al.*, 2001). SDF-1α acts via stimulation of its specific GPCR, CXCR4. The process is accompanied by shedding of TNF-α which, once released from astrocytes, acts autocrinally or paracrinally on its p55, TNF receptor type-1 (TNFR1). In turn, TNFR1 activation controls downstream signaling events, including PGE_2 production, $[Ca^{2+}]_i$ elevation, and glutamate release. The latter is blocked by the exocytosis blockers Baf A1 and tetanus neurotoxin (Fig. 1A). Importantly, we found that CXCR4 is expressed also in microglia and that

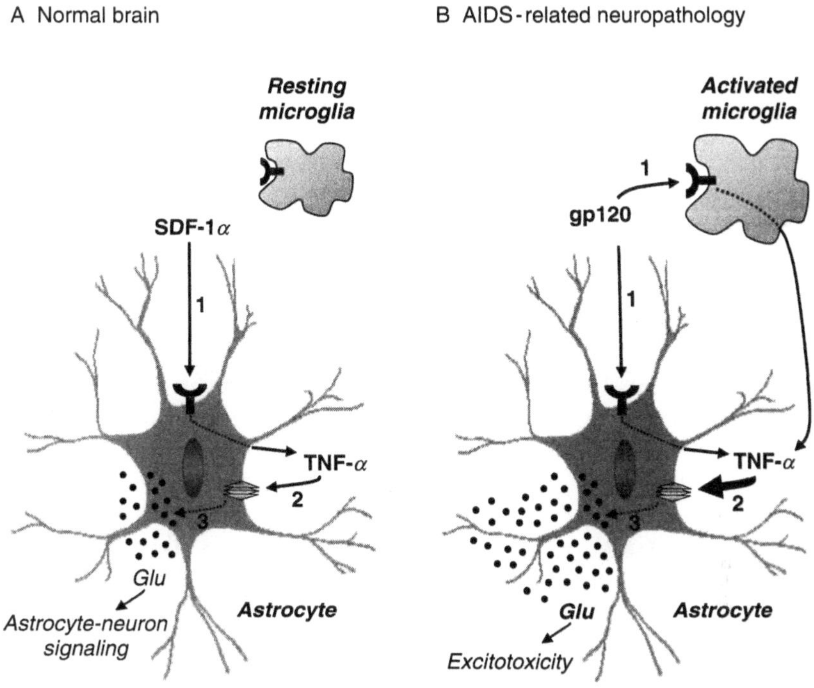

FIG. 1. Microglia-dependent transformation of CXCR4-evoked glutamate release from astrocytes into an excitotoxic pathway. (A) In physiological conditions, stimulation of the CXCR4 receptor by the endogenous ligand SDF-1α is followed by a sequence of intracellular events that triggers the shedding of TNF-α from its membrane-bound precursor (1). Once released, the cytokine acts as an autocrine/paracrine factor and stimulates TNFR1 in the same and/or in surrounding astrocytes (2). The interaction of TNF-α with its cell surface receptor initiates additional intracellular signaling events, including prostaglandin production, which act in parallel or sequence with the initial events triggered by CXCR4 stimulation, to eventually lead to Ca^{2+}-dependent glutamate (Glu) release (3). This, in turn, may have modulatory effects on synaptic activity. Microglia are in the resting state and do not participate in astrocyte-neuron signaling. (B) In AIDS-related neuropathology, microglia infected by HIV become reactive and shed viral particles, including the envelope glycoprotein gp120. This can act as an agonist on CXCR4 receptors that are present in astrocytes and also in the reactive microglia itself, thus inducing a considerably higher TNF-α release compared to physiological conditions (1). The enhanced activation of TNFR1 receptors in astrocytes (2), results in a massive and ultimately excitotoxic release of glutamate (3).

stimulation of CXCR4 in these cells results in potent TNF-α release, several-fold higher than in astrocytes, but occurring only when microglia are in a reactive state. Stimulation of microglial CXCR4 does not induce glutamate release. In the pathological brain, astrocytes and microglia often form local foci of reactive cells around the sites of lesion or infection. We mimicked this condition by adding LPS-activated microglia to astrocyte-pure cultures in about the same ratio (1:10) existing in the brain. Simultaneous stimulation of CXCR4 in contiguous reactive microglia and astrocytes resulted in a dramatic amplification of TNF-α release from both cells and, as a consequence, in strong potentiation of Ca^{2+}-dependent glutamate release from the astrocytes. In separate experiments in hippocampal cultures containing neurons, astrocytes, and microglia, we could demonstrate that the microglia- and TNF-α-dependent potentiation of astrocyte glutamate release had excitotoxic consequences, inducing slow apoptotic death of neuronal subpopulations (Fig. 1B).

Not only identification of the above CXCR4-dependent signaling cascade described a novel role for TNF-α in rapid glial communication, but also provided a novel mechanistic hypothesis (Bezzi *et al.*, 2001) for the neurodegenerative processes underlying HIV-associated neuropathology, notably the ADC (Kaul *et al.*, 2001).

Previous evidence suggested that an increasing number of neurons die by apoptosis (Bagetta *et al.*, 1996; Shi *et al.*, 1996) during progression of the disease, apparently via NMDA-mediated excitotoxicity. However, microglial cells, not neurons, are infected by penetration of the HIV-1 virus into the brain and so neuronal death is thought to occur via interactions with these cells, as well as with the astrocytes (Kaul *et al.*, 2001; Meucci and Miller, 1996; Toggas *et al.*, 1996). In this context, viral particles, including the HIV-1 coat protein gp120, shedded from infected cells, may play a key role. Indeed, the sole expression of gp120 in transgenic mice was found to reproduce several features of the ADC neuropathology (Toggas *et al.*, 1994). We discovered that a gp120 isoform, $gp120_{IIIB}$, derived from the T-tropic HIV-1 IIIB strain, acts as CXCR4 agonist in both astrocytes and microglia and, similar to the endogenous ligand SDF-1α, triggers potent TNF-α-dependent glutamate release from astrocytes. Since production and shedding of viral proteins in HIV-1-infected brains is an uncontrolled process, it is not unlikely that a consequence of it is the overstimulation of glial CXCR4 with a consequent excessive glutamate release from astrocytes eventually triggering excitotoxic neuronal cell death (Fig. 1B). Indeed, in our *in vitro* model, we obtain neuroprotection by blocking the astrocyte signaling cascade at any of the identified steps: CXCR4 receptor activation, TNF-α release, prostaglandin formation, and, finally, by scavenging released glutamate or blocking NMDA receptors (Bezzi *et al.*, 2001).

Antiretroviral therapy does not fully control AIDS-associated neuropathology while significantly prolonging the life expectancy of AIDS-affected subjects.

Therefore, there is a need for drugs limiting the development of ADC and ensuring a better quality of life. Clinical trials have been conducted with memantine, an inhibitor of the NMDA receptor channels that might prove helpful in containing excitotoxicity. Our work suggests that molecules acting at other levels along the toxic pathway could be also beneficial.

IV. Astrocytic Alterations and Ca^{2+}-Dependent Glutamate Release Dysfunction in AD

AD is a neurodegenerative disorder characterized by the progressive loss of cognitive function. One of its best-known biochemical hallmarks is the accumulation of the amyloid-β (Aβ) protein into amyloid plaques that can be found in the extracellular space of forebrain regions (Glenner and Wong, 1984). Findings suggest that astrocytes may play an important role in the clearance of the Aβ peptide and thus in preventing plaque formation (Wyss-Coray *et al.*, 2003). By incubating astrocytes with brain sections from transgenic mice expressing amyloid precursor protein, the authors showed that adult astrocytes migrate in response to chemokines present in AD lesions and are able to bind, internalize, and degrade Aβ. The mechanism of digestion is unclear, but could involve matrix metalloproteinase-9, a protease able to digest both soluble and fibrillary Aβ and expressed by astrocytes found at the border of amyloid plaques (Yan *et al.*, 2006). Interestingly, this protective action by astrocytes could be lost in AD, since endogenous astrocytes surround and contact plaques in AD brains, but appear unable to remove Aβ (Wyss-Coray *et al.*, 2003). Deficits in the astrocytic clearance of Aβ may therefore contribute to amyloid plaque formation. One of the possible causes is a defect in apolipoprotein E (Koistinaho *et al.*, 2004) that, in the mammalian brain, is mainly produced by astrocytes and represents a recognized genetic risk factor for AD. There are other indications that astrocytes are perturbed by the presence of Aβ. For example, the administration of Aβ to mixed cultures of hippocampal neurons and astrocytes causes abnormal $[Ca^{2+}]_i$ transients and mitochondrial depolarization in astrocytes long before any impairment is visible in neurons. Therefore, the subsequent neuronal death might be the result of oxidative stress generated by the astrocytic dysfunction (Abramov *et al.*, 2004). Thus, the fate of neurons in AD may be closely dependent on the maintenance of normal astrocytic function.

Chronic inflammatory glial cell reaction is well documented around Aβ plaques (for a review, see Wyss-Coray, 2006). While it is not completely understood whether inflammation plays a causative role in AD, recent data from animal models suggest that some inflammatory processes might at least accelerate the development of the disease. For instance, β-secretase, the Aβ-generating enzyme, was found to be present in reactive astrocytes from aged Tg2576 mice

(Hartlage-Rubsamen *et al.*, 2003), a transgenic AD mouse model expressing the double mutation K670N-M671L of the amyloid precursor protein APP. The proximity of these astrocytes to the plaques raises the possibility that they promote Aβ accumulation. Thus, in the transition to their reactive state, astrocytes might undergo a deleterious double transformation: loss of the ability to digest Aβ, followed by contribution to generating the toxic protein itself.

High levels of proinflammatory cytokines such as interleukin-1β, -6, and TNF-α, mostly produced by reactive microglia and astrocytes, are detected in the brain of AD subjects (Zhao *et al.*, 2003). The consequences of this phenomenon are unclear, also because proinflammatory cytokines have varied effects depending on the biological context (Wyss-Coray, 2006). However, recent work points to a toxic action of these cytokines in the neuroinflammatory reaction characterizing AD (Ralay Ranaivo *et al.*, 2006). In this study, Aβ directly injected in the mouse brain caused a reduction in the levels of the synaptic markers synaptophysin and postsynaptic density-95 (PSD-95), indicative of a synaptic defect. If the Aβ treatment was followed, starting 3 weeks later, by the daily oral administration of a selective inhibitor of proinflammatory cytokine production, the levels of the synaptic markers were restored. This positive biochemical change was matched by an improved performance of the animals in a spatial learning task, suggesting that, indeed, targeting neuroinflammation might be an effective therapeutic strategy in AD.

Glial alterations, notably increased cytokine production, may also affect the Ca^{2+}-dependent pathway of glutamate release from astrocytes. The observation that expression of both TNF-α and TNFR1 receptors is enhanced in the brain of AD patients (Del Villar and Miller, 2004; Zhao *et al.*, 2003) led our group to investigate possible alterations of the TNF-α-dependent pathway of glutamate release in PDAPP mice, a transgenic model of AD (Rossi *et al.*, 2005). We utilized aged animals (12 months old), presenting abundant amyloid plaque deposition and reactive gliosis in the forebrain, and adult animals (4 months old) with little or no amyloid deposits. Ca^{2+}-dependent glutamate release was stimulated in brain slices from PDAPP animals by direct TNF-α application. Our data indicate that the release does not originate from nerve terminals and that its pharmacological characteristics are identical to those of the process evoked by TNF-α in cultured astrocytes (Bezzi *et al.*, 2001). Contrary to our expectations and the findings in Bezzi *et al.* (2001), we detected a considerable reduction (not increase) of TNF-α-evoked glutamate release in hippocampal slices from aged PDAPP animals compared to adult animals and age-matched controls (Rossi *et al.*, 2005). The defect was region-selective as the glutamate release response from cerebellar slices of aged PDAPP mice was identical to that of controls. Interestingly, the secretory process itself appeared intact, since stimulation with PGE_2, which acts downstream of TNF-α, evoked normal glutamate release. Therefore, the alteration must take place at the level of the stimulus-secretion coupling mechanism.

An important intracellular mediator of TNF-α is the protein DENN/MADD, which binds TNFR1 and triggers activation of multiple downstream signaling pathways, including cytosolic phospholipase A_2, thus leading to arachidonic acid release and prostaglandin production. Importantly, a reduced expression of DENN/MADD was reported in the hippocampi from Alzheimer's patients (Del Villar and Miller, 2004). We found that DENN/MADD is defective also in the hippocampus (not cerebellum) of aged PDAPP mice (Rossi *et al.*, 2005), suggesting that defect of DENN/MADD expression may account for reduced glutamate release in response to TNF-α stimulation. Downregulation of DENN/MADD may result from the chronic inflammation that characterizes the slow progression of AD and might be a consequence of long-term overstimulation of TNFR1 by excessive TNF-α. This condition is very different from the *in vitro* model of acute inflammation that we utilized in the studies concerning CXCR4-evoked glutamate release. In that case, we tested glutamate release within 24 h from adding reactive microglia to astrocyte cultures. It is therefore likely that acute and chronic TNF-α overproduction cause opposite alterations of glutamate release from astrocytes.

Our present evidence in aged PDAPP mice cannot identify whether the described impairment has a pathogenetic role in AD. Therefore, at present, we can only speculate on the functional significance of these findings. Astrocytes are known to exert a critical control on the formation, maintenance, and strength of synapses by actively releasing soluble factors (Beattie *et al.*, 2002; Christopherson *et al.*, 2005; Mauch *et al.*, 2001; Ullian *et al.*, 2001). Notably, astrocyte-released TNF-α controls the strength of excitatory synapses in the hippocampus and thus is crucial for the stability and optimal performance of neuronal networks (Beattie *et al.*, 2002; Stellwagen and Malenka, 2006). In addition, Ca^{2+}-dependent glutamate release from hippocampal astrocytes modulates neuronal excitability, synchronicity, and synaptic transmission (Angulo *et al.*, 2004; Fellin *et al.*, 2004; Fiacco and McCarthy, 2004; Kang *et al.*, 1998; Liu *et al.*, 2004; Parri *et al.*, 2001; Perea and Araque, 2005). Synaptic failure in AD, probably dependent on the early formation of Aβ oligomers in the extracellular milieu, becomes apparent well before aggregation into plaques occurs and neuronal degeneration is detectable (Selkoe, 2002). Thus, although more evidence is needed, it is possible that the disruption of TNF-α-mediated glutamate release from astrocytes participates in the progressive reduction of synaptic efficacy underlying cognitive decline in AD.

V. Conclusions

The discovery that astrocytes are active partners of neurons in brain communication and possess regulated forms of transmitter release is dense of implications for understanding brain processes in health and disease. In particular, several recent findings indicate that release of gliotransmitters from astrocytes, such as the release of glutamate that we briefly summarized in this chapter, is implicated in the physiological modulation of synaptic functions. Alteration of this process, leading either to excessive release and excitotoxicity or to reduced release and possibly failure to integrate synaptic activity, may have serious consequences on the surrounding neuronal population. This suggests that understanding the biological basis of the transformations that astrocytes and microglia undergo during reactive gliosis and neuroinflammatory processes could lead to a significant progress toward the cure of neurodegenerative disorders. For instance, $[Ca^{2+}]_i$ oscillations in hippocampal astrocytes stimulate glutamate release and synchronize the activity in neighboring neurons (Angulo *et al.*, 2004; Fellin *et al.*, 2004; Parri *et al.*, 2001). However, in a model of traumatic injury, these oscillations are lost in reactive astrocytes surrounding the area of lesion (Aguado *et al.*, 2002). Ca^{2+} signaling alterations and alterations in Ca^{2+}-dependent glutamate release in reactive astrocytes appear as functionally relevant defects that should be taken into consideration when studying neurodegenerative processes. The data here reviewed call for more studies looking at glial and neuronal pathological alterations as integrated phenomena, given that glial derangements cannot be simply considered as marginal events or late reactions to neuronal injury, but rather as intrinsic components of the neurodegenerative processes.

References

Abramov, A. Y., Canevari, L., and Duchen, M. R. (2004). β-amyloid peptides induce mitochondrial dysfunction and oxidative stress in astrocytes and death of neurons through activation of NADPH oxidase. *J. Neurosci.* **24,** 565–575.

Aguado, F., Espinosa-Parrilla, J. F., Carmona, M. A., and Soriano, E. (2002). Neuronal activity regulates correlated network properties of spontaneous calcium transients in astrocytes *in situ. J. Neurosci.* **22,** 9430–9444.

Angulo, M. C., Kozlov, A. S., Charpak, S., and Audinat, E. (2004). Glutamate released from glial cells synchronizes neuronal activity in the hippocampus. *J. Neurosci.* **24,** 6920–6927.

Araque, A., Parpura, V., Sanzgiri, R. P., and Haydon, P. G. (1999). Tripartite synapses: Glia, the unacknowledged partner. *Trends Neurosci.* **22,** 208–215.

Bagetta, G., Corasaniti, M. T., Malorni, W., Rainaldi, G., Berliocchi, L., Finazzi-Agro, A., and Nistico, G. (1996). The HIV-1 gp120 causes ultrastructural changes typical of apoptosis in the rat cerebral cortex. *Neuroreport* **7,** 1722–1724.

Beattie, E. C., Stellwagen, D., Morishita, W., Bresnahan, J. C., Ha, B. K., Von Zastrow, M., Beattie, M. S., and Malenka, R. C. (2002). Control of synaptic strength by glial TNF-α. *Science* **295,** 2282–2285.

Bezzi, P., Carmignoto, G., Pasti, L., Vesce, S., Rossi, D., Rizzini, B. L., Pozzan, T., and Volterra, A. (1998). Prostaglandins stimulate calcium-dependent glutamate release in astrocytes. *Nature* **391,** 281–285.

Bezzi, P., Domercq, M., Brambilla, L., Galli, R., Schols, D., De Clercq, E., Vescovi, A., Bagetta, G., Kollias, G., Meldolesi, J., and Volterra, A. (2001). CXCR4-activated astrocyte glutamate release via TNFα: Amplification by microglia triggers neurotoxicity. *Nat. Neurosci.* **4,** 702–710.

Bezzi, P., Gundersen, V., Galbete, J. L., Seifert, G., Steinhauser, C., Pilati, E., and Volterra, A. (2004). Astrocytes contain a vesicular compartment that is competent for regulated exocytosis of glutamate. *Nat. Neurosci.* **7,** 613–620.

Chen, X., Wang, L., Zhou, Y., Zheng, L. H., and Zhou, Z. (2005). "Kiss-and-Run" Glutamate secretion in cultured and freshly isolated rat hippocampal astrocytes. *J. Neurosci.* **25,** 9236–9243.

Christopherson, K. S., Ullian, E. M., Stokes, C. C., Mullowney, C. E., Hell, J. W., Agah, A., Lawler, J., Mosher, D. F., Bornstein, P., and Barres, B. A. (2005). Thrombospondins are astrocyte-secreted proteins that promote CNS synaptogenesis. *Cell* **120,** 421–433.

Coco, S., Verderio, C., Trotti, D., Rothstein, J. D., Volterra, A., and Matteoli, M. (1997). Non-synaptic localization of the glutamate transporter EAAC1 in cultured hippocampal neurons. *Eur. J. Neurosci.* **9,** 1902–1910.

Crippa, D., Schenk, U., Francolini, M., Rosa, P., Verderio, C., Zonta, M., Pozzan, T., Matteoli, M., and Carmignoto, G. (2006). Synaptobrevin2-expressing vesicles in rat astrocytes: Insights into molecular characterization, dynamics and exocytosis. *J. Physiol.* **570,** 567–582.

Dehnes, Y., Chaudhry, F. A., Ullensvang, K., Lehre, K. P., Storm-Mathisen, J., and Danbolt, N. C. (1998). The glutamate transporter EAAT4 in rat cerebellar Purkinje cells: A glutamate-gated chloride channel concentrated near the synapse in parts of the dendritic membrane facing astroglia. *J. Neurosci.* **18,** 3606–3619.

Del Villar, K., and Miller, C. A. (2004). Down-regulation of DENN/MADD, a TNF receptor binding protein, correlates with neuronal cell death in Alzheimer's disease brain and hippocampal neurons. *Proc. Natl. Acad. Sci. USA* **101,** 4210–4215.

Domercq, M., Brambilla, L., Pilati, E., Marchaland, J., Volterra, A., and Bezzi, P. (2006). P2Y1 receptor-evoked glutamate exocytosis from astrocytes: Control by tumor necrosis factor-α and prostaglandins. *J. Biol. Chem.* **281,** 30684–30696.

Duan, S., Anderson, C. M., Keung, E. C., Chen, Y., Chen, Y., and Swanson, R. A. (2003). P2X7 receptor-mediated release of excitatory amino acids from astrocytes. *J. Neurosci.* **23,** 1320–1328.

Fellin, T., Pascual, O., Gobbo, S., Pozzan, T., Haydon, P. G., and Carmignoto, G. (2004). Neuronal synchrony mediated by astrocytic glutamate through activation of extrasynaptic NMDA receptors. *Neuron* **43,** 729–743.

Fiacco, T. A., and McCarthy, K. D. (2004). Intracellular astrocyte calcium waves *in situ* increase the frequency of spontaneous AMPA receptor currents in CA1 pyramidal neurons. *J. Neurosci.* **24,** 722–732.

Glenner, G. G., and Wong, C. W. (1984). Alzheimer's disease: Initial report of the purification and characterization of a novel cerebrovascular amyloid protein. *Biochem. Biophys. Res. Commun.* **120,** 885–890.

Goddard, D. R., Bunning, R. A., and Woodroofe, M. N. (2001). Astrocyte and endothelial cell expression of ADAM 17 (TACE) in adult human CNS. *Glia* **34,** 267–271.

Hartlage-Rubsamen, M., Zeitschel, U., Apelt, J., Gartner, U., Franke, H., Stahl, T., Gunther, A., Schliebs, R., Penkowa, M., Bigl, V., and Rossner, S. (2003). Astrocytic expression of the Alzheimer's disease β-secretase (BACE1) is stimulus-dependent. *Glia* **41,** 169–179.

Jeremic, A., Jeftinija, K., Stevanovic, J., Glavaski, A., and Jeftinija, S. (2001). ATP stimulates calcium-dependent glutamate release from cultured astrocytes. *J. Neurochem.* **77,** 664–675.

Kang, J., Jiang, L., Goldman, S. A., and Nedergaard, M. (1998). Astrocyte-mediated potentiation of inhibitory synaptic transmission. *Nat. Neurosci.* **1,** 683–692.

Kang, N., Xu, J., Xu, Q., Nedergaard, M., and Kang, J. (2005). Astrocytic glutamate release-induced transient depolarization and epileptiform discharges in hippocampal CA1 pyramidal neurons. *J. Neurophysiol.* **94,** 4121–4130.

Karwoski, C. J., Lu, H. K., and Newman, E. A. (1989). Spatial buffering of light-evoked potassium increases by retinal Muller (glial) cells. *Science* **244,** 578–580.

Kaul, M., Garden, G. A., and Lipton, S. A. (2001). Pathways to neuronal injury and apoptosis in HIV-associated dementia. *Nature* **410,** 988–994.

Kettenmann, H., and Ransom, B. R. (1995). "Neuroglia." Oxford University Press, Oxford, UK.

Kimelberg, H. K., Goderie, S. K., Higman, S., Pang, S., and Waniewski, R. A. (1990). Swelling-induced release of glutamate, aspartate, and taurine from astrocyte cultures. *J. Neurosci.* **10,** 1583–1591.

Koistinaho, M., Lin, S., Wu, X., Esterman, M., Koger, D., Hanson, J., Higgs, R., Liu, F., Malkani, S., Bales, K. R., and Paul, S. M. (2004). Apolipoprotein E promotes astrocyte colocalization and degradation of deposited amyloid-β peptides. *Nat. Med.* **10,** 719–726.

Kreft, M., Stenovec, M., Rupnik, M., Grilc, S., Krzan, M., Potokar, M., Pangrsic, T., Haydon, P. G., and Zorec, R. (2004). Properties of Ca^{2+}-dependent exocytosis in cultured astrocytes. *Glia* **46,** 437–445.

Liu, Q. S., Xu, Q., Arcuino, G., Kang, J., and Nedergaard, M. (2004). Astrocyte-mediated activation of neuronal kainate receptors. *Proc. Natl. Acad. Sci. USA* **101,** 3172–3177.

Mauch, D. H., Nagler, K., Schumacher, S., Goritz, C., Muller, E. C., Otto, A., and Pfrieger, F. W. (2001). CNS synaptogenesis promoted by glia-derived cholesterol. *Science* **294,** 1354–1357.

Meucci, O., and Miller, R. J. (1996). gp120-induced neurotoxicity in hippocampal pyramidal neuron cultures: Protective action of TGF-β1. *J. Neurosci.* **16,** 4080–4088.

Montana, V., Ni, Y., Sunjara, V., Hua, X., and Parpura, V. (2004). Vesicular glutamate transporter-dependent glutamate release from astrocytes. *J. Neurosci.* **24,** 2633–2642.

Nedergaard, M., Ransom, B., and Goldman, S. A. (2003). New roles for astrocytes: Redefining the functional architecture of the brain. *Trends Neurosci.* **26,** 523–530.

Parpura, V., Basarsky, T. A., Liu, F., Jeftinija, K., Jeftinija, S., and Haydon, P. G. (1994). Glutamate-mediated astrocyte-neuron signalling. *Nature* **369,** 744–747.

Parri, H. R., Gould, T. M., and Crunelli, V. (2001). Spontaneous astrocytic Ca^{2+} oscillations *in situ* drive NMDAR-mediated neuronal excitation. *Nat. Neurosci.* **4,** 803–812.

Pasti, L., Volterra, A., Pozzan, T., and Carmignoto, G. (1997). Intracellular calcium oscillations in astrocytes: A highly plastic, bidirectional form of communication between neurons and astrocytes *in situ*. *J. Neurosci.* **17,** 7817–7830.

Perea, G., and Araque, A. (2005). Properties of synaptically evoked astrocyte calcium signal reveal synaptic information processing by astrocytes. *J. Neurosci.* **25,** 2192–2203.

Perry, V. H., Bell, M. D., Brown, H. C., and Matyszak, M. K. (1995). Inflammation in the nervous system. *Curr. Opin. Neurobiol.* **5,** 636–641.

Pfrieger, F. W., and Barres, B. A. (1996). New views on synapse-glia interactions. *Curr. Opin. Neurobiol.* **6,** 615–621.

Ralay Ranaivo, H., Craft, J. M., Hu, W., Guo, L., Wing, L. K., Van Eldik, L. J., and Watterson, D. M. (2006). Glia as a therapeutic target: Selective suppression of human amyloid-β-induced upregulation of brain proinflammatory cytokine production attenuates neurodegeneration. *J. Neurosci.* **26,** 662–670.

Rossi, D. J., Oshima, T., and Attwell, D. (2000). Glutamate release in severe brain ischaemia is mainly by reversed uptake. *Nature* **403,** 316–321.

Rossi, D., Brambilla, L., Valori, C. F., Crugnola, A., Giaccone, G., Capobianco, R., Mangieri, M., Kingston, A. E., Bloc, A., Bezzi, P., and Volterra, A. (2005). Defective tumor necrosis factor-α-dependent control of astrocyte glutamate release in a transgenic mouse model of Alzheimer disease. *J. Biol. Chem.* **280,** 42088–42096.

Sanzgiri, R. P., Araque, A., and Haydon, P. G. (1999). Prostaglandin E_2 stimulates glutamate receptor-dependent astrocyte neuromodulation in cultured hippocampal cells. *J. Neurobiol.* **41,** 221–229.

Schipke, C. G., and Kettenmann, H. (2004). Astrocyte responses to neuronal activity. *Glia* **47,** 226–232.

Selkoe, D. J. (2002). Alzheimer's disease is a synaptic failure. *Science* **298,** 789–791.

Shi, B., De Girolami, U., He, J., Wang, S., Lorenzo, A., Busciglio, J., and Gabuzda, D. (1996). Apoptosis induced by HIV-1 infection of the central nervous system. *J. Clin. Invest.* **98,** 1979–1990.

Stellwagen, D., and Malenka, R. C. (2006). Synaptic scaling mediated by glial TNF-α. *Nature* **440,** 1054–1059.

Takano, T., Kang, J., Jaiswal, J. K., Simon, S. M., Lin, J. H., Yu, Y., Li, Y., Yang, J., Dienel, G., Zielke, H. R., and Nedergaard, M. (2005). Receptor-mediated glutamate release from volume sensitive channels in astrocytes. *Proc. Natl. Acad. Sci. USA* **102,** 16466–16471.

Toggas, S. M., Masliah, E., Rockenstein, E. M., Rall, G. F., Abraham, C. R., and Mucke, L. (1994). Central nervous system damage produced by expression of the HIV-1 coat protein gp120 in transgenic mice. *Nature* **367,** 188–193.

Toggas, S. M., Masliah, E., and Mucke, L. (1996). Prevention of HIV-1 gp120-induced neuronal damage in the central nervous system of transgenic mice by the NMDA receptor antagonist memantine. *Brain Res.* **706,** 303–307.

Tsacopoulos, M., and Magistretti, P. J. (1996). Metabolic coupling between glia and neurons. *J. Neurosci.* **16,** 877–885.

Ullian, E. M., Sapperstein, S. K., Christopherson, K. S., and Barres, B. A. (2001). Control of synapse number by glia. *Science* **291,** 657–661.

Ventura, R., and Harris, K. M. (1999). Three-dimensional relationships between hippocampal synapses and astrocytes. *J. Neurosci.* **19,** 6897–6906.

Volterra, A., and Meldolesi, J. (2005). Astrocytes, from brain glue to communication elements: The revolution continues. *Nat. Rev. Neurosci.* **6,** 626–640.

Warr, O., Takahashi, M., and Attwell, D. (1999). Modulation of extracellular glutamate concentration in rat brain slices by cystine-glutamate exchange. *J. Physiol.* **514,** 783–793.

Wyss-Coray, T. (2006). Inflammation in Alzheimer disease: Driving force, bystander or beneficial response? *Nat. Med.* **12,** 1005–1015.

Wyss-Coray, T., Loike, J. D., Brionne, T. C., Lu, E., Anankov, R., Yan, F., Silverstein, S. C., and Husemann, J. (2003). Adult mouse astrocytes degrade amyloid-β *in vitro* and *in situ*. *Nat. Med.* **9,** 453–457.

Yan, P., Hu, X., Song, H., Yin, K., Bateman, R. J., Cirrito, J. R., Xiao, Q., Hsu, F. F., Turk, J. W., Xu, J., Hsu, C. Y., Holtzman, D. M., *et al.* (2006). Matrix metalloproteinase-9 degrades amyloid-β fibrils *in vitro* and compact plaques *in situ*. *J. Biol. Chem.* **281,** 24566–24574.

Ye, Z. C., Wyeth, M. S., Baltan-Tekkok, S., and Ransom, B. R. (2002). Functional hemichannels in astrocytes: A novel mechanism of glutamate release. *J. Neurosci.* **23,** 3588–3596.

Zhang, Q., Pangrsic, T., Kreft, M., Krzan, M., Li, N., Sul, J. Y., Halassa, M., Van Bockstaele, E., Zorec, R., and Haydon, P. G. (2004a). Fusion-related release of glutamate from astrocytes. *J. Biol. Chem.* **279,** 12724–12733.

Zhang, Q., Fukuda, M., Van Bockstaele, E., Pascual, O., and Haydon, P. G. (2004b). Synaptotagmin IV regulates glial glutamate release. *Proc. Natl. Acad. Sci. USA* **101,** 9441–9446.

Zhao, M., Cribbs, D. H., Anderson, A. J., Cummings, B. J., Su, J. H., Wasserman, A. J., and Cotman, C. W. (2003). The induction of the TNF-α death domain signaling pathway in Alzheimer's disease brain. *Neurochem. Res.* **28,** 307–318.

Zonta, M., Sebelin, A., Gobbo, S., Fellin, T., Pozzan, T., and Carmignoto, G. (2003). Glutamate-mediated cytosolic calcium oscillations regulate a pulsatile prostaglandin release from cultured rat astrocytes. *J. Physiol.* **553,** 407–414.

THE HIGH-MOBILITY GROUP BOX 1 CYTOKINE INDUCES TRANSPORTER-MEDIATED RELEASE OF GLUTAMATE FROM GLIAL SUBCELLULAR PARTICLES (GLIOSOMES) PREPARED FROM *IN SITU*-MATURED ASTROCYTES

Giambattista Bonanno,[*,†] Luca Raiteri,[*] Marco Milanese,[*] Simona Zappettini,[*] Edon Melloni,[†,‡] Marco Pedrazzi,[‡] Mario Passalacqua,[†,‡] Carlo Tacchetti,[§,¶] Cesare Usai,[||] and Bianca Sparatore[†,‡]

[*]Department of Experimental Medicine, Section of Pharmacology and Toxicology University of Genoa, Italy
[†]Department of Experimental Medicine, Center of Excellence for Biomedical Research University of Genoa, Italy
[‡]Department of Experimental Medicine, Section of Biochemistry, University of Genoa, Italy
[§]Department of Experimental Medicine, Section of Human Anatomy University of Genoa, Italy
[¶]FIRC Institute of Molecular Oncology (IFOM), Milan, Italy
[||]Institute of Biophysics, National Research Council, Genoa, Italy

The multifunctional protein high-mobility group box 1 (HMGB1) is expressed in restricted areas of adult brain where it can act as a proinflammatory cytokine. We report here that HMGB1 affects CNS transmission by inducing glutamatergic release from glial (gliosomes) but not neuronal (synaptosomes) resealed subcellular particles isolated from mouse cerebellum and hippocampus. Confocal microscopy showed that gliosomes are enriched with glia-specific proteins such as GFAP and S-100, but not with neuronal proteins such as PSD-95, MAP-2, and β-tubulin III. Furthermore, gliosomes exhibit labeling neither for integrin-αM nor for myelin basic protein, specific for microglia and oligodendrocytes, respectively. The

INTERNATIONAL REVIEW OF NEUROBIOLOGY, VOL. 82
DOI: 10.1016/S0074-7742(07)82004-6

gliosomal fraction contains proteins of the exocytotic machinery coexisting with GFAP. Consistent with ultrastructural analysis, several ~30-nm nonclustered vesicles are present in the gliosome cytoplasm. Finally, gliosomes represent functional organelles that actively export glutamate when subjected to releasing stimuli, such as ionomycin or ATP, by mechanisms involving extracellular Ca^{2+} and Ca^{2+} release from intracellular stores. HMGB1-induced release of the stable glutamate analogue [^{3}H]D-aspartate and endogenous glutamate form gliosomes, whereas nerve terminals were insensitive to the protein. The HMGB1-evoked release of glutamate was independent on modifications of cytosolic Ca^{2+} concentration, but it was blocked by DL-*threo*-β-benzyloxyaspartate, suggesting the involvement of transporter-mediated release mechanisms. Moreover, dihydrokainic acid, a selective inhibitor of glutamate transporter 1 does not block the HMGB1 effect, indicating a role for the glial glutamate–aspartate transporter (GLAST) subtype in this response. HMGB1 bind to gliosomes but not to synaptosomes and can physically interact with GLAST and receptor for advanced glycation end products (RAGE). Taken together, these results suggest that the HMGB1 cytokine could act as a modulator of glutamate homeostasis in adult mammalian brain.

I. Introduction

In the last decade, exciting results have led to dramatic conceptual changes about the role of glial cells in the brain. For a long time, it was believed that glial cells provide structural, metabolic, and trophic support, while the task of information processing was attributed exclusively to neurons. An increasing number of papers suggest that glia shares at least some of the features typical of neurons, particularly those concerned with excitatory neurotransmission (reviewed in Haydon, 2001). In fact, glial cells are endowed with transporters able to remove glutamate from the extracellular space; a process even more efficient than that actuated by neurons (Bergles and Jahr, 1997; Mennerick and Zorumski, 1994). Astrocytes, due to their intimate spatial relationship with neuronal synaptic contacts, can directly respond to synaptically released messengers and, in turn, communicate via signaling substances with neurons. The disclosure of this active role of glia has led to the model of the tripartite synapse (Araque *et al.*, 1999), which affirms that a pivotal role in regulating synaptic function, strength, and plasticity is played by glial cells tightly surrounding pre- and postsynaptic neuronal elements.

Numerous stimuli have been reported to increase the intracellular Ca^{2+} concentration in astrocytes. As one of the foremost players, glutamate released from neurons can induce Ca^{2+} mobilization from intracellular stores by activating ionotropic and metabotropic receptors located on astroglial cells, which is associated with astrocytic glutamate release, through a process resembling neuronal exocytosis (reviewed in Montana *et al.*, 2006). Interestingly, exocytotic-like intracellular Ca^{2+}-driven glutamate release from astrocytes can also be elicited by substances involved in the inflammatory and immune pathways, including bradykinin (Parpura *et al.*, 1994), prostaglandins (Bezzi *et al.*, 1998), and chemokines (Bezzi *et al.*, 2001).

A recently discovered member of the cytokine family is the high-mobility group box 1 (HMGB1) protein that is widely expressed in developing nervous system and transformed cells of nerve origin, including neuroblastoma and glioma (Guazzi *et al.*, 2003; Sajithlal *et al.*, 2002; Taguchi *et al.*, 2000) and that can be actively exported by several cell types (Müller *et al.*, 2004).

In this chapter, we summarize the results obtained studying the glutamate releasing properties of HMGB1, as well as the mechanisms underlining the extracellular export of the excitatory transmitter from astrocytes by exploiting the characteristics of gliosomes, a preparation of astrocyte subcellular particles acutely isolated from the brain of adult animals. This preparation appears well suited to investigate the functional neurochemistry of mature astrocytes. In particular, we show here that release of previously taken up [^{3}H]D-aspartate ([^{3}H]D-Asp) and of endogenous glutamate can be evoked from gliosomes by HMGB1 throughout reversal of the glial glutamate–aspartate transporter (GLAST).

II. Gliosomes as a Model to Study Astrocyte Characteristics

A. Characterization of the Gliosome Preparation

Purified gliosomes and synaptosomes utilized in the experiments described here have been prepared from rat or mouse brain tissue by homogenization and purification on a discontinuous Percoll® gradient essentially according to Nakamura *et al.* (1993, 1994) with minor modifications (Stigliani *et al.*, 2006). The tissue was homogenized in 14 volumes of 0.32-M sucrose, buffered at pH 7.4 with Tris–HCl, using a glass–teflon tissue grinder (clearance 0.25 mm, 12 up–down strokes in about 1 min). The homogenate was centrifuged (5 min, 1000 × *g* at 4 °C) to remove nuclei and debris and the supernatant gently stratified on a discontinuous Percoll® gradient (2%, 6%, 10%, and 20% v/v in Tris-buffered sucrose; 3 ml of each) and centrifuged at 33,500 × *g* for 5 min at 4 °C. The layers between 2% and 6% Percoll® (gliosomal

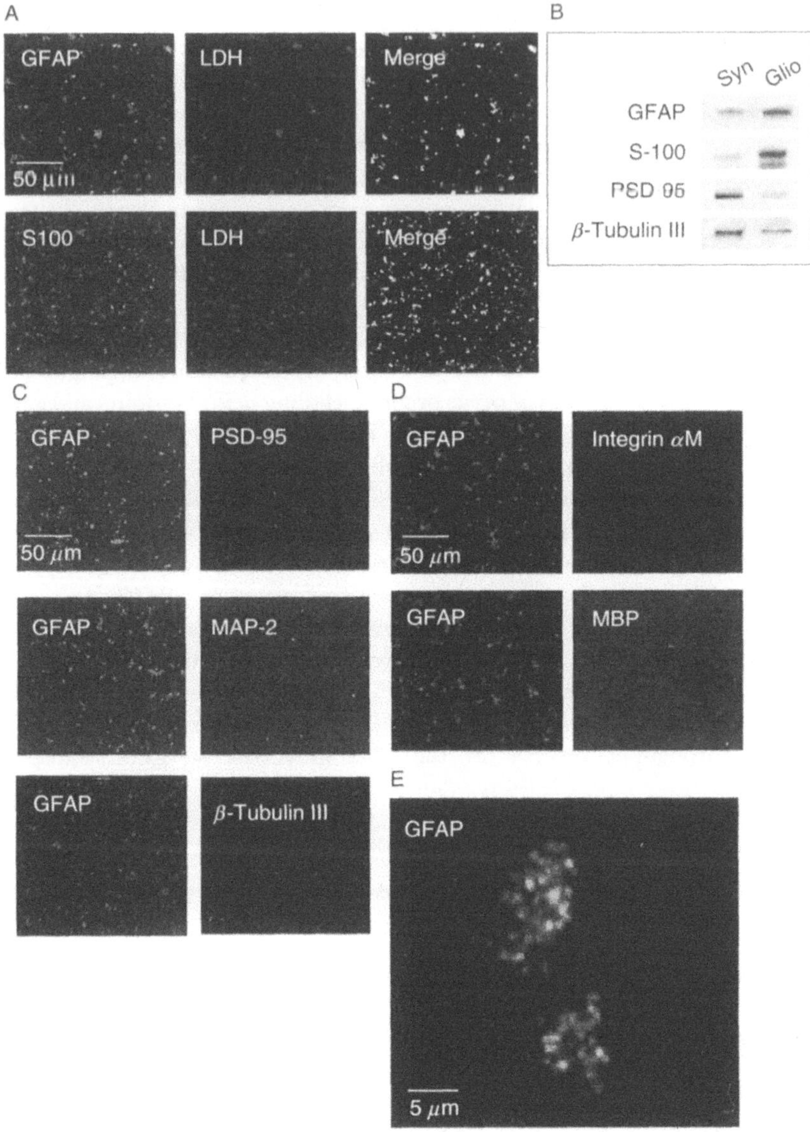

Fig. 1. Expression of glia- and neuron-specific proteins in gliosomes purified from rat cerebral cortex. (A) Identification by immunocytochemistry of GFAP, LDH, and Ca^{2+}-binding protein S-100. Gliosomes were glued onto coverslips, fixed with paraformaldehyde, permeabilized with Triton X-100 and incubated with the primary and secondary antibodies. Samples were analyzed by laser confocal microscopy. (B) Postsynaptic density protein of 95 kDa (PSD-95), GFAP, S-100, and β-tubulin III, immunoreactivity distribution evaluated by Western blotting. GFAP, S-100, PSD-95, and β-tubulin III amounts were measured on 10 μg of hippocampus synaptosomal and gliosomal

fraction) and between 10% and 20% Percoll® (synaptosomal fraction) were collected, washed by centrifugation, and resuspended in physiological medium.

Several studies have taken advantage of the characteristics of the gliosome preparation to study functional aspects of glial cells. These studies allowed identification of specific cell distribution, function, and molecular mechanisms of a number of transmitter and modulator targets, mainly membrane transporters (see for instance Daniels and Vickroy, 1999; Hirst *et al.*, 1998; Pedrazzi *et al.*, 2006; Raiteri *et al.*, 2005; Suchak *et al.*, 2003). We characterized in detail gliosomes purified from adult rat cerebral cortex, pointing out biochemical and morphological evidence in support to the concept that our gliosomal fraction is largely purified from synaptosomes (Stigliani *et al.*, 2006). The confocal microscopy images shown in Fig. 1A highlight the presence of immunoreactivity for two proteins selectively expressed in astrocytes, the glial fibrillary acidic protein (GFAP) and the Ca^{2+}-binding protein S-100, in purified gliosomes. Gliosomes were labeled with anti-GFAP or anti S-100 and with anti-lactate dehydrogenase (LDH) antibodies. Since LDH is an enzyme localized in the cytosol, it can be assumed that the LDH-positive particles represent resealed gliosomes rather than membrane debris. These experiments emphasize that the majority (about 90%) of LDH-labeled resealed particles in the gliosomal preparation were positive for GFAP or for S-100. Also the synaptosomal preparation was extensively stained by LDH, but it did not show substantial GFAP labeling, indicating low gliosome contamination (not shown). Western blot experiments substantiated that gliosomes are enriched with GFAP and S-100, while synaptosomes are enriched with the neuronal markers postsynaptic density protein of 95 kDa (PSD-95) and β-tubulin III, which are poorly represented in gliosomes (Fig. 1B). According to these data, GFAP-positive gliosomes presented only a very modest positiveness for antibodies raised against the neuronal markers PSD-95, microtubule-associated protein 2 (MAP-2), or β-tubulin III (Fig. 1C), thus supporting the idea that gliosomes represent a preparation with low synaptosomal contamination. Of note, PSD-95, MAP-2, and β-tubulin III extensively marked synaptosomes under the same experimental conditions (not shown). Moreover, GFAP-expressing gliosomal preparation did not exhibit labeling either for integrin αM or for myelin basic protein (MBP), two proteins selectively expressed in microglia and oligodendrocytes, respectively (Fig. 1D). As a control, the antibodies used in the

proteins. (C) Identification by immunocytochemistry of GFAP, PSD-95, microtuble associated protein type II (MAP-2), and β-tubulin III. Samples were analyzed by laser confocal microscopy. (D) Identification by immunocytochemistry of GFAP, integrin αM, or myelin basic protein (MBP). Samples were analyzed by laser confocal microscopy. (E) High-magnification tridimensional reconstruction of purified gliosomal particles identified by GFAP immunocytochemistry. Samples were analyzed by laser confocal microscopy. Modified from Stigliani *et al.* (2006), with the permission of Blackwell Publishing Co. (See Color Insert.)

latter confocal microscopy experiments revealed a single band of the appropriate molecular weight in Western blot experiments, using proteins extracted from microglia primary cell cultures and from white matter of 19-day-old rats (not shown). Finally, Fig. 1E shows a tridimensional reconstruction at higher-magnification of GFAP-positive gliosomal particles: these particles appear to be around 0.6-μm-diameter spheroidal organelles with a propensity to cluster in physiological medium.

B. Glutamate Release in Gliosomes

Functional experiments conducted in rat cerebral cortex showed that purified gliosomes are able to take up and release glutamate when exposed to different releasing stimuli such as ionomycin or ATP. The release of $[^3H]$D-Asp or endogenous glutamate was studied taking advantage of the uniqueness of a superfusion technique that we have used for several years, mainly with synaptosomes (Raiteri and Raiteri, 2000). The system consists of several (up to 24) parallel superfusion chambers thermostated at 37 °C in which very thin layers (mostly monolayers) of synaptosomes, plated on microporous filters, are up–down superfused in conditions in which any released compound is immediately removed by the superfusion fluid. Such a rapid removal prevents indirect effects: in particular, the changes of glutamate release observed following exposure to various agents are essentially due to direct actions on excitatory amino acid releasing particles with minimal or no involvement of neighboring elements. The fast taking away of released glutamate does not allow (1) its reuptake and, therefore, its exchange with cytosolic excitatory amino acid transporter substrates; and (2) its feedback on presynaptic targets like release-regulating receptors. If substrates just released are virtually absent from the particle biophase, release by Ca^{2+}-dependent exocytosis or by Ca^{2+}-independent transporter-mediated release can be monitored under appropriate conditions.

Ionomycin, a Ca^{2+}-selective ionophore capable to mediate Ca^{2+} influx without voltage-sensitive Ca^{2+} channel activation and previously shown to induce transmitter exocytosis from nerve terminals (Sanchez-Prieto *et al.*, 1987; Verhage *et al.*, 1991), produced a substantial stimulus-evoked release of $[^3H]$D-Asp or endogenous glutamate (Stigliani *et al.*, 2006). The release induced by ionomycin was entirely dependent on the presence of external Ca^{2+} and significantly decreased by bafilomycin-A1, a vesicle membrane V-type ATPase inhibitor (Bowman *et al.*, 1988; Floor *et al.*, 1990), which is expected to prevent the accumulation of the amino acid into vesicles (Moriyama and Futai, 1990; Roseth *et al.*, 1995).Under our experimental conditions, low concentrations of ionomycin appeared to release even higher amounts of glutamate from gliosomes than from synaptosomes. Conversely, synaptosomes were superior glutamate releasers when higher concentrations of the Ca^{2+} ionophore were applied. The dependence on Ca^{2+} and the sensitivity to bafilomycin-A1 of the ionomycin-evoked release of glutamate from gliosomes suggest that the stimulus-induced release is exocytotic in nature. In line, experiments performed with the fluorescent dye acridine orange (Zoccarato *et al.*, 1999) showed that gliosomes were able to accumulate the dye into acidic cytoplasmic organelles and that the application of ionomycin induced fusion of these organelles with the plasma membrane that was almost totally external Ca^{2+} dependent (Stigliani *et al.*, 2006).

Gliosomes were able to release glutamate also when exposed to ATP in superfusion. The ATP-induced glutamate release depended on the concentration of the nucleotide used and was abolished by the selective P2 receptor antagonist pyridoxal phosphate-6-azophenyl-2′,4′-disulfonic acid. The effect of ATP was only minimally affected in the absence of external Ca^{2+}, but it was significantly diminished by preloading gliosomes with 1,2-bis(2-aminophenoxy)ethane-*N*,*N*,*N*′, *N*′-tetraacetylmethylester (BAPTA-AM), a cytosolic Ca^{2+} chelator, suggesting the involvement of Ca^{2+} released from intra-gliosomal stores, consequent to P2 receptor activation (Stigliani *et al.*, 2006).

C. Expression of Proteins of the Release Machinery in Gliosomes

Taken together, the release and fluorescence experiments described in the preceding paragraph are consistent with the idea that gliosomes can release glutamate by an exocytotic process. Accordingly, ultrastructural analysis of the gliosomal preparation evidentiated several vesicles with a diameter of the membrane-bound area of ~30 nm and scattered within the cytoplasm and present in about 45% of the gliosomes. These vesicles were either uncoated or clathrin-coated and less numerous than in synaptosomes. Moreover, they did not show a clustered configuration, at variance to synaptosomes (Fig. 2A). Figure 2B shows that, similar to synaptosomes, the purified gliosomal fraction express the vesicular SNARE protein synaptobrevin-2 (VAMP-2) and the membrane SNARE proteins syntaxin-1 and synaptosome-associated protein of 23 kDa (SNAP-23), known to form the core complex required to execute exocytotic neurotransmitter release (Südhof, 1995). A substantial colocalization of these proteins with GFAP-positive particles could be evidentiated by the confocal experiments reported in Fig. 2C. The analysis of different image couples indicated that about 55% of GFAP-expressing particles coexpress VAMP-2 immunoreactivity and about 70% of GFAP colocalizes with both syntaxin-1 and SNAP-23. The GFAP-expressing gliosomal preparation also showed a significant (about 35%) vesicular glutamate transporter 1 (vGLUT-1) staining. Also VAMP-2 and vGLUT-1 appear to be coexpressed in gliosomes: about 65% of VAMP-2

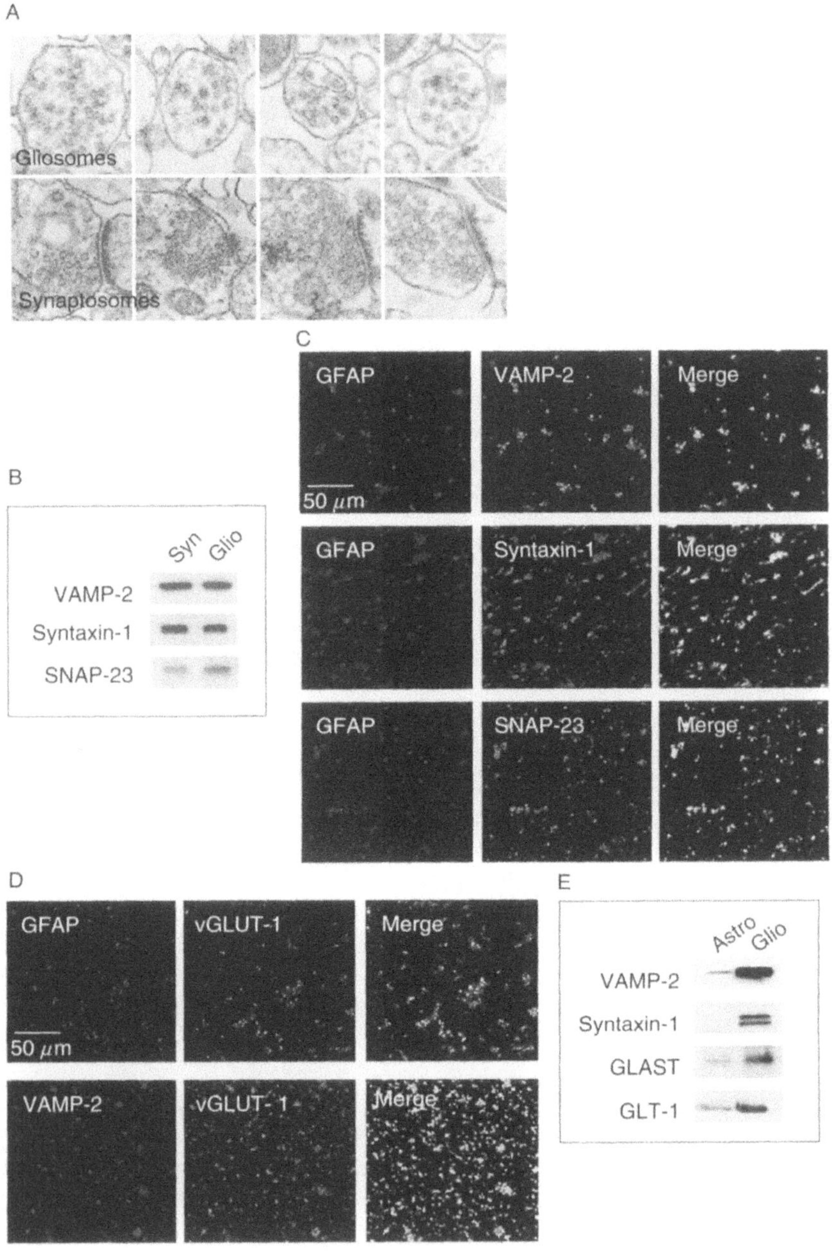

FIG. 2. Presence of cytosol vesicles and expression of proteins of the release machinery in the gliosomal fraction purified from rat cerebral cortex. (A) Electron micrographs of gliosome (upper

colocalizes with the vesicular glutamate transporter. Conversely, almost the totality of vGLUT-1 coexpresses with GFAP or VAMP-2.

Interestingly, expression of the SNARE complex proteins VAMP-2 and sintaxin-1 and of the glial-specific glutamate transporters of type 1 (GLT-1) and GLAST were much more enriched in gliosomes than in astrocytes in culture, as outlined by the Western blot experiments reported in Fig. 2E. It is reasonable to assume that during brain tissue homogenization gliosomes are formed by a process similar to that originating synaptosomes, that is, they are "pinched off" particles coming from glia cell arborizations. It has been proposed that astrocytes possess dedicated regions at the processes surrounding the synapses by which they sense the neuronal messengers for a point-to-point neuron to astrocyte communication. Astrocytes have been suggested to release transmitters from these specialized areas (Araque *et al.*, 1999; Carmignoto, 2000; Grosche *et al.*, 1999). Accordingly, a number of evidences indicate that the vesicular release sites of astrocytes might be situated at the processes rather that at the cell bodies (reviewed by Montana *et al.*, 2006). It could be proposed that the gliosomal preparation is enriched with these specific areas, where the release machinery of the glial cell should be concentrated.

III. HMGB1-Induced Glutamate Release from Gliosomes

A. Cytokine Properties of HMGB1

Cytokines are a class of small proteins acting via cell surface G-protein–coupled receptors that mediate diverse metabolic and immunological responses in other cells (Nathan, 1987). They were first identified as inflammatory mediators

panels) and synaptosome (lower panels) fractions. Purified gliosomes and synaptosomes were fixed, dehydrated, and embedded in LX112. Ultrathin sections were stained with uranyl acetate and lead citrate, and analyzed with a FEI CM10 electron microscope. (B) Synaptobrevin 2 (VAMP-2), syntaxin-1, or synaptosome-associated protein of 23 kDa (SNAP-23) immunoreactivity distribution evaluated by Western blotting. VAMP-2, syntaxin-1, and SNAP-23 amounts were measured on 10 μg of hippocampus synaptosomal and gliosomal proteins. (C) Immunocytochemical identification of GFAP, VAMP-2, syntaxin-1, and SNAP-23 immunoreactivity in gliosomes. Gliosomes were glued onto coverslips, fixed with paraformaldehyde, permeabilized with Triton X-100, and incubated with the primary and secondary antibodies. Samples were analyzed by laser confocal microscopy. (D) Immunocytochemical identification of GFAP, VAMP-2, and the vesicular glutamate transporter 1 (vGLUT-1) in gliosomes. Samples were analyzed by laser confocal microscopy. (E) VAMP-2, syntaxin-1, glutamate–aspartate transporter (GLAST), or glutamate transporter 1 (GLT-1) immunoreactivity distribution evaluated by Western blotting. VAMP-2, syntaxin-1, GLAST, and GLT-1 amounts were measured on 10 μg of proteins of cultured astrocytes or gliosomes. Modified from Stigliani *et al.* (2006), with the permission of Blackwell Publishing Co. (See Color Insert.)

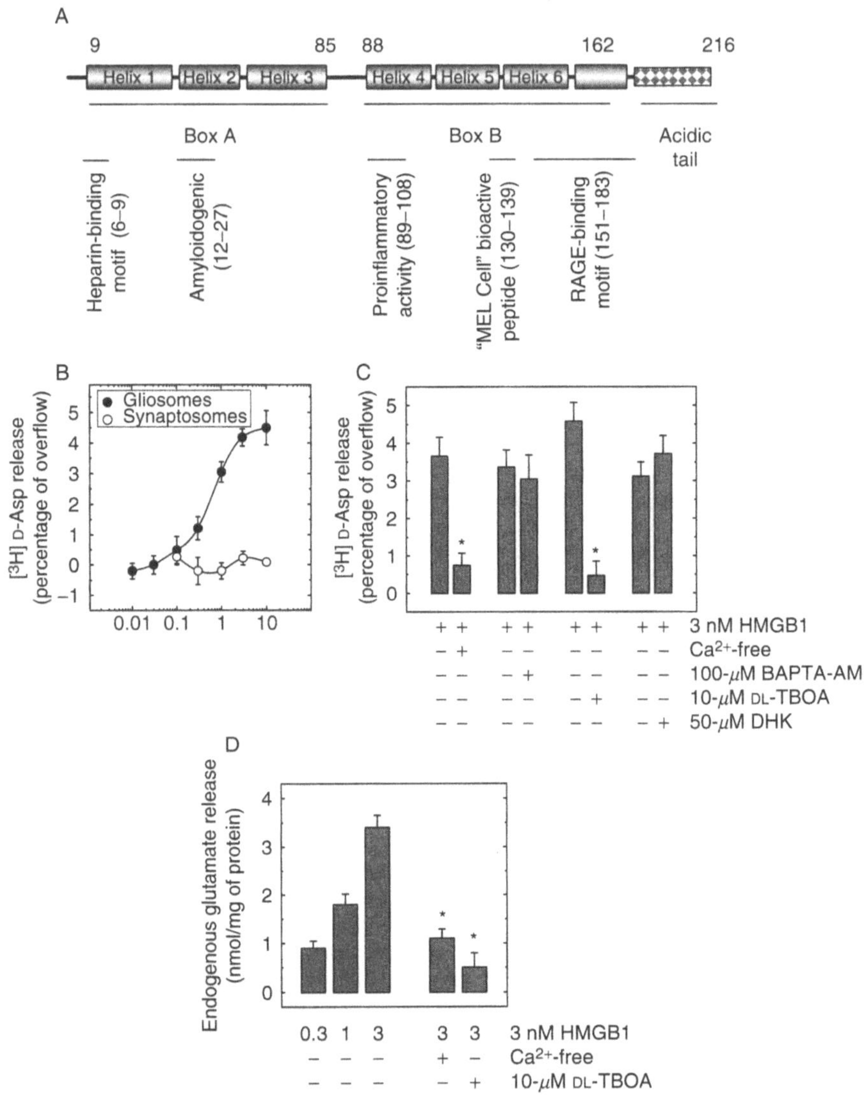

FIG. 3. Effects of HMGB1 on the release of glutamate from mouse hippocampal gliosomes. (A) Schematic representation of the HMGB1 protein domains. (B) Concentration–response curve of the HMGB1-evoked [^{3}H]D-Asp release from mouse hippocampal gliosomes and synaptosomes. Purified particles were labeled with the radioactive tracer and exposed in superfusion to various concentrations of HMGB1. Samples of superfusate were collected and counted for radioactivity. Results are expressed as percentage of overflow evoked by HMGB1. (C) Effects of Ca^{2+} omission, Ca^{2+} chelator 1,2-bis (2-aminophenoxy)ethane-*N*,*N*,*N′*,*N′*-tetraacetylmethylester (BAPTA-AM), and of the nontransportable glutamate carrier blockers DL-*threo*-β-benzyloxyaspartic acid (DL-TBOA), or DHK on the

of leukocyte chemotaxis and have been subsequently shown to possess a much larger repertoire (Rossi and Zlotnik, 2000). Various chemokines and their receptors are widely expressed in the brain, in some cases from early embryonic stages throughout adulthood (Asensio and Campbell, 1999). For this reason, it has been suggested that they may have a role in neurodegeneration disease states (Asensio and Campbell, 1999; Miller and Meucci, 1999) in addition to their inflammatory and immunological roles.

A discovery by Wang *et al.* (1999) revealed that HMGB1 is a cytokine. HMGB1 has been discovered more than 30 years ago as a nuclear-binding protein that is widely expressed and extremely conserved in mammals. HMGB1 is an architectural protein: it can bend DNA to promote nucleoprotein interactions and facilitate all sorts of DNA transactions. The protein (25 kDa) is structured in three domains: two basic HMG-box domains (box A and box B) and a long acidic C-terminal tail (Bustin, 1999; Thomas and Travers, 2001) (Fig. 3A).

HMGB1 can be actively released from a variety of cells, including macrophage, pituicytes, human primary peripheral blood mononuclear cells, and murine erythroleukemia cells (Müller *et al.*, 2004). Once released, HMGB1 can bind to cell surface receptors, such as the receptor for advanced glycation end products (RAGE), Toll-like receptor 2, and Toll-like receptor 4, to transduce signals that elicit cellular responses including fever, epithelia barrier dysfunction, chemotactic cell movement, release of proinflammatory cytokines (e.g., tumor necrosis factor and interleukin-1β), and acute inflammation (O'Connor *et al.*, 2003; Wang *et al.*, 2004). Proinflammatory effects of HMGB1, identified in specific structures of the peripheral and the central nervous system, have supported the concept that this protein may operate through direct activation of signaling cascades, commonly utilized by classic proinflammatory cytokines, or indirectly by inducing the release of these cytokines, also in the brain (Agnello *et al.*, 2002; O'Connor *et al.*, 2003; Rong *et al.*, 2004). Although HMGB1 is widely expressed in the developing central nervous system, the adult brain maintains high levels of HMGB1 only in specific areas (Guazzi *et al.*, 2003).

We report here that extracellular HMGB1 selectively modulates the release of glutamate in the central nervous system by monitoring the release of excitatory amino acid in purified synaptosomes (Gray and Whittaker, 1962) and gliosomes (Nakamura *et al.*, 1993; Stigliani *et al.*, 2006) prepared from mouse hippocampus and cerebellum.

HMGB1-evoked [^{3}H]D-Asp release from gliosomes. (D). HMGB1-evoked release of endogenous glutamate from gliosomes and effects of Ca^{2+} omission or DL-TBOA. Glutamate was measured by high performance liquid chromatography separation and fluorimetric detection. Modified from Pedrazzi *et al.* (2006), with the permission of Blackwell Publishing Co. (See Color Insert.)

B. Effect of HMGB1 on Glutamate Release from Gliosomes and Synaptosomes

The level of purification of gliosomal- and synaptosomal-enriched fractions here isolated from adult mouse cerebellum and hippocampus was studied by confocal microscopy and Western blotting, as described above in rat cerebral cortex. Studies aimed to detect the proportion of particles expressing the glial-specific marker GFAP or the neuronal markers MAP-2 and β-tubulin III indicated that no more than 15% GFAP labeling was present in the synaptosomal fraction and, conversely, that a maximum of 20% MAP-2 or β-tubulin III labeling was present in gliosomal preparations, both in hippocampus and cerebellum (Pedrazzi *et al.*, 2006).

To directly compare the ability of HMGB1 to modulate glutamate release, synaptosomes and gliosomes were prepared from adult mouse hippocampus or cerebellum. The preparations were labeled with the stable glutamate analogue [^{3}H]D-Asp and exposed to HMGB1 in superfusion. HMGB1 stimulated the release of [^{3}H]D-Asp in a concentration-dependent manner from hippocampal (Fig. 3B) or cerebellar (not shown) gliosomes. The two concentration–response curves appeared similar: the maximal potentiation of release was obtained around 3-nM HMGB1 and the EC_{50} amounted to 0.57 ± 0.043 nM and to 0.32 ± 0.041 nM in the case of hippocampal and cerebellar gliosomes, respectively. By contrast, no significant increase of [^{3}H]D-Asp efflux was observed in synaptosomes under the same experimental conditions. These findings indicate that nanomolar concentrations of extracellular HMGB1 are sufficient to elicit glutamate release from gliosomes and that this activity of HMGB1 is specifically directed to glial and not to neuronal preparations from the same regions of adult mouse brain.

C. Mechanisms of the HMGB1-Induced Release of Glutamate from Gliosomes

It is well known that neurotransmitter release may be due to Ca^{2+}-dependent or Ca^{2+}-independent mechanisms (Attwell *et al.*, 1993; Berridge, 1998; Levi and Raiteri, 1993). To determine whether the efflux [^{3}H]D-Asp promoted by HMGB1 involved vesicular exocytosis, which is known to require the entry of Ca^{2+} from the external milieu or mobilization of Ca^{2+} from internal stores to the cytosol (Berridge, 1998), the effect of HMGB1 was studied either by omitting Ca^{2+} from the extracellular medium or by using gliosomes pretreated with BAPTA-AM. As shown in Fig. 3C, omission of external Ca^{2+} largely prevented the HMGB1 effect, suggesting the occurrence of exocytotic release. Surprisingly, in the presence of external Ca^{2+}, the HMGB1-evoked [^{3}H]D-Asp release was not significantly modified in gliosomes pretreated with BAPTA-AM, thus opposing the exocytotic origin of the process. The lack of effect observed with BAPTA-AM does not look as to be due to the incapability of the Ca^{2+} chelator to remove cytosolic Ca^{2+} in the presence of 1.2-mM external Ca^{2+}. In fact, in another set of experiments, exposure of hippocampal gliosomes to ATP resulted in a stimulation of the radioactive tracer release; an effect reduced to about 60% by BAPTA-AM in the presence of external Ca^{2+}. The discrepancy between external Ca^{2+} dependency and BAPTA-AM independency of the HMGB1-evoked [^{3}H]D-Asp release prompted us to check the effect of the nontransportable blocker of glutamate uptake DL-*threo*-β-benzyloxyaspartate (DL-TBOA) on the amino acid release elicited by HMGB1. As shown in Fig. 3C, DL-TBOA completely abolished the efflux of [^{3}H]D-Asp promoted by HMGB1, suggesting that this response is due to a mechanism mediated by reversal of glutamate transporters (Attwell *et al.*, 1993; Levi and Raiteri, 1993). Interestingly, dihydrokainic acid (DHK), a selective inhibitor of GLT-1 glutamate transporter (Arriza *et al.*, 1994) did not significantly modify the HMGB1-induced [^{3}H]D-Asp overflow.

Due to occurrence of such an ambiguous scenario, these data were confirmed in a set of experiments carried out to measure endogenous glutamate release in hippocampus and the effect of HMGB1. As shown in Fig. 3D, HMGB1 also stimulated the spontaneous release of endogenous glutamate from hippocampal gliosomes. The enhancement due the cytokine was strongly reduced (about 70%) in the absence of external Ca^{2+} and abolished in the presence of the glutamate transport blocker DL-TBOA, as in the case of the radioactive tracer.

In sum, the results with [^{3}H]D-Asp and endogenous glutamate suggest that the release from hippocampal gliosomes exposed to exogenous HMGB1 involves a carrier-mediated process and concurrently is dependent on external Ca^{2+}. However, since carrier-mediated release of neurotransmitters does not depend on external Ca^{2+} (Attwell *et al.*, 1993; Levi and Raiteri, 1993), the origin of this discrepancy has been further explored.

D. HMGB1 Binding to Gliosomes

To determine whether the lack of HMGB1-induced release of glutamate, observed in the absence of external Ca^{2+}, was a consequence of a real Ca^{2+} requirement for the amino acid export or depended on altered gliosomes–HMGB1 interaction in this condition, we measured the binding of [^{125}I] HMGB1 to gliosomes in the absence or in the presence of external Ca^{2+}. As shown in Fig. 4A, the labeled protein bound to gliosomal hippocampal particles and the binding with these particles was significantly reduced when external Ca^{2+} was omitted. These results suggest that the interaction of exogenous HMGB1

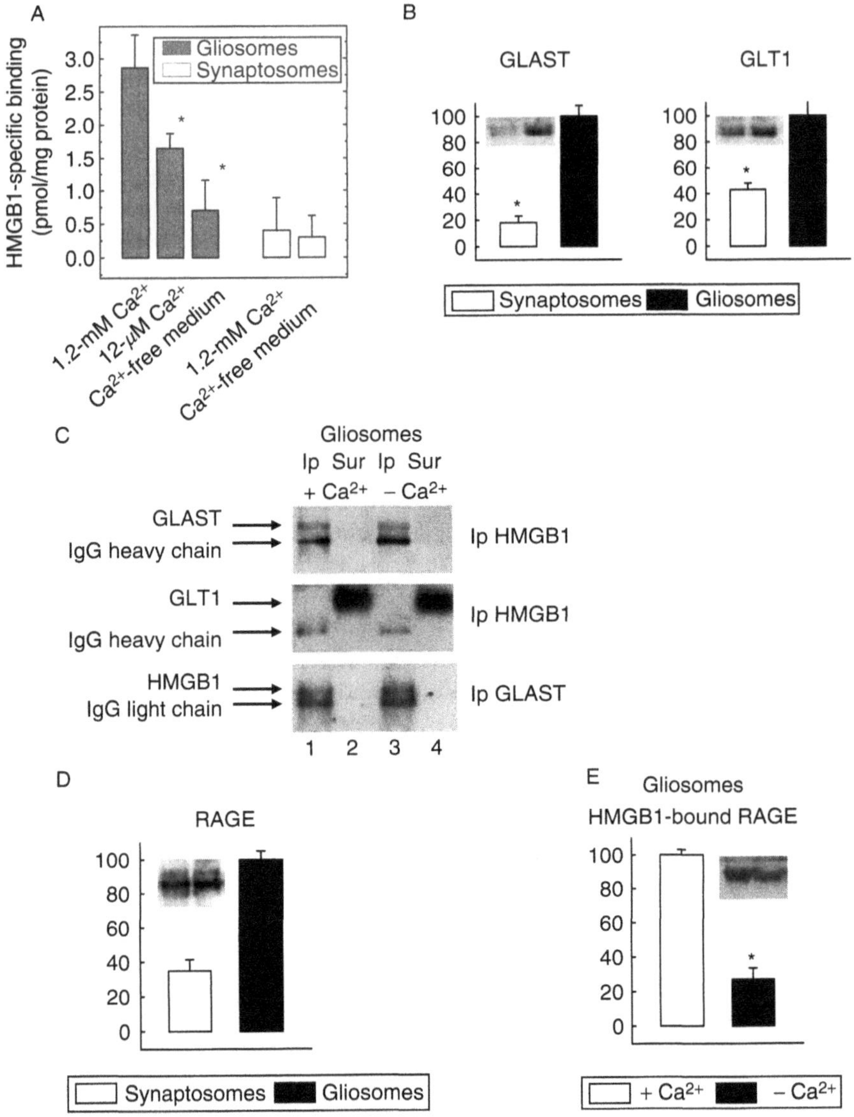

FIG. 4. Binding properties of HMGB1 to mouse hippocampal gliosomes. (A) Binding of [^{125}I]-HMGB1 to gliosomes and synaptosomes. The specific binding of the radiolabeled protein to intact gliosomes or synaptosomes was evaluated in the absence or the presence of 100-fold molar excess of unlabeled HMGB1 in a standard medium containing 1.2-mM $CaCl_2$ or 12-μM $CaCl_2$ or in a Ca^{2+}-free medium. (B) Glutamate–aspartate transporter (GLAST) or glutamate transporter 1 (GLT-1) distribution in synaptosomes and gliosomes evaluated by Western blotting. GLAST and GLT-1 amounts were measured on 10 μg proteins. (C) HMGB1-binding properties to GLAST or GLT-1 evaluated

with the gliosome surface is promoted by Ca^{2+}, even at micromolar concentrations. In contrast, synaptosomes, purified from the same brain areas, bound very low amounts of the cytokine both in the presence or in the absence of Ca^{2+} (Fig. 4A), suggesting that Ca^{2+}-activated binding sites for HMGB1 are preferentially contained in glial particles, in comparison with nerve endings.

Only two glutamate transporters have been reported to be expressed in glial cells, namely, GLAST and GLT-1 (Schmitt *et al.*, 1997; Sims and Robinson, 1999). As shown in Fig. 4B, comparison of the amounts of GLAST and GLT-1, detected by immunoblotting in gliosome and synaptosome preparations, confirmed that these transporters are exclusively (GLAST) or preferentially (GLT-1) represented in the gliosome fraction. The quantification of immunoreactive signals, obtained with different preparations, revealed a 4.5- and 1.8-fold higher amount of GLAST and GLT-1 in gliosomes than in synaptosomes, respectively. We next sought to determine whether glutamate transporters could directly interact with HMGB1 and if Ca^{2+} influenced these protein–protein bindings. Figure 4C shows that the GLAST coimmunoprecipitated with HMGB1 and that the amounts of GLAST recovered in the HMGB1 coimmunoprecipitated material in the presence (lane 1) or the absence (lane 3) of Ca^{2+} were not significantly different. By contrast, GLT-1 was detectable only in the pool of proteins not coimmunoprecipitated with HMGB1 (lanes 2 and 4). We also examined whether HMGB1 could be coimmunoprecipitated using an anti-GLAST antibody. Also in this condition, coimmunoprecipitation of HMGB1 and GLAST could be detected (Fig. 4C bottom, lanes 1 and 3), confirming the physical interaction between the two proteins. These findings indicate that HMGB1 can associate with GLAST, but not with GLT-1, and that this protein–protein interaction is not affected by Ca^{2+}.

To clarify whether the Ca^{2+}-promoted binding of extracellular HMGB1 to the gliosome membrane could be due to the possible participation of other HMGB1-binding proteins, the involvement of RAGE was tested. A report showed that brain contains several splice variants of RAGE, having transmembrane or cytoplasmic localization (Ding and Keller, 2005). We evaluated, at first, the amount of membrane-bound RAGE in gliosome and synaptosome preparations. As shown in Fig. 4D, gliosome membranes showed higher levels of

Western blotting. HMGB1 was added to detergent soluble membrane extracts and immunoprecipitation of HMGB1 (IP HMGB1) or GLAST (IP GLAST) was carried out. (D) Receptor for advanced glycosilation end product (RAGE) distribution in synaptosomes and gliosomes evaluated by Western blotting. RAGE amount was measured on 10-μg proteins. (E) HMGB1-binding properties to RAGE evaluated by Western blotting. RAGE bound to Sepharose®-HMGB1 was measured in the presence or in the absence of 1.2-mM $CaCl_2$. Modified from Pedrazzi *et al.* (2006), with the permission of Blackwell Publishing Co.

RAGE than synaptosomes: the quantification of immunoreactive signals, obtained with different preparations, revealed a 2.7- to 3.0-fold higher amount of membrane-bound RAGE in gliosomes than in synaptosomes. Since the coimmunoprecipitation strategy could not be applied, as the electrophoretic mobility of full-length transmembrane RAGE in Western blotting was superimposed to that of antibody heavy chain, to measure the effect of Ca^{2+} on HMGB1–RAGE interaction, HMGB1 was bound to Sepharose® beads and its RAGE-binding properties measured in the absence or presence of Ca^{2+} using detergent-soluble membrane fractions extracted from hippocampal gliosomes. RAGE was preferentially bound by HMGB1 in the presence of Ca^{2+} than in the absence of the ion: the amount of RAGE retained by HMGB1 in the presence of Ca^{2+} was 3.5- to 4-fold higher than in its absence.

Taken together these results suggest that HMGB1 can interact with both RAGE and GLAST on the gliosome surface but the preferential interaction of HMGB1 with gliosomes in the presence of Ca^{2+} is driven by RAGE.

IV. Concluding Remarks

We reported here evidence that purified glial derived organelles isolated from the adult rat brain, referred as to gliosomes, are able to take up and release glutamate when subjected to a variety of stimuli. Data in the literature support the view that gliosomes may represent a viable preparation that allows the study of mechanisms of transmitter release and its regulation in adult astrocytes. Besides the obvious differences between gliosomes and intact cultured astrocytes, gliosomes may have a number of advantages: they can be prepared rapidly and, most importantly, originate directly from mature brain astrocytes. Gliosomes can be obtained from animals acutely or chronically treated with drugs, from knockout or knockdown animals, from animals that represent models of brain diseases, and from fresh human brain samples of surgical origin.

In keeping with these observations, the results with HMGB1 underline that gliosomes are a vital *in vitro* glial preparation, able to respond selectively to exogenous stimuli and to reveal subtle differences from their neuronal counterpart. The present chapter, in fact, identifies extracellular HMGB1 as a stimulator of glutamate release from gliosomes. This effect was restricted to glial particles, whereas in this respect, synaptosomes from the same areas of the brain were not susceptible to HMGB1, suggesting that only gliosomes are equipped with functional mediators of HMGB1 signaling. Although the presence of Ca^{2+} in the external medium is required in order to promote association of HMGB1 with gliosomes, the release of glutamate occurs independently of the cytosolic availability of the ion; this may imply activation of carrier-mediated mechanisms. This conclusion is sustained by the ability of DL-TBOA, a glutamate uptake inhibitor that blocks all the identified glutamate transporter types (Shimamoto *et al.*, 1998), to inhibit HMGB1 releasing effect. Interestingly, DHK, a selective blocker of GLT-1, did not affect HMGB1, supporting the involvement of GLAST in the rise of glutamate release induced by HMGB1. In addition, we show here that GLAST, but not GLT-1, displays HMGB1-binding properties, suggesting that the efflux of glutamate from gliosomes requires physical association between HMGB1 and GLAST. It is worthy to note that the interaction between GLAST and HMGB1 was not affected by Ca^{2+}, in contrast, that between RAGE–HMGB1, as well as the binding of HMGB1 to gliosomal membranes, was promoted by this ion. Since the lack of extracellular Ca^{2+} also inhibited HMGB1-evoked glutamate release, this strongly suggests that binding of HMGB1 to RAGE, one major target of HMGB1 (Hori *et al.*, 1995; Rong *et al.*, 2004), plays a key role in the signaling process leading to neurotransmitter release.

It has been demonstrated that following acute injury to the brain, such as cerebral ischemia, stroke, or head trauma, glial glutamate transporters contribute to excitotoxic events by releasing glutamate into the extracellular space through a reversed operation of glutamate transport (Dunlop *et al.*, 2003; Rossi *et al.*, 2000). Moreover, it has been reported that HMGB1 level significantly increased in the brain of Alzheimer's disease patients and that coinjection of HMGB1 delayed the clearance of amyloid-β (Aβ42) and accelerated neurodegeneration in Aβ42-injected rats (Takata *et al.*, 2004). Different glial (Passalacqua *et al.*, 1998) and neuronal (Fages *et al.*, 2000) phenotypes are able to export HMGB1, but alternatively, the extracellular level of the cytokine can be increased by passive leakage from necrotic cells (Takata *et al.*, 2004). It can be hypothesized that the extracellular release of HMGB1, in brain regions of adult mammals expressing RAGE and GLAST, may be involved in neurodegenerative and neuroinflammatory processes.

Acknowledgments

This work was supported by the Italian Ministero dell'Istruzione, dell'Università e della Ricerca (PRIN 2002 and 2004), FIRB (Post-Genoma and Neuroscience Projects), and University of Genoa. We thank Ms. Maura Agate for her excellent secretarial assistance.

References

Agnello, D., Wang, H., Yang, H., Tracey, K. J., and Ghezzi, P. (2002). HMGB-1, a DNA-binding protein with cytokine activity, induces brain TNF and IL-6 production, and mediates anorexia and taste aversion. *Cytokine* **18,** 231–236.

Araque, A., Parpura, V., Sanzgiri, R. P., and Haydon, P. G. (1999). Tripartite synapses: Glia, the unacknowledged partner. *Trends Neurosci.* **22,** 208–215.

Arriza, J. L., Fairman, W. A., Wadiche, J. I., Murdoch, G. H., Kavanaugh, M. P., and Amara, S. G. (1994). Functional comparisons of three glutamate transporter subtypes cloned from human motor cortex. *J. Neurosci.* **14,** 5559–5569.

Asensio, V. C., and Campbell, I. L. (1999). Chemokines in the CNS: Plurifunctional mediators in diverse states. *Trends Neurosci.* **22,** 504–512.

Attwell, D., Barbour, B., and Szatkowski, M. (1993). Nonvesicular release of neurotransmitter. *Neuron* **11,** 401–407.

Bergles, D. E., and Jahr, C. E. (1997). Synaptic activation of glutamate transporters in hippocampal astrocytes. *Neuron* **19,** 1297–1308.

Berridge, M. J. (1998). Neuronal calcium signaling. *Neuron* **21,** 13–26.

Bezzi, P., Carmignoto, G., Pasti, L., Vesce, S., Rossi, D., Rizzini, B. L., Pozzan, T., and Volterra, A. (1998). Prostaglandins stimulate calcium-dependent glutamate release in astrocytes. *Nature* **391,** 281–285.

Bezzi, P., Domercq, M., Brambilla, L., Galli, R., Schols, D., De Clercq, E., Vescovi, A., Bagetta, G., Kollias, G., Meldolesi, J., and Volterra, A. (2001). CXCR4-activated astrocyte glutamate release via TNFa: Amplification by microglia triggers neurotoxicity. *Nat. Neurosci.* **4,** 702–710.

Bowman, E. J., Siebers, A., and Altendorf, K. (1988). Bafilomycins: A class of inhibitors of membrane ATPase from microorganisms, animal cells and plant cells. *Biochemistry* **85,** 7972–7976.

Bustin, M. (1999). Regulation of DNA-dependent activities by the functional motifs of the high-mobility-group chromosomal proteins. *Mol. Cell. Biol.* **19,** 5237–5246.

Carmignoto, G. (2000). Reciprocal communication systems between astrocytes and neurones. *Prog. Neurobiol.* **62,** 561–581.

Daniels, K. K., and Vickroy, T. W. (1999). Reversible activation of glutamate transport in rat brain glia by protein kinase C and an okadaic acid-sensitive phosphoprotein phosphatase. *Neurochem. Res.* **24,** 1017–1025.

Ding, Q., and Keller, J. N. (2005). Evaluation of rage isoforms, ligands, and signaling in the brain. *Biochim. Biophys. Acta* **1746,** 18–27.

Dunlop, J., Beal McIlvain, H., She, Y., and Howland, D. S. (2003). Impaired spinal cord glutamate transport capacity and reduced sensitivity to riluzole in a transgenic superoxide dismutase mutant rat model of amyotrophic lateral sclerosis. *J. Neurosci.* **23,** 1688–1696.

Fages, C., Nolo, R., Huttunen, H. J., Eskelinen, E., and Rauvala, H. (2000). Regulation of cell migration by amphoterin. *J. Cell Sci.* **113,** 611–620.

Floor, E., Leventhal, P. S., and Schaeffer, S. F. (1990). Partial purification and characterization of the vacuolar H^+-ATPase of mammalian synaptic vesicles. *J. Neurochem.* **55,** 1663–1670.

Gray, E. G., and Whittaker, V. P. (1962). The isolation of nerve endings from brain: An electron-microscopic study of cell fragments derived by homogenization and centrifugation. *J. Anat.* **96,** 79–87.

Grosche, J., Matyash, V., Moller, T., Verkhratsky, A., Reichenbach, A., and Kettenmann, H. (1999). Microdomains for neuron-glia interaction: Parallel fiber signaling to Bergmann glial cells. *Nat. Neurosci.* **2,** 139–143.

Guazzi, S., Strangio, A., Franzi, A. T., and Bianchi, M. E. (2003). HMGB1, an architectural chromatin protein and extracellular signaling factor, has a spatially and temporally restricted expression pattern in mouse brain. *Gene Expr. Patterns* **3,** 29–33.

Haydon, P. G. (2001). Glia: Listening and talking to the synapse. *Nat. Rev. Neurosci.* **2,** 185–193.

Hirst, W. D., Price, G. W., Rattray, M., and Wilkin, G. P. (1998). Serotonin transporters in adult rat brain astrocytes revealed by [^{3}H]5-HT uptake into glial plasmalemmal vesicles. *Neurochem. Int.* **33,** 11–22.

Hori, O., Brett, J., Slattery, T., Cao, R., Zhang, J., Chen, J. X., Nagashima, M., Lundh, E. R., Vijay, S., Nitecki, D., Morser, J., Stern, D., *et al.* (1995). The receptor for advanced glycation end products (RAGE) is a cellular binding site for amphoterin. Mediation of neurite outgrowth and co-expression of RAGE and amphoterin in the developing nervous system. *J. Biol. Chem.* **270,** 25752–25761.

Levi, G., and Raiteri, M. (1993). Carrier-mediated release of neurotransmitters. *Trends Neurosci.* **16,** 415–419.

Mennerick, S., and Zorumski, C. F. (1994). Glial contributions to excitatory neurotransmission in cultured hippocampal cells. *Nature* **368,** 59–62.

Miller, R. J., and Meucci, O. (1999). AIDS and the brain: Is there a chemokine connection? *Trends Neurosci.* **22,** 471–479.

Montana, V., Malarkey, E. B., Verderio, C., Matteoli, M., and Parpura, V. (2006). Vesicular transmitter release from astrocytes. *Glia* **54,** 700–715.

Moriyama, Y., and Futai, M. (1990). H(+)-ATPase, a primary pump for accumulation of neurotransmitters, is a major constituent of brain synaptic vesicles. *Biochem. Biophys. Res. Commun.* **173,** 443–448.

Müller, S., Ronfani, L., and Bianchi, M. E. (2004). Regulated expression and subcellular localization of HMGB1, a chromatin protein with a cytokine function. *J. Intern. Med.* **255,** 332–343.

Nakamura, Y., Iga, K., Shibata, T., Shudo, M., and Kataoka, K. (1993). Glial plasmalemmal vesicles: A subcellular fraction from rat hippocampal homogenate distinct from synaptosomes. *Glia* **9,** 48–56.

Nakamura, Y., Kubo, H., and Kataoka, K. (1994). Uptake of transmitter amino acids by glial plasmalemmal vesicles from different regions of rat central nervous system. *Neurochem. Res.* **19,** 1145–1150.

Nathan, C. F. (1987). Secretory products of macrophages. *J. Clin. Invest.* **79,** 319–326.

O'Connor, K. A., Hansen, M. K., Rachal Pugh, C., Deak, M. M., Biedenkapp, J. C., Milligan, E. D., Johnson, J. D., Wang, H., Maier, S. F., Tracey, K. J., and Watkins, L. R. (2003). Further characterization of high mobility group box 1 (HMGB1) as a proinflammatory cytokine: Central nervous system effects. *Cytokine* **24,** 254–265.

Parpura, V., Basarasky, T. A., Liu, F., Jeftinija, K., Jeftinija, S., and Haydon, P. G. (1994). Glutamate-mediated astrocyte-neuron signalling. *Nature* **369,** 744–747.

Passalacqua, M., Patrone, M., Picotti, G. B., Del Rio, M., Sparatore, B., Melloni, E., and Pontremoli, S. (1998). Stimulated astrocytes release high-mobility group 1 protein, an inducer of LAN-5 neuroblastoma cell differentiation. *Neuroscience* **82,** 1021–1028.

Pedrazzi, M., Raiteri, L., Bonanno, G., Patrone, M., Ledda, S., Passalacqua, M., Milanese, M., Melloni, E., Raiteri, M., Pontremoli, S., and Sparatore, B. (2006). Stimulation of excitatory amino acid release from adult mouse brain glia subcellular particles by high mobility group box 1 protein. *J. Neurochem.* **99,** 827–838.

Raiteri, L., and Raiteri, M. (2000). Synaptosomes still viable after 25 years of superfusion. *Neurochem. Res.* **25,** 1265–1274.

Raiteri, L., Stigliani, S., Siri, A., Passalacqua, M., Melloni, E., Raiteri, M., and Bonanno, G. (2005). Glycine taken up through GLYT1 and GLYT2 heterotransporters into glutamatergic axon terminals of mouse spinal cord elicits release of glutamate by homotransporter reversal and through anion channels. *Biochem. Pharmacol.* **69,** 158–168.

Rong, L. L., Trojaborg, W., Qu, W., Kostov, K., Yan, S. D., Gooch, C., Szabolcs, M., Hays, A. P., and Schmidt, A. M. (2004). Antagonism of RAGE suppresses peripheral nerve regeneration. *FASEB J.* **18,** 1812–1817.

Roseth, S., Fykse, E. M., and Fonnum, F. (1995). Uptake of L-glutamate into rat brain synaptic vesicles: Effect of inhibitors that bind specifically to the glutamate transporter. *J. Neurochem.* **65,** 96–103.

Rossi, D., and Zlotnik, A. (2000). The biology of chemokines and their receptors. *Annu. Rev. Immunol.* **18,** 217–242.

Rossi, D. J., Oshima, T., and Attwell, D. (2000). Glutamate release in severe brain ischemia is mainly by reversed uptake. *Nature* **403,** 316–321.

Sajithlal, G., Huttunen, H., Rauvala, H., and Munch, G. (2002). Receptor for advanced glycation end products plays a more important role in cellular survival than in neurite outgrowth during retinoic acid-induced differentiation of neuroblastoma cells. *J. Biol. Chem.* **277,** 6888–6897.

Sanchez-Prieto, J., Sihra, T. S., Evans, D., Ashton, A., Dolly, J. O., and Nicholls, D. G. (1987). Botulinum toxin A blocks glutamate exocytosis from guinea-pig cerebral cortical synaptosomes. *Eur. J. Biochem.* **165,** 675–681.

Schmitt, A., Asan, E., Püschel, B., and Kugler, P. (1997). Cellular and regional distribution of the glutamate transporter GLAST in the CNS of rats: Nonradioactive *in situ* hybridization and comparative immunocytochemistry. *J. Neurosci.* **17,** 1–10.

Shimamoto, K., Lebrun, B., Yasuda-Kamatani, Y., Sakaitani, M., Shigeri, Y., Yumoto, N., and Nakajima, T. (1998). DL-threo-beta-benzyloxyaspartate, a potent blocker of excitatory amino acid transporters. *Mol. Pharmacol.* **53,** 195–201.

Sims, K. D., and Robinson, M. B. (1999). Expression patterns and regulation of glutamate transporters in the developing and adult nervous system. *Crit. Rev. Neurobiol.* **13,** 169–197.

Stigliani, S., Zappettini, S., Raiteri, L., Passalacqua, M., Melloni, E., Venturi, C., Tacchetti, C., Diaspro, A., Usai, C., and Bonanno, G. (2006). Glia re-sealed particles freshly prepared from adult rat brain are competent for exocytotic release of glutamate. *J. Neurochem.* **96,** 656–668.

Suchak, S. K., Baloyianni, N. V., Perkinton, M. S., Williams, R. J., Meldrum, B. S., and Rattray, M. (2003). The 'glial' glutamate transporter, EAAT2 (Glt-1) accounts for high affinity glutamate uptake into adult rodent nerve endings. *J. Neurochem.* **84,** 522–532.

Südhof, T. C. (1995). The synaptic vesicle cycle: A cascade of protein-protein interactions. *Nature* **375,** 645–653.

Taguchi, A., Blood, D. C., del Toro, G., Canet, A., Lee, D. C., Qu, W., Tanji, N., Lu, Y., Lalla, E., Fu, C., Hofmann, M. A., Kislinger, T., *et al.* (2000). Blockade of RAGE-amphoterin signaling suppresses tumour growth and metastases. *Nature* **405,** 354–360.

Takata, K., Kitamura, Y., Tsuchiya, D., Kawasaki, T., Taniguchi, T., and Shimohama, S. (2004). High mobility group box protein-1 inhibits microglial A beta clearance and enhances A beta neurotoxicity. *J. Neurosci. Res.* **78,** 880–891.

Thomas, J. O., and Travers, A. A. (2001). HMG1 and 2, and related 'architectural' DNA-binding proteins. *Trends Biochem. Sci.* **26,** 167–174.

Verhage, M., McMahon, H. T., Ghijsen, W. E. J. M., Boomsma, F., Scholten, G., Wiegant, V. M., and Nicholls, D. G. (1991). Differential release of amino acids, neuropeptides, and catecholamines from isolated nerve terminals. *Neuron* **6,** 517–524.

Wang, H., Vishnubhakat, J. M., Bloom, O., Zhang, M., Ombrellino, M., Sama, A., and Tracey, K. J. (1999). Proinflammatory cytokines (tumor necrosis factor and interleukin 1) stimulate release of high mobility group protein-1 by pituicytes. *Surgery* **126,** 389–392.

Wang, H., Yang, H., and Tracey, K. J. (2004). Extracellular role of HMGB1 in inflammation and sepsis. *J. Intern. Med.* **255,** 320–331.

Zoccarato, F., Cavallini, L., and Alexandre, A. (1999). The pH-sensitive dye acridine orange as a tool to monitor exocytosis/endocytosis in synaptosomes. *J. Neurochem.* **72,** 625–633.

THE ROLE OF ASTROCYTES AND COMPLEMENT SYSTEM IN NEURAL PLASTICITY

Milos Pekny,* Ulrika Wilhelmsson,* Yalda Rahpeymai Bogestål,† and Marcela Pekna†

*Center for Brain Repair and Rehabilitation, Department of Clinical Neuroscience and Rehabilitation, Institute for Neuroscience and Physiology at Sahlgrenska Academy Göteborg University, 405 30 Göteborg, Sweden

†Department of Medical Chemistry and Cell Biology, Institute of Biomedicine at Sahlgrenska Academy, Göteborg University, 405 30 Göteborg, Sweden

In neurotrauma, brain ischemia or neurodegenerative diseases, astrocytes become reactive (which is known as reactive gliosis) and this is accompanied by an altered expression of many genes. Two cellular hallmarks of reactive gliosis are hypertrophy of astrocyte processes and the upregulation of the part of the cytoskeleton known as intermediate filaments, which are composed of nestin, vimentin, and GFAP. Our aim has been to better understand the function of reactive astrocytes in CNS diseases. Using mice deficient for astrocyte intermediate filaments ($GFAP^{-/-}Vim^{-/-}$), we were able to attenuate reactive gliosis and slow down the healing process after neurotrauma. We demonstrated the key role of reactive astrocytes in neurotrauma—at an early stage after neurotrauma, reactive astrocytes have a neuroprotective effect; at a later stage, they facilitate the formation of posttraumatic glial scars and inhibit CNS regeneration, specifically, they seem to compromise neural graft survival and integration, reduce the extent of synaptic regeneration, inhibit neurogenesis in the old age, and inhibit regeneration of severed CNS axons. We propose that reactive astrocytes are the future target for the therapeutic strategies promoting regeneration and plasticity in the brain and spinal cord in various disease conditions.

Through its involvement in inflammation, opsonization, and cytolysis, complement protects against infectious agents. Although most of the complement proteins are synthesized in CNS, the role of the complement system in the normal

INTERNATIONAL REVIEW OF NEUROBIOLOGY, VOL. 82
DOI: 10.1016/S0074-7742(07)82005-8

or ischemic CNS remains unclear. Complement activiation in the CNS has been generally considered as contributing to tissue damage. However, growing body of evidence suggests that complement may be a physiological neuroprotective mechanism as well as it may participate in maintenance and repair of the adult brain.

I. Introduction

In a striking contrast to the peripheral nervous system, the regenerative capacity of the adult brain and spinal cord (e.g., synaptic and axonal regeneration, neurogenesis) is extremely limited, despite the fact that neural stem cells are present in CNS throughout the whole life. Both the endogenous neural stem cells and neural implants grafted to replace lost neurons fail to form functional connections to the extent that would influence the clinical outcome in conditions such as neurotrauma, stroke, or neurodegenerative diseases. Moreover shortly after birth, axons in the adult mammalian CNS lose their ability to grow and regenerate following injury. We suggest that the environment, in particular astrocytes, and the immune system are important modulators of CNS regeneration.

II. Astrocytes, GFAP, and Astrocyte Intermediate Filaments

Astrocytes are the most numerous cells in the CNS, and they were implicated to be involved in many CNS pathologies such as trauma, ischemia, or neurodegenerative diseases. In response to any kind of injury in the CNS, astrocytes change their appearance and undergo a characteristic hypertrophy of their cellular processes. This phenomenon is known as reactive gliosis or astrogliosis its hallmark being upregulation of intermediate filament (IF) proteins GFAP and vimentin, reexpression of nestin as well as altered expression profiles of many proteins (Eddleston and Mucke, 1993; Hernandez *et al.*, 2002). The IFs can be considered the least understood part of the cytoskeleton. The family of IF proteins expressed in vertebrates is large (in humans 65 different IF proteins have been identified) (Herrmann and Aebi, 2004; Herrmann *et al.*, 2003), and there is a complex expression pattern of IF proteins unique for each cell type as well as during different developmental stages.

The dynamic feature of the IF network depends both on the equilibrium between filaments and unassembled subunits and the regulation of filament assembly/disassembly by phosphorylation of the head domain of the IF proteins. IFs were at first considered to be static structures primarily responsible for maintaining the cell shape (Renner *et al.*, 1981; Rueger *et al.*, 1979). However, later studies both *in vitro* (Angelides *et al.*, 1989; Nakamura *et al.*, 1991) and *in vivo* (Miller *et al.*, 1991; Vikstrom *et al.*, 1992; Wiegers *et al.*, 1991; Yoon *et al.*, 1998) revealed the rather dynamic nature of IFs and the existence of a dynamic equilibrium between the assembled filaments and the pool of soluble subunits (reviewed in Goldman *et al.*, 1999).

In vivo, IFs are often, if not always, heteropolymeric (Herrmann and Aebi, 2000). For determining the partnership in the formation of IF heteropolymers in astrocytes, transgenic mice deficient in individual IF proteins were instrumental. In nonreactive astrocytes, IFs are formed of GFAP and vimentin, while in reactive astrocytes, nestin can be found as the additional partner in the IF network (Pekny *et al.*, 1998; Table I). In addition, some reactive astrocytes, for example in neurotrauma, express another IF protein, synemin (Jing *et al.*, 2007). The studies of astrocytes lacking GFAP and/or vimentin revealed that GFAP can form IFs on its own in vimentin deficient ($Vim^{-/-}$) astrocytes, but such filaments form more compact bundles than in wild-type astrocytes, suggesting that at least a low level of vimentin is needed for normal IF formation in the astrocytes (Eliasson *et al.*, 1999; Lepekhin *et al.*, 2001; Menet *et al.*, 2001). Studies in mice deficient in GFAP ($GFAP^{-/-}$) showed that vimentin does not form IF on its own, or it does so only with a very low efficiency (McCall *et al.*, 1996; Pekny *et al.*, 1995). In contrast, the reactive $GFAP^{-/-}$ astrocytes contain IFs since vimentin can polymerize with nestin, which is expressed in reactive astrocytes (Eliasson *et al.*, 1999). GFAP does not polymerize with nestin in reactive $Vim^{-/-}$ astrocytes and consequently, the IFs contain only GFAP and exhibit the characteristic tight bundling similar to $Vim^{-/-}$ nonreactive astrocytes. In reactive astrocytes lacking both GFAP and vimentin ($GFAP^{-/-}$ $Vim^{-/-}$) no IFs are formed, and both the nestin and synemin proteins which are produced, stay in a nonfilamentous form (Eliasson *et al.*, 1999; Jing *et al.*, 2007). Nestin was proposed to facilitate phosphorylation-dependent disassembly

TABLE I

Composition of IFs in Nonreactive and Reactive Astrocytes of Wild-Type Mice and Mice Deficient in GFAP and/or Vimentin[a]

Genotype	Composition of IFs: Nonreactive astrocytes	Composition of IFs: Reactive astrocytes	Reactive astrocytes: IF amount/bundling
Wild type	GFAP, vimentin	GFAP, vimentin, nestin	Normal/normal
$GFAP^{-/-}$	No IFs (nonfilamentous vimentin)	Vimentin, nestin	Decreased/normal
$Vim^{-/-}$	GFAP	GFAP (nonfilamentous nestin)	Decreased/tight
$GFAP^{-/-}Vim^{-/-}$	No IFs	No IFs (nonfilamentous nestin)	–

[a]Some reactive astrocytes contain another IF protein, synemin (Jing *et al.*, 2007).

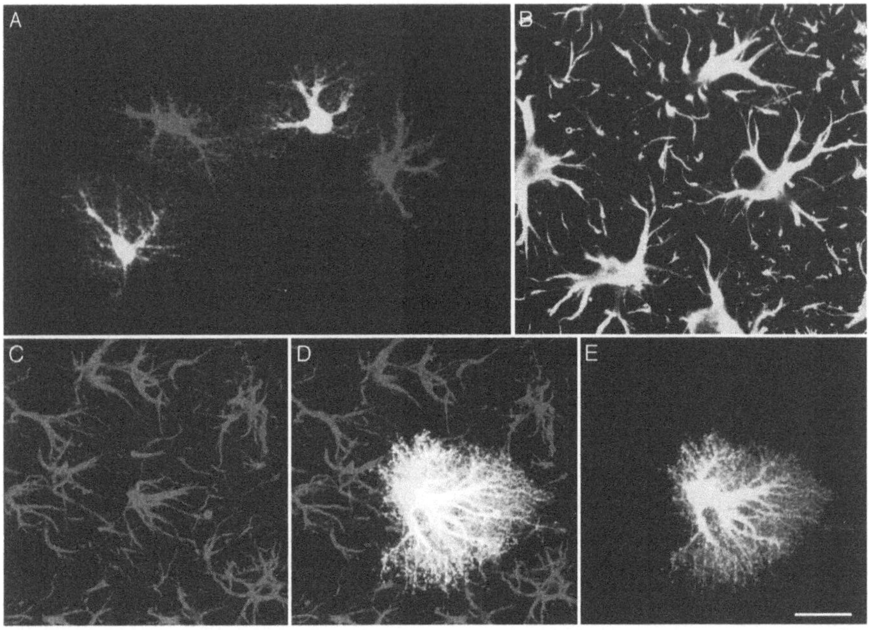

FIG. 1. Visualization of astroglial cell morphology *in vivo*. (A) Three-dimensional reconstruction of astrocytes. Astrocytes in the adult mouse hippocampus filled with two different dyes. The CNS is divided into domains and each of them is accessed by fine cellular processes of an astrocyte. (Courtesy of Wilhelmsson, Bushong, Ellisman, and Pekny.) (B) Astrocytes in the brain cortex visualized by antibodies against GFAP. (C–E) Reactive astrocytes after dye filling and three-dimensional reconstruction. Note the typical bushy appearance of astrocytes with fine cellular processes that cannot be visualized by antibodies against GFAP (compare the central astrocyte in C–E). Scale bar, 20 μm. [Reproduced from Wilhelmsson *et al.* (2004).] (See Color Insert.)

of vimentin IFs during mitosis and to play a role in the distribution of IF protein to daughter cells (Chou *et al.*, 2003).

The IF network of mature astrocytes is composed of GFAP as the major IF protein and vimentin ranging from very low to intermediate levels depending on the subpopulation of astrocytes (Pixley *et al.*, 1984; Shaw *et al.*, 1981). Mature astrocytes have fine processes extending from the main cellular processes giving each cell a characteristic bushy appearance (Fig. 1A). The IF network, however, is restricted to the main processes and the soma of astrocytes (Bushong *et al.*, 2002, 2004; Fig. 1B–E). We have shown that in denervated hippocampus or in the vicinity of cortical lesion reactive astrocytes increase the thickness of their main cellular processes but occupy a volume of tissue comparable to that of nonreactive astrocytes. Despite the hypertrophy of GFAP-containing cellular processes, the interdigitation between adjacent reactive astrocytes in denervated hippocampus remains minimal (Wilhelmsson *et al.*, 2006; Fig. 2).

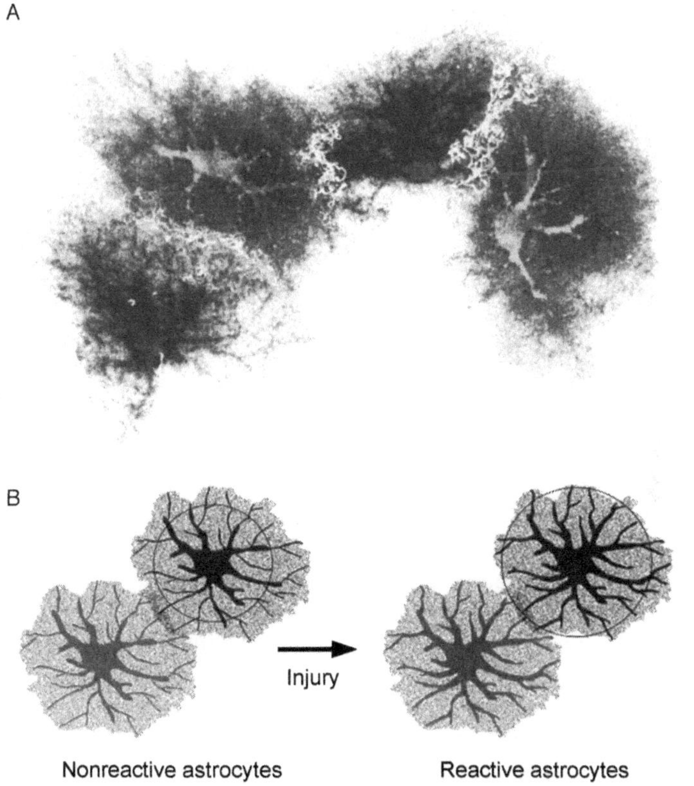

FIG. 2. The domains of nonreactive and reactive astrocytes—a concept. (A) Interdigitation of fine cellular processes in a 3D reconstruction of dye filled astrocytes in the dentate gyrus of the hippocampus. The yellow zone shows the border area where cellular processes of two adjacent astrocytes interdigitate. (B) Reactive astrocytes stay within their domains, but their main cellular processes get thicker, making them visible over a greater distance (illustrated here by the gray circles). [Reproduced from Wilhelmsson *et al.* (2006).] (See Color Insert.)

It was proposed that GFAP-positive astroglial cells are involved in the baseline neurogenesis in the adult mammalian CNS. Recent data suggest that astrocytes positively control neurogenesis in the two regions of the adult CNS, specifically in the dentate gyrus of the hippocampus and in the subventricular zone, that is, the only two CNS regions in which new neurons are generated in relatively high numbers even in the adult (Song *et al.*, 2002). It was suggested that the majority of neural stem cells in the adult CNS are at some point GFAP positive, that is, could be defined as astroglial cells (Doetsch *et al.*, 1999; Imura *et al.*, 2003; Laywell *et al.*, 2000; Morshead *et al.*, 2003). Thus, astroglial cells might be both the cell type that controls adult neurogenesis as well as the precursors to neurons that are added in the adult life.

III. Reactive Gliosis, Neurotrauma, and CNS Transplants

To assess the role of IF upregulation in reactive astrocytes in CNS injury, several trauma models were applied to mice deficient in GFAP and/or vimentin. Fine needle injury of the brain cortex and transection of the dorsal funiculus in the upper thoracic spinal cord were two of the models used. The responses of wild-type, $GFAP^{-/-}$ and $Vim^{-/-}$ mice were indistinguishable. In $GFAP^{-/-}Vim^{-/-}$ mice, however, the posttraumatic glial scarring was looser and less organized, suggesting that upregulation of IFs is an important step in astrocyte activation. These data also imply that reactive astrocytes play a role in posttraumatic healing (Pekny *et al.*, 1999; Fig. 3).

Extended healing period following CNS injury was also reported in mice in which dividing astrocytes had been ablated by GFAP-driven expression of Herpes simplex virus thymidine kinase and administration of ganciclovir (Bush *et al.*, 1999; Faulkner *et al.*, 2004). Another group used hemisections of the lower thoracic spinal cord and reported increased axonal sprouting and better functional recovery in $GFAP^{-/-}Vim^{-/-}$ mice than wild-type controls (Menet *et al.*, 2003). Two groups addressed the role of astrocyte IFs in neurite outgrowth *in vitro* (Menet *et al.*, 2000, 2001; Xu *et al.*, 1999). One group reported that $GFAP^{-/-}Vim^{-/-}$ and $GFAP^{-/-}$ astrocytes were a better substrate for the outgrowth of neurites *in vitro* than wild-type astrocytes (Menet *et al.*, 2000, 2001). The other group found comparable neurite outgrowth when neurons were cultured on wild-type and $GFAP^{-/-}$ astrocytes (Xu *et al.*, 1999). The latter finding is in agreement with the normal axonal sprouting and regeneration assessed after dorsal hemisection of the spinal cord in $GFAP^{-/-}$ mice (Wang *et al.*, 1997). Extensive axonal regeneration was reported in the severed optic nerve of young $GFAP^{-/-}Vim^{-/-}$ mice that also carried a transgene overexpressing Bcl-2 in neurons (Cho *et al.*, 2005).

Another model that was used to study the involvement of astrocytes in neurotrauma was entorhinal cortex lesion. This lesion interrupts axonal connections (known as the perforant path) between the entorhinal cortex and the projection area in the outer molecular layer of the dentate gyrus of the hippocampus (Turner *et al.*, 1998) where degenerating neurons trigger extensive reactive gliosis. The distance between these two regions allows assessment of astrocyte response, degeneration, and subsequent regeneration in the hippocampus, that is, the region that is not directly affected by the surgery. By utilizing this model, we showed that reactive astrocytes devoid of IFs ($GFAP^{-/-}Vim^{-/-}$) exhibited only limited hypertrophy of cell processes. Many processes of $GFAP^{-/-}Vim^{-/-}$ astrocytes were shorter and less straight than those of wild-type astrocytes, albeit the volume of the CNS tissue reached by a single astrocyte was comparable to wild-type mice (Wilhelmsson *et al.*, 2004). These results, along with *in vitro* data on the morphology of IF-depleted astrocytes in primary cultures (Lepekhin *et al.*, 2001),

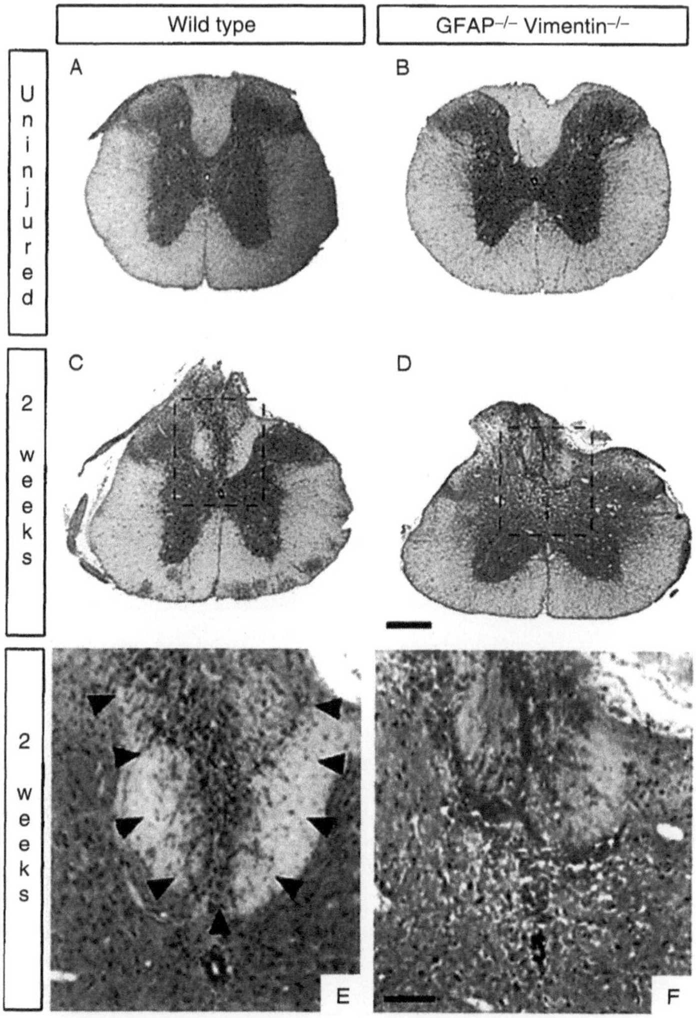

FIG. 3. Wound healing in the absence of astrocyte IF proteins after transection of the dorsal funiculus in the upper thoracic spinal cord takes longer in *GFAP*$^{-/-}$ *Vim*$^{-/-}$ than wild-type mice. H & E staining. Scale bar, 300 μm in A–D and 100 μm in E and F. [Reproduced from Pekny *et al.* (1999).] (See Color Insert.)

show a novel role for IFs in determining astrocyte morphology. In *GFAP*$^{-/-}$ *Vim*$^{-/-}$ mice, loss of neuronal synapses in the outer molecular layer of the hippocampal dentate gyrus was prominent 4 days after lesioning (Fig. 4A,B, and E). Most interestingly, there was remarkable synaptic regeneration 10 days later (at 14 days after lesions) (Fig. 4C–E). In contrast to wild type, *GFAP*$^{-/-}$ *Vim*$^{-/-}$ reactive

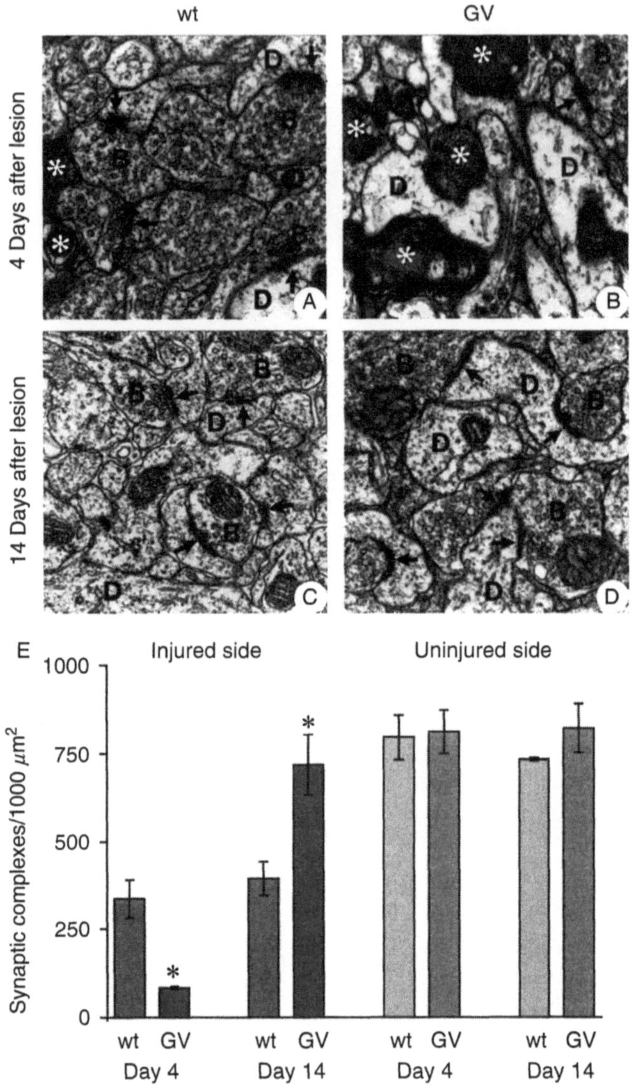

FIG. 4. Synaptic regeneration after entorhinal cortex lesion in the projection area of entorhinal cortex in the dentate gyrus of the hippocampus in *GFAP*$^{-/-}$*Vim*$^{-/-}$ (GV) and wild-type (wt) mice. At day 4 after lesioning, the synaptic loss and the signs of neurodegeneration were more prominent in *GFAP*$^{-/-}$*Vim*$^{-/-}$ than wild-type mice (A, B, and E). At day 14 after lesioning, the number of synapses in *GFAP*$^{-/-}$*Vim*$^{-/-}$, but not wild-type mice, recovered reaching the levels comparable with the uninjured hemisphere (C–E). Asterisks, degenerated axons; arrows, synaptic complexes; D, dendritic profile; B, synaptic bouton; *, $p < 0.05$. [Reproduced from Wilhelmsson *et al.* (2004).]

astrocytes did not upregulate the expression of endothelin B receptors, suggesting that the upregulation of this novel marker of reactive astrocytes (Baba, 1998; Ishikawa *et al.*, 1997; Koyama *et al.*, 1999; Peters *et al.*, 2003) is IF dependent (Wilhelmsson *et al.*, 2004). Thus, the effect of reactive astrocytes after CNS trauma seems to be twofold: reactive astrocytes play a beneficial role in the acute stage after CNS injury, however later on act as inhibitors of CNS regeneration. Support for the concept of reactive gliosis as an inhibiting factor with respect to posttraumatic repair and functional recovery was provided also by studies using transgenic mice expressing an NF-κB inhibitor in astrocytes (Brambilla *et al.*, 2005) and in mice deficient for EphA4 (Goldshmit *et al.*, 2004).

Because of their morphology and abundance in the adult CNS, astrocytes have direct physical contact with any cell that moves from one place to another. To assess the impact of astrocyte IFs on the fate of cells migrating from neural transplants, the Chen and Pekny groups transplanted dissociated retinal cells from 0- to 3-week-old donor mice that ubiquitously express enhanced green fluorescent protein (Okabe *et al.*, 1997) into the retinas of adult wild-type and $GFAP^{-/-}Vim^{-/-}$ recipients and compared the efficiency of long-term integration of such grafts in the retina (Kinouchi *et al.*, 2003). In wild-type hosts, few transplanted cells migrated from the transplantation site and few integrated into the retina. In $GFAP^{-/-}Vim^{-/-}$ hosts, however, the transplanted cells effectively moved through the retina, differentiated into neurons, integrated into the ganglion cell layer, and some of them even extended neurites about 1 mm into the optic nerve (Fig. 5A–D). The single mutants exhibited a dose effect (Fig. 5E–I). Six months after transplantation, the cells remained alive and well-integrated in the $GFAP^{-/-}Vim^{-/-}$ hosts (Kinouchi *et al.*, 2003). These results show that the absence of IFs in astroglial cells (astrocytes and Müller cells) of the retina increases the permissiveness of the retinal environment for integration of neural transplants through yet unknown mechanism. It is possible to speculate that IF depletion in astroglial cells alters their differentiation state, turning them into cells functionally similar to more immature astrocytes, and thereby also more supportive of CNS regeneration (Emsley *et al.*, 2004).

By affecting the abundance or the composition of IFs, it might be possible to alter the state of cellular differentiation and thus many cellular functions, which ultimately allow control of complex processes such as the permissiveness of the CNS for regeneration (Pekny *et al.*, 2004; Quinlan and Nilsson, 2004).

IV. The Complement System

Complement, a component of the humoral immune system, is involved in inflammation, opsonization, and cytolysis. More than 20 plasma proteins participate in the activation and regulation of complement, most of them functioning as

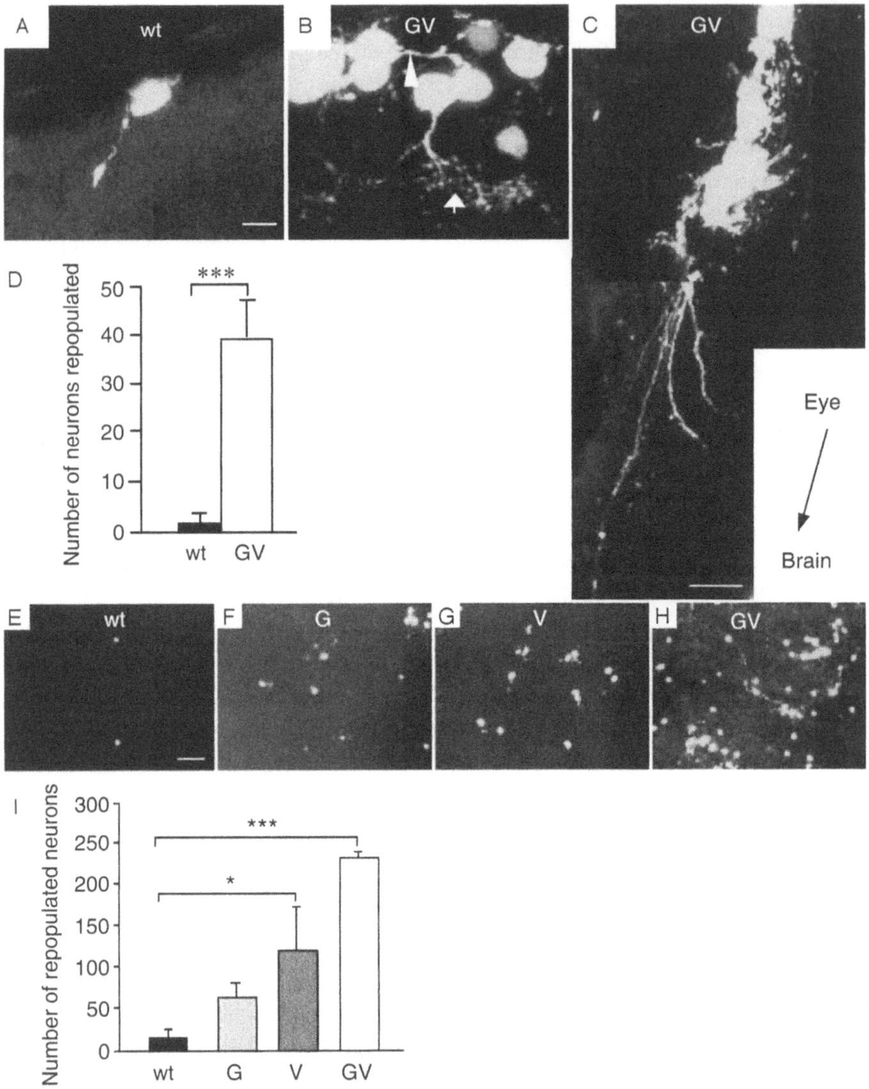

FIG. 5. Integration of neural transplants in $GFAP^{-/-}Vim^{-/-}$ mice. Retinal transplants from mice expressing enhanced green fluorescent protein integrated much better in $GFAP^{-/-}Vim^{-/-}$ (GV) than wild-type (wt) recipients (A–D). In $GFAP^{-/-}Vim^{-/-}$ recipients, transplanted cells migrated more efficiently from the transplantation site and integrated into the ganglion cell layer (GCL, D), exhibiting typical morphology of ganglion cells with axon-like process parallel to the retinal surface (arrowhead) and branched dendritic treelike structures (arrow, B). Some of these neurons even extended axons into the optic nerve (C). In single mutant recipients (G or V), the transplanted cells spread out more extensively than in wild-type but less efficiently than in $GFAP^{-/-}Vim^{-/-}$ recipients (E–I). *, $p < 0.05$; ***, $p < 0.001$. Scale bar, 5 μm in A and B, 50 μm in C, and 100 μm in E–H. Data represent mean ± SD. [Reproduced from Kinouchi *et al.* (2003).]

enzymes, enzyme inhibitors, or enzyme cofactors. In addition, there are more than 10 membrane proteins that regulate complement activation or serve as receptors for proteolytic fragments generated during activation of the cascade. Complement activation results in the formation of C3-convertase, an enzymatic complex that activates the central molecule of the cascade, the third complement component (C3). The proteolytic activation of C3 generates C3a, a small fragment with anaphylatoxic properties, and C3b, that binds to an activating surface and triggers the terminal part of the cascade, generating C5a through the proteolytic activation of C5 and culminating in the assembly of the cytolytic membrane attack complex on the target surface.

The primary site of complement protein synthesis is the liver; however, local complement production in the CNS is now well established in microglia, astrocytes, and neurons (Gasque *et al.*, 1992, 1993, 1995; Thomas *et al.*, 2000). Local expression of complement proteins by resident cells is increased following brain infection (Dandoy-Dron *et al.*, 1998; Dietzschold *et al.*, 1995; Stahel *et al.*, 1997a,b) and ischemia (Schafer *et al.*, 2000; van Beek *et al.*, 2000b). Cerebral ischemia leads also to an increased expression of receptors for the complement-derived anaphylatoxic peptides C3a and C5a (C3aR and C5aR) in the ischemic cortex in mice (van Beek *et al.*, 2000a).

Although the role of complement in normal CNS is unknown, in injury such as ischemia, complement activation has been suggested to exacerbate the inflammatory response, therefore contributing to secondary tissue damage. Systemic complement depletion reduced complement-mediated tissue damage after transient cerebral ischemia in rats (Vasthare *et al.*, 1998). Treatment with a bifunctional molecule, designed to inhibit both complement activation and selectin-mediated inflammatory cell migration, reduced the infarct volume following transient cerebral ischemia in mice (Huang *et al.*, 1999). Treatments with antibodies against complement receptor 3 or with C1 inhibitor reduced infarct size after transient cerebral ischemia (De Simoni *et al.*, 2003; Zhang *et al.*, 1995).

Notably, complement activation has been implicated in tissue regeneration. C3 mRNA and protein were specifically expressed in the blastema cell layer of the regenerating amphibian limbs but not in developing limbs (Del Rio-Tsonis *et al.*, 1998). In a study, C3 and C5 were detected in regenerating but not intact newt limb and lens (Kimura *et al.*, 2003) suggestive of specific role for complement components in the regenerative process. In addition, the newt ortholog of CD59, termed Prod 1, has been implicated in blastema positional identity during adult limb regeneration (Morais da Silva *et al.*, 2002). More importantly, complement has also been implicated in the regenerative process in higher vertebrates. In mice, C3a and C5a are critical for liver regeneration by promoting hepatocyte proliferation (Daveau *et al.*, 2004; Markiewski *et al.*, 2004;

Mastellos *et al.*, 2001; Strey *et al.*, 2003), whereas C3 activation–derived C3b/iC3b seem to contribute to the clearance of injured tissue (Markiewski *et al.*, 2004).

We have recently identified an additional and novel role for complement in the CNS. We have shown that neural stem cells *in vitro* as well as on neuroblasts *in vivo* express both C3aR and C5aR. Basal neurogenesis was impaired in C3aR deficient, C3aR antagonist treated as well as C3-deficient mice implicating signaling through C3aR as a positive regulator of adult neurogenesis (Rahpeymai *et al.*, 2006). In contrast, signaling through C5aR does not appear to be involved in this process as basal neurogensis was not affected in C5aR deficient mice (Rahpeymai Bogestål *et al.*, in press). Remarkably, the C3-deficient mice showed impaired ischemia-induced neurogenesis despite the larger infarct volume (Rahpeymai *et al.*, 2006). Thus, complement appears to function as a positive regulator of both basal and ischemia-induced adult mammalian neurogenesis. Inhibition of complement activation aggravated amyloid plaque formation and neurodegeneration in an experimental model of Alzheimer's disease (Wyss-Coray *et al.*, 2002) and exposure to C3a induced *de novo* expression of nerve growth factor, a molecule involved in neuronal growth and survival, in microglial cells *in vitro* (Heese *et al.*, 1998), both observations consistent with the involvement of intracerebral complement in brain tissue repair. Interestingly, C3a and C5a were also shown to be neuroprotective (Mukherjee and Passinetti, 2001; O'Barr *et al.*, 2001; van Beek *et al.*, 2001).

Complement activation in the CNS may have a dual role. Although it has generally been considered detrimental, it may be a physiological protective mechanism as well as participate in maintenance and repair of the adult brain. A detailed understanding of the nonimmune functions of the complement system and other components of the immune system in the normal as well as injured and diseased CNS will conceivably aid in the development of novel therapeutic strategies to promote tissue repair and to prevent or reverse neurological deficits following CNS injury or disease.

Acknowledgments

This work was supported by grants from the Swedish Research Council (projects 11548, 20116, and 5174), the Region of Västra Götaland (RUN), Swedish Stroke Foundation, Torsten and Ragnar Söderberg foundation, Heart-Lung Foundation, the Swedish Society for Medicine, W. and M. Lundgren Foundation, John and Brit Wennerström's Foundation for Neurological Research and Foundation Edit Jacobson's Donation Fund, Trygg-Hansa, Hjärnfonden, ALF Göteborg, and Åhlén-stiftelsen.

References

Angelides, K. J., Smith, K. E., and Takeda, M. (1989). Assembly and exchange of intermediate filament proteins of neurons: Neurofilaments are dynamic structures. *J. Cell Biol.* **108,** 1495–1506.

Baba, A. (1998). Role of endothelin B receptor signals in reactive astrocytes. *Life Sci.* **62,** 1711–1715.

Brambilla, R., Bracchi-Ricard, V., Hu, W. H., Frydel, B., Bramwell, A., Karmally, S., Green, E. J., and Bethea, J. R. (2005). Inhibition of astroglial nuclear factor kappaB reduces inflammation and improves functional recovery after spinal cord injury. *J. Exp. Med.* **202,** 145–156.

Bush, T. G., Puvanachandra, N., Horner, C. H., Polito, A., Ostenfeld, T., Svendsen, C. N., Mucke, L., Johnson, M. H., and Sofroniew, M. V. (1999). Leukocyte infiltration, neuronal degeneration, and neurite outgrowth after ablation of scar-forming, reactive astrocytes in adult transgenic mice. *Neuron* **2,** 297–308.

Bushong, E. A., Martone, M. E., Jones, Y. Z., and Ellisman, M. H. (2002). Protoplasmic astrocytes in CA1 stratum radiatum occupy separate anatomical domains. *J. Neurosci.* **22,** 183–192.

Bushong, E. A., Martone, M. E., and Ellisman, M. H. (2004). Maturation of astrocyte morphology and the establishment of astrocyte domains during postnatal hippocampal development. *Int. J. Dev. Neurosci.* **22,** 73–86.

Cho, K. S., Yang, L., Lu, B., Feng Ma, H., Huang, X., Pekny, M., and Chen, D. F. (2005). Re-establishing the regenerative potential of central nervous system axons in postnatal mice. *J. Cell Sci.* **118,** 863–872.

Chou, Y. H., Khuon, S., Herrmann, H., and Goldman, R. D. (2003). Nestin promotes the phosphorylation-dependent disassembly of vimentin intermediate filaments during mitosis. *Mol. Biol. Cell* **14,** 1468–1478.

Dandoy-Dron, F., Guillo, F., Benboudjema, L., Deslys, J. P., Lasmezas, C., Dormont, D., Tovey, M. G., and Dron, M. (1998). Gene expression in scrapie. Cloning of a new scrapie-responsive gene and the identification of increased levels of seven other mRNA transcripts. *J. Biol. Chem.* **273,** 7691–7697.

Daveau, M., Benard, M., Scotte, M., Schouft, M.-T., Hiron, M., Francois, A., Salier, J.-P., and Fontaine, M. (2004). Expression of a functional C5a receptor in regenerating hepatocytes and its involvement in a proliferative signalling pathway in rat. *J. Immunol.* **173,** 3418–3424.

De Simoni, M. G., Storini, C., Barba, M., Catapano, L., Arabia, A. M., Rossi, E., and Bergamaschini, L. (2003). Neuroprotection by complement (C1) inhibitor in mouse transient brain ischemia. *J. Cereb. Blood Flow Metab.* **23,** 232–239.

Del Rio-Tsonis, K., Tsonis, P. A., Zarkadis, I. K., Tsagas, A. G., and Lambris, J. D. (1998). Expression of the third component of complement, C3, in regenerating limb blastema cells of urodeles. *J. Immunol.* **161,** 6819–6824.

Dietzschold, B., Schwaeble, W., Schafer, M. K., Hooper, D. C., Zehng, Y. M., Petry, F., Sheng, H., Fink, T., Loos, M., Koprowski, H., and Weihe, E. (1995). Expression of C1q, a subcomponent of the rat complement system, is dramatically enhanced in brains of rats with either Borna disease or experimental allergic encephalomyelitis. *J. Neurol. Sci.* **130,** 11–16.

Doetsch, F., Caille, I., Lim, D. A., Garcia-Verdugo, J. M., and Alvarez-Buylla, A. (1999). Subventricular zone astrocytes are neural stem cells in the adult mammalian brain. *Cell* **97,** 703–716.

Eddleston, M., and Mucke, L. (1993). Molecular profile of reactive astrocytes—implications for their role in neurologic disease. *Neuroscience* **54,** 15–36.

Eliasson, C., Sahlgren, C., Berthold, C. H., Stakeberg, J., Celis, J. E., Betsholtz, C., Eriksson, J. E., and Pekny, M. (1999). Intermediate filament protein partnership in astrocytes. *J. Biol. Chem.* **274,** 23996–24006.

Emsley, J. G., Arlotta, P., and Macklis, J. D. (2004). Star-cross'd neurons: Astroglial effects on neural repair in the adult mammalian CNS. *Trends Neurosci.* **27,** 238–240.

Faulkner, J. R., Herrmann, J. E., Woo, M. J., Tansey, K. E., Doan, N. B., and Sofroniew, M. V. (2004). Reactive astrocytes protect tissue and preserve function after spinal cord injury. *J. Neurosci.* **24,** 2143–2155.

Gasque, P., Julen, N., Ischenko, A. M., Picot, C., Mauger, C., Chauzy, C., Ripoche, J., and Fontaine, M. (1992). Expression of complement components of the alternative pathway by glioma cell lines. *J. Immunol.* **149,** 1381–1387.

Gasque, P., Ischenko, A., Legoedec, J., Mauger, C., Schouft, M. T., and Fontaine, M. (1993). Expression of the complement classical pathway by human glioma in culture. A model for complement expression by nerve cells. *J. Biol. Chem.* **268,** 25068–25074.

Gasque, P., Fontaine, M., and Morgan, B. P. (1995). Complement expression in human brain. Biosynthesis of terminal pathway components and regulators in human glial cells and cell lines. *J. Immunol.* **154,** 4726–4733.

Goldman, R. D., Chou, Y. H., Prahlad, V., and Yoon, M. (1999). Intermediate filaments: Dynamic processes regulating their assembly, motility, and interactions with other cytoskeletal systems. *FASEB J.* **13**(Suppl. 2), S261–S265.

Goldshmit, Y., Galea, M. P., Wise, G., Bartlett, P. F., and Turnley, A. M. (2004). Axonal regeneration and lack of astrocytic gliosis in EphA4-deficient mice. *J. Neurosci.* **24,** 10064–10073.

Heese, K., Hock, C., and Otten, U. (1998). Inflammatory signals induce neurotropin expression in human microglial cells. *J. Neurochem.* **70,** 699–707.

Hernandez, M. R., Agapova, O. A., Yang, P., Salvador-Silva, M., Ricard, C. S., and Aoi, S. (2002). Differential gene expression in astrocytes from human normal and glaucomatous optic nerve head analyzed by cDNA microarray. *Glia* **38,** 45–64.

Herrmann, H., and Aebi, U. (2000). Intermediate filaments and their associates: Multi-talented structural elements specifying cytoarchitecture and cytodynamics. *Curr. Opin. Cell Biol.* **12,** 79–90.

Herrmann, H., and Aebi, U. (2004). Intermediate filaments: Molecular structure, assembly mechanism, and integration into functionally distinct intracellular scaffolds. *Annu. Rev. Biochem.* **73,** 749–789.

Herrmann, H., Hesse, M., Reichenzeller, M., Aebi, U., and Magin, T. M. (2003). Functional complexity of intermediate filament cytoskeletons: From structure to assembly to gene ablation. *Int. Rev. Cytol.* **223,** 83–175.

Huang, J., Kim, L. J., Mealey, R., Marsh, H. C., Jr., Zhang, Y., Tenner, A. J., Connolly, E. S., Jr., and Pinsky, D. J. (1999). Neuronal protection in stroke by an sLex-glycosylated complement inhibitory protein. *Science* **285,** 595–599.

Imura, T., Kornblum, H. I., and Sofroniew, M. V. (2003). The predominant neural stem cell isolated from postnatal and adult forebrain but not early embryonic forebrain expresses GFAP. *J. Neurosci.* **23,** 2824–2832.

Ishikawa, N., Takemura, M., Koyama, Y., Shigenaga, Y., Okada, T., and Baba, A. (1997). Endothelins promote the activation of astrocytes in rat neostriatum through ET(B) receptors. *Eur. J. Neurosci.* **9,** 895–901.

Jing, R., Wilhelmsson, U., Goodwill, W., Li, L., Pan, Y., Pekny, M., and Skalli, O. (2007). Synemin is expressed in reactive astrocytes in neurotrauma and interacts differentially with vimentin and GFAP intermediate filament networks. *J. Cell Sci.* **12,** 1267–1277.

Kimura, Y., Madhavan, M., Call, M. K., Santiago, W., Tsonis, P. A., Lambris, J. D., and Del Rio-Tsonis, K. (2003). Expression of complement 3 and complement 5 in newt limb and lens regeneration. *J. Immunol.* **170,** 2331–2339.

Kinouchi, R., Takeda, M., Yang, L., Wilhelmsson, U., Lundkvist, A., Pekny, M., and Chen, D. F. (2003). Robust neural integration from retinal transplants in mice deficient in GFAP and vimentin. *Nat. Neurosci.* **6,** 863–868.

Koyama, Y., Takemura, M., Fujiki, K., Ishikawa, N., Shigenaga, Y., and Baba, A. (1999). BQ788, an endothelin ET(B) receptor antagonist, attenuates stab wound injury-induced reactive astrocytes in rat brain. *Glia* **26,** 268–271.

Laywell, E. D., Rakic, P., Kukekov, V. G., Holland, E. C., and Steindler, D. A. (2000). Identification of a multipotent astrocytic stem cell in the immature and adult mouse brain. *Proc. Natl. Acad. Sci. USA* **97,** 13883–13888.

Lepekhin, E. A., Eliasson, C., Berthold, C. H., Berezin, V., Bock, E., and Pekny, M. (2001). Intermediate filaments regulate astrocyte motility. *J. Neurochem.* **79,** 617–625.

Markiewski, M. M., Mastellos, D., Tudoran, R., DeAngelis, R. A., Strey, C. W., Franchini, S., Wetsel, R. A., Erdei, A., and Lambris, J. D. (2004). C3a and C3b activation products of the third component of complement (C3) are critical for normal liver recovery after toxic injury. *J. Immunol.* **173,** 747–754.

Mastellos, D., Papadimitriou, J. C., Franchini, S., Tsonis, P. A., and Lambris, J. D. (2001). A novel role of complement: Mice deficient in the fifth component of complement (C5) exhibit impaired liver regeneration. *J. Immunol.* **166,** 2479–2486.

McCall, M. A., Gregg, R. G., Behringer, R. R., Brenner, M., Delaney, C. L., Galbreath, E. J., Zhang, C. L., Pearce, R. A., Chiu, S. Y., and Messing, A. (1996). Targeted deletion in astrocyte intermediate filament (GFAP) alters neuronal physiology. *Proc. Natl. Acad. Sci. USA* **93,** 6361–6366.

Menet, V., Gimenez, Y. R. M., Sandillon, F., and Privat, A. (2000). GFAP null astrocytes are a favorable substrate for neuronal survival and neurite growth. *Glia* **31,** 267–272.

Menet, V., Gimenez y Ribotta, M., Chauvet, N., Drian, M. J., Lannoy, J., Colucci-Guyon, E., and Privat, A. (2001). Inactivation of the glial fibrillary acidic protein gene, but not that of vimentin, improves neuronal survival and neurite growth by modifying adhesion molecule expression. *J. Neurosci.* **21,** 6147–6158.

Menet, V., Prieto, M., Privat, A., and Gimenez y Ribotta, M. (2003). Axonal plasticity and functional recovery after spinal cord injury in mice deficient in both glial fibrillary acidic protein and vimentin genes. *Proc. Natl. Acad. Sci. USA* **100,** 8999–9004.

Miller, R. K., Vikstrom, K., and Goldman, R. D. (1991). Keratin incorporation into intermediate filament networks is a rapid process. *J. Cell Biol.* **113,** 843–855.

Morais da Silva, S., Gates, P. B., and Brockers, J. P. (2002). The newt ortholog of CD59 is implicated in proximodistal identity during amphibian limb regeneration. *Dev. Cell* **3,** 547–555.

Morshead, C. M., Garcia, A. D., Sofroniew, M. V., and van Der Kooy, D. (2003). The ablation of glial fibrillary acidic protein-positive cells from the adult central nervous system results in the loss of forebrain neural stem cells but not retinal stem cells. *Eur. J. Neurosci.* **18,** 76–84.

Mukherjee, P., and Passinetti, G. M. (2001). Complement anaphylatoxin C5a neuroprotects through mitogen-activated protein kinase-dependent inhibition of caspase 3. *J. Neurochem.* **77,** 43–49.

Nakamura, Y., Takeda, M., Angelides, K. J., Tada, K., Hariguchi, S., and Nishimura, T. (1991). Assembly, disassembly, and exchange of glial fibrillary acidic protein. *Glia* **4,** 101–110.

O'Barr, S. A., Caguioa, J., Gruol, D., Perkins, G., Ember, J. A., Hugli, T., and Cooper, N. R. (2001). Neuronal expression of a functional receptor for the C5a complement activation fragment. *J. Immunol.* **166,** 4154–4162.

Okabe, M., Ikawa, M., Kominami, K., Nakanishi, T., and Nishimune, Y. (1997). 'Green mice' as a source of ubiquitous green cells. *FEBS Lett.* **407,** 313–319.

Pekny, M., Leveen, P., Pekna, M., Eliasson, C., Berthold, C. H., Westermark, B., and Betsholtz, C. (1995). Mice lacking glial fibrillary acidic protein display astrocytes devoid of intermediate filaments but develop and reproduce normally. *EMBO J.* **14,** 1590–1598.

Pekny, M., Eliasson, C., Chien, C. L., Kindblom, L. G., Liem, R., Hamberger, A., and Betsholtz, C. (1998). GFAP-deficient astrocytes are capable of stellation *in vitro* when cocultured with neurons and exhibit a reduced amount of intermediate filaments and an increased cell saturation density. *Exp. Cell Res.* **239,** 332–343.

Pekny, M., Johansson, C. B., Eliasson, C., Stakeberg, J., Wallen, A., Perlmann, T., Lendahl, U., Betsholtz, C., Berthold, C. H., and Frisen, J. (1999). Abnormal reaction to central nervous system injury in mice lacking glial fibrillary acidic protein and vimentin. *J. Cell Biol.* **145,** 503–514.

Pekny, M., Pekna, M., Wilhelmsson, U., and Chen, D. F. (2004). Response to Quinlan and Nilsson: Astroglia sitting at the controls? *Trends Neurosci.* **27,** 243–244.

Peters, C. M., Rogers, S. D., Pomonis, J. D., Egnazyck, G. F., Keyser, C. P., Schmidt, J. A., Ghilardi, J. R., Maggio, J. E., and Mantyh, P. W. (2003). Endothelin receptor expression in the normal and injured spinal cord: Potential involvement in injury-induced ischemia and gliosis. *Exp. Neurol.* **180,** 1–13.

Pixley, S. K., Kobayashi, Y., and de Vellis, J. (1984). A monoclonal antibody against vimentin: Characterization. *Brain Res.* **317,** 185–199.

Quinlan, R., and Nilsson, M. (2004). Reloading the retina by modifying the glial matrix. *Trends Neurosci.* **27,** 241–242.

Rahpeymai, Y., Hietala, M. A., Wilhelmsson, U., Fotheringham, A. P., Davies, I., Nilsson, A.-K., Zwirner, J., Wetsel, R. A., Gerard, C., Pekny, M., and Pekna, M. (2006). Complement: A novel factor in basal and ischemia-induced neurogenesis. *EMBO J.* **25,** 1364–1374.

Rahpeymai Bogeståł, Y., Barnum, S. R., Smith, P. L. P., Mattisson, V., Pekny, M., and Pekna, M. (2007). Signaling through C5aR is not involved in basal neurogenesis.. *J. Neurosci. Res.* (in press).

Renner, W., Franke, W. W., Schmid, E., Geisler, N., Weber, K., and Mandelkow, E. (1981). Reconstitution of intermediate-sized filaments from denatured monomeric vimentin. *J. Mol. Biol.* **149,** 285–306.

Rueger, D. C., Huston, J. S., Dahl, D., and Bignami, A. (1979). Formation of 100 A filaments from purified glial fibrillary acidic protein *in vitro*. *J. Mol. Biol.* **135,** 53–68.

Schafer, M. K., Schwaeble, W. J., Post, C., Salvati, P., Calabresi, M., Sim, R. B., Petry, F., Loos, M., and Weihe, E. (2000). Complement C1q is dramatically up-regulated in brain microglia in response to transient global cerebral ischemia. *J. Immunol.* **164,** 5446–5452.

Shaw, G., Osborn, M., and Weber, K. (1981). An immunofluorescence microscopical study of the neurofilament triplet proteins, vimentin and glial fibrillary acidic protein within the adult rat brain. *Eur. J. Cell Biol.* **26,** 68–82.

Song, H., Stevens, C. F., and Gage, F. H. (2002). Astroglia induce neurogenesis from adult neural stem cells. *Nature* **417,** 39–44.

Stahel, P. F., Frei, K., Fontana, A., Eugster, H. P., Ault, B. H., and Barnum, S. R. (1997a). Evidence for intrathecal synthesis of alternative pathway complement activation proteins in experimental meningitis. *Am. J. Pathol.* **151,** 897–904.

Stahel, P. F., Nadal, D., Pfister, H. W., Paradisis, P. M., and Barnum, S. R. (1997b). Complement C3 and factor B cerebrospinal fluid concentrations in bacterial and aseptic meningitis. *Lancet* **349,** 1886–1887.

Strey, C. W., Markiewski, M., Mastellos, D., Tudoran, R., Spruce, L. A., Greenbaum, L. E., and Lambris, J. D. (2003). The proinflammatory mediators C3a and C5a are essential for liver regeneration. *J. Exp. Med.* **198,** 913–923.

Thomas, A., Gasque, P., Vaudry, D., Gonzalez, B., and Fontaine, M. (2000). Expression of a complete and functional complement system by human neuronal cells *in vitro*. *Int. Immunol.* **12,** 1015–1023.

Turner, D. A., Buhl, E. H., Hailer, N. P., and Nitsch, R. (1998). Morphological features of the entorhinal-hippocampal connection. *Prog. Neurobiol.* **55,** 537–562.

van Beek, J., Bernaudin, M., Petit, E., Gasque, P., Nouvelot, A., MacKenzie, E. T., and Fontaine, M. (2000a). Expression of receptors for complement anaphylatoxins C3a and C5a following permanent focal cerebral ischemia in the mouse. *Exp. Neurol.* **161,** 373–382.

van Beek, J., Chan, P., Bernaudin, M., Petit, E., MacKenzie, E. T., and Fontaine, M. (2000b). Glial responses, clusterin, and complement in permanent focal cerebral ischemia in the mouse. *Glia* **31,** 39–50.

van Beek, J., Nicole, O., Ali, C., Ischenko, A., MacKenzie, E. T., Buisson, A., and Fontaine, M. (2001). Complement anaphylatoxin C3a is selectively protective against NMDA-induced neuronal cell death. *Neuroreport* **12,** 289–293.

Vasthare, U. S., Barone, F. C., Sarau, H. M., Rosenwasser, R. H., DiMartino, M., Young, W. F., and Tuma, R. F. (1998). Complement depletion improves neurological function in cerebral ischemia. *Brain Res. Bull.* **45,** 413–419.

Vikstrom, K. L., Lim, S. S., Goldman, R. D., and Borisy, G. G. (1992). Steady state dynamics of intermediate filament networks. *J. Cell Biol.* **118,** 121–129.

Wang, X., Messing, A., and David, S. (1997). Axonal and nonneuronal cell responses to spinal cord injury in mice lacking glial fibrillary acidic protein. *Exp. Neurol.* **148,** 568–576.

Wiegers, W., Honer, B., and Traub, P. (1991). Microinjection of intermediate filament proteins into living cells with and without preexisting intermediate filament network. *Cell. Biol. Int. Rep.* **15,** 287–296.

Wilhelmsson, U., Li, L., Pekna, M., Berthold, C. H., Blom, S., Eliasson, C., Renner, O., Bushong, E., Ellisman, M., Morgan, T. E., and Pekny, M. (2004). Absence of glial fibrillary acidic protein and vimentin prevents hypertrophy of astrocytic processes and improves post-traumatic regeneration. *J. Neurosci.* **24,** 5016–5021.

Wilhelmsson, U., Bushong, E. A., Price, D. L., Smarr, B. L., Phung, V., Terada, M., Ellisman, M. H., and Pekny, M. (2006). Redefining the concept of reactive astrocytes as cells that remain within their unique domains upon reaction to injury. *Proc. Natl. Acad. Sci. USA* **103,** 17513–17518.

Wyss-Coray, T., Yan, F., Lin, A. H., Lambris, J. D., Alexander, J. J., Quigg, R. J., and Masliah, E. (2002). Prominent neurodegeneration and increased plaque formation in complement-inhibited Alzheimer's mice. *Proc. Natl. Acad. Sci. USA* **99,** 10837–10842.

Xu, K., Malouf, A. T., Messing, A., and Silver, J. (1999). Glial fibrillary acidic protein is necessary for mature astrocytes to react to beta-amyloid. *Glia* **25,** 390–403.

Yoon, M., Moir, R. D., Prahlad, V., and Goldman, R. D. (1998). Motile properties of vimentin intermediate filament networks in living cells. *J. Cell Biol.* **143,** 147–157.

Zhang, Z. G., Chopp, M., Tang, W. X., Jiang, N., and Zhang, R. L. (1995). Post-ischemic treatment (2–4 h) with anti-CD11b and anti-CD18 monoclonal antibodies are neuroprotective after transient (2 h) cerebral ischemia in the rat. *Brain Res.* **698,** 79–85.

NEW INSIGHTS INTO THE ROLES OF METALLOPROTEINASES IN NEURODEGENERATION AND NEUROPROTECTION

A. J. Turner* and N. N. Nalivaeva*,†

*Institute of Molecular and Cellular Biology, Faculty of Biological Sciences, University of Leeds Leeds LS2 9JT, United Kingdom

†I. M. Sechenov Institute of Evolutionary Physiology and Biochemistry, Russian Academy of Sciences, 194223 St. Petersburg, Russia

Proteolytic enzymes constitute around 2% of the human genome and are involved in many stages of cell development from fertilization to death (apoptosis). The identification of many novel proteases from genome-sequencing programs has suggested them as potential new therapeutic targets. In addition, several well-characterized metallopeptidases were recently shown to possess new biological roles in neuroinflammation and neurodegeneration. As a result of these studies, metabolism of the neurotoxic and inflammatory amyloid peptide (Aβ) is considered as a physiologically relevant process with several metallopeptidases being suggested for the role of amyloid-degrading enzymes. These include the neprilysin (NEP) family of metalloproteinases (including its homologue endothelin-converting enzyme), insulin-degrading enzyme, angiotensin-converting enzyme, plasmin, and, possibly, some other enzymes. NEP also has a role in metabolism of sensory and inflammatory neuropeptides such as tachykinins and neurokinins. The existence of natural enzymatic mechanisms for removal of amyloid peptides has extended the therapeutic avenues in Alzheimer's disease (AD) and neurodegeneration. The proteolytic events underlying AD are highly compartmentalized in the cell and formation of amyloid peptide from its precursor molecule APP (amyloid precursor protein) takes place both within intracellular compartments and in the plasma membrane, especially in lipid raft domains. Degradation of amyloid peptide by metallopeptidases can also be both intra- and extracellular

INTERNATIONAL REVIEW OF
NEUROBIOLOGY, VOL. 82
DOI: 10.1016/S0074-7742(07)82006-X

depending on the activity of membrane-bound enzymes and their soluble partners. Soluble forms of proteases can be secreted or released from the cell surface through the activity of "sheddases"—another group of proteolytic enzymes involved in key cellular regulatory functions. The activity of proteases involved in amyloid metabolism depends on numerous factors (e.g., genetic, environmental, age), and some conditions (e.g., hypoxia and ischemia) shift the balance of amyloid metabolism toward accumulation of higher concentrations of $A\beta$. In this regard, regulation of the activity of amyloid-degrading enzymes should be considered as a viable strategy in neuroprotection.

I. Introduction

Proteolysis represents one of the key processes underlying biological events from fertilization through development to death. In the human genome there are more than 500 proteases and homologues (almost 2% of the genome), many of whose physiological roles are yet to be identified and which may provide potential therapeutic targets in the treatment of human disease. At present there are 70 known human hereditary diseases caused by mutations in protease-coding genes (http://www.uniovi.es/degradome/). Abnormal functioning of these genes are implicated in such pathologies as inflammatory diseases, cancer, cardiovascular diseases, and neurodegeneration. Understanding of the role of these enzymes and their evolutionary conservation is important for design of appropriate drugs and therapeutic strategies. Other eukaryotic species whose genomes have been deciphered to date also possess a very high number of protease and protease-like genes which in the case of *Drosophila melanogaster* and mouse are even higher than in human (Rawlings *et al.*, 2006). The number of species in which proteases have now been identified is approaching 3000 (http://merops.sanger.ac.uk/). Cysteine-, serine-, and metallopeptidases represent the major classes of peptidases; among these three classes, the metallopeptidase class is more consistently represented with the number of genes varying from 183 in *Drosophila* to 198 in mice (in human, 186). Two other classes have much higher diversity with the range in the number of genes for cysteine proteases from 76 in *Drosophila* to 153 in mice (Puente *et al.*, 2003). For the serine peptidases this range is even more pronounced: from 100 in *Caenorhabditis elegans* to 309 in *Drosophila.* There are a smaller number of aspartic proteinases, although some are of key therapeutic importance. This more conservative representation of metallopeptidases in the genomes of various species may reflect roles of these enzymes in more generic cellular functions and their unique catalytic properties which required water and metal ions for activation of the proteolytic (hydrolytic) process, which were carefully preserved in the course of evolution.

The metallopeptidases can be divided into 12 clans according to the type and number of metal ions required for their activity (Rawlings and Barrett, 2004). This chapter will focus on two distinct zinc metallopeptidase families belonging to one of these clans, namely MA, whose representatives require zinc for their activity and contain one or two histidine residues in the zinc binding motif. The first of these metallopeptidase families is the M13 family represented by neprilysin [or neutral endopeptidase (NEP)] and the second is the M2 family represented by angiotensin-converting enzyme [or peptidyl dipeptidase (ACE)]. Members of each of these families have served as important drug targets, particularly in cardiovascular disease and, more recently, have provided insight into mechanisms involved in neurodegeneration and neuroinflammation, especially from the point of view of processing of the amyloid precursor protein (APP) and its products in Alzheimer's disease (AD).

The discovery of ACE2 in our laboratory (Tipnis *et al.*, 2000), as a result of genomics approaches to the identification of zinc metalloproteinases, together with its critical role in cardiac and lung development and function (Crackower *et al.*, 2002; Donoghue *et al.*, 2000) and as the cell surface receptor for the severe-acute respiratory syndrome (SARS) *Coronavirus* (Li *et al.*, 2003), emphasizes the validity of this strategy for identification of novel therapeutic targets. ACE2 was also suggested to play an important role in the brain renin–angiotensin system being widely expressed in various brain areas but restricted mostly to neuronal cells (Doobay *et al.*, 2007).

II. The NEP Family

NEP, or neprilysin as it is now known, is a cell surface membrane-bound glycoprotein and zinc peptidase also known as CD10 or common acute lymphoblastic leukemia antigen (CALLA). It was originally identified as a major antigen of renal membranes over 30 years ago and, at that time, was implicated in the metabolism of insulin. However, as found later, NEP degrades only the insulin B chain *in vitro* and not the intact insulin dimer, which suggests that NEP does not have any physiological role in degradation of insulin. Moreover, another zinc metallopeptidase, insulin-degrading enzyme (IDE; insulysin), was subsequently discovered which appeared to fulfill the physiologically relevant function of insulin degradation. Whereas the highest concentrations of NEP are in the renal microvillar membrane, the first clues to its physiological roles came from studies on the metabolism of neuropeptides [especially the enkephalins and substance P (SP)] in the central nervous system where NEP is several orders of magnitude less abundant (Malfroy *et al.*, 1978; Matsas *et al.*, 1984; Relton *et al.*, 1983). It is now accepted that NEP functions to turn off neuropeptide signals at the synapse in an analogous fashion to the hydrolysis of acetylcholine by acetylcholinesterase at cholinergic synapses (see Turner and Tanzawa, 1997 for review). Evidence *in vitro* has clearly demonstrated that synaptic membranes efficiently degrade enkephalins and SP and that NEP was the primary enzyme responsible for these events (Matsas *et al.*, 1983). Combined with the data obtained *in vivo* in rodents using potent and selective NEP inhibitors, such as phosphoramidon and thiorphan (Roques *et al.*, 1980), key roles were established for NEP in the central nervous system. Subsequently, renal NEP was shown to be the principal enzyme inactivating the vasodilator, atrial natriuretic peptide (Kenny and Stephenson, 1988), which has led to much investment in the development of NEP inhibitors, either alone or in combination with ACE inhibitors (vasopeptidase inhibitors), as drugs in the treatment of hypertension, congestive heart failure, and renal disease (Bralet and Schwartz, 2001), although concerns over their safety have been raised (Quaschning, 2005).

NEP also plays important roles in other peripheral tissues such as in chemoreception and in potentiation of the response of the carotid body to hypoxia via degradation of its preferred substrate, SP (Kumar *et al.*, 1990, 2000). NEP was also detected in the liver, lungs, muscles, fat deposits, bones, the vertebrae, articular cartilages, and synoviae (Sales *et al.*, 1991). In the bones, NEP plays a role in regulation of osteoblast and osteoclast metabolism mediated by both hormones and local bone peptide factors (Howell *et al.*, 1993; Ruchon *et al.*, 2000). In the skeletal muscles, NEP participates in regeneration of muscle fibers and there are data on its role associated with hereditary muscle disorders (Broccolini *et al.*, 2006).

The cloning of NEP also revealed its identity with the CALLA or CD10 and has implicated NEP in cancer mechanisms (Letarte *et al.*, 1988; Tran-Paterson *et al.*, 1989). For example, in human prostate cancer, NEP is dramatically downregulated (Papandreou *et al.*, 1998) allowing mitogenic peptides such as bombesin and endothelin to drive androgen-independent cell division in the prostate. The survey of the expression of NEP and the NEP homologue, endothelin-converting enzyme 1 (ECE-1), in a range of prostate cancers demonstrated that there is a certain balance between the levels of expression of NEP and ECE-1 which determines the level of malignancy of the cells and that upregulation of ECE-1 expression in metastatic cells may be indicative of its role in metastatic progression (Usmani *et al.*, 2002). Later it was shown that NEP and ECE-1 act as mediators of prostate cancer invasion via a stromal–epithelial interaction (Dawson *et al.*, 2004). This and related observations have led to suggestions that the dysregulation of the balance of NEP and ECE can lead to the disease and selective reexpression of NEP in prostate cells may provide a novel approach to the treatment of prostate cancer. This is an excellent example where two homologous peptidases counterbalance each other's actions in physiology and pathology.

NEP cleaves a wide range of substrates including SP and other tachykinin peptides, in particular neurokinin B (Fig. 1), which makes it an important player

Proinflammatory peptides cleaved by metalloproteinases

- Substance P (NEP, ACE)
- Bradykinin (NEP, ACE, ECE)
- CGRP (NEP)
- Neurokinins (NEP, ACE)
- Endothelin (NEP)
- Angiotensin II (ACE2)
- Interleukin-1β (NEP)
- Insulin (IDE)
- Insulin B chain (NEP)
- Aβ (NEP, ECE, ACE, IDE)

FIG. 1. List of proinflammatory peptides cleaved by metalloproteinases.

in the arena of inflammation (Hooper and Turner, 1985). SP is released from sensory nerves inducing neurogenic inflammation and NEP, via SP degradation, limits its effects and those of other proinflammatory peptides. There are several studies reporting the role of NEP in peripheral inflammation, for example in the skin (Scholzen and Luger, 2004; Scholzen *et al.*, 2001). Moreover, NEP but not ACE was shown to be the most important for CGRP degradation in human skin, and NEP inhibitors facilitated neurogenic inflammation in the skin (Kramer *et al.*, 2005). NEP levels were also found to be significantly decreased in the plasma and monocytes of patients with juvenile idiopathic arthritis (Simonini *et al.*, 2005). In the mouse model of intestinal inflammation caused by nematode infection NEP was demonstrated to downregulate the early onset of inflammation (Barbara *et al.*, 2003). Inhibition of NEP was found to exacerbate both experimental pancreatitis and the associated lung injury (Day *et al.*, 2005), and pretreatment with recombinant human NEP ameliorated this injury in NEP$^{-/-}$ transgenic mice (Lightner *et al.*, 2002). In this connection, upregulation of NEP is considered as a potential therapeutic approach for pancreatitis-associated lung injury, which can also be true in the case of other inflammatory diseases.

Although the role of NEP in peripheral inflammation is well characterized, there are much less data on the role of NEP in inflammatory processes in the brain. However, the data on the localization and properties of NEP in the nervous system of insects provide evidence for an evolutionarily conserved role for NEP in the inactivation of tachykinin-related peptides in the brain (Isaac and Nassel, 2003; Isaac *et al.*, 2002).

It was demonstrated that NEP might play an important role in the pathogenesis of AD due to its capacity to cleave the neurotoxic and inflammatory amyloid-β (Aβ) peptide (Iwata *et al.*, 2000) that is a primary trigger for the

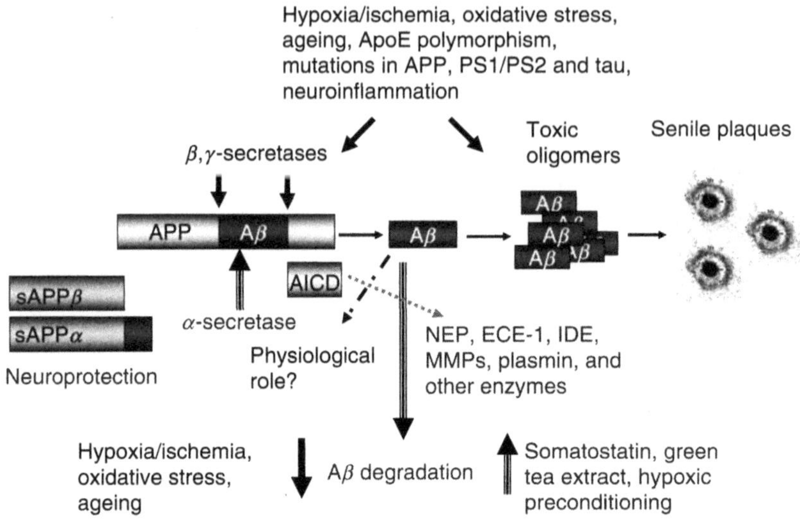

FIG. 2. Amyloidogenic processing of APP and formation of amyloid (Aβ) peptide, fibrils, and plaques. Main avenues in neurodegeneration and neuroprotection. APP, amyloid precursor protein; sAPPα, soluble product of α-secretase cleavage; sAPPβ, soluble product of β-secretase cleavage; AICD, C-terminal fragment of APP, product of γ-secretase cleavage. (See Color Insert.)

development of this disease (Hardy and Higgins, 1992). It is now well documented that Aβ is formed from a large precursor molecule called APP via two consecutive cleavages (Fig. 2). The first of these cleavages occurs in the extracellular or lumenal domain and is mediated by a membrane-bound aspartic protease termed β-secretase (BACE) (Vassar and Citron, 2000). It releases a large soluble fragment of sAPPβ and the residual membrane-bound fragment. The latter is cleaved by a γ-secretase protease complex, at residues 40–42 (termed γ-site) or at residues 48–52 (termed ε-site) within the transmembrane domain. The presenilin-dependent γ-secretase and γ-site proteolytic activities are dependent on a multimeric complex of at least four different membrane proteins, including presenilin-1 (PS1) or presenilin-2 (PS2), nicastrin, Aph-1, and Pen-2 (Francis *et al.*, 2002; Yu *et al.*, 2000). In these complexes, the presenilins have been proposed as a novel type of transmembrane aspartic protease bearing the catalytic core of the γ-secretase (Wolfe *et al.*, 1999). Whereas the cleavage at the γ-site generates Aβ, the subsequent cleavage at the ε-site generates a cytosolic fragment referred to as ICD (Passer *et al.*, 2000) or AICD (*βA*PP *I*ntra*C*ellular *D*omain). The exact role of AICD remains unclear but it has been suggested to act as a functional transcriptional regulator (Cao and Sudhof, 2001) in combination with the regulatory proteins Fe65 and the chromatin-associated histone acetyltransferase, Tip60. There were reports that AICD and nicastrin regulate expression of NEP but this is controversial and still has to be proved experimentally in animal models (Hass and Yankner, 2005; Hebert *et al.*, 2006; Pardossi-Piquard *et al.*, 2005, 2006). An additional component of the γ-secretase complex, termed TMP21, which is a member of the p24 cargo protein family, appears to differentially regulate γ-secretase cleavage without affecting ε-secretase activity (Chen *et al.*, 2006). The proteolytic events involved in processing of APP are highly compartmentalized in the cell taking place both within intracellular compartments and in the plasma membrane. The rate limiting reaction of APP cleavage by β-secretase (BACE) was shown to take place especially in lipid raft domains enriched with cholesterol and glycosphingolipids and targeting BACE to lipid rafts increased production of Aβ (Cordy *et al.*, 2003). Depletion of cell cholesterol by lovastatin resulted in a decrease in both of sAPPβ and Aβ levels in the cell culture model. This explains the positive effect of statins on the development of AD pathology observed in patients with statin treatment (Sparks *et al.*, 2006).

Under normal conditions, Aβ occurs as a soluble fragment, the concentration of which is normally tightly controlled below the threshold for its self-aggregation into β sheet fibrils (Burdick *et al.*, 1992). Until recently, production of Aβ in the brain and other tissues was thought to be an irreversible process, leading in the case of their disruption, to amyloidogenic diseases. However, in the last few years, neprilysin and several other proteases (IDE, ECE-1 and ECE-2, plasmin) have been found to be capable of degrading Aβ *in vitro* and *in vivo* (for review see Carson and Turner, 2002; Turner *et al.*, 2004). The sites of Aβ cleaved by known proteinases are shown in Fig. 3. Pathological downregulation of these enzymes and, in particular of NEP, could predispose to accumulation of Aβ and the development of AD (Apelt *et al.*, 2003; Nalivaeva *et al.*, 2004). In particular, in mice deficient in NEP or ECE-1, amyloid deposits are seen to deposit at significant levels in the brain (Eckman *et al.*, 2003; Hersh *et al.*, 2002). Furthermore, injection of amyloid-β peptide into the brains of rodents significantly enhances the concentrations of NEP mRNA and protein, suggesting the operation of a regulatory feedback mechanism to protect neurons from toxic damage (Mohajeri *et al.*, 2002). NEP levels appear to be reduced in high-plaque-bearing areas of human brain in AD and in cerebral amyloid angiopathy (Carpentier *et al.*, 2002; Yasojima *et al.*, 2001) but no association has been detected to date between polymorphisms in the NEP gene and AD (Lilius *et al.*, 2003). Clinical data suggest that ischemia and stroke predispose to development of AD (Snowdon *et al.*, 1997), and it was shown that hypoxia and ischemia lead to a decrease of NEP and ECE expression (Fisk *et al.*, 2006; Nalivaeva *et al.*, 2004). Thus, a possible therapeutic approach for treatment of AD might be a chronic upregulation of these proteinases (Fig. 4), either pharmacologically or through a gene therapy approach (Marr *et al.*, 2003; Saito *et al.*, 2005; Turner *et al.*, 2004).

Maintenance of cellular concentrations of NEP, which is rather widely distributed in the body, is critical to peptide homeostasis and its up- or downregulation can lead to a range of pathological conditions including those of

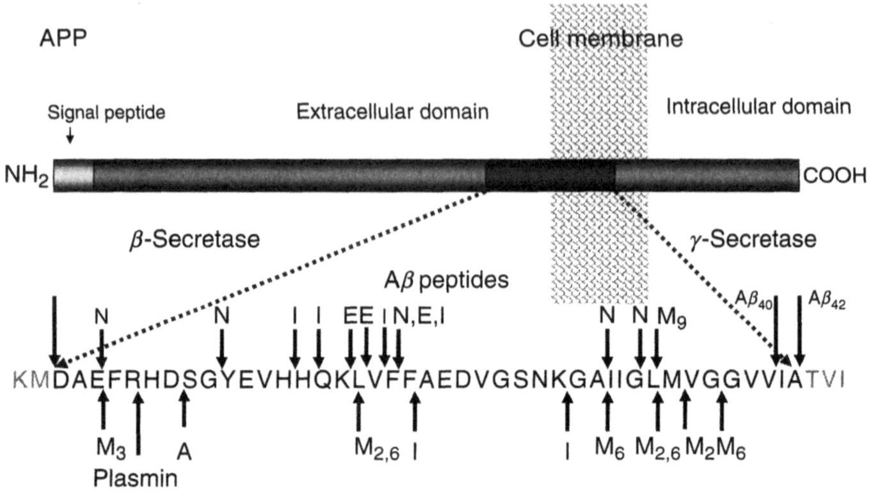

FIG. 3. Cleavage sites of Aβ-degrading enzymes. A, angiotensin-converting enzyme; E, ECE-1; I, IDE; N, NEP; M, matrix metalloproteinases (MMPs: M3, MMP-3; M2 and M6, MMP-2 and MMP-6; M9, MMP-9).

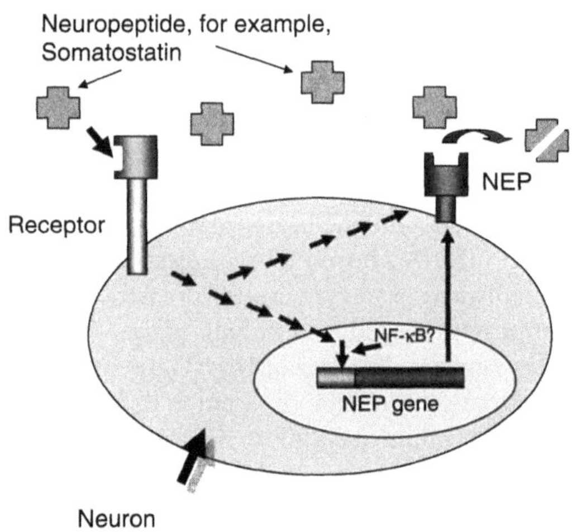

FIG. 4. Scheme of upregulation of NEP gene and its possible involvement in neuroprotection [adapted from Saito *et al.* (2003)].

cardiovascular, neurodegenerative, and tumorigenic origins. NEP was, for quite some years, a lone mammalian zinc peptidase apparently mechanistically similar to the bacterial enzyme, thermolysin, both enzymes being potently inhibited by phosphoramidon. The structural solution of the catalytic domain of human NEP complexed with phosphoramidon (Oefner *et al.*, 2000) reveals both similarities and differences with thermolysin, and the restricted access to the active site highlights why the enzyme acts exclusively as an oligopeptidase rather than a proteinase, unlike thermolysin.

A homologue of NEP, namely neprilysin 2 (NEP2), was discovered and in the brain shown to be restricted mainly to developing and differentiated fields of the CNS. Unlike NEP and ECE-1, which are broadly expressed in the CNS and periphery, NEP2 was found to be almost exclusively expressed only in selected populations of neurons and in the spinal cord. The only peripheral areas where expression of NEP2 was detected were the pituitary and choroid plexuses. NEP2 was also found capable of degrading Aβ and its distinct localization from NEP suggests that, together with ECE-1, it may be better poised to catabolize Aβ as it is more abundantly expressed in the areas relevant to AD pathology (Facchinetti *et al.*, 2003; Thomas *et al.*, 2005).

The human genome is now known to contain at least seven NEP-related enzymes (summarized in Turner *et al.*, 2001), of which the best characterized is ECE-1, which catalyzes the final step in the biosynthesis of the potent vasoconstrictor peptide, endothelin-1 (ET-1; Matsumura *et al.*, 1990; Xu *et al.*, 1994). Several of the NEP-like enzymes are, as yet, orphan peptidases with no recognized peptide substrates. Novel strategies are urgently needed to allow the identification of physiologically relevant substrates for such newly identified proteases.

III. The NEP Homologue ECE-1

ECE-1 was first purified from rat lung (Takahashi *et al.*, 1993) but then was also found in a variety of tissues. It is most abundant in endothelial cells but is also expressed by exocrine cells, smooth muscle cells, neurons, and glia in the brain (Barnes and Turner, 1999; Barnes *et al.*, 1997; Takahashi *et al.*, 1995). To date, four isoforms of human ECE-1 differing only in a part of their N-terminal cytoplasmic region but which cleave big ETs with similar efficiencies have been characterized: named ECE-1a, ECE-1b, ECE-1c, and ECE-1d (Schweizer *et al.*, 1997; Valdenaire *et al.*, 1999). Although the relative levels of the isoform mRNA species vary between human tissues, ECE-1c mRNA is generally the predominant isoform message. There are distinct subcellular localizations for the four isoforms: whereas ECE-1a, ECE-1c, and ECE-1d proteins are localized mainly at the cell surface, ECE-1b was found to be intracellular and showed significant colocalization with a marker protein for the *trans*-Golgi network (Schweizer *et al.*, 1997). There are no significant differences in the catalytic properties between them, so it has been suggested that intracellular ECE-1 localized in Golgi and vesicles might be involved in processing of big ET whereas cell surface ECE-1 may metabolize other regulatory peptides (Turner *et al.*, 1998).

ECE-1 and its product ET-1 have been shown to be involved in such inflammatory conditions as asthma (Zhang *et al.*, 2004), chronic rhinitis with its expression in the nasal epithelium and mucosa being much higher in the case of rhinitis than in controls (Furukawa *et al.*, 1996), and idiopathic pulmonary fibrosis (Saleh *et al.*, 1997). In the latter case, an increased release of ET-1 was accompanied by increased levels of ECE-1 mRNA. Using normal bronchial epithelial cell culture, these authors have demonstrated that proinflammatory cytokines (IL-1α, IL-1β) induced a significant increase in ET-1 release and mRNA expression, while TNF-α stimulated expression of ECE-1 mRNA. In the model of acute gastric infection caused by administration of *Helicobacter pylori* lipopolysaccharide, it was also shown that the levels of ET-1 and expression of ECE-1 were significantly increased in the gastric mucosa of infected cells on day 4 but then significantly reduced (down to 60%) by day 10 following reduced gastric inflammation (Slomiany *et al.*, 2000). Developing this work further, the authors have discovered that ET-1 has also an effect on leptin production in the gastric mucosa as a consequence of ET(A) receptor activation (Slomiany and Slomiany, 2005).

Together with ET-1, ECE-1 is abundantly present in human arteries and is involved in chronic inflammation in human atherosclerosis. Upregulation of the ECE-1–ET-1 system was shown to be closely linked to the presence of chronic inflammation at the very early stages of plaque evolution and, thus, has been suggested as a target in atherosclerosis therapy (Ihling *et al.*, 2001, 2004). Plasma levels of ECE-1 were also shown to reflect the severity of ischemic complications after subarachnoid hemorrhage. The higher levels of plasma ECE-1 resulted in reduced big ET-1 and increased ET-1/big ET-1 ratio in patients who experienced symptomatic delayed cerebral ischemia, compared with other patients (Juvela, 2002). However, the levels of ECE-1 expression in rat brain cortex hemispheres and hippocampus was found to be decreased after 15-min global ischemia and restored to control levels after reperfusion (Nalivaeva *et al.*, 2004), which might reflect an adaptive reaction of the brain to increased blood supply to the affected areas.

Although ECE-1 has been regarded as a highly specific endopeptidase, it was demonstrated to be able to hydrolyze, apart from big ET-1, a number of other biologically active peptides, such as bradykinin, SP, neurotensin, angiotensin I, and insulin B chain; it is not yet clear whether any of these, or other peptides, are physiological substrates of ECE-1 (Hoang and Turner, 1997; Johnson *et al.*, 1999). It was demonstrated that ECE-1 can also degrade amyloid peptide Aβ, which made it an important player in the arena of AD (Eckman *et al.*, 2001). Overexpression of ECE-1 in Chinese hamster ovary cells, lacking endogenous ECE activity, was found to reduce extracellular Aβ concentration by up to 90% and this effect was completely abolished by treatment with a metalloproteinase inhibitor phosphoramidon. Recombinant soluble ECE-1 was shown to hydrolyze synthetic Aβ_{40} and Aβ_{42} *in vitro* at multiple sites. Comparing ECE heterozygous and knockout mice, Eckman *et al.* (2003) showed that the concentration of Aβ peptides in the brain of these animals was elevated in comparison with control littermates in a gene-dependent manner.

IV. The ACE Family

ACE was originally identified 50 years ago as a "hypertensin-converting enzyme" (Skeggs *et al.*, 1956) and its primary substrate was identified as angiotensin I, which it converts into the vasoconstrictor angiotensin II. In parallel, it inactivates the vasodilator and inflammatory peptide bradykinin. Hence, inhibition of ACE has a powerful effect in reduction of blood pressure and the enzyme has therefore been a major cardiovascular target for many years. The catalytic activity of ACE is primarily as a "peptidyl dipeptidase," removing dipeptides from the C-terminus of a susceptible peptide substrate. In the hydrolysis of some peptides (e.g., SP, luliberin), ACE can act as an endopeptidase, although with much lower catalytic efficiency (Turner and Hooper, 2002). Mammalian ACE exists as two distinct forms, arising from the use of alternative promoters. The simplest form is germinal or testicular ACE, which is essential for male fertility and which carries a single zinc-binding and catalytic domain. Elsewhere in the body, the somatic form of ACE is duplicated and carries two active sites. The analysis of ACE X-ray structures has revealed that ACE most closely resembles a neurotensin-degrading zinc endopeptidase known as neurolysin rather than NEP, or carboxypeptidase A, on whose structure the design of ACE inhibitors was originally based (Hooper and Turner, 2003).

While seeking novel expressed sequence tags encoding zinc metallopeptidases, we identified and cloned the first human homologue of ACE (ACEH), which is now more commonly referred to as ACE2 (Tipnis *et al.*, 2000). ACE2 is also a type I integral membrane peptidase showing 40% identity and 61% similarity with ACE and conserving the critical active site residues. It contains a single catalytic domain like testicular ACE, and it is most abundantly expressed in kidney, heart, and lung (Donoghue *et al.*, 2000; Tipnis *et al.*, 2000). Physiologically, the principal role of ACE2 is thought to be the conversion of angiotensin 2 to angiotensin (1–7), which opposes the actions of ACE2. Hence ACE and ACE2 act as counterbalances in metabolism in the renin–angiotensin system. Clues to the roles of ACE2 have come from the development of mice deficient in the ACE2 gene (Crackower *et al.*, 2002). These mice have severe cardiac contractility defects, increased plasma angiotensin II levels, and an upregulation of cardiac hypoxia-related genes, implicating ACE2 as an essential regulator of heart function. Intriguingly, a double knockout in mice of both the ACE and ACE2 genes is able to rescue the cardiac defect seen with the ACE2-deficient mice (Crackower *et al.*, 2002). ACE2 mRNA and protein levels are substantially reduced in the kidney in diabetic rats, suggesting that the enzyme may have a role in the development of diabetic complications (Tikellis *et al.*, 2004). It has been shown that ACE2 is widely distributed throughout the brain but is mainly localized to the cytoplasm of neuronal cells in the brain and not present in glia. Moreover, ACE2 levels appear to be highly regulated by the renin–angiotensin system, suggesting its involvement in this system in the brain. ACE2 expression in the brain structures involved in the control of cardiovascular function suggests that it may have a role in the central regulation of blood pressure and hypertension (Doobay *et al.*, 2007).

The most remarkable discovery in relation to ACE2 biology has been the demonstration that the enzyme functions as the receptor for the SARS virus (Li *et al.*, 2003) and numerous subsequent studies have confirmed and extended this observation. Thus, the structure of the SARS *Coronavirus* spike receptor-binding domain complexed with receptor has been determined (Li *et al.*, 2005) and compounds blocking spike protein and ACE2 interaction discovered, for example emodin (Ho *et al.*, 2007). ACE2 also appears to play a critical role in protection against acute lung injury from SARS infection, or other causes of acute respiratory distress syndrome (Imai *et al.*, 2005).

One feature that both ACE (Hooper *et al.*, 1987) and ACE2 (Lambert *et al.*, 2005) share is the ability to be shed from the plasma membrane by cleavage within the juxtamembrane region releasing the bulk of the protein, including the intact catalytic domain, into the extracellular medium. This process is common to a growing number of membrane proteins of diverse characteristics (see Hooper *et al.*, 1997 for review), an event which is generally receptor-regulated and sensitive to inhibition by a group of hydroxamate metalloproteinase inhibitors such as batimastat. This has led to the identification of members of the ADAMs (a disintegrin and metalloproteinase) family of zinc proteinases, typified by tumor necrosis factor-α converting enzyme ("TACE"; ADAM17) as candidate shedding enzymes (Allinson *et al.*, 2004). Our data suggest that NEP might be also shed from the cell surface by a similar mechanism (Fisk, L., Nalivaeva, N. N., and Turner, A. J., unpublished data). However, the detailed molecular mechanism of this phenomenon has still to be elucidated.

ACE inhibitors have been shown to reduce development of diabetes, improve surrogate markers of inflammation, and reduce cardiovascular disease and renal disease (McFarlane *et al.*, 2003). Increasing evidence indicates that systemic inflammation and neuroinflammation are central features in cerebrovascular disease and that hypertension, through the vasoactive peptides angiotensin and ET-1, promotes and accelerates the atherosclerotic process via inflammatory mechanisms (Di Napoli and Papa, 2005). It was demonstrated that in humans the product of ACE activity angiotensin II induces IL-6 production through a mineralocorticoid receptor-dependent mechanism (Luther *et al.*, 2007). This suggests that ACE inhibitors might also be considered as targets in neuroprotection.

An ACE polymorphism was demonstrated to be associated with AD in the Japanese population (Hu *et al.*, 1999), and later it was found that ACE can indeed hydrolyze Aβ *in vitro* and reduce accumulation of Aβ in cell cultures (Hu *et al.*, 2001; Oba *et al.*, 2005). However, Eckman *et al.* (2006) have demonstrated that *in vivo* ACE does not have a physiological role in clearing Aβ and it is cleaved in the brain mostly by NEP and ECE-1 since ACE-deficient mice did not demonstrate accumulation of Aβ while deficit of NEP or ECE-1 activity resulted in additive increases in brain Aβ levels.

V. Ischemia/Hypoxia and Ageing as Factors Affecting Metalloproteinases

It is becoming more obvious that neurodegeneration and development of AD can be promoted by cardiovascular lesions, ischemia and stroke (Hofman *et al.*, 1997; Kalaria, 2000). Indeed, the analysis of various autopsy series demonstrates that 60–90% of AD cases exhibit variable cerebrovascular pathology. Hypertension has also been suggested as a risk factor in development of AD (for review see Skoog and Gustafson, 2006). Taking into account that NEP and ECE-1 are involved in vascular functions and can contribute to amyloid metabolism, we have analyzed levels of expression of ECE-1 in the brain cortex of rats after 15-min global ischemia and found a significant decrease of NEP and ECE-1 protein levels in brain hemispheres and both hippocampi which returned to normal levels after 2-h reperfusion (Nalivaeva *et al.*, 2004). We have also demonstrated that NEP and ECE-1 levels were lower in the cortex and striatum of rats subjected to prenatal hypoxia (7% O_2, 3 h, 13th day of gestation) (Nalivaeva *et al.*, 2003). Preconditioning to mild (15% O_2) hypoxia before the episodes of acute hypoxia had a protective effect restoring the levels of NEP and ECE-1 in rat brain structures analyzed during the first month after birth. Using human neuroblastoma NB7 cells, we have also demonstrated that hypoxia (1.0–2.5% O_2) resulted in a decrease in expression of ECE-1 (Fisk *et al.*, 2006) and also in expression of NEP both at the protein and mRNA levels at and NEP activity (Fisk *et al.*, 2007). Moreover, both hypoxia and ischemia resulted in an increased production of sAPPβ and reduced amount of sAPPα. These data allowed us to conclude that hypoxic and ischemic conditions in the brain might lead to a shift of amyloid metabolism toward formation of higher levels of Aβ due to an increased rate of β-secretase reaction and reduced activity of such amyloid-degrading enzymes as NEP and ECE-1 (Fig. 4).

Analyzing the effect of chronic hypoxia (1%, 24 h) on expression of NEP estimated by the method of real-time PCR in rat primary cortical neurons and astrocytes, we have found that NEP mRNA levels were downregulated (by 20%) under hypoxia in neurons and upregulated in astrocytes (Fig. 5).

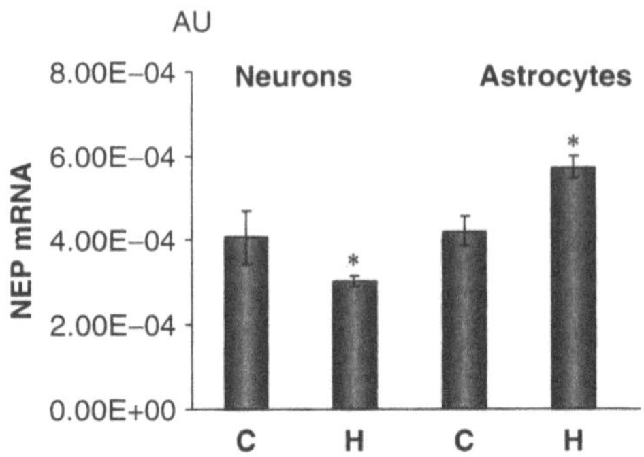

FIG. 5. Effect of hypoxia on the level of NEP mRNA expression in primary cortical neurons and astrocytes. NEP mRNA detected by real-time PCR (related to GAPDH mRNA). C, control; H, hypoxia (1% O_2, 24 h, $n = 12$). *$p < 0.05$ compared to control.

Previously, Apelt *et al.* (2003) reported NEP mRNA upregulation in amyloid plaque-surrounding reactive astrocytes in transgenic Tg2576 mice that produce human amyloid-β peptides from birth and develop amyloid-β plaques which may suggest a role of plaque-mediated astrogliosis in Aβ degradation. In our experiments, mRNA levels were also upregulated in hypoxic astrocytes, which might be an adaptive reaction of astrocytes to pathological conditions (Fisk *et al.*, 2007).

It is important to note that expression of NEP with ageing was found to decrease in the cortex and hippocampus of rats, while its levels in the striatum (where amyloid deposits have not been reported in AD brains) were as high as at the end of the first month after birth (Nalivaeva *et al.*, 2004). Decreased levels of another Aβ-degrading enzyme, IDE, have also been reported in the brain of aged mice (Caccamo *et al.*, 2005) and in rat brain structures (Nalivaeva-Turner *et al.*, 2006). Moreover, our experiments demonstrated that in the brain of rats with experimental type II diabetes IDE levels were lower than in control animals (Kochkina *et al.*, 2006). These data suggest that age-related deficit of amyloid-degrading enzymes might be one of the factors leading to the development of the sporadic form of AD and upregulation of NEP and other Aβ-degrading enzymes become one of the possible therapeutic targets in neurodegeneration and AD.

Several physiological and pharmaceutical ways of upregulation of the neprilysin gene have been suggested. One of the approaches, suggested by Saito *et al.* (2003), involves NEP substrates as molecules activating expression of the NEP gene via a positive feedback mechanism and as yet unknown signaling pathways (Fig. 4). Analyzing various NEP substrates in this experimental paradigm, these authors (Saito *et al.*, 2005) were able to demonstrate that only somatostatin was capable to stimulate NEP activity in primary cortical neurons. Since somatostatin levels in the brain decrease with age these results indicate that age-related downregulation of somatostatin expression could be one of the triggers for Aβ accumulation leading to late-onset sporadic AD. In our experiments using human neuroblastoma cells expressing NEP and ECE-1, we were not able to detect somatostatin-dependent upregulation of NEP mRNA or activity while ECE-1 was significantly upregulated by this peptide in a dose-dependent manner (Fisk, L., Nalivaeva, N. N., and Turner, A. J., unpublished data).

Another physiological pathway of regulation of NEP gene has been suggested by Checler and colleagues who demonstrated that the C-terminal product of APP cleavage by γ-secretase (AICD) was able to transactivate the NEP gene promoter activating NEP expression and that presenilin molecules (PS1 and PS2) and the protein nicastrin in the γ-secretase complex were essential for this effect (Pardossi-Piquard *et al.*, 2005, 2006).

Among chemical compounds that can upregulate NEP expression, green tea extract (EFLA85942) has been demonstrated as effective in human neuroblastoma SK-N SH cells (Ayoub and Melzig, 2006; Melzig and Janka, 2003). According to our observations, the active compound of the green tea extract, namely, polyphenol (−)-epigallocatechin-3-gallate stimulates in a dose-dependent manner both NEP expression at protein level and its activity in human neuroblastoma NB7 cells (Fisk, L., Nalivaeva, N. N., and Turner, A. J., unpublished data).

VI. Conclusions

In this chapter, several examples have demonstrated the role of metalloproteinases in neurodegeneration, neuroprotection, and neuroinflammation. This was mostly shown for NEP, ECE-1, and ACE; however, there are other enzymes whose targeting might be beneficial for prevention of neurological disorders. Among them are the enzymes of amyloidogenic processing of APP by β-secretase (and subsequently by γ-secretase) and nonamyloidogenic APP processing by α-secretase (for review see Cordy *et al.*, 2006; Neve, 2003). Although both α- and β-secretases produce neurotrophic fragments of APP (sAPPα and sAPPβ) and are believed to play a role in normal functioning of neuronal cells (Thornton *et al.*, 2006; Turner *et al.*, 2003), the α-secretase pathway prevents production of Aβ and thus is regarded as neuroprotective. Downregulation of α-secretase or upregulation of β-secretase activity will lead to changes in the balance of production of Aβ toward its accumulation and there are emerging experimental data that this might be the case in ischemic or hypoxic brain (Nalivaeva *et al.*, 2004; Wen *et al.*, 2004). There are also some data that the same compounds can activate both α-secretase and NEP activity, for example green tea extracts (Levites *et al.*, 2003; Melzig and Janka, 2003), suggesting that neuroprotective therapies might target several points in pathogenic processes. However, there also might be situations when one compound upregulates one neuroprotective pathway and downregulates another. For example, the phorbol ester PMA was shown to activate the α-secretase pathway (Zhu *et al.*, 2001) but inhibit expression of ECE-1 (Fisk *et al.*, 2006). This observation implies the necessity of differential analysis of the effects of potentially effective neuroprotective drugs on various enzymes participating in amyloid metabolism.

Metalloproteinases represent important therapeutic targets not only in neurodegeneration but also in cardiovascular diseases and prostate cancer, and understanding their intricate interrelationship is still one of the most fascinating areas of modern molecular and cell biology.

References

Allinson, T., Parkin, E., Turner, A. J., and Hooper, N. M. (2004). The role of ADAM10 and ADAM17 in the ectodomain shedding of angiotensin converting enzyme and the amyloid precursor protein. *Eur. J. Biochem.* **271,** 2539–2547.

Apelt, J., Ach, K., and Schliebs, R. (2003). Aging-related down-regulation of neprilysin, a putative β-amyloid-degrading enzyme, in transgenic Tg2576 Alzheimer-like mouse brain is accompanied by an astroglial upregulation in the vicinity of β-amyloid plaques. *Neurosci. Lett.* **339,** 183–186.

Ayoub, S., and Melzig, M. F. (2006). Induction of neutral endopeptidase (NEP) activity of SK-N-SH cells by natural compounds from green tea. *J. Pharm. Pharmacol.* **58,** 495–501.

Barbara, G., De Giorgio, R., Stanghellini, V., Corinaldesi, R., Cremon, C., Gerard, N., Gerard, C., Grady, E. F., Bunnett, N. W., Blennerhassett, P. A., and Collins, S. M. (2003). Neutral endopeptidase (EC 3. 4. 24. 11) downregulates the onset of intestinal inflammation in the nematode infected mouse. *Gut* **52,** 1457–1464.

Barnes, K., and Turner, A. J. (1999). Endothelin converting enzyme is located on β-actin filaments in smooth muscle cells. *Cardiovasc. Res.* **42,** 814–822.

Barnes, K., Walkden, B. J., Wilkinson, T. C., and Turner, A. J. (1997). Expression of endothelin-converting enzyme in both neuroblastoma and glial cell lines and its localization in rat hippocampus. *J. Neurochem.* **68,** 570–577.

Bralet, J., and Schwartz, J.-C. (2001). Vasopeptidase inhibitors: An emerging class of cardiovascular drugs. *Trends Pharmacol. Sci.* **22,** 106–109.

Broccolini, A., Gidaro, T., Morosetti, R., Gliubizzi, C., Servidei, T., Pescatori, M., Tonali, P. A., Ricci, E., and Mirabella, M. (2006). Neprilysin participates in skeletal muscle regeneration and is accumulated in abnormal muscle fibres of inclusion body myositis. *J. Neurochem.* **96,** 777–789.

Burdick, D., Soreghan, B., Kwon, M., Kosmoski, J., Knauer, M., Henschen, A., Yates, J., Cotman, C., and Glabe, C. (1992). Assembly and aggregation properties of synthetic Alzheimer's A4/amyloid peptide analogs. *J. Biol. Chem.* **267,** 546–554.

Caccamo, A., Oddo, S., Sugarman, M. C., Akbari, Y., and LaFerla, F. M. (2005). Age- and region-dependent alterations in Aβ-degrading enzymes: Implications for Aβ-induced disorders. *Neurobiol. Aging* **26,** 645–654.

Cao, X., and Sudhof, T. C. (2001). A transcriptionally active complex of APP with Fe65 and histone acetyltransferase Tip60. *Science* **293,** 115–120.

Carpentier, M., Robitaille, Y., DesGroseillers, L., Boileau, G., and Marcinkiewicz, M. (2002). Declining expression of neprilysin in Alzheimer disease vasculature: Possible involvement in cerebral amyloid angiopathy. *J. Neuropathol. Exp. Neurol.* **61,** 849–856.

Carson, J. A., and Turner, A. J. (2002). β-amyloid catabolism: Roles for neprilysin (NEP) and other metallopeptidases? *J. Neurochem.* **81,** 1–8.

Chen, F., Hasegawa, H., Schmitt-Ulms, G., Kawarai, T., Bohm, C., Katayama, T., Gu, Y., Sanjo, N., Glista, M., Rogaeva, E., Wakutani, Y., Pardossi-Piquard, R., *et al.* (2006). TMP21 is a presenilin complex component that modulates γ-secretase but not ε-secretase activity. *Nature* **440,** 1208–1212.

Cordy, J. M., Hussain, I., Dingwall, C., Hooper, N. M., and Turner, A. J. (2003). Exclusively targeting β-secretase to lipid rafts by GPI-anchor addition up-regulates beta-site processing of the amyloid precursor protein. *Proc. Natl. Acad. Sci. USA* **100,** 11735–11740.

Cordy, J. M., Hooper, N. M., and Turner, A. J. (2006). The involvement of lipid rafts in Alzheimer's disease. *Mol. Membr. Biol.* **23,** 111–122.

Crackower, M. A., Sarao, R., Oudit, G. Y., Yagil, C., Kozieradzki, I., Scanga, S. E., Oliveira-dos-Santos, A. J., da Costa, J., Zhang, L., Pei, Y., Scholey, J., Ferrario, C. M., *et al.* (2002). Angiotensin-converting enzyme 2 is an essential regulator of heart function. *Nature* **417,** 799–802.

Dawson, L. A., Maitland, N. J., Turner, A. J., and Usmani, B. A. (2004). Stromal-epithelial interactions influence prostate cancer cell invasion by altering the balance of metallopeptidase expression. *Br. J. Cancer* **90,** 1577–1582.

Day, A. L., Wick, E., Jordan, T. H., Jaffray, C. E., Bunnett, N. W., Grady, E. F., and Kirkwood, K. S. (2005). Neutral endopeptidase determines the severity of pancreatitis-associated lung injury. *J. Surg. Res.* **128,** 21–27.

Di Napoli, M., and Papa, F. (2005). Inflammation, blood pressure, and stroke: An opportunity to target primary prevention? *Curr. Hypertens. Rep.* **7,** 44–51.

Doobay, M. F., Talman, L. S., Obr, T. D., Tian, X., Davisson, R. L., and Lazartigues, E. (2007). Differential expression of neuronal ACE2 in transgenic mice with overexpression of the brain renin-angiotensin system. *Am. J. Physiol. Regul. Integr. Comp. Physiol.* **292,** R373–381.

Donoghue, M., Hsieh, F., Baronas, E., Godbout, K., Gosselin, M., Stagliano, N., Donovan, M., Woolf, B., Robison, K., Jeyaseelan, R., Breitbart, R. E., and Acton, S. (2000). A novel angiotensin-converting enzyme-related carboxypeptidase (ACE2) converts angiotensin I to angiotensin 1–9. *Circ. Res.* **87,** 1–9.

Eckman, E. A., Reed, D. K., and Eckman, C. B. (2001). Degradation of the Alzheimer's amyloid β peptide by endothelin-converting enzyme. *J. Biol. Chem.* **276,** 24540–24548.

Eckman, E. A., Watson, M., Marlow, L., Sambamurti, K., and Eckman, C. B. (2003). Alzheimer's disease β-amyloid peptide is increased in mice deficient in endothelin-converting enzyme. *J. Biol. Chem.* **278,** 2081–2084.

Eckman, E. A., Adams, S. K., Troendle, F. J., Stodola, B. A., Kahn, M. A., Fauq, A. H., Xiao, H. D., Bernstein, K. E., and Eckman, C. B. (2006). Regulation of steady-state β-amyloid levels in the brain by neprilysin and endothelin-converting enzyme but not angiotensin-converting enzyme. *J. Biol. Chem.* **281,** 30471–30478.

Facchinetti, P., Rose, C., Schwartz, J. C., and Ouimet, T. (2003). Ontogeny, regional and cellular distribution of the novel metalloprotease neprilysin 2 in the rat: A comparison with neprilysin and endothelinconverting enzyme-1. *NeuroScience* **118,** 627–639.

Fisk, L., Nalivaeva, N. N., and Turner, A. J. (2006). Regulation of endothelin-converting enzyme-1 expression in human neuroblastoma cells. *Exp. Biol. Med.* **231,** 1048–1053.

Fisk, L., Nalivaeva, N. N., Boyle, J. P., Peers, C. S., and Turner, A. J. (2007). Effects of hypoxia and oxidative stress on expression of neprilysin in human neuroblastoma cells and rat cortical neurons and astro cytes. *Neurochem. Res.* May 7, 2007 [Epub ahead of print].

Francis, R., McGrath, G., Zhang, J., Ruddy, D. A., Sym, M., Apfeld, J., Nicoll, M., Maxwell, M., Hai, B., Ellis, M. C., Parks, A. L., Xu, W., *et al.* (2002). Aph-1 and pen-2 are required for Notch pathway signaling, γ-secretase cleavage of APP, and presenilin protein accumulation. *Dev. Cell* **3,** 85–97.

Furukawa, K., Saleh, D., Bayan, F., Emoto, N., Kaw, S., Yanagisawa, M., and Giaid, A. (1996). Co-expression of endothelin-1 and endothelin-converting enzyme-1 in patients with chronic rhinitis. *Am. J. Respir. Cell Mol. Biol.* **14,** 248–253.

Hardy, J. A., and Higgins, G. A. (1992). Alzheimer's disease: The amyloid cascade hypothesis. *Science* **256,** 184–185.

Hass, M. R., and Yankner, B. A. (2005). A γ-secretase-independent mechanism of signal transduction by the amyloid precursor protein. *J. Biol. Chem.* **280,** 36895–36904.

Hebert, S. S., Serneels, L., Tolia, A., Craessaerts, K., Derks, C., Filippov, M. A., Muller, U., and De Strooper, B. (2006). Regulated intramembrane proteolysis of amyloid precursor protein and regulation of expression of putative target genes. *EMBO Rep.* **7,** 739–745.

Hersh, L. B., Yu, J., Zhu, H., Gao, J. T., Mucke, L., and Kindy, M. S. (2002). Neprilysin regulation of amyloid deposition in Alzheimer's disease. *J. Neurochem.* **81**(Suppl. 1), 67.

Ho, T. Y., Wu, S. L., Chen, J. C., Li, C. C., and Hsiang, C. Y. (2007). Emodin blocks the SARS coronavirus spike protein and angiotensin-converting enzyme 2 interaction. *Antiviral Res.* **74,** 92–101.

Hoang, M. V., and Turner, A. J. (1997). Novel activity for endothelin-converting enzyme: Hydrolysis of bradykinin. *Biochem. J.* **327,** 23–26.

Hofman, A., Ott, A., Breteler, M. M., Bots, M. L., Slooter, A. J., Van Harskamp, F., van Duijn, C. N., Van Broeckhoven, C., and Grobbee, D. E. (1997). Atherosclerosis, apolipoprotein E, and prevalence of dementia and Alzheimer's disease in the Rotterdam Study. *Lancet* **349,** 151–154.

Hooper, N. M., and Turner, A. J. (1985). Neurokinin B is hydrolysed by synaptic membranes and by endopeptidase-24.11 (enkephalinase) but not by angiotensin converting enzyme. *FEBS Lett.* **190,** 133–136.

Hooper, N. M., and Turner, A. J. (2003). An ACE structure. *Nat. Struct. Biol.* **10,** 155–157.

Hooper, N. M., Keen, J., Pappin, D. J. C., and Turner, A. J. (1987). Pig kidney angiotensin converting enzyme. Purification and characterization of amphipathic and hydrophilic forms of the enzyme establishes C-terminal anchorage to the plasma membrane. *Biochem. J.* **247,** 85–93.

Hooper, N. M., Karran, E. H., and Turner, A. J. (1997). Membrane protein secretases. *Biochem. J.* **321,** 265–279.

Howell, S., Caswell, A. M., Kenny, A. J., and Turner, A. J. (1993). Membrane peptidases on human osteoblast-like cells in culture: Hydrolysis of calcitonin and hormonal regulation of endopeptidase-24.11. *Biochem. J.* **290,** 159–164.

Hu, J., Miyatake, F., Aizu, Y., Nakagawa, H., Nakamura, S., Tamaoka, A., Takahashi, R., Urakami, K., and Shoji, M. (1999). Angiotensin-converting enzyme genotype is associated with Alzheimer disease in the Japanese population. *Neurosci. Lett.* **277,** 65–67.

Hu, J., Igarashi, A., Kamata, M., and Nakagawa, H. (2001). Angiotensin-converting enzyme degrades Alzheimer amyloid β-peptide (Aβ); retards Aβ aggregation, deposition, fibril formation; and inhibits cytotoxicity. *J. Biol. Chem.* **276,** 47863–47868.

Ihling, C., Szombathy, T., Bohrmann, B., Brockhaus, M., Schaefer, H. E., and Loeffler, B. M. (2001). Coexpression of endothelin-converting enzyme-1 and endothelin-1 in different stages of human atherosclerosis. *Circulation* **104,** 864–869.

Ihling, C., Bohrmann, B., Schaefer, H. E., Technau-Ihling, K., and Loeffler, B. M. (2004). Endothelin-1 and endothelin converting enzyme-1 in human atherosclerosis–novel targets for pharmacotherapy in atherosclerosis. *Curr. Vasc. Pharmacol.* **2,** 249–258.

Imai, Y., Kuba, K., Rao, S., Huan, Y., Guo, F., Guan, B., Yang, P., Sarao, R., Wada, T., Leong-Poi, H., Crackower, M. A., Fukamizu, A., *et al.* (2005). Angiotensin-converting enzyme 2 protects from severe acute lung failure. *Nature* **436,** 112–116.

Isaac, R. E., and Nassel, D. R. (2003). Identification and localization of a neprilysin-like activity that degrades tachykinin-related peptides in the brain of the cockroach, *Leucophaea maderae*, and locust, *Locusta migratoria*. *J. Comp. Neurol.* **457,** 57–66.

Isaac, R. E., Parkin, E. T., Keen, J. N., Nassel, D. R., Siviter, R. J., and Shirras, A. D. (2002). Inactivation of a tachykinin-related peptide: Identification of four neuropeptide-degrading enzymes in neuronal membranes of insects from four different orders. *Peptides* **23,** 725–733.

Iwata, N., Tsubuki, S., Takaki, Y., Watanabe, K., Seikiguchi, M., Hosoki, E., Kawashima-Morishima, M., Lee, H.-J., Hama, E., Sekine-Aizawa, Y., and Saido, T. C. (2000). Identification of the major Aβ1–42-degrading catabolic pathway in brain parenchyma: Suppression leads to biochemical and pathological deposition. *Nat. Med.* **6,** 143–150.

Johnson, G. D., Stevenson, T., and Ahn, K. (1999). Hydrolysis of peptide hormones by endothelin-converting enzyme-1. A comparison with neprilysin. *J. Biol. Chem.* **274,** 4053–4058.

Juvela, S. (2002). Plasma endothelin and big endothelin concentrations and serum endothelin-converting enzyme activity following aneurysmal subarachnoid hemorrhage. *J. Neurosurg.* **97,** 1287–1293.

Kalaria, R. N. (2000). The role of cerebral ischemia in Alzheimer's disease. *Neurobiol. Aging* **21,** 321–330.

Kenny, A. J., and Stephenson, S. L. (1988). Role of endopeptidase-24.11 in the inactivation of atrial natriuretic peptide. *FEBS Lett.* **232,** 1–8.

Kochkina, E. G., Plesneva, S. A., Zhuravin, I. A., Turner, A. J., and Nalivaeva, N. N. (2006). Study of insulin-degrading enzyme expression in the brain and peripheral organs of hypoxic and diabetic rats 11th meeting of Czech and Slovak Neurochemical Society, Martin, Slovakia, 54.

Kramer, H. H., Schmidt, K., Leis, S., Schmelz, M., Sommer, C., and Birklein, F. (2005). Inhibition of neutral endopeptidase (NEP) facilitates neurogenic inflammation. *Exp. Neurol.* **195,** 179–184.

Kumar, G. K., Runold, M., Ghai, R. D., Cherniack, N. S., and Prabhakar, N. R. (1990). Occurrence of neutral endopeptidase activity in the cat carotid body and its significance in chemoreception. *Brain Res.* **517,** 341–343.

Kumar, G. K., Yu, R. K., Overholt, J. L., and Prabhakar, N. R. (2000). Role of substance P in neutral endopeptidase modulation of hypoxic response of the carotid body. *Adv. Exp. Med. Biol.* **475,** 705–713.

Lambert, D. W., Yarski, M., Warner, F. J., Thornhill, P., Parkin, E. T., Smith, A. I., Hooper, N. M., and Turner, A. J. (2005). Tumor necrosis factor-alpha convertase (ADAM17) mediates regulated ecto-domain shedding of the severe-acute respiratory syndrome-coronavirus (SARS-CoV) receptor, angiotensin-converting enzyme-2 (ACE2). *J. Biol. Chem.* **280,** 30113–30119.

Letarte, M., Vera, S., Tran, R., Addis, J. B., Onizuka, R. J., Quackenbush, E. J., Jongeneel, C. V., and McInnes, R. R. (1988). Common acute lymphocytic leukemia antigen is identical to neutral endopeptidase. *J. Exp. Med.* **168,** 1247–1253.

Levites, Y., Amit, T., Mandel, S., and Youdim, M. B. (2003). Neuroprotection and neurorescue against Aβ toxicity and PKC-dependent release of nonamyloidogenic soluble precursor protein by green tea polyphenol (–)-epigallocatechin-3-gallate. *FASEB J.* **17,** 952–954.

Li, F., Li, W., Farzan, M., and Harrison, S. C. (2005). Structure of SARS coronavirus spike receptor-binding domain complexed with receptor. *Science* **309,** 1864–1868.

Li, W., Moore, M. J., Vasilieva, N., Sui, J., Wong, S. K., Berne, M. A., Somasundaran, M., Sullivan, J. L., Luzuriaga, K., Greenough, T. C., Choe, H., and Farzan, M. (2003). Angiotensin-converting enzyme 2 is a functional receptor for the SARS coronavirus. *Nature* **426,** 450–454.

Lightner, A. M., Jordan, T. H., Bunnett, N. W., Grady, E. F., and Kirkwood, K. S. (2002). Recombinant human neutral endopeptidase ameliorates pancreatic elastase-induced lung injury. *Surgery* **132,** 193–199.

Lilius, L., Forsell, C., Axelman, K., Winblad, B., Graff, C., and Tjernberg, L. (2003). No association between polymorphisms in the neprilysin promoter region and Swedish Alzheimer's disease patients. *Neurosci. Lett.* **337,** 111–113.

Luther, J. M., Gainer, J. V., Murphey, L. J., Yu, C., Vaughan, D. E., Morrow, J. D., and Brown, N. J. (2007). Angiotensin II induces interleukin-6 in humans through a mineralocorticoid receptor-dependent mechanism. *Hypertension* **48,** 1050–1057.

Malfroy, B., Swerts, J. P., Guyon, A., Roques, B. P., and Schwartz, J.-C. (1978). High-affinity enkephalin-degrading peptidase in brain is increased after morphine. *Nature* **276,** 523–526.

Marr, R. A., Rockenstein, E., Mukherjee, A., Kindy, M. S., Hersh, L. B., Gage, F. H., Verma, I. M., and Masliah, E. (2003). Neprilysin gene transfer reduces human amyloid pathology in transgenic mice. *J. Neurosci.* **23,** 1992–1996.

Matsas, R., Fulcher, I. S., Kenny, A. J., and Turner, A. J. (1983). Substance P and [Leu]enkephalin are hydrolyzed by an enzyme in pig caudate synaptic membranes that is identical with the endopeptidase of kidney microvilli. *Proc. Natl. Acad. Sci. USA* **80,** 3111–3115.

Matsas, R., Turner, A. J., and Kenny, A. J. (1984). The metabolism of neuropeptides. The hydrolysis of peptides, including enkephalins, tachykinins and their analogues, by endopeptidase-24.11. *Biochem. J.* **223,** 433–440.

Matsumura, Y., Ikegawa, R., Takaoka, M., and Morimoto, S. (1990). Conversion of porcine big endothelin to endothelin by an extract from the porcine aortic endothelial cells. *Biochem. Biophys. Res. Commun.* **167,** 203–210.

McFarlane, S. I., Kumar, A., and Sowers, J. R. (2003). Mechanisms by which angiotensin-converting enzyme inhibitors prevent diabetes and cardiovascular disease. *Am. J. Cardiol.* **91,** 30H–37H.

Melzig, M. F., and Janka, M. (2003). Enhancement of neutral endopeptidase activity in SK-N-SH cells by green tea extract. *Phytomedicine* **10,** 494–498.

Mohajeri, M. H., Wollmer, M. A., and Nitsch, R. M. (2002). Aβ42-induced increase in neprilysin is associated with prevention of amyloid plaque formation *in vivo*. *J. Biol. Chem.* **277,** 35460–35465.

Nalivaeva, N. N., Fisk, L., Canet Aviles, R. M., Plesneva, S. A., Zhuravin, I. A., and Turner, A. J. (2003). Effects of prenatal hypoxia on expression of amyloid precursor protein and metallopeptidases in the rat brain. *Lett. Peptide Sci.* **10,** 455–462.

Nalivaeva, N. N., Fisk, L., Kochkina, E. G., Plesneva, S. A., Zhuravin, I. A., Babusikova, E., Dobrota, D., and Turner, A. J. (2004). Effect of hypoxia/ischemia and hypoxic preconditioning/reperfusion on expression of some amyloid-degrading enzymes. *Ann. NY Acad. Sci.* **1035,** 21–33.

Nalivaeva-Turner, N. N., Fisk, L., Kochkina, E. G., Makova, N., Plesneva, S. A., Zhuravin, I. A., and Turner, A. J. (2006). Developmental dynamics of amyloid-degrading enzymes in the rat brain under normal and hypoxic conditions. Proceedings of the 5th FENS Forum, Vienna, Austria, July 8–12, 2006. *FENS Abstr.* **3,** A130.30.

Neve, R. L. (2003). A new wrestler in the battle between α- and β-secretases for cleavage of APP. *Trends Neurosci.* **26,** 461–463.

Oba, R., Igarashi, A., Kamata, M., Nagata, K., Takano, S., and Nakagawa, H. (2005). The N-terminal active centre of human angiotensin-converting enzyme degrades Alzheimer amyloid β-peptide. *Eur. J. Neurosci.* **21,** 733–740.

Oefner, C., D'Arcy, A., Nennig, M., Winkler, F. K., and Dale, G. E. (2000). Structure of human neutral endopeptidase (Neprilysin) complexed with phosphoramidon. *J. Mol. Biol.* **296,** 341–349.

Papandreou, C. N., Usmani, B., Geng, Y., Bogenrieder, T., Freeman, R., Wilk, S., Finstad, C. L., Reuter, V. E., Powell, C. T., Scheinberg, D., Magill, C., Scher, H. I., *et al.* (1998). Neutral endopeptidase 24.11 loss in metastatic human prostate cancer contributes to androgen-independent progression. *Nat. Med.* **4,** 50–57.

Pardossi-Piquard, R., Petit, A., Kawarai, T., Sunyach, C., Alves da Costa, C., Vincent, B., Ring, S., D'Adamio, L., Shen, J., Muller, U., St. George-Hyslop, P., and Checler, F. (2005). Presenilin-dependent transcriptional control of the Aβ-degrading enzyme neprilysin by intracellular domains of APP and APLP. *Neuron* **46,** 541–554.

Pardossi-Piquard, R., Dunys, J., Yu, G., St. George-Hyslop, P., Alves da Costa, C., and Checler, F. (2006). Neprilysin activity and expression are controlled by nicastrin. *J. Neurochem.* **97,** 1052–1056.

Passer, B., Pellegrini, L., Russo, C., Siegel, R. M., Lenardo, M. J., Schettini, G., Bachmann, M., Tabaton, M., and D'Adamio, L. (2000). Generation of an apoptotic intracellular peptide by gamma-secretase cleavage of Alzheimer's amyloid protein precursor. *J. Alzheimers Dis.* **2,** 289–301.

Puente, X. S., Sanchez, L. M., Overall, C. M., and Lopez-Otin, C. (2003). Human and mouse proteases: A comparative genomic approach. *Nat. Rev. Genet.* **4,** 544–558.

Quaschning, T. (2005). Vasopeptidase inhibition for blood pressure control: Emerging experience. *Curr. Pharm. Des.* **11,** 3293–3299.

Rawlings, N. D., and Barrett, A. J. (2004). Intruduction: Metallopeptidases and their clans. *In* "Handbook of Proteolytic Enzymes" (A. J. Barrett, N. D. Rawlings, and J. F. Woessner, Eds.), pp. 231–268. Elsevier Academic Press, London.

Rawlings, N. D., Morton, F. R., and Barrett, A. J. (2006). *MEROPS*: The peptidase database. *Nucleic Acids Res.* **34,** D270–D272.

Relton, J. M., Gee, N. S., Matsas, R., Turner, A. J., and Kenny, A. J. (1983). Purification of endopeptidase-24.11 ('enkephalinase') from pig brain by immunoadsorbent chromatography. *Biochem. J.* **215,** 519–523.

Roques, B. P., Fournié-Zaluski, M.-C., Soroca, E., Lecomte, J.-M., Malfroy, B., Llorens-Cortes, C., and Schwartz, J.-C. (1980). The enkephalinase inhibitor thiorphan shows antinociceptive activity in mice. *Nature* **288,** 286–288.

Ruchon, A. F., Marcinkiewicz, M., Ellefsen, K., Basak, A., Aubin, J., Crine, P., and Boileau, G. (2000). Cellular localization of neprilysin in mouse bone tissue and putative role in hydrolysis of osteogenic peptides. *J. Bone Miner. Res.* **15,** 1266–1274.

Saito, T., Takaki, Y., Iwata, N., Trojanowski, J., and Saido, T. C. (2003). Alzheimer's disease, neuropeptides, neuropeptidase, and amyloid-β peptide metabolism. *Sci. Aging Knowledge Environ.* **3,** E1.

Saito, T., Iwata, N., Tsubuki, S., Takaki, Y., Takano, J., Huang, S. M., Suemoto, T., Higuchi, M., and Saido, T. C. (2005). Somatostatin regulates brain amyloid-β peptide A42 through modulation of proteolytic degradation. *Nat. Med.* **11,** 434–439.

Saleh, D., Furukawa, K., Tsao, M. S., Maghazachi, A., Corrin, B., Yanagisawa, M., Barnes, P. J., and Giaid, A. (1997). Elevated expression of endothelin-1 and endothelin-converting enzyme-1 in idiopathic pulmonary fibrosis: Possible involvement of proinflammatory cytokines. *Am. J. Respir. Cell Mol. Biol.* **16,** 187–193.

Sales, N., Dutriez, I., Maziere, B., Ottaviani, M., and Roques, B. P. (1991). Neutral endopeptidase 24.11 in rat peripheral tissues: Comparative localization by '*ex vivo*' and '*in vitro*' autoradiography. *Regul. Pept.* **33,** 209–222.

Scholzen, T. E., and Luger, T. A. (2004). Neutral endopeptidase and angiotensin-converting enzyme—key enzymes terminating the action of neuroendocrine mediators. *Exp. Dermatol.* **13**(Suppl. 4), 22–26.

Scholzen, T. E., Steinhoff, M., Bonaccorsi, P., Klein, R., Amadesi, S., Geppetti, P., Lu, B., Gerard, N. P., Olerud, J. E., Luger, T. A., Bunnett, N. W., Grady, E. F., *et al.* (2001). Neutral endopeptidase terminates substance P-induced inflammation in allergic contact dermatitis. *J. Immunol.* **166,** 1285–1291.

Schweizer, A., Valdenaire, O., Nelböck, P., Deuschle, U., Dumas, M., Edwards, J. B., Stumpf, J. G., and Löffler, B. M. (1997). Human endothelin converting enzyme (ECE-1): Three isoforms with distinct subcellular localizations. *Biochem. J.* **328,** 871–877.

Simonini, G., Azzari, C., Gelli, A. M., Giani, T., Calabri, G. B., Leoncini, G., Del Rosso, A., Generini, S., Cimaz, R., Cerinic, M. M., and Falcini, F. (2005). Neprilysin levels in plasma and synovial fluid of juvenile idiopathic arthritis patients. *Rheumatol. Int.* **25,** 336–340.

Skeggs, L. T., Kahn, J. R., and Shumway, N. P. J. (1956). The preparation and function of the hypertensin-converting enzyme. *Exp. Med.* **103,** 295–299.

Skoog, I., and Gustafson, D. (2006). Update on hypertension and Alzheimer's disease. *Neurol. Res.* **28,** 605–611.

Slomiany, B. L., and Slomiany, A. (2005). Endothelin-1-dependent leptin induction in gastric mucosal inflammatory responses to *Helicobacter pylori* lipopolysaccharide. *Biochem. Biophys. Res. Commun.* **336,** 1106–1111.

Slomiany, B. L., Piotrowski, J., and Slomiany, A. (2000). Up-regulation of endothelin-converting enzyme-1 in gastric mucosal inflammatory responses to *Helicobacter pylori* lipopolysaccharide. *Biochem. Biophys. Res. Commun.* **267,** 801–805.

Snowdon, D. A., Greiner, L. H., Mortimer, J. A., Riley, K. P., Greiner, P. A., and Markesbery, W. R. (1997). Brain infarction and the clinical expression of Alzheimer disease. The Nun Study. *JAMA* **277,** 813–817.

Sparks, D. L., Sabbagh, M., Connor, D., Soares, H., Lopez, J., Stankovic, G., Johnson-Traver, S., Ziolkowski, C., and Browne, P. (2006). Statin therapy in Alzheimer's disease. *Acta Neurol. Scand. Suppl.* **185,** 78–86.

Takahashi, M., Matsushita, Y., Iijima, Y., and Tanzawa, K. (1993). Purification and characterization of endothelin converting enzyme from rat lung. *J. Biol. Chem.* **268,** 21394–21398.

Takahashi, M., Fukuda, K., Shimada, K., Barnes, K., Turner, A. J., Ikeda, M., Koike, M., Yamamoto, Y., and Tanzawa, K. (1995). Localization of rat endothelin converting enzyme to vascular endothelial cells and some secretory cells. *Biochem. J.* **311,** 657–665.

Thomas, J. E., Rylett, C. M., Carhan, A., Bland, N. D., Bingham, R. J., Shirras, A. D., Turner, A. J., and Isaac, R. E. (2005). *Drosophila melanogaster* NEP2 is a new soluble member of the neprilysin family of endopeptidases with implications for reproduction and renal function. *Biochem. J.* **386,** 357–366.

Thornton, E., Vink, R., Blumbergs, P. C., and Van Den Heuvel, C. (2006). Soluble amyloid precursor protein α reduces neuronal injury and improves functional outcome following diffuse traumatic brain injury in rats. *Brain Res.* **1094,** 38–46.

Tikellis, C., Johnston, C. I., Forbes, J. M., Burns, W. C., Burrell, L. M., Risvanis, J., and Cooper, M. E. (2004). Characterization of renal angiotensin-converting enzyme 2 in diabetic nephropathy. *Hypertension* **41,** 392–397.

Tipnis, S. R., Hooper, N. M., Hyde, R., Karran, E., Christie, G., and Turner, A. J. (2000). A human homolog of angiotensin-converting enzyme. Cloning and functional expression as a captopril-insensitive carboxypeptidase. *J. Biol. Chem.* **275,** 33238–33243.

Tran-Paterson, R., Willard, H. F., and Letarte, M. (1989). The common acute lymphoblastic leukemia antigen (neutral endopeptidase-3.4.24.11) gene is located on human chromosome 3. *Cancer Genet. Cytogenet.* **42,** 129–134.

Turner, A. J., and Hooper, N. M. (2002). The angiotensin-converting enzyme gene family: Genomics and pharmacology. *Trends Pharmacol. Sci.* **23,** 177–183.

Turner, A. J., and Tanzawa, K. (1997). Mammalian membrane metallopeptidases: NEP, ECE, KELL, and PEX. *FASEB J.* **11,** 355–364.

Turner, A. J., Barnes, K., Schweizer, A., and Valdenaire, O. (1998). Isoforms of endothelinconverting enzyme: Why and where? *Trends Pharmacol. Sci.* **19,** 483–486.

Turner, A. J., Isaac, R. E., and Coates, D. (2001). The neprilysin (NEP) family of zinc metalloendopeptidases: Genomics and function. *Bioessays* **23,** 261–269.

Turner, A. J., Fisk, L., and Nalivaeva, N. N. (2004). Targeting amyloid-degrading enzymes as therapeutic strategies in neurodegeneration. *Ann. NY Acad. Sci.* **1035,** 1–20.

Turner, P. R., O'Connor, K., Tate, W. P., and Abraham, W. C. (2003). Roles of amyloid precursor protein and its fragments in regulating neural activity, plasticity and memory. *Prog. Neurobiol.* **70,** 1–32.

Usmani, B. A., Harden, B., Maitland, N. J., and Turner, A. J. (2002). Differential expression of neutral endopeptidase-24.11 (neprilysin) and endothelin-converting enzyme in human prostate cancer cell lines. *Clin. Sci. (Lond.)* **103**(Suppl. 48), 314S–317S.

Valdenaire, O., Lepailleur-Enouf, D., Edidy, G., Thouard, A., Barrer, A., Vranckx, R., Tougard, C., and Michel, J. B. (1999). A fourth isoform of endothelin-converting enzyme (ECE-1) is generated from an additional promoter. *Eur. J. Biochem.* **264,** 341–349.

Vassar, R., and Citron, M. (2000). Aβ-generating enzymes: Recent advances in β- and γ-secretase research. *Neuron* **27,** 419–422.

Wen, Y., Onyewuchi, O., Yang, S., Liu, R., and Simpkins, J. W. (2004). Increased β-secretase activity and expression in rats following transient cerebral ischemia. *Brain Res.* **1009,** 1–8.

Wolfe, M. S., Xia, W., Ostaszewski, B. L., Diehl, T. S., Kimberly, W. T., and Selkoe, D. J. (1999). Two transmembrane aspartates in presenilin-1 required for presenilin endoproteolysis and γ-secretase activity. *Nature* **398,** 513–517.

Yasojima, K., Akiyama, H., McGeer, E. G., and McGeer, P. L. (2001). Reduced neprilysin in high plaque areas of Alzheimer brain: A possible relationship to deficient degradation of β-amyloid peptide. *Neurosci. Lett.* **297,** 97–100.

Xu, D., Emoto, N., Giaid, A., Slaughter, C., Kaw, S., de Wit, D., and Yanagisawa, M. (1994). ECE-1: A membrane-bound metalloprotease that catalyzes the proteolytic activation of big endothelin-1. *Cell* **78,** 473–845.

Yu, G., Nishimura, M., Arawaka, S., Levitan, D., Zhang, L., Tandon, A., Song, Y. Q., Rogaeva, E., Chen, F., Kawarai, T., Supala, A., Levesque, L., *et al.* (2000). Nicastrin modulates presenilin-mediated notch/glp-1 signal transduction and APP processing. *Nature* **407,** 48–54.

Zhang, Y., Adner, M., and Cardell, L. O. (2004). Interleukin-1β attenuates endothelin B receptor-mediated airway contractions in a murine *in vitro* model of asthma: Roles of endothelin converting enzyme and mitogen-activated protein kinase pathways. *Clin. Exp. Allergy* **34,** 1480–1487.

Zhu, G., Wang, D., Lin, Y. H., McMahon, T., Koo, E. H., and Messing, R. O. (2001). Protein kinase Cϵ suppresses Aβ production and promotes activation of α-secretase. *Biochem. Biophys. Res. Commun.* **285,** 997–1006.

RELEVANCE OF HIGH-MOBILITY GROUP PROTEIN BOX 1 TO NEURODEGENERATION

Silvia Fossati and Alberto Chiarugi

Department of Preclinical and Clinical Pharmacology, University of Florence, Florence, Italy

High-mobility group protein box 1 (HMGB1), also know as amphoterin, is a nonhistone nuclear protein with well-established functions in gene transcription and homeostasis regulation into the cell nucleus. Interestingly, the protein can be passively released in the extracellular space during necrosis, whereas retained into the nucleus by cells undergoing apoptosis. Also, immune cells actively release HMGB1 on stimulation. Emerging evidence undoubtedly demonstrates that extracellular HMGB1 engages membrane receptors on different cells signaling proliferation, differentiation, cytoprotection, as well as immune activation. Consistently, numerous reports point to HMGB1 as a novel player in disease pathogenesis in peripheral organs. This chapter provides an appraisal of the emerging roles of HMGB1 in neuropathology and the neuroinflammatory response, highlighting the relevance of HMGB1-blocking agents as innovative therapeutic tools to be harnessed for neuroprotection.

I. Introduction

High-mobility group protein box 1 (HMGB1) is a protein with key roles in maintenance of nuclear homeostasis. Surprisingly, a large body of experimental evidence demonstrates that HMGB1 is also endowed with extracellular signaling functions on various cell types, including those of the immune system. In light of this property, the protein has been included into the "alarmin" family, a term used by

INTERNATIONAL REVIEW OF NEUROBIOLOGY, VOL. 82
DOI: 10.1016/S0074-7742(07)82007-1

Oppenheim and coworkers (Yang *et al.*, 2007), to identify a group of endogenous factors, also known as "endokines," which, once released in the extracellular space, interact with membrane receptors on immune cells to activate the inflammatory response. In addition to HMGB1, innate immune mediators included in the alarmin family are IL-1α, S-100 proteins, cytosolic calcium-binding proteins, heat shock proteins (HSPs), defensins, cathelicidins, eosinophil-derived neurotoxin, and hepatoma-derived growth factor (HDGF) (Fig. 1).

Numerous studies report that HMGB1 is secreted under different physiological and pathological conditions thereby inducing proliferation, cell migration, and inflammation. Surprisingly enough, although the protein was first isolated from the rat brain, to date the majority of information on HMGB1's functions has been gathered from studies focused on peripheral organs. Yet, several lines of evidence indicate different roles for HMGB1 in brain development and degeneration. Hence, the aim of the present chapter is to provide an appraisal of current knowledge on HMBG1's actions in the CNS, underlying the relevance of the protein as a target for innovative neuroprotective strategies.

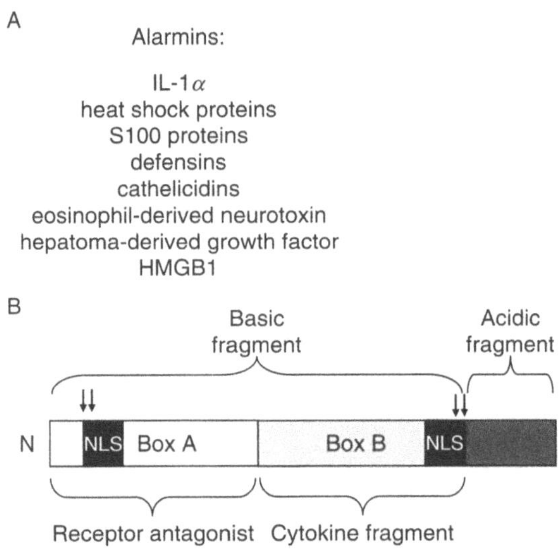

FIG. 1. Classification and structure. (A) HMGB1 belongs to the growing family of alarmins, a list of proteins, also known as endokines, which, once released from the cell, show both cytokine and chemokine properties. Specifically, alarmins are released chemotactic activators. (B) Schematic structure of HMGB1. The acidic and basic fragments are shown. The basic fragment comprises Box A and B which can function as a RAGE receptor antagonist and cytokine, respectively. The two nuclear localization signals (NLSs) are depicted. Arrows represent the hyperacetylation sites (arrows) targeted by acetyltransferases and responsible for masking of NLS and cytoplasmic accumulation of HMGB1.

II. Structure and Nuclear Functions

HMGB1, also called HMG1 or amphoterin, is a 215-residue protein isolated more than 30 years ago (Goodwin and Johns, 1973). As for its structure (see Fig. 1), HMGB1 is organized into three different regions encompassing two DNA-binding domains called box A and B typically rich in basic residues and an acidic C-terminal tail (Landsman and Bustin, 1993). Further studies showed that box A can function as an antagonist of full-length HMGB1, while box B is endowed with cytokine activity. Besides box A and B, two nuclear localization signals (NLSs) are present in the protein structure at amino acids 28–44 and 179–185.

It was soon appreciated that HMGB1 is highly expressed in the nucleus of almost every cell type. Since its first identification, a great deal of efforts had been directed at elucidating HMGB1's functions. Evidence for its high nuclear concentration (in the micromolar range), evolutionary sequence conservation and lack of sequence-specific DNA-binding activity suggested that the protein should play an essential as well as general role in maintenance of nuclear homeostasis. The protein binds to the minor groove of linear DNA (Yu *et al.*, 1977) and bends the double helix up to 90° or more, thereby altering chromatin architecture and allowing recruitment of transcription-regulating factors. One of the first hints about the functional relevance of HMGB1 to transcription came with the study by Giese *et al.* (1992). By investigating the properties of a structural domain called HMG in the transcription-regulating protein lymphoid enhancer factor-1 (LEF-1), the authors demonstrated that the domain under scrutiny bends DNA and facilitates assembly of supramolecular nucleoprotein complexes (Giese *et al.*, 1991). Given that this domain is homologous to HMGB1, data suggested that the latter is also involved in remodeling of chromatin architecture and transcription. Subsequent work confirmed this assumption. For instance, relevance of HMGB1 to transcription is well evidenced by the study of McKinney and associates (McKinney and Prives, 2002), reporting that the weak interaction between p53 and linear DNA is turned into an efficient binding once HMGB1 bends the nucleic acid. Then, dissociation of HMGB1 from the DNA element promotes p53 transcriptional activity. In light of the its central role in transcription, it is not surprising that mice deficient for HMGB1 die 24 h after birth, and that fibroblasts null for HMGB1 show general impairment of gene expression activation (Calogero *et al.*, 1999). At present, a large body of evidence clearly points to HMGB1 as a key regulator of enhanceosome organization, transcription factor recruitment, and overall activation of the basal transcriptional machinery (Allain *et al.*, 1999; Ellwood *et al.*, 2000; Verrijdt *et al.*, 2002).

III. Cytokine Functions

HMGB1 can be released from the nucleus to the extracellular milieu in response to different stimuli and via several mechanisms, depending on the cell type and the pathophysiological conditions.

While necrotic cells passively release the protein following nuclear and plasma membrane rupture, cells undergoing apoptosis retain HMGB1 into the nucleus strongly associated with condensed chromatin (Scaffidi *et al.*, 2002). The molecular mechanisms underlying nuclear retention of HMGB1 during apoptosis are still in part unclear, but it seems that both acetylation and phosphorylation are involved. Remarkably, released HMGB1 triggers activation of bystander immune cells (i.e., macrophages, dendritic cells, natural killer cells, and microglia) and inflammation. Conversely, it has been postulated that the ability of apoptotic cells to retain HMGB1 in the nucleus is a strategy to prevent development of the inflammatory response (Bell *et al.*, 2006) (Fig. 2).

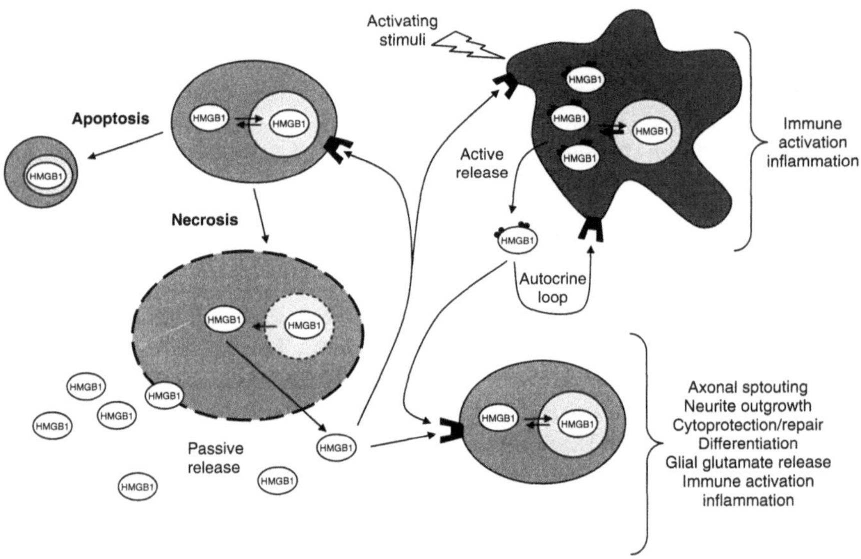

FIG. 2. The pleiotypic effects of HMGB1. Under normal conditions HMGB1 nucleocytoplasmic shuttling participates to maintenance of nuclear homeostasiss. In apoptotic cells, HMGB1 is sequestered into the nucleus strictly bound to chromatin because of phosphorylation and acetylation events (see text). Conversely, HMGB1 is passively released by necrotic cells into the extracellular space, thereby allowing its interaction with different membrane receptors (see text) on bystander cells. Activated immune cells can actively release HMGB1 on hyperacetylation (black dots). HMGB1 receptor activation is then responsible for a large array of beneficial and detrimental effects encompassing cytoprotection, differentiation, cell migration, as well as immune activation and inflammation.

Besides passive release during necrosis, HMGB1 can be actively secreted by several cell types. Among these, activated immune cells constitute a major source of extracellular HMGB1 in inflamed tissues (Lotze and Tracey, 2005). Although the mechanisms contributing to active HMGB1 release are still under investigation, there is general consensus that protein hyperacetylation on lysine residues of its two NLSs plays a major role in HMGB1 active extrusion. It has been reported, indeed, that HMGB1 shuttles from the nucleus to the cytosol and vice versa, with a predominant nuclear localization under resting conditions. Once hyperacetylated (it is still unclear the relative importance of acetyltransferase activation and/or deacetylase inhibition), the NLS is masked and HMGB1 nuclear import is prevented, thereby leading to the cytoplasmic accumulation of the protein. The latter through mechanisms in part delineated is then loaded into secretory lysosomes and exocytosed (Bonaldi *et al.*, 2003). In addition to acetylation, HMGB1 also undergoes different posttranslational modifications such as methylation, glycosilation, poly(ADP-ribosyl)ation, phosphorylation, and oxidation, which likely regulate its extracellular release and biological functions (Hoppe *et al.*, 2006; Ulloa and Messmer, 2006).

As mentioned above, HMGB1 behaves as an inflammatory mediator. Indeed, immune cells not only are the major source of actively secreted HMGB1 but also display high sensitivity to extracellular HMGB1. Thus, the abilities of HMGB1 to activate immune cells and that of the latter to release HMGB1 establish a proinflammatory, paracrine, and/or autocrine vicious circle, which sustains and promotes the immune response within different tissues (Fig. 2). Notably, classic inflammatory mediators such as TNF-α, IL-1β, INF-γ, and bacterial lipopolysaccharides (LPS) also prompt HMGB1 release from immune cells. From a functional point of view, therefore, nuclear HMGB1 behaves as a prototypical cytokine once extracellularly released and plays an important role in developing and maintenance of the inflammatory response. It is worth noting, however, that HMGB1 release occurs considerably later than secretion of other classical proinflammatory mediators (TNF-α, IL-1β). For instance, in a model of septic shock, HMGB1 serum levels increase at 12–18 h whereas TNF-α concentrations reach a peak at 2 h (Yang *et al.*, 2004). As a whole, these findings, along with the notions that HMGB1 serum levels are higher in septic patients destined to die than in survivors and that HMGB1 antibodies protect mice from septic shock demonstrate that the protein is a delayed mediator of inflammation of pathological relevance.

As for the signaling mechanisms adopted by extracellular HMGB1, several receptors able to bind the protein exist on the plasma membrane of different cells. These receptors encompass RAGE (receptor for advanced glycation end products) and TLR (Toll-like receptor)-2 and -4 (Park *et al.*, 2004). Signal transduction through RAGE leads to activation of kinases, such as ERK1/2 and p38, which in turn lead to activation of transcription factors including nuclear factor-κB (NF-κB). On the contrary, HMGB1 binding on TLR-2 or -4 promotes transcriptional activity of NF-κB by activating the kinase MyD88 (myeloid differentiation primary response protein 88) (Park *et al.*, 2003). How activation of different receptors mediates the pleiotypic effects of HMGB1 is still unclear. The report that this protein mainly engages TLR-4 on primary human blood monocytes and macrophages, whereas TLR-2 mediates HMGB1's effects in macrophage-like cells such as RAW 264.7 or CHO (Bell *et al.*, 2006; Yu *et al.*, 2006), hints that HMGB1-dependent signaling is differently mediated in primary cell cultures and neoplastic cell lines.

IV. Role of HMGB1 in CNS (DYS)Function

Findings point to HMGB1 as an active mediator of liver, lung, gut, as well as joint disorders. The protein's role in diseases of the CNS is less appreciated. It is worthy of note that the extracellular presence of HMGB1 has been first discovered in the brain (Bianchi, 2007). Specifically, in 1987 the group of Rauvala, in an effort to isolate novel neurotrophic factors, identified a protein, at that time named p30, which was endowed with potent neurite outgrowth promoting activity. Interestingly, the extracellular protein was rich in basic residues, a feature typical of DNA-binding factors, but appeared associated with the plasma membrane (Rauvala and Pihlaskari, 1987). Further studies than revealed that p30, also called "amphoterin" because of the concomitant presence of positively and negatively charged domains, was indeed HMGB1 (Merenmies *et al.*, 1991). Since its discovery, a great deal of interest focused on HMGB1's roles in CNS functioning. Studies on rodent brain development suggested a role for HMGB1 in regulating proper neuronal migration and sprouting. HMGB1 is differently expressed in the mouse brain at various steps of development and tends to be downregulated in the adult animal (Guazzi *et al.*, 2003). In keeping with this, the expression of the HMGB1-binding factor sulfoglucuronyl carbohydrate as well as that of HMGB1-binding receptor RAGE colocalize at precise regions of the developing mouse cortex and cerebellum in a timely restricted manner (Chou *et al.*, 2004). More interestingly, this tripartite interaction seems not to occur in the adult cerebellum because of reduced expression of both HMGB1 and the glucuronyl carbohydrate. RAGE expression levels are not affected in the adult mouse brain but anti-RAGE antibodies inhibit neurite outgrowth and neuronal migration (Chou *et al.*, 2004). These findings strengthen the relevance of RAGE and its ligand HMGB1 to brain development. In keeping with this, HMGB1 functionally link the immune and endocrine system at the hypothalamic level. It has been reported that proinflammatory cytokines such as IL-1β and TNF-α trigger release of HMGB1 from pituicytes. Other inflammatory mediators including INF-γ do not prompt HMGB1 release but enhance that caused by TNF-α

(Wang *et al.*, 1999). Given the proinflammatory properties of HMGB1 (see above), these findings suggest that the protein contributes to establish a positive feedback loop in the inflammatory response within the brain. Further evidence demonstrating that HMGB1 regulates the neuroendocrine response to immune stimuli comes with the study showing that intracerebroventricular injection of HMGB1 increases brain levels of TNF-α and IL-1β in the mouse and induces anorexia and loss of body weight (Agnello *et al.*, 2002). Subsequent work builds on these findings showing that rats develop fever and show increased TNF-α and IL-1β contents in various brain regions on intracerebroventricular injection of HMGB1 (O'Connor *et al.*, 2003). Interestingly, allodynia occurs in rats receiving HMGB1 in the subarachnoid space at the lumbosacral level (O'Connor *et al.*, 2003). Overall, data point to a key role of HMGB1 in triggering production of proinflammatory mediators in the rodent brain as well as in participating to the establishment of the array of signs and symptoms (the so-called "sickness behavior") that are typically associated with activation of the neuroimmune response.

The finding that HMGB1 regulates neurotransmitter release open new vistas on the protein's roles in brain functioning. Pedrazzi *et al.* (2006) report that nanomolar concentrations of HMGB1 trigger release of [^{3}H]-D-aspartate as well as that of endogenous glutamate from glial subcellular particles (gliosomes) isolated *in vitro*. Of note, the protein does not affect excitatory amino acid release from synaptosomes. In search for the mechanisms underlying the effect of HMGB1 on gliosomes, the authors provide evidence that HMGB1-induced excitatory amino acid release is due to a complex and still in part unclear interaction between the protein, RAGE, and the glial glutamate transporter GLAST. The authors also report that Ca^{2+} regulates HMGB1-dependent stimulation of glutamate release by facilitating the protein's binding to RAGE (Pedrazzi *et al.*, 2006). Of course, it is premature attributing a defined role to HMGB1 in the modulation of glutamate release from astrocytes. It is worthy of note, however, that the protein can be released from neural cells both passively in conditions of necrotic cell death, and actively from astrocytes challenged with forskolin or desamethasone (Passalacqua *et al.*, 1998). Thus, in light of the emerging relevance of glial cells to the regulation of excitatory neurotransmission, these findings suggest that HMGB1-dependent modulation of glutamate release is of pathophysiological significance.

Several reports indicate that, beyond the involvement of HMGB1 in maintenance of brain homeostasis, the protein actively participates to neuropathology. Takata *et al.* (2003) reported that nanomolar concentrations calf thymus-purified HMGB1 binds amyloid fibrils, stabilizes their oligomerization, and inhibits amyloid phagocytosis by cultured rat microglia. Of note, brains from patients affected by Alzheimer's disease show increased levels of HMGB1 which colocalizes with microglia at the level of senile plaques (Takata *et al.*, 2003). In a subsequent study from the same laboratory, HMGB1 immunoreactivity increases in the hippocampi of kainic acid- or β-amyloid-microinjected mice. More interestingly, injection of β-amyloid together with HMGB1 delays fibril clearance and increases their neurotoxic potential (Takata *et al.*, 2004). Altogether, these findings suggest that HMGB1 promotes amyloid neurotoxicity and actively participates to neurodegeneration during Alzheimer's disease.

A large body of information indicates that numerous mechanisms concur to ischemic brain injury and that different neurotoxic events, more or less causally concatenated, occur at different times within the core and ischemic penumbra (Dirnagl *et al.*, 1999). Among these, delayed inflammatory reaction is a well-established event, and there is general consensus that it plays an active role in progression of ischemic brain injury. A still unresolved and relevant issue is that of identifying the triggers of the ischemic neuroinflammatory response. Of course, cellular necrosis has been considered a major source of immune-activating substances, but the mediators responsible for recruitment of peripheral leukocytes as well as activation of resident microglia remain in large part to be unequivocally identified at the molecular level. Also, factors linking the acute to the delayed phase of ischemic neurodegeneration wait to be understood (Lo *et al.*, 2003). In this regard, evidence that HMGB1 is released by necrotic cells along with the notion that the protein behaves as a cytokine able to activate immune cells and prompt release of inflammatory mediators allows to speculate an involvement of HMGB1 in the inflammatory reaction within the ischemic brain. The work by Kim and associates clearly indicates that, at least in the rodent brain, this hypothesis holds true. The authors report that HMGB1 levels decrease in the ischemic hemisphere and increase in the cerebrospinal fluid and serum after brain ischemia in the rat. Importantly, reducing HMGB1 expression by shRNA reduces ischemia-dependent microglia activation and induction of inflammatory cytokines/enzymes (TNF-α, IL-1β, COX-2, and iNOS) in the ischemic brain. Downregulation of HMGB1 brain levels correlates with diminished infarct volumes (Kim *et al.*, 2006). Additional evidence indicating that HMGB1 is responsible for activation of the neuroinflammatory reaction on neurodegeneration comes from experiments showing that cultured neurons undergoing excitotoxicity release HMGB1 in the incubating media. Of note, microglia exposed to these media release proinflammatory mediators, whereas media preadsorbed with anti-HMGB1 loose their proinflammatory activity. Further, recombinant HMGB1 activates cultured microglia and boosts INF-γ-dependent microglia activation (Kim *et al.*, 2006). These results taken together, point to a key role of HMGB1 in triggering neuroinflammation and injury following brain ischemia. Remarkably, the report by Goldstain *et al.* (2006) showing that plasma levels of HMGB1 are elevated up to tenfold (from 16.8 ± 10 to 218 ± 18 ng/ml) in stroke patients (Goldstein *et al.*, 2006) corroborates the hypothesis that this protein is also of pathogenetic relevance to stroke in humans.

V. Conclusions

A wealth of evidence indicates that the nonhistone chromatin-binding protein HMGB1 leads a double life. It plays a major role in homeostatic regulation of nuclear functions, but, once released, behaves as a cytokine with multiple signaling properties (Ulloa and Messmer, 2006). The latter range from beneficial effects such as myocardial regeneration after infarction (Limana *et al.*, 2005) and regulation of CNS development to harmful events, including triggering of immune activation and lethal septic shock. Likely, additional HMGB1 functions will be added to this list shortly. Of course, because of its role in the pathogenesis of different disorders, a great deal of efforts is being directed at developing potent inhibitors of HMGB1's signaling. Potential drugs candidates might target the HMGB1 signaling pathway at different levels in order to prevent HMGB1 release, scavenge the protein once released, or inhibit HMGB1 binding to its receptors. Commercially available anti-HMGB1 drugs are ethyl pyruvate (Ulloa *et al.*, 2002), ethacrynic acid (Killeen *et al.*, 2006), and glicizzyrric acid (Sakamoto *et al.*, 2001). Efficient inhibition of HMGB1 signaling has been also achieved by using HMGB1 box A as a decoy agent or anti-HMGB1 antibodies (Lotze and Tracey, 2005). Unfortunately, however, to date there are no specific and powerful HMGB1 inhibitors with an acceptable degree of selectivity and pharmacokinetic profile. Yet, the great interest of several drug and biotech companies in HMGB1-blocking agents should warrant a successful development in the near future. Hopefully, these efforts will help elucidating the relevance of HMGB1 signaling cascade to the pathophysiology of several human disorders, as well as providing new tolls of significance to therapeutic interventions.

References

Agnello, D., Wang, H., Yang, H., Tracey, K. J., and Ghezzi, P. (2002). HMGB-1, a DNA-binding protein with cytokine activity, induces brain TNF and IL-6 production, and mediates anorexia and taste aversion. *Cytokine* **18,** 231–236.

Allain, F. H., Yen, Y. M., Masse, J. E., Schultze, P., Dieckmann, T., Johnson, R. C., and Feigon, J. (1999). Solution structure of the HMG protein NHP6A and its interaction with DNA reveals the structural determinants for non-sequence-specific binding. *EMBO J.* **18,** 2563–2579.

Bell, C. W., Jiang, W., Reich, C. F., III, and Pisetsky, D. S. (2006). The extracellular release of HMGB1 during apoptotic cell death. *Am. J. Physiol. Cell Physiol.* **291,** C1318–C1325.

Bianchi, M. E. (2007). Interview with Dr. Heikki Rauvala regarding pivotal advance: Analysis of proinflammatory activity of highly purified eukaryotic recombinant HMGB1 (amphoterin). *J. Leukoc. Biol.* **81,** 1–5.

Bonaldi, T., Talamo, F., Scaffidi, P., Ferrera, D., Porto, A., Bachi, A., Rubartelli, A., Agresti, A., and Bianchi, M. E. (2003). Monocytic cells hyperacetylate chromatin protein HMGB1 to redirect it towards secretion. *EMBO J.* **22,** 5551–5560.

Calogero, S., Grassi, F., Aguzzi, A., Voigtlander, T., Ferrier, P., Ferrari, S., and Bianchi, M. E. (1999). The lack of chromosomal protein Hmg1 does not disrupt cell growth but causes lethal hypoglycaemia in newborn mice. *Nat. Genet.* **22,** 276–280.

Chou, D. K., Zhang, J., Smith, F. I., McCaffery, P., and Jungalwala, F. B. (2004). Developmental expression of receptor for advanced glycation end products (RAGE), amphoterin and sulfoglucuronyl (HNK-1) carbohydrate in mouse cerebellum and their role in neurite outgrowth and cell migration. *J. Neurochem.* **90,** 1389–1401.

Dirnagl, U., Iadecola, C., and Moskowitz, M. A. (1999). Pathobiology of ischaemic stroke: An integrated view. *Trends Neurosci.* **22,** 391–397.

Ellwood, K. B., Yen, Y. M., Johnson, R. C., and Carey, M. (2000). Mechanism for specificity by HMG-1 in enhanceosome assembly. *Mol. Cell. Biol.* **20,** 4359–4370.

Giese, K., Amsterdam, A., and Grosschedl, R. (1991). DNA-binding properties of the HMG domain of the lymphoid-specific transcriptional regulator LEF-1. *Genes Dev.* **5,** 2567–2578.

Goldstein, R. S., Gallowitsch-Puerta, M., Yang, L., Rosas-Ballina, M., Huston, J. M., Czura, C. J., Lee, D. C., Ward, M. F., Bruchfeld, A. N., Wang, H., Lesser, M. L., Church, A. L., *et al.* (2006). Elevated high-mobility group box 1 levels in patients with cerebral and myocardial ischemia. *Shock* **25,** 571–574.

Goodwin, G. H., and Johns, E. W. (1973). Isolation and characterisation of two calf-thymus chromatin non-histone proteins with high contents of acidic and basic amino acids. *Eur. J. Biochem.* **40,** 215–219.

Guazzi, S., Strangio, A., Franzi, A. T., and Bianchi, M. E. (2003). HMGB1, an architectural chromatin protein and extracellular signalling factor, has a spatially and temporally restricted expression pattern in mouse brain. *Gene Expr. Patterns* **3,** 29–33.

Hoppe, G., Talcott, K. E., Bhattacharya, S. K., Crabb, J. W., and Sears, J. E. (2006). Molecular basis for the redox control of nuclear transport of the structural chromatin protein Hmgb1. *Exp. Cell Res.* **312,** 3526–3538.

Killeen, M. E., Englert, J. A., Stolz, D. B., Song, M., Han, Y., Delude, R. L., Kellum, J. A., and Fink, M. P. (2006). The phase 2 enzyme inducers ethacrynic acid, DL-sulforaphane, and oltipraz inhibit lipopolysaccharide-induced high-mobility group box 1 secretion by RAW 264.7 cells. *J. Pharmacol. Exp. Ther.* **316,** 1070–1079.

Kim, J. B., Sig, C. J., Yu, Y. M., Nam, K., Piao, C. S., Kim, S. W., Lee, M. H., Han, P. L., Park, J. S., and Lee, J. K. (2006). HMGB1, a novel cytokine-like mediator linking acute neuronal death and delayed neuroinflammation in the postischemic brain. *J. Neurosci.* **26,** 6413–6421.

Landsman, D., and Bustin, M. (1993). A signature for the HMG-1 box DNA-binding proteins. *Bioessays* **15,** 539–546.

Limana, F., Germani, A., Zacheo, A., Kajstura, J., Di Carlo, A., Borsellino, G., Leoni, O., Palumbo, R., Battistini, L., Rastaldo, R., Muller, S., Pompilio, G., *et al.* (2005). Exogenous high-mobility group box 1 protein induces myocardial regeneration after infarction via enhanced cardiac C-kit+ cell proliferation and differentiation. *Circ. Res.* **97,** e73–e83.

Lo, E. H., Dalkara, T., and Moskowitz, M. A. (2003). Mechanisms, challenges and opportunities in stroke. *Nat. Rev. Neurosci.* **4,** 399–415.

Lotze, M. T., and Tracey, K. J. (2005). High-mobility group box 1 protein (HMGB1): Nuclear weapon in the immune arsenal. *Nat. Rev. Immunol.* **5,** 331–342.

McKinney, K., and Prives, C. (2002). Efficient specific DNA binding by p53 requires both its central and C-terminal domains as revealed by studies with high-mobility group 1 protein. *Mol. Cell. Biol.* **22,** 6797–6808.

Merenmies, J., Pihlaskari, R., Laitinen, J., Wartiovaara, J., and Rauvala, H. (1991). 30-kDa heparin-binding protein of brain (amphoterin) involved in neurite outgrowth. Amino acid sequence and localization in the filopodia of the advancing plasma membrane. *J. Biol. Chem.* **266,** 16722–16729.

O'Connor, K. A., Hansen, M. K., Rachal, P. C., Deak, M. M., Biedenkapp, J. C., Milligan, E. D., Johnson, J. D., Wang, H., Maier, S. F., Tracey, K. J., and Watkins, L. R. (2003). Further characterization of high mobility group box 1 (HMGB1) as a proinflammatory cytokine: Central nervous system effects. *Cytokine* **24,** 254–265.

Park, J. S., Arcaroli, J., Yum, H. K., Yang, H., Wang, H., Yang, K. Y., Choe, K. H., Strassheim, D., Pitts, T. M., Tracey, K. J., and Abraham, E. (2003). Activation of gene expression in human neutrophils by high mobility group box 1 protein. *Am. J. Physiol. Cell Physiol.* **284,** C870–C879.

Park, J. S., Svetkauskaite, D., He, Q., Kim, J. Y., Strassheim, D., Ishizaka, A., and Abraham, E. (2004). Involvement of toll-like receptors 2 and 4 in cellular activation by high mobility group box 1 protein. *J. Biol. Chem.* **279,** 7370–7377.

Passalacqua, M., Patrone, M., Picotti, G. B., Del Rio, M., Sparatore, B., Melloni, E., and Pontremoli, S. (1998). Stimulated astrocytes release high-mobility group 1 protein, an inducer of LAN-5 neuroblastoma cell differentiation. *Neuroscience* **82,** 1021–1028.

Pedrazzi, M., Raiteri, L., Bonanno, G., Patrone, M., Ledda, S., Passalacqua, M., Milanese, M., Melloni, E., Raiteri, M., Pontremoli, S., and Sparatore, B. (2006). Stimulation of excitatory amino acid release from adult mouse brain glia subcellular particles by high mobility group box 1 protein. *J. Neurochem.* **99,** 827–838.

Rauvala, H., and Pihlaskari, R. (1987). Isolation and some characteristics of an adhesive factor of brain that enhances neurite outgrowth in central neurons. *J. Biol. Chem.* **262,** 16625–16635.

Sakamoto, R., Okano, M., Takena, H., and Ohtsuki, K. (2001). Inhibitory effect of glycyrrhizin on the phosphorylation and DNA-binding abilities of high mobility group proteins 1 and 2 *in vitro*. *Biol. Pharm. Bull.* **24,** 906–911.

Scaffidi, P., Misteli, T., and Bianchi, M. E. (2002). Release of chromatin protein HMGB1 by necrotic cells triggers inflammation. *Nature* **418,** 191–195.

Takata, K., Kitamura, Y., Kakimura, J., Shibagaki, K., Tsuchiya, D., Taniguchi, T., Smith, M. A., Perry, G., and Shimohama, S. (2003). Role of high mobility group protein-1 (HMG1) in amyloid-beta homeostasis. *Biochem. Biophys. Res. Commun.* **301,** 699–703.

Takata, K., Kitamura, Y., Tsuchiya, D., Kawasaki, T., Taniguchi, T., and Shimohama, S. (2004). High mobility group box protein-1 inhibits microglial Abeta clearance and enhances Abeta neurotoxicity. *J. Neurosci. Res.* **78,** 880–891.

Ulloa, L., and Messmer, D. (2006). High-mobility group box 1 (HMGB1) protein: Friend and foe. *Cytokine Growth Factor Rev.* **17,** 189–201.

Ulloa, L., Ochani, M., Yang, H., Tanovic, M., Halperin, D., Yang, R., Czura, C. J., Fink, M. P., and Tracey, K. J. (2002). Ethyl pyruvate prevents lethality in mice with established lethal sepsis and systemic inflammation. *Proc. Natl. Acad. Sci. USA* **99,** 12351–12356.

Verrijdt, G., Haelens, A., Schoenmakers, E., Rombauts, W., and Claessens, F. (2002). Comparative analysis of the influence of the high-mobility group box 1 protein on DNA binding and transcriptional activation by the androgen, glucocorticoid, progesterone and mineralocorticoid receptors. *Biochem. J.* **361,** 97–103.

Wang, H., Vishnubhakat, J. M., Bloom, O., Zhang, M., Ombrellino, M., Sama, A., and Tracey, K. J. (1999). Proinflammatory cytokines (tumor necrosis factor and interleukin 1) stimulate release of high mobility group protein-1 by pituicytes. *Surgery* **126,** 389–392.

Yang, D., Chen, Q., Yang, H., Tracey, K. J., Bustin, M., and Oppenheim, J. J. (2007). High mobility group box-1 (HMGB1) protein induces the migration and activation of human dendritic cells and acts as an alarmin. *J. Leukoc. Biol.* **81,** 59–66.

Yang, H., Ochani, M., Li, J., Qiang, X., Tanovic, M., Harris, H. E., Susarla, S. M., Ulloa, L., Wang, H., DiRaimo, R., Czura, C. J., Wang, H., *et al.* (2004). Reversing established sepsis with antagonists of endogenous high-mobility group box 1. *Proc. Natl. Acad. Sci. USA* **101,** 296–301.

Yu, M., Wang, H., Ding, A., Golenbock, D. T., Latz, E., Czura, C. J., Fenton, M. J., Tracey, K. J., and Yang, H. (2006). HMGB1 signals through toll-like receptor (TLR) 4 and TLR2. *Shock* **26,** 174–179.

Yu, S. S., Li, H. J., Goodwin, G. H., and Johns, E. W. (1977). Interaction of non-histone chromosomal proteins HMG1 and HMG2 with DNA. *Eur. J. Biochem.* **78,** 497–502.

EARLY UPREGULATION OF MATRIX METALLOPROTEINASES FOLLOWING REPERFUSION TRIGGERS NEUROINFLAMMATORY MEDIATORS IN BRAIN ISCHEMIA IN RAT

Diana Amantea,* Rossella Russo,* Micaela Gliozzi,[†] Vincenza Fratto,[†] Laura Berliocchi,*[,‡] G. Bagetta,* G. Bernardi,[§] and M. Tiziana Corasaniti[†,‡]

*Department of Pharmacobiology, UCHAD Section of Neuropharmacology of Normal and Pathological Neuronal Plasticity, University of Calabria, 87036 Rende, Italy
[†]Department of Pharmacobiological Sciences, Faculty of Pharmacy, University "Magna Graecia" of Catanzaro, 88100 Catanzaro, Italy
[‡]Mondino-Tor Vergata Center for Experimental Neuropharmacology, University of Rome Tor Vergata, 00133 Rome, Italy
[§]CERC-Fondazione S. Lucia IRCCS, University of Rome Tor Vergata, 00133 Rome, Italy

Abnormal expression of matrix metalloproteinases (MMPs) has been implicated in the pathophysiology of neuroinflammatory processes that accompany most central nervous system disease. In particular, early upregulation of the gelatinases MMP-2 and MMP-9 has been shown to contribute to disruption of the blood–brain barrier and to death of neurons in ischemic stroke. *In situ* zymography reveals a significant increase in gelatinolytic MMPs activity in the ischemic brain hemisphere after 2-h middle cerebral artery occlusion (MCAo) followed by 2-h reperfusion in rat. Accordingly, gel zymography demonstrates that expression and activity of MMP-2 and MMP-9 are enhanced in cortex and striatum ipsilateral to the ischemic insult. The latter effect appears to be instrumental for development

DOI: 10.1016/S0074-7742(07)82008-3

0074-7742/07 $35.00

of delayed brain damage since administration of a broad spectrum, highly specific MMPs inhibitor, GM6001, but not by its negative control, results in a significant (50%) reduction in ischemic brain volume. Increased gelatinase activity in the ischemic cortex coincides with elevation (166% vs sham) of mature interleukin-1β (IL-1β) after 2-h reperfusion and this does not appear to implicate a caspase-1-dependent processing of pro(31 kDa)-IL-1β to yield mature (17 kDa) IL-1β. More importantly, when administered at a neuroprotective dose GM6001 abolishes the early IL-1β increase in the ischemic cortex and reduces the cleavage of the cytokine proform supporting the deduction that MMPs may initiate IL-1β processing.

In conclusion, development of tissue damage that follows transient ischemia implicates a crucial interplay between MMPs and mediators of neuroinflammation (e.g., IL-1β), and this further underscores the therapeutic potential of MMPs inhibitors in the treatment of stroke.

I. Introduction

Interleukin-1β (IL-1β) is a proinflammatory cytokine that has been identified as an important mediator of neurodegeneration induced by excitatory or traumatic brain injury and, most notably, by experimental cerebral ischemia in rodents (Rothwell, 2003). Induction of IL-1β mRNA has been shown in rats following either permanent (Buttini *et al.*, 1994; Liu *et al.*, 1993) or transient (Wang *et al.*, 1994) middle cerebral artery occlusion (MCAo). Accordingly, IL-1β protein increases very early following permanent MCAo (Davies *et al.*, 1999; Legos *et al.*, 2000) and peaks within hours of reperfusion in transient focal ischemic models in rodents (Hara *et al.*, 1997b; Zhang *et al.*, 1998). Intracerebral injection of neutralizing anti-IL-1β antibody to rats reduces ischemic brain damage (Yamasaki *et al.*, 1995) and both intracerebroventricular and systemic administration of IL-1receptor antagonist (IL-1ra) markedly reduce brain damage induced by focal stroke, further implicating IL-1β in ischemic pathophysiology (Garcia *et al.*, 1995; Mulcahy *et al.*, 2003; Relton and Rothwell, 1992; Relton *et al.*, 1996).

IL-1β is synthesized as a precursor molecule, pro-IL-1β, which is cleaved and converted into the mature, biologically active, form of the cytokine by caspase-1, formerly referred to as interleukin-1β converting enzyme (ICE; Black *et al.*, 1988; Howard *et al.*, 1991; Thornberry *et al.*, 1992). Inhibition of caspase-1 by Ac-YVAD-cmk affords neuroprotection in rodent models of permanent (Rabuffetti *et al.*, 2000) or transient (Hara *et al.*, 1997b) MCAo (tMCAo), and evidence from knockout mice indicates that caspase-1 is important in the development of cerebral ischemic damage (Friedlander and Yuan, 1998; Schielke *et al.*, 1998).

However, to date, it is not clear whether neuroprotection yielded by caspase-1 preferring inhibitors is mediated by reduced IL-1β production or by interference with the death process.

Although *in vitro* studies have clearly established the role of ICE in the maturation of IL-1β, evidence from ICE-deficient mice suggests that cytokine activation might also involve other mechanisms (Fantuzzi *et al.*, 1997). Matrix metalloproteinases (MMPs) have been suggested to contribute to the biological processing of IL-1β as they have been shown to be involved in both the maturation and inactivation of the cytokine *in vitro* (Ito *et al.*, 1996; Schönbeck *et al.*, 1998).

MMPs are zinc-dependent endopeptidases, classically recognized as matrix-degrading enzymes implicated in tissue remodeling during development, wound healing, and angiogenesis. MMPs are expressed as zymogens that are activated on disruption of the zinc–thiol interaction between the catalytic site and the prodomain. MMPs cleave protein components of the extracellular matrix (ECM) such as collagen, proteoglycan, and laminin, but also process a number of cell surface and soluble proteins including receptors, cytokines, and chemokines (Sternlicht and Werb, 2001). In addition to their physiological roles, MMPs are markedly upregulated in the central nervous system (CNS) in response to injury and have been implicated in the propagation and regulation of neuroinflammatory processes that accompany most CNS disease (Cunningham *et al.*, 2005; Rosenberg, 2002). Evidence suggests that abnormal MMP activity plays a role in the pathophysiology of cerebral ischemia. In particular, the gelatinases MMP-2 and MMP-9 become activated following focal brain ischemia and participate to the disruption of the blood–brain barrier (BBB) and hemorrhagic transformation following injury both in animal models (Asahi *et al.*, 2000; Heo *et al.*, 1999; Romanic *et al.*, 1998; Rosenberg *et al.*, 1998) and stroke patients (Horstmann *et al.*, 2003; Rosell *et al.*, 2006). Treatment with MMP inhibitors or MMP neutralizing antibodies has been shown to decrease infarct volume and prevent BBB disruption after permanent and tMCAo in rodents (Asahi *et al.*, 2000; Gasche *et al.*, 2001; Romanic *et al.*, 1998). MMP-9, but not MMP-2 (Asahi *et al.*, 2001a), gene knockout is associated with reduced infarct size and less BBB damage in mouse models of ischemic stroke (Asahi *et al.*, 2000, 2001b). This large body of evidence suggests that MMPs and, most notably, the gelatinases MMP-2 and MMP-9 might contribute to the development of ischemic brain damage, though the underlying mechanisms need to be discovered. Various studies have emphasized the role of MMPs and their endogenous inhibitors (TIMPs) in the regulation of neuronal cell death through the modulation of excitotoxicity (Jourquin *et al.*, 2003), anoikis (Gu *et al.*, 2002), calpain activity (Copin *et al.*, 2005), death receptor activation (Wetzel *et al.*, 2003), neurotrophic factor bioavailability (Lee *et al.*, 2001), and production of neurotoxic products (Gu *et al.*, 2002; Zhang *et al.*, 2003a). This suggests that, in addition to ECM degradation, MMPs might elicit some direct, pathogenic effects, which contribute to brain tissue damage under various neuropathological conditions.

Here, we suggest that cytokine maturation may represent a likely mechanism by which MMPs contribute to ischemic neuronal death and postulate that MMPs might be involved in the early increase of IL-1β occurring following tMCAo in rat. In fact, increased gelatinolytic, namely, MMP-2 and MMP-9, activity is observed as early as 2 h following reperfusion, and this is coincident with a significant increase of IL-1β levels in the ischemic cortex. More interestingly, pharmacological inhibition of gelatinolytic activity prevents elevation of cortical IL-1β levels and results in significant neuroprotection.

II. Methods

A. Focal Cerebral Ischemia and Drug Treatments

Adult male Wistar rats (Charles River, Calco, Como, Italy) were housed under controlled environmental conditions with ambient temperature of 22 °C, relative humidity of 65%, and 12-h light:12-h dark cycle, with free access to food and water.

Brain ischemia was induced by MCAo in rats weighing 280–320 g by intraluminal filament, using the relatively noninvasive technique previously described by Longa *et al.* (1989). Briefly, rats were anesthetized with 5% isoflurane in air and were maintained with the lowest acceptable concentration of the anesthetic (1.5–2%). Body temperature was measured with a rectal probe and was kept at 37 °C during the surgical procedure with a heating pad. Under an operating microscope, the external and internal right carotid arteries were exposed through a neck incision. The external carotid artery was cut approximately 3 mm above the common carotid artery (CCA) bifurcation and a silk suture was tied loosely around the external carotid stump. A silicone-coated nylon filament (diameter, 0.28 mm) was then inserted into the external carotid artery and gently advanced into the internal carotid artery, approximately 18 mm from the carotid bifurcation, until mild resistance was felt, thereby indicating occlusion of the origin of the middle cerebral artery in the Willis circle. The silk suture was tightened around the intraluminal filament to prevent bleeding. The wound was then sutured and anesthesia discontinued. Sham rats were exposed to the same surgical procedure without occlusion of MCA.

To allow reperfusion, rats were briefly reanesthetized with isoflurane, and the nylon filament was withdrawn 2 h after MCAo. After the discontinuation of isoflurane and wound closure, the animals were allowed to awake and were kept in their cages with free access to food and water.

N-[(2*R*)-2-(Hydroxamidocarbonylmethyl)-4-methylpenthanoyl]-L-tryptophan methylamide (GM6001; also known as Galardin), a potent broad-range inhibitor of MMPs (Levy *et al.*, 1998), and its negative control (N-*t*-butoxycarbonyl-L-leucyl-L-tryptophan methylamide, GM6001 negative control), obtained from Calbiochem (La Jolla, CA), were dissolved in DMSO and administered in a volume of 2 μl through the external carotid artery (i.a.), 15 min prior to MCAo.

All the experimental procedures were in accordance to the guidelines of the European Community Council Directive 86/609, included in the D.M. 116/1992 of the Italian Ministry of Health.

B. Neuropathology and Quantification of Ischemic Damage

Cerebral infarct volume was evaluated 22 h after reperfusion in rats subjected to 2-h MCAo. Rats were sacrificed by decapitation and the brains were rapidly removed. Eight serial sections from each brain were cut at 2-mm intervals from the frontal pole using a rat brain matrix (Harvard Apparatus, Massachusetts). To measure ischemic damage, brain slices were stained in a solution containing 2% 2,3,5-triphenyltetrazolium chloride (TTC) in saline, at 37 °C. After 10-min incubation, the slices were transferred to 10% neutral buffered formaldehyde and stored at 4 °C prior to analysis. Images of TTC-stained sections were captured using a digital scanner and analyzed using an image analysis software (ImageJ, version 1.30). The infarct volume (mm^3) was calculated by summing the infarcted area (unstained) of the eight sections and multiplying by the interval thickness between sections as previously described (Li *et al.*, 2000).

C. IL-1β ELISA

Immunoreactive IL-1β levels were analyzed in individual brain cortical tissue homogenates by an established, rat-specific, sandwich ELISA previously described (Corasaniti *et al.*, 2001; Hagan *et al.*,1996), using an immunoaffinity-purified polyclonal sheep anti-rat IL-1β-coating antibody (1 μg/ml) and a biotinylated, immunoaffinity-purified polyclonal sheep anti-rat IL-1β-detecting antibody (1:1000 dilution), kindly provided by Dr. Stephen Poole (National Institute of Biological Standards and Controls, NIBSC, Hertfordshire, United Kingdom). Poly-horseradish peroxidase-conjugated streptavidin (CLB, Amsterdam, the Netherlands) was used at 1:5000 dilution and the color was developed by using the chromogen o-phenylenediamine. Optical densities (OD) were read at 492 nm by using an automated plate reader (Multiscan MS, Labsystems, Helsinki, Finland) and cytokine levels were calculated by interpolation from a standard curve obtained from recombinant rat IL-1β (0.0–1000 pg/ml). Data were corrected for protein concentration and the results expressed as picogram of IL-1β per milligram of protein.

D. Western Blotting

For Western blotting analysis of mature and pro-IL-1β immunoreactivities, 20 μg of proteins, from the same aliquots used for ELISA assay, were resolved by 15% sodium dodecyl sulfate-polyacrylamide gel electrophoresis (SDS-PAGE), transferred to nitrocellulose membranes (Optitran BA-S 83, Schleicher & Schuell Bioscience, Dassel, Germany), and probed overnight at 4 °C with a polyclonal sheep anti-rat antibody, which recognizes both the precursor and mature forms of this cytokine (S1002/BM, NIBSC; 7.5 μg/ml). The membranes were then incubated with horseradish peroxidase-conjugated anti-sheep IgG (1:5000 dilution; Chemicon International, Inc., Temecula, CA) for 1 h at room temperature. Immunoreactivity was visualized by chemiluminescent detection (Amersham Biosciences, GE Healthcare, Milan, Italy).

E. *In Situ* Zymography

In situ zymography with the MMP fluorogenic substrate DQ-gelatin-FITC (Molecular Probes, Eugene, OR) was performed on OCT-embedded fresh brain cryostat-cut sections obtained from rat sacrificed after 2-h MCAo followed by 2-h reperfusion, as previously described (Gu *et al.*, 2002). Briefly, rat brains were dissected out, immediately embedded in OCT (Tissue-Tek, Pennsylvania) and frozen on dry ice; 15-μm-thick coronal sections were cut using a cryostat, air-dried for 1 h at room temperature, then rehydrated in PBS, and incubated overnight at 37 °C with the quenched fluorogenic substrate DQ-gelatin-FITC (40 μg/ml in PBS). The excess of fluorogenic substrate was washed out by three washing steps, 5 min each, in PBS. Sections were fixed with 2% paraformaldehyde in PBS for 5 min and nuclei counterstained with propidium iodide (0.5 μg/ml) for 20 min at room temperature. Mounted slides were examined by confocal microscopy to detect the green fluorescence due to gelatinolytic activity.

F. Gel Zymography

MMP-2 (gelatinase A) and MMP-9 (gelatinase B) gelatinolytic activities were detected by gelatin gel zymography (Gu *et al.*, 2005). Individual brain striatal, cortical, and hippocampal tissue samples ($n = 3$ per experimental group) were homogenized in ice-cold Tris-buffered saline (TBS), containing 150-mM NaCl, 5-mM $CaCl_2$, 0.05% Brij35, pH 7.6, 0.02% NaN_3, 1% Triton X-100, 100-μM PMSF, and a protease inhibitor cocktail (Sigma, Milan), and centrifuged at 14,000 × *g* for 20 min at 4 °C. Supernatants were subjected to affinity precipitation with gelatin-conjugated Sepharose beads (Gelatine-Sepharose 4B, Amersham Biosciences, GE Healthcare, Milan, Italy), overnight at 4 °C. The bound proteins were eluted from the beads in TBS containing 10% DMSO by shaking for 1 h at 4 °C. Ten microliters of extracted brain samples were diluted (1:1) in a nonreducing loading buffer (0.0625-M Tris-HCl, pH 6.8, 2% SDS, 10% glycerol, 0.25% bromophenol blue) and subjected to electrophoresis through a 10% SDS polyacrylamide gel copolymerized with 0.1% gelatin. After electrophoretic separation, the gel was incubated (2 × 30 min) in 2.5% Triton X-100 to remove SDS, washed two times, 15 min each, with water and then incubated for 40 h at 37 °C in a developing buffer containing 50-mM Tris–HCl, pH 7.78; 10-mM $CaCl_2$ and 0.02% NaN_3. Following incubation, the gel was stained for 1 h with 0.25% Coomassie Brilliant Blue R-250 diluted with methanol (50%) and acetic acid (10%) and finally destained in a solution of acetic acid:methanol:water (1:3:6). Enzyme activity attributed to MMP-2 and MMP-9 was visualized (on the basis of molecular weight) in the gelatin-containing zymograms as clear bands against a blue background.

G. Fluorimetric Caspase-1 Activity Assay

Individual brain cortical tissue samples ($n = 4$ per experimental group) were rapidly dissected out and homogenized in ice-cold lysis buffer (50-mM HEPES, pH 7.4, 150-mM NaCl, 5-mM $MgCl_2$, 5-mM EDTA, 0.1% CHAPS, 5-mM DTT, 10 μg/ml pepstatin A, 10-μg/ml leupeptin, 10-μg/ml aprotinin); following 10 min incubation on ice, samples were centrifuged at 12,000 × *g* for 10 min at 4 °C and protein concentration in supernatants was determined by the DC protein assay (Bio-Rad Laboratories, Milan, Italy). Brain cortical supernatants were diluted in assay buffer (100-mM HEPES, pH 7.4, 5-mM EDTA, 0.1% CHAPS, 5-mM dithiothreitol, 10% glycerol) to a final concentration of 1.2 μg protein/μl and incubated in triplicate in a 96-well clear-bottom plate with the fluorogenic substrate acetyl-Trp-Glu-His-Asp-7-amino-4-methylcoumarin (Ac-WEHD-AMC; 10 μM; Bachem) (Rano *et al.*, 1997). Production of fluorescent-free AMC, released by caspase-1 activity, was monitored over 60 min at 37 °C using a microplate fluorometer (Victor^2 multilabel counter, Perkin-Elmer Life Sciences; excitation, 355 nm; emission, 460 nm). Specific contribution of caspase-1 activity in each brain extract was determined by preincubating parallel sample aliquots with the caspase-1 preferring inhibitor acetyl-Trp-Glu-His-Asp-aldehyde (Ac-WEHD-CHO; 10 μM; Bachem) (Garcia-Calvo *et al.*, 1998) for 10 min at 37 °C prior to the addition of the caspase substrate; the difference between the substrate cleavage activity in the absence and presence of Ac-WEHD-CHO was regarded as specific caspase-1 activity. The increase in fluorescence was linear for 40 min after addition of the fluorogenic substrate. Data were analyzed by linear regression within the linear range of the enzymatic reaction and the fluorescence units were converted to micromoles of AMC by using a standard curve generated with free AMC (Calbiochem); the results were expressed as micromoles of free AMC released per minute per milligram of protein and reported as mean ± SEM.

H. Statistical Analysis

Data are expressed as mean ± SEM and statistical analysis performed by ANOVA followed by Dunnett's or Tukey's post hoc tests using the Prism 3 program (GraphPAD Software for Science, San Diego, CA). Differences were considered statistically significant when $p < 0.05$.

III. Results

In situ zymography revealed a considerable increase of total gelatinolytic MMP activity in the brain of rats subjected to 2-h MCAo followed by 2-h reperfusion (Fig. 1). This early increase in gelatinolytic activity was evident in the brain areas supplied by the MCA, including the striatum (E and G) and the periinfarct motor cortex (M and O) as compared to the corresponding regions of sham-operated animals (A, C, I, and K). MMPs broadly colocalized with nuclear DNA staining as shown by the merging of the *in situ* zymography signal with propidium iodide staining (D, H, L, and P).

Consistent with these findings, gel zymography assay demonstrated that both MMP-2 and MMP-9 levels were increased in the ipsilateral, ischemic cortex of rats subjected to tMCAo, as compared to the contralateral, nonischemic hemisphere; whereas no difference was found in the cortical samples from sham-operated animals (Fig. 2A). As shown in Fig. 2B, enhanced gelatinolytic activity corresponding to MMP-2 and MMP-9 was also observed in the striatum with less significant changes in the hippocampus, a brain region only marginally affected by the ischemic insult produced by MCAo (Fig. 2B). On the basis of molecular weight, MMP-9 increases predominantly in the dimeric (250-kDa band) and latent (90- to 95-kDa band) "pro" forms, whereas enhanced levels of MMP-2 following ischemia/reperfusion occurred mainly in the active forms (65- to 67-kDa band). The levels of MMP-9 were low, in comparison to MMP-2, in both ischemic and nonischemic hemispheres and throughout the brain regions examined, as shown by the weaker gelatinolytic bands on the zymography gels.

To determine whether administration of the broad spectrum MMP inhibitor, GM6001, affects brain gelatinolytic activity following brain ischemia, we performed gel zymography on striatal tissue samples from rats exposed to 2-h MCAo followed by 2-h reperfusion. As shown in Fig. 3, intra-arterial administration of GM6001 (5 μg/rat) prevented the increase of MMP-2 and MMP-9 activity in the striatum ipsilateral to the ischemic insult, otherwise observed after pretreatment with the GM6001 negative, inactive, control.

Inhibition of MMPs by GM6001 resulted in a significant reduction of MCAo-induced brain damage as revealed 22 h after reperfusion by the TTC-staining

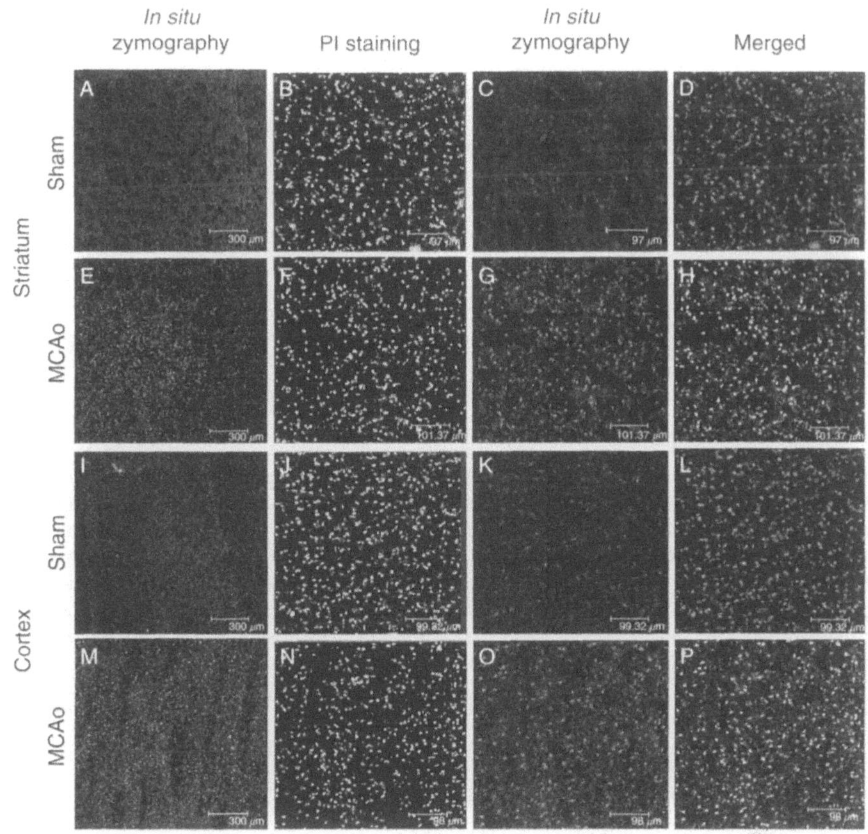

FIG. 1. Increased MMP gelatinolytic activity after transient focal cerebral ischemia. Typical *in situ* zymography with the MMP fluorogenic substrate DQ-gelatin FITC showing the robust increase of MMP gelatinolytic activity in the striatum (E and G) and cortex (M and O) after 2-h occlusion of MCA followed by 2-h reperfusion compared with sham-operated control (A, C, I, and K). Activity appeared as a green fluorescent product and developed after incubation of coronal sections (10-μm in thickness) with the fluorogenic substrate DQ-gelatin FITC (panels A, C, E, G, I, K, M, and O). *In situ* zymography merged with nuclear DNA staining with propidium iodide (PI) dye (red plus green in panels D, H, L, and P) shows that increased gelatinolytic activity (green fluorescence in panels G and O) broadly colocalizes with DNA staining (red fluorescence in panels F and N) in ischemic striatum (H) and cortex (P). Scale bar: 97–101 μm in B–D, F, G, H, J, K, L, N, O, and P; 300 μm in A, E, I, and M. (See Color Insert.)

technique. As illustrated in Fig. 4, administration of GM6001 negative control (GM Neg, 5 μg/rat, i.a.), given 15 min before tMCAo, resulted in an extensive brain damage, comparable to vehicle-treated control (data not shown), involving the cortex and the striatum (infarct volume: 502.6 ± 18.4 mm^3); this was significantly reduced by administration of 5 μg of GM6001 (infarct volume:

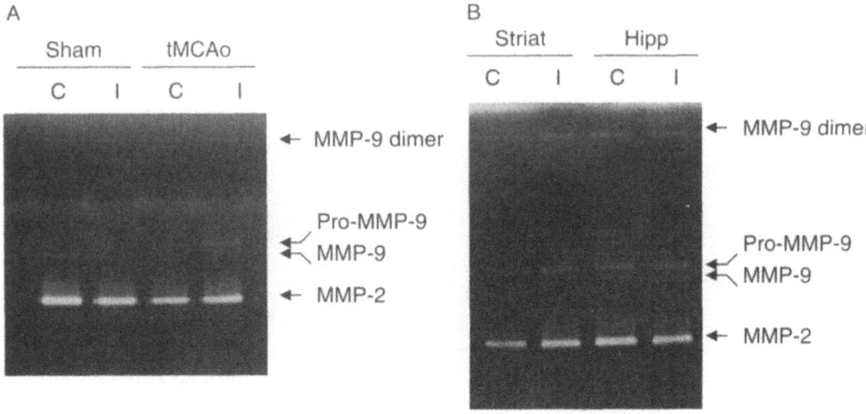

FIG. 2. Increased MMP-9 and MMP-2 after transient focal cerebral ischemia. (A) Typical gel zymography of brain cortical homogenates, subjected to affinity precipitation with gelatin-conjugated Sepharose beads, after 2-h occlusion of MCA followed by 2-h reperfusion. Representative zymogram gel showing elevation of MMP-9 and MMP-2 in the ipsilateral (I), ischemic side as compared to contralateral (C), nonischemic side in a rat subjected to tMCAo but not in a sham-operated animal. (B) Typical gel zymography of striatal and hippocampal brain homogenates, subjected to affinity precipitation with gelatin-conjugated Sepharose beads, after 2-h transient cerebral ischemia and 2-h reperfusion. Representative zymogram gel showing elevation of MMP-9 and MMP-2 in the ipsilateral (I), ischemic side as compared to contralateral (C), nonischemic side.

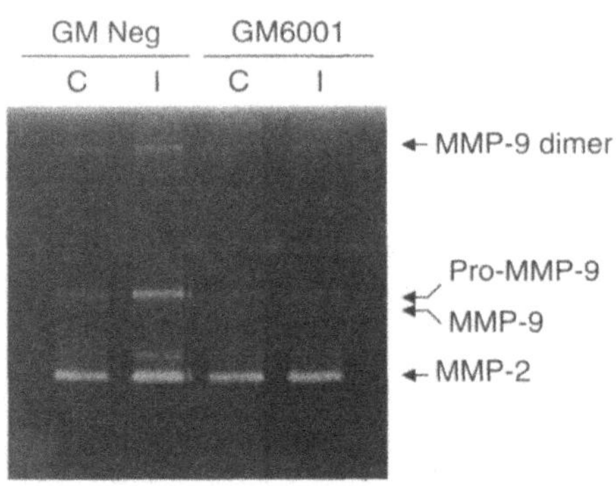

FIG. 3. GM6001 decreases MMP-9 and MMP-2 activity in the brain of rats subjected to transient focal cerebral ischemia. Rats pretreated with the broad-spectrum metalloproteinase inhibitor, GM6001 (5 μg/rat i.a., 15 min before MCAo), but not with the inactive, negative control of GM6001 (GM Neg; 5 μg/rat i.a.), show decreased MMP-9 and MMP-2 activity after ischemia (2-h MCAo followed by 2-h reperfusion) as assessed by gel zymography of brain striatal homogenates subjected to affinity precipitation with gelatin-conjugated Sepharose beads.

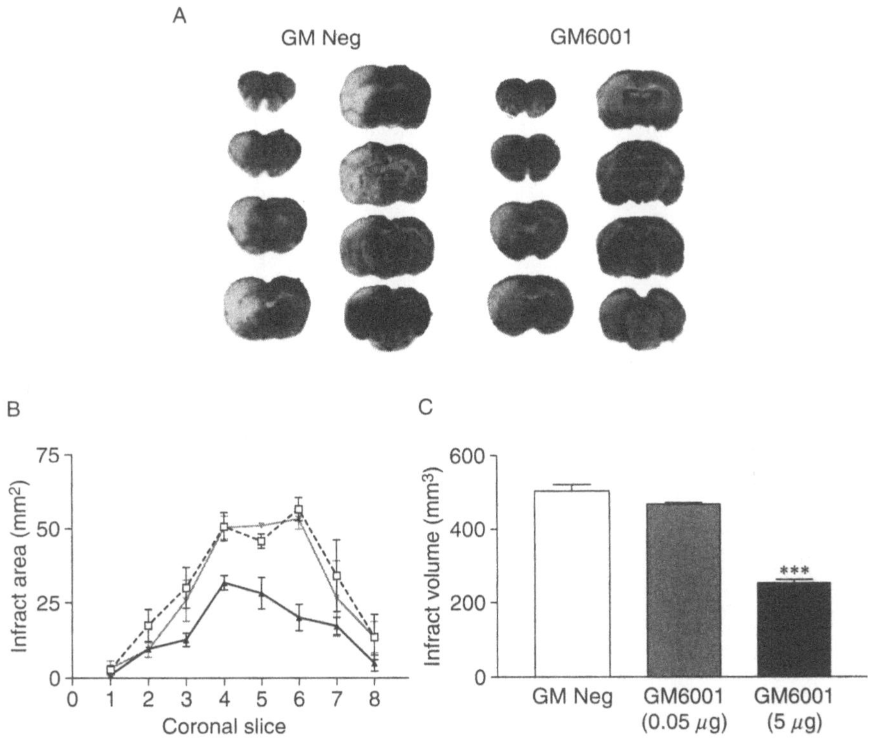

FIG. 4. GM6001 protects against brain damage after transient focal cerebral ischemia. (A) Histological evidence for GM6001-mediated neuroprotection against brain damage produced by tMCAo. After 2-h MCA and 22-h reperfusion, eight consecutive coronal sections from each brain were cut at 2-mm intervals from the frontal pole and incubated in TTC, which stains viable tissue red but not infracted areas. GM6001 or its negative control (GM Neg) were dissolved in DMSO and administered (5 μg/rat, 2-μl injection volume) into the external carotid artery 15 min before MCAo. Quantification of infarct area (B) and volume (C) by TTC staining revealed significant neuroprotection by GM6001 (5 μg/rat) as compared to its negative control. Infarct volumes were determined after 2h MCAo followed by 22-h reperfusion by summing the infarcted area of the eight TTC-stained sections and multiplying by the interval thickness between sections. Values are expressed as mean ± SEM and analyzed by ANOVA followed by Dunnett's post hoc test ($n = 3$ rats per experimental group). *** denotes $p < 0.001$ versus GM Neg. (See Color Insert.)

251.9 ± 9.5 mm^3, $p < 0.001$ vs GM Neg), whereas a lower dose (0.05 μg) resulted ineffective (infarct volume: 467.0 ± 3.9 mm^3).

Under our present experimental conditions, ELISA assay of IL-1β revealed significantly increased cytokine levels in the ipsilateral cortex of rats subjected to MCAo followed by 2-h reperfusion (Fig. 5A), whereas a less marked increase was detected following 1-h reperfusion (data not shown). More importantly, cytokine

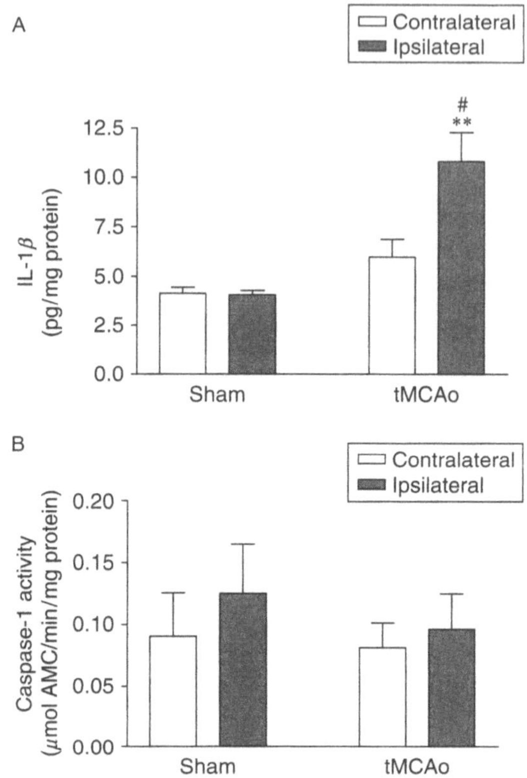

FIG. 5. Transient focal cerebral ischemia increases IL-1β levels but not caspase-1 activity in the ischemic cortex of rat. (A) Immunoreactive IL-1β levels are increased in the ipsilateral (I), ischemic cortex of rats ($n = 6$) subjected to 2-h occlusion of MCA followed by 2-h reperfusion as compared to sham-operated animals ($n = 4$). Immunoreactive IL-1β levels were assayed in individual brain cortical tissue samples by an established, rat specific, sandwich ELISA. IL-1β levels were corrected for protein concentration and the results expressed as pg of IL-1β per milligram of protein. Data are expressed as mean $\pm$ SEM ($n = 4$–6 per group). The resulting means were evaluated statistically for differences using ANOVA followed by Tukey-Kramer test for multiple comparisons. # denotes $p < 0.05$ versus control, contralateral cortex (C); ** denote $p < 0.01$ versus sham (contralateral and ipsilateral). (B) Caspase-1 activity does not increase in the ischemic cortex of rats subjected to transient occlusion (2 h) of MCA followed by 2-h reperfusion. Caspase-1 activity was determined by measuring cleavage of the fluorogenic substrate Ac-WEHD-AMC in individual cortical homogenates obtained from sham-operated and ischemic rats ($n = 4$ per group). Data are expressed as micromoles of free AMC released per minute per milligram of protein and reported as mean $\pm$ SEM.

production was not associated with enhanced caspase-1 activity as determined by measuring cleavage of the fluorogenic substrate Ac-WEHD-AMC (Fig. 5B).

To evaluate whether the observed elevation MMPs activity could contribute to IL-1β processing and maturation, we used a pharmacological approach. After pretreatment with the broad spectrum MMPs inhibitor, GM6001, IL-1β levels measured in the cortex ipsilateral to the ischemic insult (2-h MCAo followed by 2-h reperfusion) were comparable with levels detected in the contralateral nonischemic brain tissue. By contrast, administration of GM6001 negative control did not affect the increase of IL-1β observed in the cortex ipsilateral to the ischemic insult (Fig. 6A). Western blotting analysis confirmed that mature (17-kDa band) IL-1β levels were significantly reduced in the ischemic cortex from rats treated with GM6001 as compared to GM6001 negative control (Fig. 6B). By contrast, pro-IL-1β (33- to 31-kDa band) levels were not affected by GM6001 treatment. Moreover, we detected a strong band (about 28 kDa) corresponding to a cleavage product of the 33- to 31-kDa pro-IL-1β which was less intense in the cortex of rats treated with GM6001 as compared to GM negative control (Fig. 6B).

IV. Discussion

Despite the numerous reports implicating IL-1β in the neurodegeneration associated with ischemic insult, the role of caspase-1 in cytokine maturation following injury is still controversial. Here we demonstrate that IL-1β increases very early following tMCAo and this appears to occur independently from caspase-1 activation. Interestingly, we found that inhibition of MMPs activity prevented cytokine enhancement in the ischemic cortex, thus suggesting that MMPs might contribute to the early IL-1β maturation following a transient ischemic insult.

The role of MMPs in the pathophysiology of ischemic stroke has been widely investigated particularly regarding their ability to degrade the neurovascular matrix, thus leading to BBB damage, hemorrhage, and anoikis-like cell death triggered by disruption of cell-matrix homeostasis (see Cunningham *et al.*, 2005; Rosenberg, 2002). An early and progressive increase of MMP-9 expression has been detected particularly in endothelial cells in the ischemic hemisphere following tMCAo (Planas *et al.*, 2001; Zhao *et al.*, 2006) and has been implicated in the early stages of tissue injury produced by the ischemic insult (Romanic *et al.*, 1998). Although the involvement of MMP-2 during the early stages after stroke has been questioned (Asahi *et al.*, 2001a), there is evidence for a transient enhancement of MMP-2 early after temporary ischemia, which has been suggested to contribute to the effects of reperfusion and to the early opening of the BBB (Planas *et al.*, 2001; Rosenberg *et al.*, 1998). At later stages following tMCAo, a redistribution of gelatinases in the neurovascular unit does occur, with a massive increase of MMP-2 expression in microglia/macrophages (Planas *et al.*, 2001; Rosenberg *et al.*, 2001) and MMP-9 in astrocytes and neurons (Zhao *et al.*, 2006). The former effect has been suggested to facilitate migration of macrophages into the ischemic lesion, whereas the latter might represent an endogenous mechanism involved in

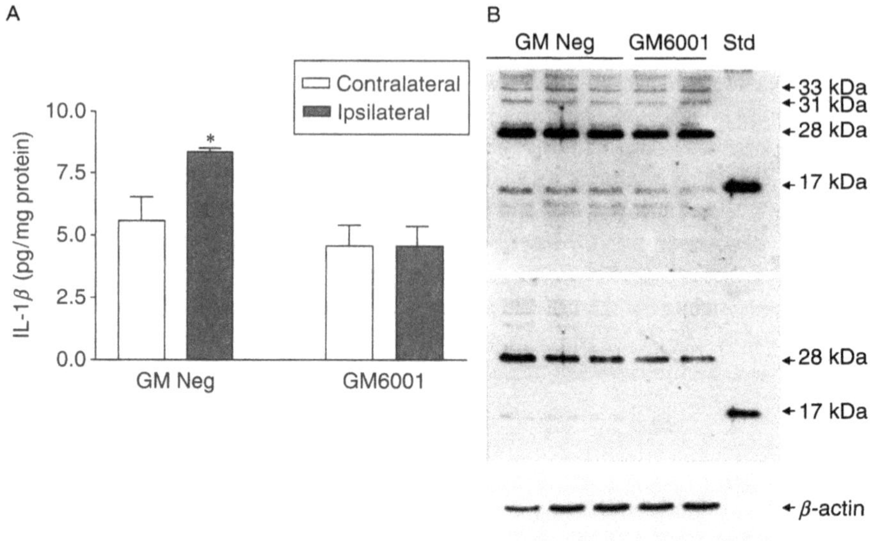

FIG. 6. GM6001 abrogates the increase of cortical IL-1β induced by tMCAo and reduces the expression of pro-IL-1β cleavage products. (A) GM6001 (5 μg/rat i.a.), but not the inactive form of the metalloproteinase inhibitor, GM6001 negative control (GM Neg; 5 μg/rat i.a.), abrogates increase of cortical IL-1β levels in the ischemic cortex of rats undergone tMCAo (2h/2h reperfusion). Immunoreactive IL-1β levels were assayed in individual brain cortical tissue samples ($n = 3$ per group) by an established, rat specific, sandwich ELISA. IL-1β levels were corrected for protein concentration and the results expressed as picograms of IL-1β per milligram of protein. Data are expressed as mean $\pm$ SEM values ($n = 3$ per group). The resulting means were evaluated statistically for differences using ANOVA followed by Tukey-Kramer test for multiple comparisons. * denotes $p < 0.05$ versus GM6001. (B) Pro-IL-1β immunoreactivity of ipsilateral, ischemic brain cortical homogenates obtained from rats subjected to tMCAo (2 h plus 2 h reperfusion) and pretreated with GM6001 (5 μg/rat i.a.) or with its negative control (GM Neg; 5 μg/rat i.a.) was assessed by Western blotting analysis using an antibody specific for pro-IL-1β. The antibody detected a strong band of about 28 kDa which has been previously described as a cleavage product of the 33- to 31-kDa pro-IL-1β, together with less intense bands of 33 and 31 kDa and a band of 17 kDa corresponding to mature IL-1β as determined by the use of recombinant mouse IL-1β employed as a standard (Std). As compared to animals pretreated with the inactive form of GM6001, a reduction of the intensity of IL-1β is evident in the cortex of GM6001-treated animals (upper panel in B) and this is associated to a reduction of the intensity of the 28-kDa cleavage product band as better shown in middle panel in B which represents the same gel as in upper panel but at a lower exposure time.

the resolution phase. In fact, inhibition of MMP-9 between 7 and 14 days after stroke has been reported to result in a substantial reduction in the number of neurons and new vessels implicated in neurovascular remodeling (Zhao *et al.*, 2006). These findings underscore the complexity of MMPs activity during tissue injury, ranging from detrimental effects during the early phases after stroke to beneficial roles at later stages (Yong, 2005).

Here we demonstrate that increased gelatinolytic activity is detected *in situ* as early as 2 h following reperfusion both in the cortex and striatum of rats exposed to 2-h MCAo. This is confirmed by gel zymography revealing that active MMP-2, as well as pro- and dimeric forms of MMP-9 are significantly increased in the brain regions affected by the ischemic insult. Previous studies have demonstrated that latent MMP-9 (dimer and propeptide forms) levels increase after focal (Asahi *et al.*, 2000; Gu *et al.*, 2002, 2005; Romanic *et al.*, 1998; Rosenberg *et al.*, 1996) and global cerebral ischemia in rodents (Magnoni *et al.*, 2004; Rivera *et al.*, 2002). Accordingly, here we observed a gelatinolytic band around 250 kDa, which increases significantly following brain ischemia/reperfusion. Similar gelatinolytic activity (200- to 250-kDa band) was reported in the hippocampus of rats subjected to global brain ischemia (Rivera *et al.*, 2002) and in the brain of mice treated with LPS (Pagenstecher *et al.*, 2000). These high-molecular-weight bands may correspond to homodimers produced intracellularly during the maturation process of the enzyme or to complexes of dimerized MMP-9, favored by increases in the gelatinase B/TIMP-1 ratio (Dubois *et al.*, 1998; Goldberg *et al.*, 1992), or to complexes of gelatinase B and integrins (Brooks *et al.*, 1996).

It is widely accepted that uncontrolled expression of MMPs may result in tissue damage and inflammation, though the mechanisms involved are still poorly understood, especially concerning those leading to neuronal cell death. Activated MMP-9 has been demonstrated to directly induce neuronal apoptosis both *in vitro* and *in vivo* after focal cerebral ischemia/reperfusion (Gu *et al.*, 2002). Some studies have suggested that MMP-9 contributes to neuronal cell death via proteolysis of basement membrane proteins, including laminin (Asahi *et al.*, 2001b; Castellanos *et al.*, 2003; Chen *et al.*, 2003; Gu *et al.*, 2005; Horstmann *et al.*, 2003). MMPs are also related to activation and release of several bioactive molecules such as cytokines, chemokines, and growth factors (see Parks *et al.*, 2004). Here we suggest that brain tissue damage produced by MMPs in the ischemic brain might be linked to cleavage and activation of neuroinflammatory mediators. In fact, increased gelatinolytic activity is associated with a rise of mature IL-1β production in the ischemic cortex and, most importantly, inhibition of MMPs activity by GM6001 prevents the early increase of IL-1β yielded by the ischemic insult in the cortex and this results in significant neuroprotection.

There is *in vitro* evidence that MMPs activate pro-IL-1β proteolitically and cleave the activated form of IL-1β to an inactive form, thereby providing both positive and negative regulation (Ito *et al.*, 1996; Schönbeck *et al.*, 1998). Our results extend these findings and suggest that MMPs contribute to maturation of IL-1β also *in vivo*, under pathophysiological conditions such as brain ischemia, by a mechanism independent from caspase-1 activation. Increased IL-1β generated via mechanisms other than caspase-1 has also been reported in the brain of rats exposed to kainic acid or 3,4-methylenedioxymethamphetamine (Eriksson *et al.*, 1999; O'Shea *et al.*, 2005) and in an animal model of neuroAIDS (Corasaniti *et al.*, 2005), further emphasizing that, under certain circumstances, cytokine activation might involve proteases other than caspase-1 (Fantuzzi *et al.*, 1997).

Although ICE has been strongly implicated in the development of ischemic brain damage (Hara *et al.*, 1997a,b; Schielke *et al.*, 1998; Yang *et al.*, 1999), we failed to detect increase in caspase-1 activity during the early stages after reperfusion. In animal models of MCAo, there is evidence that caspase-1 activity is enhanced shortly after the induction of ischemia (Benchoua *et al.*, 2001) and ICE-like protease activity increases few hours after reperfusion (Hara *et al.*, 1997a). Nevertheless, some studies have revealed that caspase-1 displays a biphasic activation after the ischemic insult (Benchoua *et al.*, 2001), whereas others have failed to detect any change in the expression of ICE after permanent MCAo (Asahi *et al.*, 1997). Therefore, we cannot exclude that under our experimental conditions caspase-1 may be activated at some stage following tMCAo, thus contributing to tissue damage through mechanisms that do not necessarily involve IL-1β production. In fact, caspase-1 is able to directly process pro-caspase-3 to its active form (Tewari *et al.*, 1995), and it is also involved in the processing of IL-18 (Akita *et al.*, 1997). Moreover, there is evidence that caspase-1 is an apical caspase in neuronal cell death pathways, being involved in Bid cleavage and caspase-3 activation following an ischemic insult (Benchoua *et al.*, 2001; Zhang *et al.*, 2003b).

In conclusion, our data suggest that early production of IL-1β following ischemic stroke is not associated with caspase-1 activation. By contrast, MMPs appear to contribute to cytokine processing thus representing a crucial upstream signal for the induction of neuroinflammatory responses. This strengthens the involvement of MMPs in the development of tissue damage during the early phases after stroke and emphasizes their potential as useful pharmacological targets for the treatment of ischemia-induced neurodegeneration.

Acknowledgments

Partial finantial support from Italian Ministry of University and Research (PRIN prot. 2004053099_004 to G.B.) is gratefully acknowledged. Mr. Guido Fico is gratefully acknowledged for skillful technical assistance.

References

Akita, K., Ohtsuki, T., Nukada, Y., Tanimoto, T., Namba, M., Okura, T., Takakura-Yamamoto, R., Torigoe, K., Gu, Y., Su, M. S., Fujii, M., Satoh-Itoh, M., *et al.* (1997). Involvement of caspase-1 and caspase-3 in the production and processing of mature human interleukin 18 in monocytic THP.1 cells. *J. Biol. Chem.* **272,** 26595–26603.

Asahi, M., Asahi, K., Jung, J. C., del Zoppo, G. J., Fini, M. E., and Lo, E. H. (2000). Role for matrix metalloproteinase 9 after focal cerebral ischemia: Effects of gene knockdown and enzyme inhibition with BB-94. *J. Cereb. Blood Flow Metab.* **20,** 1681–1689.

Asahi, M., Hoshimaru, M., Uemura, Y., Tokime, T., Kojima, M., Ohtsuka, T., Matsuura, N., Aoki, T., Shibahara, K., and Kikuchi, H. (1997). Expression of interleukin-1 beta converting enzyme gene family and bcl-2 gene family in the rat brain following permanent occlusion of the middle cerebral artery. *J. Cereb. Blood Flow Metab.* **17,** 11–18.

Asahi, M., Sumii, T., Fini, M. E., Itohara, S., and Lo, E. H. (2001a). Matrix metalloproteinase 2 gene knockout has no effect on acute brain injury after focal ischemia. *Neuroreport* **12,** 3003–3007.

Asahi, M., Wang, X., Mori, T., Sumii, T., Jung, J. C., Moskowitz, M. A., Fini, M. E., and Lo, E. H. (2001b). Effects of matrix metalloproteinase-9 gene knock-out on the proteolysis of blood-brain barrier and white matter components after cerebral ischemia. *J. Neurosci.* **21,** 7724–7732.

Benchoua, A., Guegan, C., Couriaud, C., Hosseini, H., Sampaio, N., Morin, D., and Onteniente, B. (2001). Specific caspase pathways are activated in the two stages of cerebral infarction. *J. Neurosci.* **21,** 7127–7134.

Black, R. A., Kronheim, S. R., Cantrell, M., Deeley, M. C., March, C. J., Prickett, K. S., Wignall, J., Conlon, P. J., Cosman, D., Hopp, T. P., and Mochizuki, D. Y. (1988). Generation of biologically active interleukin-1β by proteolytic cleavage of the inactive precursor. *J. Biol. Chem.* **263,** 9437–9442.

Brooks, P. C., Stromblad, S., Sanders, L. C., von Schalscha, T. L., Aimes, R. T., Stetler-Stevenson, W. G., Quigley, J. P., and Cheresh, D. A. (1996). Localization of matrix metalloproteinase MMP-2 to the surface of invasive cells by interaction with integrin alpha v beta 3. *Cell* **85,** 683–693.

Buttini, M., Sauter, A., and Boddeke, H. (1994). Induction of interleukin-1beta mRNA after focal ischemia in the rat. *Mol. Brain Res.* **23,** 126–134.

Castellanos, M., Leira, R., Serena, J., Pumar, J. M., Lizasoain, I., Castillo, J., and Davalos, A. (2003). Plasma metalloproteinase-9 concentration predicts hemorrhagic transformation in acute ischemic stroke. *Stroke* **34,** 40–46.

Chen, Z. L., Indyk, J. A., and Strickland, S. (2003). The hippocampal laminin matrix is dynamic and critical for neuronal survival. *Mol. Biol. Cell* **14,** 2665–2676.

Copin, J.-C., Goodyear, M. C., Gidday, J. M., Shah, A. R., Gascon, E., Dayer, A., Morel, D. M., and Gasche, Y. (2005). Role of matrix metalloproteinases in apoptosis after transient focal cerebral ischemia in rats and mice. *Eur. J. Neurosci.* **22,** 1597–1608.

Corasaniti, M. T., Turano, P., Bilotta, A., Malorni, W., Stringaro, A. R., Nisticò, R., Finazzi-Agrò, A., and Bagetta, G. (2001). Evidence that increases of mitochondrial immunoreactive IL-1β by HIV-1 gp120 implicate in situ cleavage of pro-IL-1β in the neocortex of rat. *J. Neurochem.* **78,** 611–618.

Corasaniti, M. T., Russo, R., Amantea, D., Gliozzi, M., Siviglia, E., Stringaro, A. R., Malori, W., Melino, G., and Bagetta, G. (2005). Neuroprotection by the caspase-1 inhibitor Ac-YVAD-(acyloxy)mk in experimental neuroAIDS is independent from IL-1β generation. *Cell Death Differ.* **12,** 999–1001.

Cunningham, L. A., Wetzel, M., and Rosenberg, G. A. (2005). Multiple roles for MMPs and TIMPs in cerebral ischemia. *Glia* **50,** 329–339.

Davies, C. A., Loddick, S. A., Toulmond, S., Stroemer, R. P., Hunt, J., and Rothwell, N. J. (1999). The progression and topographic distribution of interleukin-1 beta expression after permanent middle cerebral artery occlusion in the rat. *J. Cereb. Blood Flow Metab.* **19,** 87–98.

Dubois, B., Peumans, W. J., Van Damme, E. J., Van Damme, J., and Opdenakker, G. (1998). Regulation of gelatinase B (MMP-9) in leukocytes by plant lectins. *FEBS Lett.* **427,** 275–278.

Eriksson, C., Van Dam, A. -M., Lucassen, P. J., Bol, J. G. J., Winblad, B., and Schultzberg, M. (1999). Immunohistochemical localization of interleukin-1β, interleukin-1 receptor antagonist and interleukin-1β converting enzyme/caspase-1 in the rat brain after peripheral administration of kainic acid. *Neuroscience* **93,** 915–930.

Fantuzzi, G., Ku, G., Harding, M. W., Livingston, D. J., Sipe, J. D., Kuida, K., Flavell, R. A., and Dinarello, C. A. (1997). Response to local inflammation of IL-1β-converting enzyme-deficient mice. *J. Immunol.* **158,** 1818–1824.

Friedlander, R. M., and Yuan, J. (1998). ICE, neuronal apoptosis and neurodegeneration. *Cell Death Differ.* **5,** 823–831.

Garcia, J. H., Liu, K. F., and Relton, J. K. (1995). Interleukin-1 receptor antagonist decreases the number of necrotic neurons in rats with middle cerebral artery occlusion. *Am. J. Pathol.* **147,** 1477–1486.

Garcia-Calvo, M., Peterson, E. P., Leiting, B., Ruel, R., Nicholson, D. W., and Thornberry, N. A. (1998). Inhibition of human caspases by peptide-based and macromolecular inhibitors. *J. Biol. Chem.* **273,** 32608–32613.

Gasche, Y., Copin, J. C., Sugawara, T., Fujimura, M., and Chan, P. H. (2001). Matrix metalloproteinase inhibition prevents oxidative stress associated blood-brain barrier disruption after transient focal cerebral ischemia. *J. Cereb. Blood Flow Metab.* **21,** 1393–1400.

Goldberg, G. I., Strongin, A., Coolier, I. E., Genrich, L. T., and Marmer, B. L. (1992). Interaction of 92-kDa type IV collagenase with the tissue inhibitor of metalloproteases prevents dimerization, complex formation with interstitial collagenase, and activation of the proenzyme with stromelysin. *J. Biol. Chem.* **267,** 4583–4591.

Gu, Z., Kaul, M., Yan, B., Kridel, S. J., Cui, J., Strongin, A., Smith, J. W., Liddington, R. C., and Lipton, S. A. (2002). S-nitrosylation of matrix metalloproteinases: Signalling pathway to neuronal cell death. *Science* **297,** 1186–1190.

Gu, Z., Cui, J., Brown, S., Fridman, R., Mobashery, S., Strongin, A. Y., and Lipton, S. A. (2005). A highly specific inhibitor of matrix metalloproteinase-9 rescues laminin from proteolysis and neurons from apoptosis in transient focal cerebral ischemia. *J. Neurosci.* **25,** 6401–6408.

Hagan, P., Poole, S., Bristow, A. F., Tilders, F., and Silverstein, F. S. (1996). Intracerebral NMDA injection stimulates production of interleukin-1β in perinatal rat brain. *J. Neurochem.* **67,** 2215–2218.

Hara, H., Fink, K., Endres, M., Friedlander, R. M., Gagliardini, V., Yuan, J., and Moskowitz, M. A. (1997a). Attenuation of transient focal cerebral ischemic injury in transgenic mice expressing a mutant ICE inhibitory protein. *J. Cereb. Blood Flow Metab.* **17,** 370–375.

Hara, H., Friedlander, R. M., Gagliardini, V., Ayata, C., Fink, K., Huang, Z., Shimizu-Sasamata, M., Yuan, J., and Moskowitz, M. A. (1997b). Inhibition of interleukin 1β converting enzyme family proteases reduces ischemic and excitotoxic neuronal damage. *Proc. Natl. Acad. Sci. USA* **94,** 2007–2012.

Heo, J. H., Lucero, J., Abumiya, T., Koziol, J. A., Copeland, B. R., and del Zoppo, G. J. (1999). Matrix metalloproteinases increase very early during experimental focal cerebral ischemia. *J. Cereb. Blood Flow Metab.* **19,** 624–633.

Horstmann, S., Kalb, P., Koziol, J., Gardner, H., and Wagner, S. (2003). Profiles of matrix metalloproteinases, their inhibitors, and laminin in stroke patients: Influence of different therapies. *Stroke* **34,** 2165–2170.

Howard, A. D., Kostura, M. J., Thornberry, N., Ding, G. J. F., Limjuco, G., Weidner, J., Salley, J. P., Hogquist, K. A., Chaplin, D. D., Mumford, R. A., Schmidt, J. A., and Tocci, M. J. (1991). IL-1-converting enzyme requires aspartic acid residues for processing of the IL-1β precursor at two distinct sites and does not cleave 31-kDa IL-1α. *J. Immunol.* **147,** 2964–2969.

Ito, A., Mukaiyama, A., Itoh, Y., Nagase, H., Thøgersen, I. B., Enghild, J. J., Sasaguri, Y., and Mori, Y. (1996). Degradation of interleukin 1β by matrix metalloproteinases. *J. Biol. Chem.* **271,** 14657–14660.

Jourquin, J., Tremblay, E., Decanis, N., Charton, G., Hanessian, S., Chollet, A. M., Le Diguardjer, T., Khrestchatisky, M., and Rivera, S. (2003). Neuronal activity-dependent increase of net matrix metalloproteinase activity is associated with MMP-9 neurotoxicity after kainate. *Eur. J. Neurosci.* **18,** 1507–1517.

Lee, R., Kermani, P., Teng, K. K., and Hempstead, B. L. (2001). Regulation of cell survival by secreted proneurotrophins. *Science* **294,** 1945–1948.

Legos, J. J., Whitmore, R. G., Erhardt, J. A., Parsons, A. A., Tuma, R. F., and Barone, F. C. (2000). Quantitative changes in interleukin proteins following focal stroke in the rat. *Neurosci. Lett.* **282,** 189–192.

Levy, D. E., Lapierre, F., Liang, W., Ye, Y., Lange, C. W., Li, X., Grobelny, D., Casabonne, M., Tyrrell, D., Holme, K., Nadzan, A., and Galardy, R. E. (1998). Matrix metalloproteinase inhibitors: A structure-activity study. *J. Med. Chem.* **41,** 199–223.

Li, H., Colbourne, F., Sun, P., Zhao, Z., Buchan, A. M., and Iadecola, C. (2000). Caspase inhibitors reduce neuronal injury after focal but not global cerebral ischemia in rats. *Stroke* **31,** 176–182.

Liu, T., McDonnell, P. C., Young, P. R., White, R. F., Siren, A. L., Hallenbeck, J. M., Barone, F. C., and Feurerstein, G. Z. (1993). Interleukin-1-beta messenger RNA expression in ischemic rat cortex. *Stroke* **24,** 1746–1751.

Longa, E. Z., Weinstein, P. R., Carlson, S., and Cummins, R. (1989). Reversible middle cerebral artery occlusion without craniotomy in rats. *Stroke* **20,** 84–91.

Magnoni, S., Baker, A., George, S. J., Duncan, W. C., Kerr, L. E., McCulloch, J., and Horsburgh, K. (2004). Differential alterations in the expression and activity of matrix metalloproteinases 2 and 9 after transient cerebral ischemia in mice. *Neurobiol. Dis.* **17,** 188–197.

Mulcahy, N., Ross, J., Rothwell, N. J., and Loddick, S. A. (2003). Delayed administration of interleukin-1 receptor antagonist protects against transient cerebral ischemia in the rat. *Br. J. Pharmacol.* **140,** 471–476.

O'Shea, E., Sanchez, V., Orio, L., Escobedo, I., Green, A. R., and Colado, M. I. (2005). 3,4-methylenedioxymethamphetamine increases prointerleukin-1β production and caspase-1 protease activity in frontal cortex, but not in hypothalamus, of Dark Agouti rats: Role of inteleukin-1β in neurotoxicity. *Neuroscience* **135,** 1095–1105.

Pagenstecher, A., Stadler, A. K., Kincaid, C. L., Volk, B., and Campbell, I. L. (2000). Regulation of matrix metalloproteinases and their inhibitor genes in lipopolysaccaride-induced endotoxemia in mice. *Am. J. Pathol.* **157,** 197–210.

Parks, W. C., Wilson, C. L., and Lopez-Boado, Y. S. (2004). Matrix metalloproteinases as modulators of inflammation and innate immunity. *Nat. Rev. Immunol.* **4,** 617–629.

Planas, A. M., Sole, A., and Justicia, C. (2001). Expression and activation of matrix metalloproteinase-2 and -9 in rat brain after transient focal cerebral ischemia. *Neurobiol. Dis.* **8,** 834–846.

Rabuffetti, M., Sciorati, C., Tarozzo, G., Clementi, E., Manfredi, A. A., and Beltramo, M. (2000). Inhibition of caspase-1-like activity by Ac-Tyr-Val-Ala-Asp-Chloromethyl ketone induces long-lasting neuroprotection in cerebral ischemia through apoptosis reduction and decrease of proinflammatory cytokines. *J. Neurosci.* **20,** 4398–4404.

Rano, T. A., Timkey, T., Peterson, E. P., Rotonda, J., Nicholson, D. W., Becker, J. W., Chapman, K. T., and Thornberry, N. A. (1997). A combinatorial approach for determining protease specificities: Application to interleukin-1beta converting enzyme (ICE). *Chem. Biol.* **4,** 149–155.

Relton, J. K., and Rothwell, N. J. (1992). Interleukin-1 receptor antagonist inhibits ischaemic and excitotoxic neuronal damage in the rat. *Brain Res. Bull.* **29,** 242–246.

Relton, J. K., Martin, D., Thompson, R. C., and Russell, D. A. (1996). Peripheral administration of interleukin-1 receptor antagonist inhibits brain damage after focal cerebral ischemia in the rat. *Exp. Neurol.* **138,** 206–213.

Rivera, S., Ogier, C., Jourquin, J., Timsit, S., Szklarczyk, A. W., Miller, K., Gearing, A. J. H., Kaczmarek, L., and Khrestchatisky, M. (2002). Gelatinase B and TIMP-1 are regulated in a cell- and time-dependent manner in association with neuronal death and glial reactivity after global forebrain ischemia. *Eur. J. Neurosci.* **15,** 19–32.

Romanic, A. M., White, R. F., Arleth, A. J., Ohlstein, E. H., and Barone, F. C. (1998). Matrix metalloproteinase expression increases after cerebral focal ischemia in rats: Inhibition of matrix metalloproteinase-9 reduces infarct size. *Stroke* **29,** 1020–1030.

Rosell, A., Ortega-Aznar, A., Alvarez-Sabin, J., Fernandez-Cadenas, I., Ribo, M., Molina, C. A., Lo, E. H., and Montanter, J. (2006). Increased brain expression of matrix metalloproteinase-9 after ischemic and hemorrhagic human stroke. *Stroke* **37,** 1399–1406.

Rosenberg, G. A. (2002). Matrix metalloproteinases in neuroinflammation. *Glia* **39,** 279–291.

Rosenberg, G. A., Navratil, M., Barone, F., and Feuerstein, G. (1996). Proteolytic cascade enzymes increase in focal cerebral ischemia in rat. *J. Cereb. Blood Flow Metab.* **16,** 360–366.

Rosenberg, G. A., Estrada, E. Y., and Dencoff, J. E. (1998). Matrix metalloproteinases and TIMPs are associated with blood-brain barrier opening after reperfusion in rat brain. *Stroke* **29,** 2189–2195.

Rosenberg, G. A., Cunningham, L. A., Wallace, J., Alexander, S., Estrada, E. Y., Grossetete, M., Razhagi, A., Miller, K., and Gearing, A. (2001). Immunohistochemistry of matrix metalloproteinases in reperfusion injury to rat brain: Activation of MMP-9 linked to stromelysin-1 and microglia in cell cultures. *Brain Res.* **893,** 104–112.

Rothwell, N. (2003). Interleukin-1 and neuronal injury: Mechanisms, modification, and therapeutic potential. *Brain Behav. Immun.* **17,** 152–157.

Schielke, G. P., Yang, G. Y., Shivers, B. D., and Lorris Betz, A. (1998). Reduced ischemic brain injury in interleukin-1β converting enzyme-deficient mice. *J. Cereb. Blood Flow Metab.* **18,** 180–185.

Schönbeck, U., Mach, F., and Libby, P. (1998). Generation of biologically active IL-1β by matrix metalloproteinases: A novel caspase-1-independent pathway of IL-1β processing. *J. Immunol.* **161,** 3340–3346.

Sternlicht, M. D., and Werb, Z. (2001). How matrix metalloproteinases regulate cell behaviour. *Annu. Rev. Cell Dev. Biol.* **17,** 463–516.

Tewari, M., Quan, L. T., O'Rourke, K., Desnoyers, S., Zeng, Z., Beidler, D. R., Poirier, G. G., Salvesen, G. S., and Dixit, V. M. (1995). Yama/CPP32 beta, a mammalian homolg of CED-3, is a CrmA-inhibitable protease that cleaves the death substrate poly(ADP-ribose) polymerase. *Cell* **81,** 801–809.

Thornberry, N. A., Bull, H. G., Calaycay, J. R., Chapman, K. T., Howard, A. D., Kostura, M. J., Miller, D. K., Molineaux, S. M., Weidner, J. R., Aunins, J., Elliston, K. O., Ayala, J. M., *et al.* (1992). A novel heterodimeric cysteine protease is required for interleukin-1β processing in monocytes. *Nature* **356,** 768–774.

Wang, X. K., Yue, T. L., Barone, F. C., White, R. F., Gagnon, R. C., and Feuerstein, G. Z. (1994). Concomitant cortical expression of TNF-α and IL-1β mRNAs follows early response gene expression in transient focal ischemia. *Mol. Chem. Neuropathol.* **23,** 103–114.

Wetzel, M., Rosenberg, G. A., and Cunningham, L. A. (2003). Tissue inhibitor of metalloproteinase-3 and matrix metalloproteinase-3 regulate neuronal sensitivity to doxorubicin-induced apoptosis. *Eur. J. Neurosci.* **18,** 1050–1060.

Yamasaki, Y., Matsuura, N., Shizuhara, H., Onodera, H., and Itoyama, Y. K. K. (1995). Interleukin-1 as a pathogenetic mediator of ischemic brain damage in the rats. *Stroke* **26,** 676–681.

Yang, G. Y., Schielke, G. P., Gong, C., Mao, Y., Ge, H. L., Liu, X. H., and Betz, A. L. (1999). Expression of tumor necrosis factor-alpha and intercellular adhesion molecule-1 after focal cerebral ischemia in interleukin-1β converting enzyme deficient mice. *J. Cereb. Blood Flow Metab.* **19,** 1109–1117.

Yong, V. W. (2005). Metalloproteinases: Mediators of pathology and regeneration in the CNS. *Nat. Rev. Neurosci.* **6,** 931–944.

Zhang, Z., Chopp, M., Goussev, A., and Powers, C. (1998). Cerebral vessels express interleukin 1 beta after focal cerebral ischemia. *Brain Res.* **784,** 210–217.

Zhang, K., McQuibban, G. A., Silva, C., Butler, G. S., Johnston, J. B., Holden, J., Clark-Lewis, I., Overall, C. M., and Power, C. (2003a). HIV-induced metalloproteinase processing of the chemokine stromal cell derived factor-1 causes neurodegeneration. *Nat. Neurosci.* **6,** 1064–1071.

Zhang, W. H., Wang, X., Narayanan, M., Zhang, Y., Huo, C., Reed, J. C., and Friedlander, R. M. (2003b). Fundamental role of the Rip2/caspase-1 pathway in hypoxia and ischemia-induced neuronal cell death. *Proc. Natl. Acad. Sci. USA* **100,** 16012–7.

Zhao, B.-Q., Wang, S., Kim, H.-Y., Storrie, H., Rosen, B. R., Mooney, D. J., Wang, X., and Lo, E. H. (2006). Role of matrix metalloproteinases in delayed cortical responses after stroke. *Nat. Med.* **12,** 441–445.

THE (ENDO)CANNABINOID SYSTEM IN MULTIPLE SCLEROSIS AND AMYOTROPHIC LATERAL SCLEROSIS

Diego Centonze,[1,*,†] Silvia Rossi,[1,*,†] Alessandro Finazzi-Agrò,[‡] Giorgio Bernardi,[*,†] and Mauro Maccarrone[†,§]

*Neurological Clinics, Department of Neurosciences, University of Rome Tor Vergata, Rome, Italy

†European Center for Brain Research, (CERC)/IRCCS S. Lucia Foundation, Rome, Italy

‡Department of Experimental Medicine and Biochemical Sciences, University of Rome Tor Vergata, Rome, Italy

§Department of Biomedical Sciences, University of Teramo, Teramo, Italy

Alterations of the endocannabinoid system (ECS) have been recently implicated in a number of neuroinflammatory and neurodegenerative conditions so that the pharmacological modulation of cannabinoid (CB) receptors and/or of the enzymes controlling synthesis, transport, and degradation of these substances has emerged as a valuable option to treat neurological diseases. Here, we describe the current knowledge concerning the rearrangement of ECS in a primarily inflammatory disorder of the central nervous system such as multiple sclerosis (MS), and in a primarily degenerative condition such as amyotrophic lateral sclerosis (ALS). Furthermore, the data supporting a therapeutic role of agents modulating CB receptors

[1]These authors equally contributed to the present work.

INTERNATIONAL REVIEW OF NEUROBIOLOGY, VOL. 82
DOI: 10.1016/S0074-7742(07)82009-5

or endocannabinoid tone in these disorders will also be reviewed. Complex changes of ECS take place in both diseases, influencing crucial aspects of their pathophysiology and clinical manifestations. Neuroinflammation, microglial activation, oxidative stress, and excitotoxicity are variably combined in MS and in ALS and can be modulated by endocannabinoids or by drugs targeting the ECS.

I. Introduction

The involvement of endocannabinoid system (ECS) in the pathophysiology of inflammatory and degenerative disorders of the central nervous system (CNS) has recently emerged, along with the therapetic potential of compounds targeting cannabinoid (CB) receptors.

Central neurons and immune cells possess the biochemical machinery required for endocannabinoid synthesis, transport, and degradation, and respond to the stimulation of CB receptors by changing their physiological activity. In neurons, the mechanisms that regulate the interaction of endocannabinoid signaling with other neurotransmitter system are still poorly understood, but evidence exists that cannabinoid may regulate both excitatory and inhibitory transmission through pre- and postsynaptic mechanisms. In immune cells, the activation of CB receptors significantly alters their metabolic activity and responses during inflammation.

Aim of the present chapter is to describe the ECS, and then to present the current knowledge on the involvement of this system in the pathophysiology of multiple sclerosis (MS) and of amyotrophic lateral sclerosis (ALS), paradigmatic inflammatory, and degenerative diseases of the CNS, respectively. The data supporting a therapeutic role of pharmacological agents modulating CB receptors or endocannabinoid tone in these disorders will also be reviewed.

II. The ECS

Two arachidonate derivatives, *N*-arachidonoylethanolamine (AEA) and 2-arachidonoylglycerol (2-AG), are the most biologically active endocannabinoids described to date (Bari *et al.*, 2006; Piomelli, 2003). AEA is generated starting from membrane phosphoglycerides and phosphatidylethanolamine (Ligresti *et al.*, 2005). Conversely, the biological activity of AEA at CB receptors is terminated by its removal from the extracellular space through cellular uptake by a purported high-affinity transporter (AMT). Once taken up by cells, AEA is substrate for fatty acid amide hydrolase (FAAH), which breaks the amide bond and releases arachidonic acid and ethanolamine (McKinney and Cravatt, 2005).

Like AEA, 2-AG is produced by cleavage of lipid precursors (Ligresti *et al.*, 2005). A biosynthetic pathway provides for 2-AG formation through the hydrolysis of phospholipids by a specific diacylglycerol lipase (DAGL). Then, 2-AG is subjected to a rapid transport across the plasma membrane and can be degradated by FAAH or, more efficiently, by a specific monoacylglycerol lipase (MAGL).

AEA and 2-AG bind to and activate type-1 (CB_1) and type-2 (CB_2) receptors (Howlett *et al.*, 2002; Pertwee and Ross, 2002). CB_1 is localized mainly in the CNS (Egertova *et al.*, 2003), but is also expressed in peripheral tissues including the immune cells (Klein *et al.*, 2003; Maccarrone *et al.*, 2001; Nong *et al.*, 2001). CB_2 is predominantly expressed in the periphery, but is also present in the brain (Nunez *et al.*, 2004; Van Sickle *et al.*, 2005). The binding to these receptors triggers a number of signaling pathways, which include the inhibition of adenylyl cyclase and the activation of mitogen-activated protein kinase (Maccarrone *et al.*, 2006). In addition to these receptors, AEA and 2-AG are able to activate non-CB_1/non-CB_2 receptors (Breivogel *et al.*, 2001) and/or a purported CB_3 (or GPR55) receptor (Drmota *et al.*, 2004). Furthermore AEA, but not 2-AG, behaves as a ligand to type-1 vanilloid receptor (now called transient receptor potential vanilloid 1, TRPV1) Since TRPV1 is expressed in peripheral sensory fibers and also in several nuclei of the CNS (Marinelli *et al.*, 2003), the endovanilloid activity of AEA may play a physiological control of brain function.

III. ECS in MS

MS is an inflammatory demyelinating disorder, but the development of neuroimaging and histological techniques has recently stressed the concept that neurodegenerative processes are also important events in this disorder. Accordingly, neuronal damage and axonal loss are common and abundant in MS, and affect both inflammatory lesions and normally appearing white matter (Bjartmar *et al.*, 2001; De Stefano *et al.*, 2002; Filippi and Rocca, 2005; Filippi *et al.*, 2003; Kuhlmann *et al.*, 2002; Peterson *et al.*, 2001). Furthermore, neuronal loss in gray matter significantly contributes to brain atrophy seen even in the early phases of MS (Bakshi *et al.*, 2002; Bermel *et al.*, 2003; Peterson *et al.*, 2001). Despite this, however, it is still a valid concept that the inflammatory-demyelinating component of MS predominates in the early phases of the disease, whereas neuronal and axonal damage, although precocious, reach critical levels after several years, causing long-term disability (Bruck and Stadelmann, 2003; De Stefano *et al.*, 2002; Lassmann, 2003; Stevenson *et al.*, 1998).

Increasing evidence supports the involvement of the ECS not only in the inflammatory process typical of MS, but also in the neurodegeneration related to this disease. Accordingly, MS significantly alters the ECS, since increased levels of AEA have been found in tissues from inflammatory lesions of patients with active or silent MS (Eljaschewitsch *et al.*, 2006). Endocannabinoids are also increased in the spinal cord of a mouse model of MS (Baker *et al.*, 2001), while both AEA and 2-AG are normal (Witting *et al.*, 2006) or downregulated in the brain of other models of MS (Cabranes *et al.*, 2005).

Several studies have provided solid experimental evidence that endocannabinoids might also influence the expression of disabling clinical symptoms of the disease (Baker *et al.*, 2000, 2001, 2003; Malfitano *et al.*, 2005), providing a convincing support to previous reports affirming the beneficial effects against spasticity, dystonia, tremor, ataxia, bladder control, and pain in MS patients self-medicating with marijuana (Consroe, 1998; Fernandez-Ruiz *et al.*, 2002; Pertwee, 2002). The demonstration that central endocannabinoid levels increase in spastic mice with experimental MS, and that pharmacological stimulation of CB receptors inhibits spasticity in these animals (Baker *et al.*, 2000, 2001, 2003) offered a rationale for testing the effect of orally administered cannabinoid agonists in MS spastic patients (Pertwee, 2002; Zajicek *et al.*, 2003). The encouraging results obtained so far, however, need to be validated with long-term trials. Also, the reported beneficial effects of cannabinoids in the control of tremor, pain, and bladder dysfunction need further validation with clinical trials involving a large sample of MS patients (Killestein and Uitdehaag, 2005; Pertwee, 2002).

A. Interplay Between Inflammatory Process and Cannabinoids in MS

Convincing evidence exists in support to a role for the ECS in the inflammatory process of experimental MS. For example, treatment with cannabinoid agonists significantly delays and attenuates the clinical signs of experimentally induced MS in animals, as well as the inflammatory infiltrates in the spinal cord (Lyman *et al.*, 1989; Wirguin *et al.*, 1994).

The anti-inflammatory properties of cannabinoids against MS pathophysiology seem to be associated with a reduced number of activated microglia in the spinal cord, decreased expression of major histocompatibility complex class II, and decreased number of CD4+ T cells, as described in models of MS. Decreased synthesis of proinflammatory cytokines [such as TNF-α, interleukin (IL)-1β, and IL-6] are other important effects of these compounds in mice with experimental MS (Croxford and Miller, 2003). Not only exogenous cannabinoids, but also endocannabinoids provide an effective therapy for experimental MS, since inhibition of AMT, aimed at elevating AEA levels, has been shown to downregulate macrophage function, to inhibit microglial activation, and to reduce NO production and major histocompatibility complex class II antigen expression in the CNS (Mestre *et al.*, 2005; Ortega-Gutierrez *et al.*, 2005).

The immune cells involved in the pathophysiology of MS express functional CB_2 receptors at higher levels than CB_1 receptors (Galiegue *et al.*, 1995; Yiangou *et al.*, 2006). Stimulation of CB_2 receptors mediates complex inhibitory actions in these cells, accounting for the protective effects of cannabinoid agonists in MS and experimental MS. In-line with this, activation of CB_2 receptors inhibits the production of several inflammatory cytokines, cell proliferation and chemiotaxis, and limits nitric oxide release. All these factors are critical determinants for the maintenance of inflammation in MS and experimental MS (Ehrhart *et al.*, 2005; Eljaschewitsch *et al.*, 2006; Pertwee, 2002; Walter and Stella, 2004). It should be mentioned, however, that activated microglia and macrophages express not only CB_2 but also CB_1 receptors, suggesting that both receptor subtypes might be important for the anti-inflammatory effects of cannabinoids in MS (Facchinetti *et al.*, 2003; Walter and Stella, 2004). A report also allowed to extend to non-CB_1 and non-CB_2 receptors the inhibitory action of cannabinoids on astrocyte activation, suggesting a role for these receptors in the anti-inflammatory effects of exogenous and endogenous cannabinoids in MS and experimental MS (Curran *et al.*, 2005).

B. Interplay Between Neurodegeneration Mechanisms and Cannabinoids in MS

A portion of the neurodegenerative damage seen in MS results from the release of toxic cytokines and free radicals by activated microglia or other immune cells invading the CNS. Thus endocannabinoids, by inhibiting inflammation, also exert an indirect neuroprotective action in MS (Eljaschewitsch *et al.*, 2006).

Cannabinoids, however, exert direct neuroprotective effects by limiting glutamate release and excitotoxic damage in several neurodegenerative disorders by acting through CB_1 receptors. Excitotoxicity plays a crucial role also in MS-associated neuronal damage, suggesting that pharmacological stimulation of CB_1 may contrast neurodegenerative damage also in this disorder. Accordingly, glutamate clearance is altered in experimental MS (Hardin-Pouzet *et al.*, 1997; Ohgoh *et al.*, 2002) as well as in MS (Vallejo-Illarramendi *et al.*, 2006; Werner *et al.*, 2001), and glutamate levels are significantly higher in the CSF (Sarchielli *et al.*, 2003; Stover *et al.*, 1997) and in the brains of MS patients (Srinivasan *et al.*, 2005). Finally, glutamate receptor antagonists exert beneficial effects in EAE (Bolton and Paul, 1997; Pitt *et al.*, 2000; Smith *et al.*, 2000; Wallstrom *et al.*, 1996) and in MS (Plaut, 1987) by limiting not only oligodendrocyte but also neuronal damage (Pitt *et al.*, 2000; Smith *et al.*, 2000). The antiexcitotoxic properties of endocannabinoids in experimental MS may contribute to explain the finding that mice

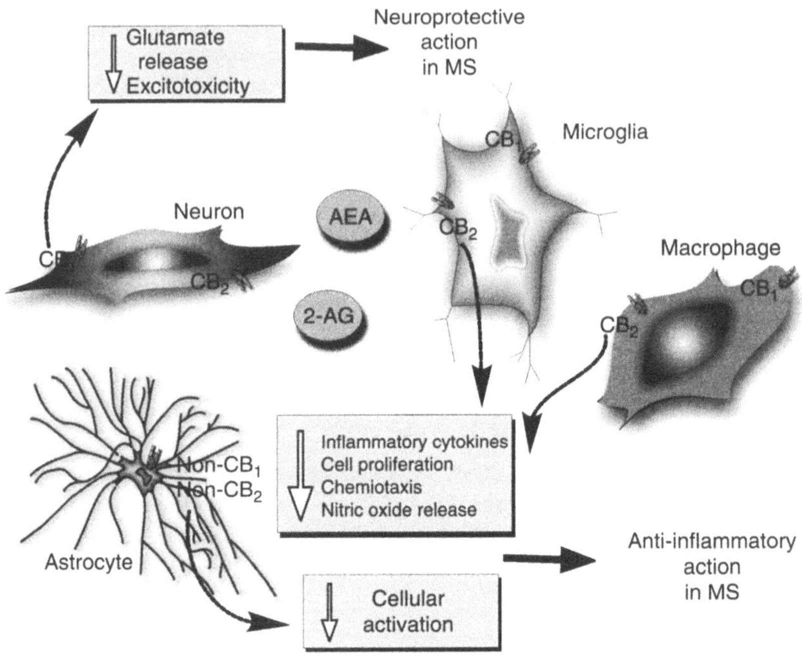

FIG. 1. Representation of the interplay between (endo)cannabinoids and the pathogenic events of MS. The stimulation of CB receptors by AEA and 2-AG in distinct cell types results in a series of effects potentially useful to contrast both inflammation and neurodegeneration in MS.

lacking CB_1 receptors do not tolerate excitotoxic insults and develop more severe neurodegeneration following EAE induction (Pryce *et al.*, 2003).

As a final consideration, a study has shown that cell damage in experimental MS does not lead, unlike other brain diseases, to enhancement of endocannabinoid levels, an effect associated with poor neuroprotection (Witting *et al.*, 2006). Of note, the functionality of CB receptors seems to be intact in experimental MS (Pertwee, 2002; Witting *et al.*, 2006), providing additional support to the use of agonists of these receptors for the treatment of MS (Fig. 1).

IV. ECS in ALS

ALS is a severe neurodegenerative disorder that selectively damages upper and lower motorneurons (Bruijn *et al.*, 2004; Nirmalananthan and Greensmith, 2005). Most cases of ALS are sporadic, but about 10% are familial (Bruijn *et al.*, 2004; Nirmalananthan and Greensmith, 2005). The underlying cause of the sporadic form of ALS remains unclear, while progresses in understanding the mechanisms of disease have been made into the familial forms.

Glutamate excitotoxicity, oxidative stress, microglial activation, and neuroinflammation have all been implicated in the pathogenesis of ALS (Bendotti and Carrì, 2004; Bruijn *et al.*, 2004; Nirmalananthan and Greensmith, 2005; Rao and Weiss, 2004). Interestingly, these pathological events are potentially modulated by endocannabinoids, possibly explaining the neuroprotective effects of increasing endocannabinoid levels in ALS animal models. It has been found, in fact, that pharmacological agonists of CB receptors and elevated levels of AEA, obtained through genetic ablation of FAAH, exerted robust anti-inflammatory and neuroprotective effects, delaying disease progression in SOD1 mice, a reliable animal model of ALS (Bilsland *et al.*, 2006; Raman *et al.*, 2004; Weydt *et al.*, 2005). Increased endocannabinoid levels in models of ALS are likely to play a neuroprotective role by stimulating various CB receptors. Accordingly, activation of CB_1 receptors reduces glutamate release from presynaptic terminals, and produce protection against neuronal death induced by glutamate through pre- and postsynaptic mechanisms (Shen and Thayer, 1998). Furthermore, stimulation of CB_1 receptors attenuates kainate-induced excitotoxicity through CB_1 receptors, when added to primary spinal cord cultures (Abood *et al.*, 2001). Finally, non-CB_1 and non-CB_2 receptors might also be involved in the neuroprotective effects of endocannabinoids against excitotoxic damage since glutamate neurotoxicity and ROS-induced cell death is also prevented in a CB_1 and CB_2 receptor independent manner (Hampson *et al.*, 1998). Finally, a report has shown that the neuroprotective effects observed in ALS mice augmentation of endocannabinoid levels were largely dependent on CB_2 receptor stimulation (Bilsland *et al.*, 2006).

A. Interplay Between Glutamate Transmission and Endocannabinoids in ALS

Disruption of extracellular glutamate homeostasis has been claimed to play a crucial role in ALS-associated excitotoxic damage. In-line with this, abnormal glutamate metabolism has been demonstrated both in patients with sporadic ALS and in animal models of the disease (Doble, 1999; Maragakis and Rothstein, 2001; Vermeiren *et al.*, 2006), while dysfunction of glutamate transporters has been shown to induce or exacerbate ALS symptoms and neurodegeneration (Doble, 1999). Endocannabinoids are synthesized and released by neurons in an activity-dependent manner (Chevaleyre *et al.*, 2006; Piomelli, 2003), raising the possibility that motorneurons subjected to hyperstimulation of glutamate receptors may constitute an important source of these substances during excitotoxic processes. According to this idea, increased levels of both AEA and 2-AG have been found in the spinal cord of SOD1 transgenic mice (Witting *et al.*, 2004).

In ALS mice, increased endocannabinoid levels might also be secondary to altered activity of metabotropic glutamate receptors 5 (mGlu5 receptor). In fact, regulation of glutamate transporter activity by these receptors is lost in experimental ALS (Vermeiren *et al.*, 2006) and stimulation of these receptors also induces endocannabinoid release (Doherty and Dingledine, 2003). It is therefore conceivable that increased endocannabinoid production in ALS models may represent a compensatory mechanism aimed at counteracting the loss of mGlu5 receptor function.

B. Interplay Between Oxidative Stress and Endocannabinoids in ALS

Glutamate is known to induce the formation of reactive oxygen species, which inhibit glutamate uptake, thereby potentiating excitotoxic damage and oxidative stress in several pathological conditions including ALS (Rao and Weiss, 2004; Rao *et al.*, 2003). Mitochondrial abnormalities have been found in ALS mouse models and in human ALS (Kong and Xu, 1998; Wong *et al.*, 1995), and amplify ROS production that follows AMPA (α-amino-3-hydroxy-5-methyl-4-isoxazole propionic acid) glutamate receptor stimulation in motorneurons (Carriedo *et al.*, 2000; Von Lewinski and Keller, 2005). Based on these considerations, it is conceivable that the neuroprotective effects on ALS motorneurons exerted by (endo)cannabinoids are partially dependent on the limitation of oxidative damage associated with glutamate receptor hyperstimulation.

C. Interplay Between Microglial Activation and Endocannabinoids in ALS

Studies on postmortem spinal cord tissues of ALS patients showed that the number of activated microglia is significantly greater in these patients than in controls (Henkel *et al.*, 2004; Kawamata *et al.*, 1992). Furthermore, the presence of activated microglia in the brain of living ALS patients has been confirmed by positron emission tomography (Turner *et al.*, 2004). Similar findings have been documented in early studies on tissues of ALS mouse models (Almer *et al.*, 1999; Hall *et al.*, 1998).

The evidence that expression of mutant SOD1 in neurons alone was insufficient to cause motorneuron degeneration indicates that altered activity of microglial, and possibly of other nonneuronal cells, is required for ALS pathophysiology. In this line, microglia from ALS mice secretes more actively cytokines potentially toxic for neighboring neurons (Weydt *et al.*, 2004), while pharmacological suppression of microglial activation on neurons exposed to the cerebrospinal fluid of ALS patients exerts clear prosurvival effects (Tikka *et al.*, 2002).

It is conceivable that the neuroprotective effects of CB_2 receptor stimulation in ALS models depend on the modulation of the activity of microglial cells

(Bilsland *et al.*, 2006; Ehrhart *et al.*, 2005). Accordingly, CB_2 receptors are scarcely or not at all expressed in neurons, while they are abundant in microglia (Facchinetti *et al.*, 2003), and are upregulated in microglial cells of ALS patients (Yiangou *et al.*, 2006). (Endo)cannabinoids have been shown to reduce microglial activation, as well as expression and release of proinflammatory cytokines from microglia via CB_2 receptors (Klegeris *et al.*, 2003; Puffenbarger *et al.*, 2000), providing a plausible explanation for their protective action in ALS.

D. Interplay Between Inflammatory Mediators and Endocannabinoids in ALS

The upregulation of COX-2 in ALS and the resulting production of proinflammatory mediators are believed to be responsible, at least in part, for the induction and the maintenance of inflammation seen during the progression of the disease (Drachman *et al.*, 2002; Klivenyi *et al.*, 2004; Pompl *et al.*, 2003). COX-2 activity is increased in spinal cords of ALS patients (Yasojima *et al.*, 2001; Yiangou *et al.*, 2006) and in SOD1 mice (Almer *et al.*, 2001), where a correlation between the intensity of inflammation and disease progression has been observed (Alexianu *et al.*, 2001; Hall *et al.*, 1998). Furthermore, pharmacological inhibition of COX-2 prolongs survival of SOD1 mice (Drachman *et al.*, 2002; Klivenyi *et al.*, 2004; Pompl *et al.*, 2003), and a clinical trial exploring the effect of celecoxib, inhibitor of COX-2, on ALS patients is currently ongoing.

Endocannabinoids can be selectively metabolized by COX-2 (Kozak and Marnett, 2002), implying that the toxic effect of a hyperactive COX-2 during ALS can be mediated by the reduction of endocannabinoid levels and the inhibition of their neuroprotective effects. These data indicate that at least some of the protective effects of COX-2 inhibition in ALS might be mediated by a reduction of endocannabinoid degradation, which mitigates excitotoxic damage and inflammatory response. In this respect, it is interesting to note that COX-2 inhibition is indeed able to elicit endocannabinoid-mediated depression of glutamate release through the stimulation of CB_1 receptors (Slanina and Schweitzer, 2005) (Fig. 2).

V. Conclusions

MS and ALS are remarkably distinct pathological disorders, both causing rearrangements of the ECS. Neuroinflammation, microglial activation, oxidative stress, and excitotoxicity are variably involved in the pathophysiology of MS and of ALS and are influenced by endocannabinoids or by drugs targeting the ECS. The pharmacological modulation of ECS in central neurons and/or in peripheral

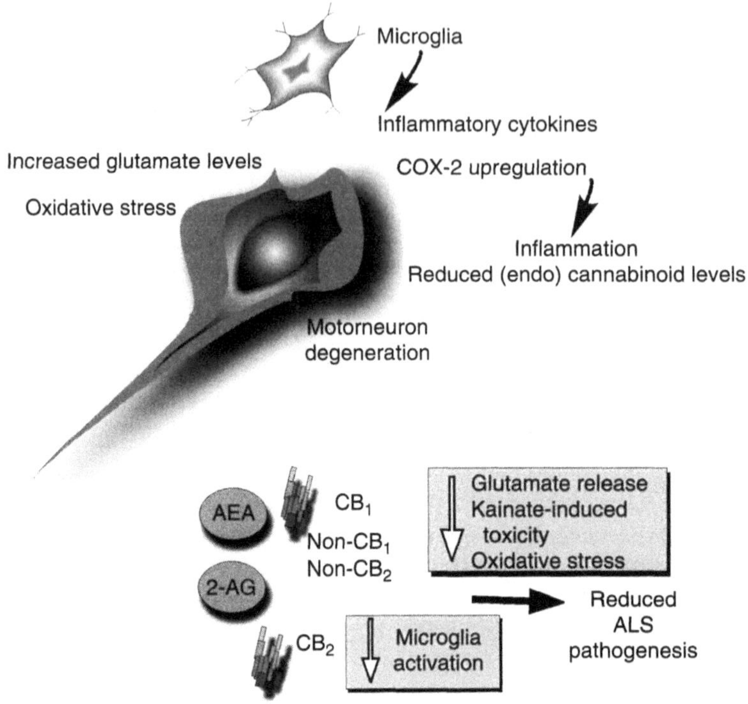

FIG. 2. Representation of the interplay between (endo)cannabinoids and the pathogenic events of ALS. The stimulation of CB receptors by AEA and 2-AG in distinct cell types results in a series of effects potentially useful to contrast ALS pathogenesis.

immune cells might represent a useful option to treat inflammatory and degenerative diseases of the CNS.

Acknowledgments

This investigation was supported by Ministero dell'Università e della Ricerca (PRIN 2005, FIRB 2006) to D.C. and M.M., by Ministero della Salute to D.C. (grants 2006), by Agenzia Spaziale Italiana (DCMC and MoMa projects) to A.F.-A., G.B., and M.M., and by Fondazione TERCAS (Research Programs 2004 and 2005) to M.M.

References

Abood, M. E., Rizvi, G., Sallapudi, N., and McAllister, S. D. (2001). Activation of the CB1 cannabinoid receptor protects cultured mouse spinal neurons against excitotoxicity. *Neurosci. Lett.* **309,** 197–201.

Alexianu, M. E., Kozovska, M., and Appel, S. H. (2001). Immune reactivity in a mouse model of familial ALS correlates with disease progression. *Neurology* **57,** 1282–1289.

Almer, G., Vukosavic, S., Romero, N., and Przedborski, S. (1999). Inducible nitric oxide synthase up-regulation in a transgenic mouse model of familial amyotrophic lateral sclerosis. *J. Neurochem.* **72,** 2415–2425.

Almer, G., Guegan, C., Teismann, P., Naini, A., Rosoklija, G., Hays, A. P., Chen, C., and Przedborski, S. (2001). Increased expression of the pro-inflammatory enzyme cyclooxygenase-2 in amyotrophic lateral sclerosis. *Ann. Neurol.* **49,** 176–185.

Baker, D., Pryce, G., Croxford, J. L., Brown, P., Pertwee, R. G., Huffman, J. W., and Layward, L. (2000). Cannabinoids control spasticity and tremor in a multiple sclerosis model. *Nature* **404,** 84–87.

Baker, D., Pryce, G., Croxford, J. L., Brown, P., Pertwee, R. G., Makriyannis, A., Khanolkar, A., Layward, L., Fezza, F., Bisogno, T., and Di Marzo, V. (2001). Endocannabinoids control spasticity in a multiple sclerosis model. *FASEB J.* **15,** 300–302.

Baker, D., Pryce, G., Giovannoni, G., and Thompson, A. J. (2003). The therapeutic potential of cannabis. *Lancet Neurol.* **2,** 291–298.

Bakshi, R., Benedict, R. H., Bermel, R. A., Caruthers, S. D., Puli, S. R., Tjoa, C. W., Fabiano, A. J., and Jacobs, L. (2002). T2 hypointensity in the deep gray matter of patients with multiple sclerosis: A quantitative magnetic resonance imaging study. *Arch. Neurol.* **59,** 62–68.

Bari, M., Battista, N., Fezza, F., Gasperi, V., and Maccarrone, M. (2006). New insights into endocannabinoid degradation and its therapeutic potential. *Mini-Rev. Med. Chem.* **6,** 109–120.

Bendotti, C., and Carrì, M. T. (2004). Lessons from models of SOD1-linked familial ALS. *Trends Mol. Med.* **10,** 393–400.

Bermel, R. A., Innus, M. D., Tjoa, C. W., and Bakshi, R. (2003). Selective caudate atrophy in multiple sclerosis: A 3D MRI parcellation study. *Neuroreport* **14,** 335–339.

Bilsland, L. G., Dick, J. R., Pryce, G., Petrosino, S., Di Marzo, V., Baker, D., and Greensmith, L. (2006). Increasing cannabinoid levels by pharmacological and genetic manipulation delay disease progression in SOD1 mice. *FASEB J.* **20,** 1003–1005.

Bjartmar, C., Kinkel, R. P., Kidd, G., Rudick, R. A., and Trapp, B. D. (2001). Axonal loss in normal-appearing white matter in a patient with acute MS. *Neurology* **57,** 1248–1252.

Bolton, C., and Paul, C. (1997). MK-801 limits neurovascular dysfunction during experimental allergic encephalomyelitis. *J. Pharmacol. Exp. Ther.* **282,** 397–402.

Breivogel, C. S., Griffin, G., Di Marzo, V., and Martin, B. R. (2001). Evidence for a new G protein-coupled cannabinoid receptor in mouse brain. *Mol. Pharmacol.* **60,** 155–163.

Bruck, W., and Stadelmann, C. (2003). Inflammation and degeneration in multiple sclerosis. *Neurol. Sci.* **24,** S265–S267.

Bruijn, L. I., Miller, T. M., and Cleveland, D. W. (2004). Unraveling the mechanisms involved in motor neuron degeneration in ALS. *Annu. Rev. Neurosci.* **27,** 723–749.

Cabranes, A., Venderova, K., de Lago, E., Fezza, F., Sanchez, A., Mestre, L., Valenti, M., Garcia-Merino, A., Ramos, J. A., Di Marzo, V., and Fernandez-Ruiz, J. (2005). Decreased endocannabinoid levels in the brain and beneficial effects of agents activating cannabinoid and/or vanilloid receptors in a rat model of multiple sclerosis. *Neurobiol. Dis.* **20,** 207–217.

Carriedo, S. G., Sensi, S. L., Yin, H. Z., and Weiss, J. H. (2000). AMPA exposures induce mitochondrial Ca(2+) overload and ROS generation in spinal motor neurons *in vitro*. *J. Neurosci.* **20,** 240–250.

Chevaleyre, V., Takahashi, K. A., and Castello, P. E. (2006). Endocannabinoid-mediated synaptic plasticity in the CNS. *Annu. Rev. Neurosci.* **29,** 37–76.

Consroe, P. (1998). Brain cannabinoid systems as targets for the therapy of neurological disorders. *Neurobiol. Dis.* **6,** 534–551.

Croxford, J. L., and Miller, S. D. (2003). Immunoregulation of a viral model of multiple sclerosis using the synthetic cannabinoid R(+)WIN55,212-2,212. *J. Clin. Invest.* **111,** 1231–1240.

Curran, N. M., Griffin, B. D., O'Toole, D., Brady, K. J., Fitzgerald, S. N., and Moynagh, P. N. (2005). The synthetic cannabinoid R(+)WIN 55,212-2 inhibits the interleukin-1 signaling pathway in human astrocytes in a cannabinoid receptor-independent manner. *J. Biol. Chem.* **280,** 35797–35806.

De Stefano, N., Narayanan, S., Francio, S. J., Smith, S., Mortella, M., Tartaglia, M. C., Bartolozzi, M. L., Guidi, L., Federico, A., and Arnold, D. L. (2002). Diffuse axonal and tissue injury in patients with multiple sclerosis with low cerebral lesion load and no disability. *Arch. Neurol.* **59,** 1565–1571.

Doble, A. (1999). The role of excitotoxicity in neurodegenerative disease: Implications for therapy. *Pharmacol. Ther.* **81,** 163–221.

Doherty, J., and Dingledine, R. (2003). Functional interactions between cannabinoid and metabotropic glutamate receptors in the central nervous system. *Curr. Opin. Pharmacol.* **3,** 46–53.

Drachman, D. B., Frank, K., Dykes-Hoberg, M., Teismann, P., Almer, G., Przedborski, S., and Rothstein, J. D. (2002). Cyclooxygenase 2 inhibition protects motor neurons and prolongs survival in a transgenic mouse model of ALS. *Ann. Neurol.* **52,** 771–778.

Drmota, T., Greasley, P., and Groblewski, T. (2004). Screening assays for cannabinoid-ligand-type modulators of GPR55. Patent WO 2004/074844 A1.

Egertova, M., Cravatt, B. F., and Elphick, M. R. (2003). Comparative analysis of fatty acid amide hydrolase and CB1 cannabinoid receptor expression in the mouse brain: Evidence of a widespread role for fatty acid amide hydrolase in regulation of endocannabinoid signaling. *Neuroscience* **119,** 481–496.

Ehrhart, J., Obregon, D., Mori, T., Hou, H., Sun, N., Bai, Y., Klein, T., Fernandez, F., Tan, J., and Shytle, R. D. (2005). Stimulation of cannabinoid receptor 2 (CB2) suppresses microglial activation. *J. Neuroinflammation* **2,** 29–32.

Eljaschewitsch, E., Witting, A., Mawrin, C., Lee, T., Schmidt, P. M., Wolf, S., Hoertnagl, H., Raine, C. S., Schneider-Stock, R., Nitsch, R., and Ullrich, O. (2006). The endocannabinoid anandamide protects neurons during CNS inflammation by induction of MKP-1 in microglial cells. *Neuron* **49,** 67–79.

Facchinetti, F., Del Giudice, E., Furegato, S., Passarotto, M., and Leon, A. (2003). Cannabinoids ablate release of TNFalpha in rat microglial cells stimulated with lypopolysaccharide. *Glia* **41,** 161–168.

Fernandez-Ruiz, J., Lastres-Becker, I., Cabranes, A., Gonzalez, S., and Ramos, J. A. (2002). Endocannabinoids and basal ganglia functionality. *Prostaglandins Leukot. Essent. Fatty Acids* **66,** 257–267.

Filippi, M., and Rocca, M. A. (2005). MRI evidence for multiple sclerosis as a diffuse disease of the central nervous system. *J. Neurol.* **252,** S16–S24.

Filippi, M., Bozzali, M., Rovaris, M., Gonen, O., Kesavadas, C., Ghezzi, A., Martinelli, V., Grossman, R. I., Scotti, G., Comi, G., and Falini, A. (2003). Evidence for widespread axonal damage at the earliest clinical stage of multiple sclerosis. *Brain* **126,** 433–437.

Galiegue, S., Mary, S., Marchand, J., Dussossoy, D., Carriere, D., Carayon, P., Bouaboula, M., Shire, D., Le Fur, G., and Casellas, P. (1995). Expression of central and peripheral cannabinoid receptors in human immune tissues and leukocyte subpopulations. *Eur. J. Biochem.* **232,** 54–61.

Hall, E. D., Andrus, P. K., Oostveen, J. A., Fleck, T. J., and Gurney, M. E. (1998). Relationship of oxygen radical-induced lipid peroxidative damage to disease onset and progression in a transgenic model of familial ALS. *J. Neurosci. Res.* **53,** 66–77.

Hampson, A. J., Grimaldi, M., Axelrod, J., and Wink, D. (1998). Cannabidiol and (–)Delta9-tetrahydrocannabinol are neuroprotective antioxidants. *Proc. Natl. Acad. Sci. USA* **95,** 8268–8273.

Hardin-Pouzet, H., Krakowski, M., Bourbonniere, L., Didier-Bazes, M., Tran, E., and Owens, T. (1997). Glutamate metabolism is down-regulated in astrocytes during experimental allergic encephalomyelitis. *Glia* **20,** 79–85.

Henkel, J. S., Engelhardt, J. I., Siklos, L., Simpson, E. P., Kim, S. H., Pan, T., Goodman, J. C., Siddique, T., Beers, D. R., and Appel, S. H. (2004). Presence of dendritic cells, MCP-1, and activated microglia/macrophages in amyotrophic lateral sclerosis spinal cord tissue. *Ann. Neurol.* **55,** 221–235.

Howlett, A. C., Barth, F., Bonner, T. I., Cabral, G., Casellas, P., Devane, W. A., Felder, C. C., Herkenham, M., Mackie, K., Martin, B. R., Mechoulam, R., and Pertwee, R. G. (2002). International union of pharmacology. XXVII. Classification of cannabinoid receptors. *Pharmacol. Rev.* **54,** 161–202.

Kawamata, T., Akiyama, H., Yamada, T., and McGeer, P. L. (1992). Immunologic reactions in amyotrophic lateral sclerosis brain and spinal cord tissue. *Am. J. Pathol.* **140,** 691–707.

Killestein, J., and Uitdehaag, B. M. (2005). Cannabinoids in multiple sclerosis: Urgent need for long term trials. *J. Neurol. Neurosurg. Psychiatr.* **76,** 1612–1614.

Klegeris, A., Bissonnette, C. J., and McGeer, P. L. (2003). Reduction of human monocytic cell neurotoxicity and cytokine secretion by ligands of the cannabinoid-type CB2 receptor. *Br. J. Pharmacol.* **139,** 775–786.

Klein, T. W., Newton, C., Larsen, K., Lu, L., Perkins, I., Nong, L., and Friedman, H. (2003). The cannabinoid system and immune modulation. *J. Leukoc. Biol.* **74,** 486–496.

Klivenyi, P., Kiaei, M., Gardian, G., Calingasan, N. Y., and Beal, M. F. (2004). Additive neuroprotective effects of creatine and cyclooxygenase 2 inhibitors in a transgenic mouse model of amyotrophic lateral sclerosis. *J. Neurochem.* **88,** 576–582.

Kong, J., and Xu, Z. (1998). Massive mitochondrial degeneration in motor neurons triggers the onset of amyotrophic lateral sclerosis in mice expressing a mutant SOD1. *J. Neurosci.* **18,** 3241–3250.

Kozak, K. R., and Marnett, L. J. (2002). Oxidative metabolism of endocannabinoids. *Prostaglandins Leukot. Essent. Fatty Acids* **66,** 211–220.

Kuhlmann, T., Lingfeld, G., Bitsch, A., Schuchardt, J., and Bruck, W. (2002). Acute axonal damage in multiple sclerosis is most extensive in early disease stages and decreases over time. *Brain* **125,** 2202–2212.

Lassmann, H. (2003). Axonal injury in multiple sclerosis. *J. Neurol. Neurosurg. Psychiatr.* **74,** 695–697.

Ligresti, A., Cascio, M. G., and Di Marzo, V. (2005). Endocannabinoid metabolic pathways and enzymes. *Curr. Drug Targets CNS Neurol. Disord.* **4,** 615–623.

Lyman, W. D., Sonett, J. R., Brosnan, C. F., Elkin, R., and Bornstein, M. B. (1989). Delta 9-tetrahydrocannabinol: A novel treatment for experimental autoimmune encephalomyelitis. *J. Neuroimmunol.* **23,** 73–81.

Maccarrone, M., Valensise, H., Bari, M., Lazzarin, N., Romanini, C., and Finazzi-Agrò, A. (2001). Progesterone up-regulates anandamide hydrolase in human lymphocytes: Role of cytokines and implications for fertility. *J. Immunol.* **166,** 7183–7189.

Maccarrone, M., Fiori, A., Bari, M., Granata, F., Gasperi, V., De Stefano, M. E., Finazzi-Agrò, A., and Strom, R. (2006). Regulation by cannabinoid receptors of anandamide transport across the blood-brain barrier and through other endothelial cells. *Thromb. Haemost.* **95,** 117–127.

Malfitano, A. M., Matarese, G., and Bifulco, M. (2005). From cannabis to endocannabinoids in multiple sclerosis: A paradigm of central nervous system autoimmune diseases. *Curr. Drug Targets CNS Neurol. Disord.* **4,** 667–675.

Maragakis, N. J., and Rothstein, J. D. (2001). Glutamate transporters in neurologic disease. *Arch. Neurol.* **58,** 365–370.

Marinelli, S., Di Marzo, V., Berretta, N., Matias, I., Maccarrone, M., Bernardi, G., and Mercuri, N. B. (2003). Presynaptic facilitation of glutamatergic synapses to dopaminergic neurons of the rat substantia nigra by endogenous stimulation of vanilloid receptors. *J. Neurosci.* **23,** 3136–3144.

McKinney, M. K., and Cravatt, B. F. (2005). Structure and function of fatty acid amide hydrolase. *Annu. Rev. Biochem.* **74,** 411–432.

Mestre, L., Correa, F., Arevalo-Martin, A., Molina-Holgado, E., Valenti, M., Ortar, G., Di Marzo, V., and Guaza, C. (2005). Pharmacological modulation of the endocannabinoid system in a viral model of multiple sclerosis. *J. Neurochem.* **92,** 1327–1339.

Nirmalananthan, N., and Greensmith, L. (2005). Amyotrophic lateral sclerosis: Recent advances and future therapies. *Curr. Opin. Neurol.* **18,** 712–719.

Nong, L., Newton, C., Friedman, H., and Klein, T. W. (2001). CB1 and CB2 receptor mRNA expression in human peripheral blood mononuclear cells (PBMC) from various donor types. *Adv. Exp. Med. Biol.* **493,** 229–233.

Nunez, E., Benito, C., Pazos, M. R., Barbachano, A., Fajardo, O., Gonzalez, S., Tolon, R. M., and Romero, J. (2004). Cannabinoid CB2 receptors are expressed by perivascular microglial cells in the human brain: An immunohistochemical study. *Synapse* **53,** 208–213.

Ohgoh, M., Hanada, T., Smith, T., Hashimoto, T., Ueno, M., Yamanishi, Y., Watanabe, M., and Nishizawa, Y. (2002). Altered expression of glutamate transporters in experimental autoimmune encephalomyelitis. *J. Neuroimmunol.* **125,** 170–178.

Ortega-Gutierrez, S., Molina-Holgado, E., Arevalo-Martin, A., Correa, F., Viso, A., Lopez-Rodriguez, M. L., Di Marzo, V., and Guaza, C. (2005). Activation of the endocannabinoid system as therapeutic approach in a murine model of multiple sclerosis. *FASEB J.* **19,** 1338–1340.

Pertwee, R. G. (2002). Cannabinoids and multiple sclerosis. *Pharmacol. Ther.* **95,** 165–174.

Pertwee, R. G., and Ross, R. A. (2002). Cannabinoid receptors and their ligands. *Prostaglandins Leukot. Essent. Fatty Acids* **66,** 101–121.

Peterson, J. W., Bo, L., Mork, S., Chang, A., and Trapp, B. D. (2001). Transected neurites, apoptotic neurons, and reduced inflammation in cortical multiple sclerosis lesions. *Ann. Neurol.* **50,** 389–400.

Piomelli, D. (2003). The molecular logic of endocannabinoid signalling. *Nat. Rev. Neurosci.* **4,** 873–884.

Pitt, D., Werner, P., and Raine, C. S. (2000). Glutamate excitotoxicity in a model of multiple sclerosis. *Nat. Med.* **6,** 67–70.

Plaut, G. S. (1987). Effectiveness of amantadine in reducing relapses in multiple sclerosis. *J. R. Soc. Med.* **80,** 91–93.

Pompl, P. N., Ho, L., Bianchi, M., McManus, T., Qin, W., and Pasinetti, G. M. (2003). A therapeutic role for cyclooxygenase-2 inhibitors in a transgenic mouse model of amyotrophic lateral sclerosis. *FASEB J.* **17,** 725–727.

Pryce, G., Ahmed, Z., Hankey, D. J., Jackson, S. J., Croxford, J. L., Pocock, J. M., Ledent, C., Petzold, A., Thompson, A. J., Giovannoni, G., Cuzner, M. L., and Baker, D. (2003). Cannabinoids inhibit neurodegeneration in models of multiple sclerosis. *Brain* **126,** 2191–2202.

Puffenbarger, R. A., Boothe, A. C., and Cabral, G. A. (2000). Cannabinoids inhibit LPS-inducible cytokine mRNA expression in rat microglial cells. *Glia* **29,** 58–69.

Raman, C., McAllister, S. D., Rizvi, G., Patel, S. G., Moore, D. H., and Abood, M. E. (2004). Amyotrophic lateral sclerosis: Delayed disease progression in mice by treatment with a cannabinoid. *Amyotroph. Lateral Scler. Other Motor Neuron Disord.* **5,** 33–39.

Rao, S. D., and Weiss, J. H. (2004). Excitotoxic and oxidative cross-talk between motor neurons and glia in ALS pathogenesis. *Trends Neurosci.* **27,** 17–23.

Rao, S. D., Yin, H. Z., and Weiss, J. H. (2003). Disruption of glial glutamate transport by reactive oxygen species produced in motor neurons. *J. Neurosci.* **23,** 2627–2633.

Sarchielli, P., Greco, L., Floridi, A., and Gallai, V. (2003). Excitatory amino acids and multiple sclerosis: Evidence from cerebrospinal fluid. *Arch. Neurol.* **60,** 1082–1088.

Shen, M., and Thayer, S. A. (1998). Cannabinoid receptor agonists protect cultured rat hippocampal neurons from excitotoxicity. *Mol. Pharmacol.* **54,** 459–462.

Slanina, K. A., and Schweitzer, P. (2005). Inhibition of cyclooxygenase-2 elicits a CB1-mediated decrease of excitatory transmission in rat CA1 hippocampus. *Neuropharmacology* **49,** 653–659.

Smith, T., Groom, A., Zhu, B., and Turski, L. (2000). Autoimmune encephalomyelitis ameliorated by AMPA antagonists. *Nat. Med.* **6,** 62–66.

Srinivasan, R., Sailasuta, N., Hurd, R., Nelson, S., and Pelleetier, D. (2005). Evidence of elevated glutamate in multiple sclerosis using magnetic resonance spectroscopy at 3 T. *Brain* **128,** 1016–1025.

Stevenson, V. L., Leary, S. M., Losseff, N. A., Parker, G. J., Barker, G. J., Husmani, Y., Miller, D. H., and Thompson, A. J. (1998). Spinal cord atrophy and disability in MS: A longitudinal study. *Neurology* **51,** 234–238.

Stover, J. F., Pleines, U. E., Morganti-Kossmann, M. C., Kossmann, T., Lowitzsch, K., and Kempski, O. S. (1997). Neurotransmitters in cerebrospinal fluid reflect pathological activity. *Eur. J. Clin. Invest.* **27,** 1038–1043.

Tikka, T. M., Vartiainen, N. E., Goldsteins, G., Oja, S. S., Andersen, P. M., Marklund, S. L., and Koistinaho, J. (2002). Minocycline prevents neurotoxicity induced by cerebrospinal fluid from patients with motor neurone disease. *Brain* **125,** 722–731.

Turner, M. R., Cagnin, A., Turkheimer, F. E., Miller, C. C., Shaw, C. E., Brooks, D. J., Leigh, P. N., and Banati, R. B. (2004). Evidence of widespread cerebral microglial activation in amyotrophic lateral sclerosis: An [11C](R)-PK11195 positron emission tomography study. *Neurobiol. Dis.* **15,** 601–609.

Vallejo-Illarramendi, A., Domercq, M., Perez-Cerda, F., Ravid, R., and Matute, C. (2006). Increased expression and function of glutamate transporters in multiple sclerosis. *Neurobiol. Dis.* **21,** 154–164.

Van Sickle, M. D., Duncan, M., Kingsley, P. J., Mouihate, A., Urbani, P., Mackie, K., Stella, N., Makriyannis, A., Piomelli, D., Davison, J. S., Marnett, L. J., Di Marzo, V., *et al.* (2005). Identification and functional characterization of brainstem cannabinoid CB2 receptors. *Science* **310,** 329–332.

Vermeiren, C., Hemptinne, I., Vanhoutte, N., Tilleux, S., Maloteaux, J. M., and Hermans, E. (2006). Loss of metabotropic glutamate receptor-mediated regulation of glutamate transport in chemically activated astrocytes in a rat model of amyotrophic lateral sclerosis. *J. Neurochem.* **96,** 719–731.

Von Lewinski, F., and Keller, B. U. (2005). Ca^{2+}, mitochondria and selective motoneuron vulnerability: Implications for ALS. *Trends Neurosci.* **28,** 494–500.

Wallstrom, E., Diener, P., Ljungdahl, A., Khademi, M., Nilsson, C. G., and Olsson, T. (1996). Memantine abrogates neurological deficits, but not CNS inflammation, in Lewis rat experimental autoimmune encephalomyelitis. *J. Neurol. Sci.* **137,** 89–96.

Walter, L., and Stella, N. (2004). Cannabinoids and neuroinflammation. *Br. J. Pharmacol.* **141,** 775–785.

Werner, P., Pitt, D., and Raine, C. S. (2001). Multiple sclerosis: Altered glutamate homeostasis in lesions correlates with oligodendrocyte and axonal damage. *Ann. Neurol.* **50,** 169–180.

Weydt, P., Yuen, E. C., Ransom, B. R., and Moller, T. (2004). Increased cytotoxic potential of microglia from ALS-transgenic mice. *Glia* **48**(2), 179–182.

Weydt, P., Hong, S., Witting, A., Moller, T., Stella, N., and Kliot, M. (2005). Cannabinol delays symptom onset in SOD1 (G93A) transgenic mice without affecting survival. *Amyotroph. Lateral Scler. Other Motor Neuron Disord.* **6,** 182–184.

Wirguin, I., Mechoulam, R., Breuer, A., Schezen, E., Weidenfeld, J., and Brenner, T. (1994). Suppression of experimental autoimmune encephalomyelitis by cannabinoids. *Immunopharmacology* **28,** 209–214.

Witting, A., Weydt, P., Hong, S., Kliot, M., Moller, T., and Stella, N. (2004). Endocannabinoids accumulate in spinal cord of SOD1 G93A transgenic mice. *J. Neurochem.* **89,** 1555–1557.

Witting, A., Chen, L., Cudaback, E., Straiker, A., Walter, L., Rickman, B., Moller, T., Brosnan, C., and Stella, N. (2006). Experimental autoimmune encephalomyelitis disrupts endocannabinoid-mediated neuroprotection. *Proc. Natl. Acad. Sci. USA* **103,** 6362–6367.

Wong, P. C., Pardo, C. A., Borchelt, D. R., Lee, M. K., Copeland, N. G., Jenkins, N. A., Sisodia, S. S., Cleveland, D. W., and Price, D. L. (1995). An adverse property of a familial ALS-linked SOD1 mutation causes motor neuron disease characterized by vacuolar degeneration of mitochondria. *Neuron* **14,** 1105–1116.

Yasojima, K., Tourtellotte, W. W., McGeer, E. G., and McGeer, P. L. (2001). Marked increase in cyclooxygenase-2 in ALS spinal cord: Implications for therapy. *Neurology* **57,** 952–956.

Yiangou, Y., Facer, P., Durrenberger, P., Chessell, I. P., Naylor, A., Bountra, C., Banati, R. R., and Anand, P. (2006). COX-2, CB2 and P2X7-immunoreactivities are increased in activated microglial cells/macrophages of multiple sclerosis and amyotrophic lateral sclerosis spinal cord. *BMC Neurol.* **6,** 12.

Zajicek, J., Fox, P., Sanders, H., Wright, D., Vickery, J., Nunn, A., and Thompson, A., and UK MS Research Group (2003). Cannabinoids for treatment of spasticity and other symptoms related to multiple sclerosis (CAMS study): Multicentre randomised placebo-controlled trial. *Lancet* **362,** 517–526.

CHEMOKINES AND CHEMOKINE RECEPTORS: MULTIPURPOSE PLAYERS IN NEUROINFLAMMATION

Richard M. Ransohoff,*,† LiPing Liu,* and Astrid E. Cardona*

*Neuroinflammation Research Center, Department of Neurosciences, Lerner Research Institute, Cleveland Clinic, Cleveland, Ohio 44195, USA

†Mellen Center for MS Treatment and Research, Cleveland Clinic, Cleveland Ohio 44195, USA

Chemokines were detected by virtue of chemotactic effects toward neutrophils in the late 1970s. During subsequent decades, it has become clear that their primordial role in vertebrate biology was to facilitate organogenesis, with particularly important functions in the central nervous system (CNS). In common with other developmentally relevant factors, chemokines and their G-protein–coupled receptors continue to be expressed in the adult CNS as neuromodulators. In our progress toward chemokine receptor blockade for treatment of inflammatory and infectious diseases, the CNS physiology of the chemokine system will need to be a material consideration. In some cases, the dual functions of the chemokine system in the periphery and in the CNS offer unique possibilities for disease treatment.

INTERNATIONAL REVIEW OF NEUROBIOLOGY, VOL. 82
DOI: 10.1016/S0074-7742(07)82010-1

I. Introduction

A. The Chemokine System Plays an Essential Role in Leukocyte Trafficking and Immunity

The chemokine system is composed of approximately 50 peptides and 20 receptors in humans, with close homologues in other mammalian species (Charo and Ransohoff, 2006). The chemokine ligand superfamily is divided into subgroups, of which the largest are the CC chemokines (28 members) and the CXC chemokines (16 members). Chemokine subgroup members, encoded in multigene arrays, are functionally related and signal to corresponding families of chemokine receptors (e.g., CXC chemokine action is mediated by CXC chemokine receptors). Chemokine receptors are G-protein–coupled receptors (GPCRs) and act specifically through pertussis toxin-sensitive $G_{\alpha i}$ components. GPCRs are drug targets, and the biotech/pharmaceutical industry has mounted substantial efforts to modulate chemokine receptor activity, heightening the medical importance of understanding how chemokines regulate inflammatory disease. First identified by their eponymous function of chemoattraction *in vitro*, chemokines are now recognized to mediate a wide array of leukocyte functions during inflammation and immunity.

B. Chemokines Regulate Cell–Cell Interactions

Perhaps most significantly, chemokines generate signals for integrin clustering and conformational changes, which lead to high-affinity and high-avidity interactions with cell adhesion molecules (CAMs). Chemokine signaling is therefore essential both for leukocyte-endothelial recognition (which regulates leukocyte trafficking) and for T cell interactions with antigen presenting cells (APCs), during lymphocyte recognition of antigen. Leukocyte extravasation requires two chemokine-mediated signals: the first signal induces leukocyte *arrest* through binding leukocyte integrins (such as LFA-1) to CAMs (such as ICAM-1) on endothelium. The second signal causes cytoskeletal reorganization and *chemotaxis*. For example, CXCR2 is the monocyte "arrest receptor" while CCR2 is the "chemotaxis receptor," and CXCR2 and CCR2 both play nonredundant functions for monocyte entry into vascular fatty streak lesions during atherogenesis (Huo *et al.*, 2001; Rollins, 2001). In the context of leukocyte trafficking, chemokines and their receptors are grouped as homeostatic (chemokines expressed constitutively in organs such as in lymph nodes and spleen, with receptors on leukocytes homing to those organs) or inflammatory (chemokines induced on demand at sites of inflammation, with receptors on infiltrating leukocytes) (Charo and Ransohoff, 2006). Homeostatic chemokine receptors such as CCR7 and CXCR5 are implicated in developmental organogenesis for lymphoid organs. Inflammatory chemokine receptors such as CCR2 are essential for responses to a wide variety of infectious and inflammatory challenges.

C. Nonsignaling Chemokine "Interceptors" Help Localize Chemokines Within Tissues

Chemokine receptor–like molecules that lack G-protein coupling have important functions in chemokine biology (Rot and von Andrian, 2004). The nonsignaling receptors are closely related to chemokine receptors, being "seven-spanning" plasma membrane components. The best characterized are Duffy antigen receptor for chemokines (DARC) and D6, each of which efficiently binds and internalizes numerous chemokines; based on being *inte*rnalizing re*ceptors*, they have been termed "interceptors" (Nibbs *et al.*, 2003). DARC, expressed on erythrocytes and venular endothelium, exerts a critical proinflammatory function: on postcapillary endothelial cells, DARC binds inflammatory CXC and CC chemokines abluminally, and internalizes them for transcytosis and immobilization on the lumenal aspect of the capillary, to signal to rolling leukocytes. D6 was described more recently. Encoded at human 3p21, near a cluster of chemokine receptor genes, D6 binds at least 12 inflammatory CC chemokines, internalizes them, and targets them for degradation through constitutive recycling between plasma membrane and endocytic vesicles at an extraordinarily rapid rate. Topical phorbol ester or intradermal complete Freund's adjuvant (CFA) caused remarkably enhanced and sustained inflammation of the skin of $D6^{-/-}$ mice due to persistently elevated local chemokine levels (Jamieson *et al.*, 2005). In this context, D6 is a "scavenger receptor," with functional similarity to IL1RA and plays a major anti-inflammatory role. D6 is expressed constitutively on lymphatic endothelium and, inducibly, on circulating leukocytes. The functional importance of D6 expression on leukocytes versus lymphatic endothelium has not been determined. We speculated that $D6^{-/-}$ mice might show worsened EAE, if chemokine clearance in the CNS were impaired. Unexpectedly, $D6^{-/-}$ mice did not generate efficient encephalitogenic responses to immunization with myelin oligodendrocyte glycoprotein $(MOG)_{35–55}$ peptide/CFA and were relatively resistant to EAE (Liu *et al.*, 2006). These findings (Liu *et al.*, 2006) are the first to delineate an impaired immune response due to enhanced inflammation at the immunization site. The further understanding of this observation may clarify peripheral autoimmune responses.

D. On Beyond Immunity: Chemokines Regulate Development and Physiology of the Nervous System

Beginning in 1998, it was discovered that $CXCR4^{-/-}$ mice, which died immediately after birth due to cardiac anomalies, harbored extensive defects in neuronal localization and development, with prominent malpositioning of neurons of the cerebellum, dentate gyrus, trigeminal ganglia, dorsal root ganglia, and cortical interneurons as well as aberrant initial trajectory of spinal motor axons (Lazarini *et al.*, 2003). The same year (1998), we collaborated with Bob Miller

(CWRU School of Medicine) to show that chemokine CXCL1 synergized with platelet-derived growth factor (PDGF) to stimulate proliferation of oligodendrocyte progenitor cells (OPCs) (Robinson *et al.*, 1998). We later showed that CXCR2 and its ligand CXCL1 helped determine both positioning and numbers of oligodendrocytes in the developing spinal cord, by acting as an "arrest receptor" and proliferative signal for OPCs (Tsai *et al.*, 2002). These reports about CXCR4 and CXCR2 sparked phylogenetic research, culminating in the proposal that CXC chemokines emerged at the dawn of vertebrate evolution, to pattern the nervous system (Huising *et al.*, 2003). This concept helps to explain the initially perplexing observation that chemokine receptors continue to be expressed on adult neural cells and mediate functions during normal physiology (Tran and Miller, 2003). Together, these observations led to the suggestion that chemokines and their receptors, along with neurotransmitters and neuropeptides, constitute a "Third System" of transmitters (Adler and Rogers, 2005). This "Third System" proposal is in-line with our hypothesis that specific chemokine receptors regulate EAE both by virtue of expression on leukocytes and on CNS cells. As examples, our current research addresses how EAE pathogenesis is affected by CXCR2 (on inflammatory monocytes in the circulation and on OPCs in the CNS) and by CX3CR1 (on regulatory NK cells in the circulation and on microglia in the CNS).

E. Chemokines are Selective Leukocyte Chemoattractants in EAE and are Implicated in Disease Pathogenesis

During EAE, the cellular infiltrate within CNS tissues includes CD4+ and CD8+ lymphocytes, monocytes, natural killer (NK) and NK-T cells, and B cells. Trafficking of some but not all cell types to the inflamed CNS of animals with EAE has been attributed to specific chemokines and their receptors. Importantly, disruption of chemokine-mediated leukocyte trafficking also alters disease expression in EAE. However, significant issues remain unresolved, particularly in regard to T cells. For example, no chemokine receptor is known to play a nonredundant, essential role for migration of lymphocytes to the CNS during EAE. Below, we describe progress in identifying how chemokine receptors regulate the migration of specific leukocyte populations to the inflamed CNS of mice with EAE.

1. *CCL2 and CCR2 Mediate Monocyte Accumulation and Tissue Injury in the CNS During EAE*

Classical depletion studies showed that monocyte-derived macrophages are critical for EAE pathogenesis. We found that $CCL2^{-/-}$ mice with EAE exhibited extremely mild disease, with near-complete lack of infiltrating monocytes and minimal demyelination (Huang *et al.*, 2001). Work by others identified CCR2 as the receptor responsible for CCL2 action in EAE (Fife *et al.*, 2000;

Izikson *et al.*, 2000). Aggressive immunization with high antigen concentration could drive an atypical EAE in $CCR2^{-/-}$ mice, but the infiltrating myeloid population comprised neutrophils, underlining the essential role of CCR2 for monocytic inflammation of the CNS in EAE (Gaupp *et al.*, 2003). Additional work by others showed that modulation of CCL2/CCR2 signaling with neutralizing antibodies or by naked DNA immunization also suppressed EAE, with particular efficacy for relapsing disease (Mahad and Ransohoff, 2003). We then performed extensive studies of transgenic mice that overexpressed CCL2 in the CNS, and found complex effects, which were both age and concentration dependent. The clinical importance of this work is suggested by reports of altered concentrations of CCL2 in the CSF of patients. Specifically, multiple sclerosis (MS) patients exhibit reduced amounts of CSF CCL2, particularly during active disease, whereas most neuroinflammatory diseases (such as AIDS dementia) feature very high CSF CCL2 concentrations. Our clinical and laboratory studies suggested that this surprising finding might be mediated through consumption of CCL2 by CCR2+ migrating cells (Mahad *et al.*, 2006).

2. *CX3CR1, Fractalkine, and NK Cells*

NK cells mediate regulatory effects in EAE, as suggested a decade ago, by showing exacerbated EAE in mice treated with depleting anti-NK1.1 antibodies (Zhang *et al.*, 1997). Studies using human NK cells showed that CX3CR1 is a chemotactic receptor for this subset. Our experiments with $CX3CR1^{-/-}$ mice demonstrated that this receptor is essential for recruitment of NK cells, but not NK-T cells, T lymphocytes, or monocytes in EAE tissues (Huang *et al.*, 2006). This work is described below.

II. Fractalkine and Fractalkine Receptor (CX3CR1) Govern Regulatory NK Accumulation and Microglial Activation During Neuroinflammation

A. Fractalkine is Essential for Accumulation of Regulatory NK Cells in the Inflamed CNS During EAE

1. *Background*

Fractalkine and CX3CR1 are relatively unique among chemokines and chemokine receptors in that their primary sites of expression are within the CNS, where high levels of message and protein are found under physiological conditions (Harrison, 2002). Fractalkine is a transmembrane chemokine expressed constitutively in the CNS by neurons. In its transmembrane form, fractalkine functions as an adhesion molecule, and fractalkine/CX3CR1 interactions are equivalent in adhesive potency with integrin/CAM binding (Haskell *et al.*, 1999). In peripheral tissues (but not CNS), fractalkine is expressed on endothelium and mediates monocyte recruitment during atherogenesis (Charo, 2001). Small quantities of fractalkine are continually released by proteolysis from neuronal membranes into the extracellular fluid (ECF) of the healthy brain, probably through the action of ADAM10 (Hundhausen *et al.*, 2003). On excitotoxic or inflammatory challenge to neurons, soluble fractalkine in the ECF increases dramatically, without a corresponding increase in fractalkine mRNA. Inducible proteolysis of fractalkine is mediated by ADAM17 (which is the TNF-α converting enzyme, TACE) (Garton *et al.*, 2001). Therefore, availability of soluble fractalkine during CNS inflammation is regulated at the posttranslational level by proteolysis, not by inducible transcription and increased mRNA availability. CX3CR1 is expressed by monocytes, CD8+ T cells, NK, and NK-T cells in the circulation, by some DCs, and by CNS microglia.

2. *Severe EAE in CX3CR1$^{-/-}$ Mice*

We studied EAE in CX3CR1$^{-/-}$ mice in which the CX3CR1 coding region had been replaced by green fluorescent protein (GFP) in Dan Littman's laboratory [Howard Hughes Medical Institutes, NYU Medical School (Jung *et al.*, 2000)] so that cells from heterozygous mice were CX3CR1$^{+/GFP}$ and retained receptor function while cells from knockout mice were CX3CR1$^{GFP/GFP}$. Cells expressing the CX3CR1-GFP reporter could be conveniently identified using flow cytometry or histochemistry in vibratome sections. Previously, the Charo Lab (UCSF), using different line of mice, showed that CX3CR1$^{-/-}$ mice on the B6/129 background developed EAE slightly more rapidly and with moderately increased severity, as compared with B6 controls (Haskell *et al.*, 2001). We extended these findings by extensive back-crossing to B6 and direct comparison of CX3CR1$^{-/-}$ and CX3CR1$^{+/+}$ mice (Huang *et al.*, 2006). We observed significantly worse EAE in CX3CR1$^{-/-}$ mice as compared with littermate controls. The most salient finding was nonremitting, spastic paralytic disease, which correlated with hemorrhagic inflammatory lesions in the spinal cords of CX3CR1$^{-/-}$ mice with EAE. To address whether altered T cell priming could underlie this phenotype, we evaluated recall proliferation and cytokine production by primed LN cells from CX3CR1$^{-/-}$ and CX3CR1$^{+/+}$ mice, and found no differences (Huang *et al.*, 2006). We asked whether lack of CX3CR1 might affect trafficking of an effector or regulatory subset to the CNS. To address this issue, we studied CNS-infiltrating cells GFP+ from CX3CR1$^{-/-}$ and CX3CR1$^{+/+}$ mice with EAE. All GFP+ mononuclear cell subsets were markedly enriched in the CNS as compared with peripheral blood (PB) of both heterozygous and knockout mice with EAE. We found a striking and selective deficiency of GFP+/NK1.1+ cells in the CNS of knockout mice with EAE, as compared with wild-type littermates, while all other cell populations (both GFP+ and GFP−) were approximately equivalent. These results suggested that fractalkine-CX3CR1 signaling might be selectively required for recruitment of NK1.1 cells to the CNS. These results were intriguing as antibody-mediated depletion of NK1.1+ cells had been shown a decade before to produce a similar phenotype of severe nonremitting EAE with high mortality (Zhang *et al.*, 1997). Taken in this context, our observations suggested that CX3CR1 deficiency might lead to severe EAE due to defective recruitment of regulatory NK cells. We confirmed that depletion with anti-NK1.1 antibodies caused severe, nonremitting EAE in CX3CR1$^{+/-}$ mice, with mortality rates and disease scores equivalent to those seen in CX3CR1$^{-/-}$ mice (Huang *et al.*, 2006). However, supportive functional assays, such as adoptive transfer of NK cells to reverse or ameliorate EAE in knockout mice, were not feasible, given the cell numbers required for such experiments and the low percentage of NK cells in the circulation.

3. *Severe EAE in CX3CR1$^{-/-}$ Mice Correlated with Impaired Recruitment of NK1.1+/TCRβ− Cells to the Inflamed CNS*

To extend our initial findings, a genetic approach was used. We collaborated with Fu-Dong Shi (Barrow Neurological Institute), an expert in the study of NK cells, to generate a colony of CX3CR1$^{+/-}$ mice lacking one copy of CD1d, a restricting element that is required for production of NKT cells that recognize glycolipid antigens. Mice were intercrossed to yield CX3CR1$^{+/-}$, CX3CR1$^{-/-}$, CD1d$^{-/-}$/CX3CR1$^{+/-}$, and CD1d$^{-/-}$/CX3CR1$^{-/-}$ mice, which were immunized with MOG/CFA and sacrificed at the peak of EAE for analysis of CNS infiltrates. The results showed selective deficiency of true NK cells in the CNS of CX3CR1$^{-/-}$ mice with EAE. CD1d+/CX3CR1$^{-/-}$ mice showed approximately 50% deficiency for CNS GFP+/NK1.1/TCRβ− cells at the peak of EAE. Findings were equally striking in CX3CR1$^{-/-}$/CD1d$^{-/-}$ mice, with a deficiency of >75% in CNS GFP+/NK1.1/TCRβ− cells. We also used CD1d/α−galactosylceramide tetramers for positive identification of NKT cells, and showed no difference between CX3CR1$^{-/-}$ and CX3CR1$^{+/-}$ mice in the CNS accumulation of these cells during EAE. As expected, tetramer-positive cells were absent from CD1d$^{-/-}$ mice. There were no other significant differences in CNS cell populations between CX3CR1$^{+/-}$ and CX3CR1$^{-/-}$ mice with EAE. These findings also established that we could validly compare EAE severity between CX3CR1$^{-/-}$ mice with or without NKT cells (i.e., CD1d$^{+/+}$ or CD1d$^{-/-}$), to begin to decipher the role of NK cells in the severe-EAE phenotype of CX3CR1$^{-/-}$ mice. Analysis of disease severity in all four genotypes showed that CX3CR1$^{-/-}$ mice experienced equally severe EAE, whether or not CD1d was also present. Further, CX3CR1$^{+/+}$ mice experienced equivalent severity of EAE in the presence or absence of CD1d. In these studies, NKT cells restricted by CD1d did not play an evident role in EAE severity.

From these results, it was not certain whether NK cells might exhibit a global deficiency in the migratory response to inflammatory stimuli. This question was addressed by collaborative studies with Christine Biron (Brown University), a leading investigator of cytokine and chemokine function in viral infection models, with particularly extensive studies of murine *Cytomegalovirus* (mCMV) hepatitis. Dr. Biron previously showed that recruitment of NK cells to the livers of mCMV-infected mice was critical for efficacious host defense. Mice lacking CX3CR1 were infected with mCMV and NK cell accumulation in the liver was monitored daily during the first 3 days postinfection. Both CX3CR1$^{+/+}$ and CX3CR1$^{-/-}$ mice showed massive accumulation of hepatic NK1.1+/TCRβ− cells ($\sim 1 \times 10^7$ per gram of tissue), with no genotype-related differences (Huang *et al.*, 2006). These observations excluded the possibility of a global NK cell migratory defect in CX3CR1$^{-/-}$ mice.

4. *Summary*

We showed that CX3CR1$^{-/-}$ mice exhibit an EAE phenotype of severe, frequently fatal, and nonremitting disease with spastic paralysis and common hemorrhagic lesions. This disease phenotype is accompanied by selective failure of accumulation of NK cells in the CNS and can be mimicked by depleting NK cells from CX3CR1$^{+/-}$ mice. Interestingly, reduced numbers of circulating CX3CR1+ NK cells were recently reported in MS patients with active disease. We propose the hypothesis that the CX3CR1$^{-/-}$ EAE phenotype stems, at least in part, because CX3CR1+ NK cells exert regulatory influences in the CNS of mice with EAE.

B. CX3CR1 is a Critical Inhibitor of Microglial Neurotoxicity

1. *CX3CR1 Deficient Mice Exhibited Increased Microglial Activation Associated with Increased Neuronal Damage After Induction of Systemic Inflammation*

The deficient accumulation of NK cells within CNS tissues is a contributing factor for EAE disease severity in CX3CR1$^{-/-}$ mice. However, it remains unclear whether CX3CR1 deficiency in microglia also contributes to the observed worsening of EAE. It is not straightforward to define the roles of microglia during EAE as macrophages derived from microglia or from infiltrating peripheral monocytes are morphologically and phenotypically indistinguishable. Therefore, we used intraperitoneal (i.p.) injections of lipopolysaccharide (LPS) to induce a systemic inflammatory reaction, without disrupting the BBB or recruiting hematogenous leukocytes into the CNS (Rivest, 2003). CX3CR1$^{+/-}$ mice showed moderate morphological activation of CNS microglia, while CX3CR1$^{-/-}$ mice exhibited intense and widespread microglial activation throughout gray matter (Cardona *et al.*, 2006). Scattered, infrequent neurons with cytoplasmic annexin V positivity (an indicator of cellular stress) were found throughout the gray matter of both knockout and heterozygous mice following control saline injections, without increase in

$CX3CR1^{+/-}$ mice after LPS injections. By contrast, annexin V-positive neurons were significantly more numerous in LPS-injected $CX3CR1^{-/-}$ mice as compared with all other groups.

2. *Neurotoxicity of* $CX3CR1^{-/-}$ *Microglia in Multiple Models of Neurodegeneration*

In additional experiments, we explored the possibility that microglial neurotoxicity, manifest in the absence of CX3CR1, would enhance the severity of neurodegenerative diseases (Cardona *et al.*, 2006). We characterized the responses of $CX3CR1^{+/+}$, $CX3CR1^{-/-}$, and $CX3CL1^{-/-}$ mice to administration of the dopaminergic neurotoxin MPTP, a model of parkinsonism. In response to MPTP, mice lacking CX3CR1 or the ligand CX3CL1/fractalkine (Cook *et al.*, 2001) exhibited a dramatic reduction in TH-IR neurons at 2 and 7 days postinjection when compared to wild-type littermates. Furthermore in the $SOD1^{G93A}$ transgenic model of amyotrophic lateral sclerosis, absence of CX3CR1 caused increased microglial activation, enhanced neuronal damage, and poorer neurobehavioral outcomes. Findings in the MPTP model and the $SOD1^{G93A}$ model were documented with extensive quantitative analysis, as presented in our published report (Cardona *et al.*, 2006).

3. *Cell-Autonomous Neurotoxic Effect of CX3CR1-Deficient Microglia Demonstrated by Adoptive Transfer Experiments*

To examine mechanisms of microglial neurotoxicity *in vivo*, we developed an adoptive transfer protocol, using purified populations of GFP-labeled activated microglial cells, which were isolated from LPS-injected $CX3CR1^{-/-}$ or $CX3CR1^{+/-}$ mice, and stereotactically placed into the frontal cortex of wild-type recipient mice (Cardona *et al.*, 2006).

Within 36-h posttransfer, microglia from $CX3CR1^{+/-}$ mice, as previously described for other myeloid cells, migrated widely throughout the CNS parenchyma, preferentially in white matter tracts (not shown) and subventricular zones. Examination of injection sites showed strikingly divergent phenotypes for $CX3CR1^{+/-}$ as compared with $CX3CR1^{-/-}$microglial cells. In wild-type recipients of $CX3CR1^{-/-}$ microglia, we readily identified terminal deoxynucleotidyl transferase biotin-dUTP labeling nick end (TUNEL)-positive neurons within 50 μm of the microglial aggregate. $CX3CR1^{+/-}$ cells were no longer present at the injection site, while, in marked contrast, microglial cells from $CX3CR1^{-/-}$ mice remained localized within 50 μm of the injection site.

In additional experiments using intracranial adoptive transfer (Cardona *et al.*, 2006), we performed quantitative analysis and showed that the CNS white matter of wild-type recipients of $CX3CR1^{-/-}$ microglia contained significantly fewer migrating cells (0.5 ± 0.5 migrating microglia/high-power field compared to 11 ± 2 cells in wild-type recipients of $CX3CR1^{+/-}$ microglia, $p = 0.02$). Furthermore, significantly fewer apoptotic neurons were detected in the CNS of wild-type recipients of CX3CR1$^{+/-}$ microglial cells (34 ± 0.7 apoptotic neurons in recipients of CX3CR1$^{-/-}$ microglia vs 0.5 ± 0.2 apoptotic neurons in recipients of CX3CR1$^{+/-}$ microglia, $p = 0.03$). These findings demonstrated cell-autonomous neurotoxicity *in vivo*, mediated by CX3CR1$^{-/-}$ microglia in the CNS of wild-type mice (Cardona *et al.*, 2006).

4. Ex Vivo *Gene Expression Profiling of Microglia After LPS Administration*

The ability to isolate GFP+ microglial cells of high purity, and to identify them *in vivo*, provided a basis for evaluating putative mediators of this neurotoxicity. To provide insight into mechanisms of microglial neurotoxicity, we performed expression profiling, using high-density oligonucleotide microarrays. Mice of differing genotypes were injected with LPS or with saline (as negative controls) by the same protocol as used to induce systemic inflammation. Four hours after the last injection, microglia were purified rapidly, RNA was prepared and analyzed *ex vivo* on microarrays. Each chip required pooled RNA from microglia from six to eight mice, and each analysis of LPS-treated mice was performed at least three times to permit application of statistical methods. Proof of principle for this approach came from initial anonymous hierarchical clustering, which showed a clear distinction between control and LPS-injected mice without genotype-related variation. *Further analysis readily distinguished between CX3CR1$^{-/-}$ and CX3CR1$^{+/-}$ mice in the microglial response to systemic inflammation evoked by i.p. LPS injections.* Repeat hierarchical clustering, using only samples from LPS-injected animals of CX3CR1$^{-/-}$ ($n = 3$ chips; 18–24 mice) or CX3CR1$^{+/+}$ ($n = 4$; 24–32 mice) genotype, showed that microglia from CX3CR1-deficient mice expressed distinct genes from those expressed by wild-type mice. Pathway analysis showed that genes overexpressed by microglial cells from CX3CR1$^{-/-}$ mice were associated with inflammation and oxidative stress, while transcripts more abundant in cells from CX3CR1$^{+/+}$ mice included anti-inflammatory and neuroprotective components. Increased levels of IL-1α in CX3CR1$^{-/-}$ microglia were confirmed by ribonuclease protection assay, and additional IL-1 pathway genes, such as Myd88, were also highly expressed in CX3CR1$^{-/-}$ microglia. These results were particularly interesting as IL-1 is a cytokine, which has been implicated in neurodegeneration *in vivo* (Rothwell, 2003).

5. *IL-1 is a Downstream Mediator of Microglial Neurotoxicity in CX3CR1$^{-/-}$ Mice*

These microarray data prompted an interventional study in which activated microglia from CX3CR1$^{-/-}$ mice were adoptively transferred along with carrier protein (control) or an equal mass of IL-1RA. Strikingly, IL-1RA partially reversed the neurotoxic phenotype of CX3CR1$^{-/-}$ microglia in this assay: some microglial cells were observed to migrate postinjection, and apoptotic neurons were not detected near the injection site (Cardona *et al.*, 2006). Compatible results were obtained from adoptive transfer of activated CX3CR1$^{-/-}$

microglia into the cortices of IL1R1$^{-/-}$ mice: we observed a moderate degree of microglial migration through the white matter of recipients, and no apoptotic neurons were observed near the injection site.

6. *Summary*

Our preliminary results showed that mice lacking CX3CR1 develop very severe, frequently fatal, nonremitting EAE with spastic paralysis and hemorrhagic spinal cord lesions. We described two distinct functions for fractalkine-CX3CR1 signaling in neuroinflammation. In EAE, CX3CR1 is required for recruitment of putative regulatory NK cells to the inflamed CNS (Huang *et al.*, 2006). Following induction of systemic inflammation with LPS, CX3CR1 is responsible for restraining microglial neurotoxicity (Cardona *et al.*, 2006). Important unresolved questions include the relative significance of CX3CR1 action toward microglia as compared with NK cells in EAE. It is also uncertain what roles are played by membrane-associated versus soluble fractalkine. The functional relevance of the common human polymorphic variant receptor CX3CR1$^{I249/M280}$ for neurodegeneration remains incompletely defined.

III. CXCR2 Regulates Both Monocyte Infiltration and Oligodendrocyte-Mediated Tissue Repair in EAE

In collaboration with Bob Miller (CWRU School of Medicine), we have been studying how CXCL1 signaling to CXCR2 contributes to development of the oligodendrocyte lineage, since 1997. Our studies led to EAE experiments. CXCR2 is expressed both on inflammatory myeloid cells, such as monocytes and neutrophils, and on OPCs (Tsai *et al.*, 2002). CXCR2 mediates monocyte arrest under flow conditions (Smith *et al.*, 2005). On OPCs, CXCR2 governs both migratory arrest and (in the presence of PDGF) proliferative responses during development (Robinson *et al.*, 1998; Tsai *et al.*, 2002). These findings are summarized below.

A. CXCL1 Acts Through CXCR2 to Arrest Migrating OPCs

In the developing rodent spinal cord, OPCs carry out a precisely orchestrated program of migration, proliferation, and differentiation (Miller, 2002). Factors that regulate this developmental program have been partially elucidated with the central finding that PDGF is necessary, but not sufficient, for all phases of this intricate process (Miller, 2002). We evaluated the action of CXCL1 toward OPCs to address the possibility that chemotactic properties might localize OPCs in the developing spinal cord. OPC migration can be modeled *in vitro* by incubation with PDGF. *Unexpectedly, we found that exposure of OPCs to CXCL1 caused migration arrest, virtually abrogating the migratory response to PDGF* (Tsai *et al.*, 2002). OPC migration arrest was mediated through CXCR2, expressed both *in vitro* and *in vivo* by OPCs. CXCL1-treated OPCs arrested *in vitro* due to increased interaction with laminin substrate, which is known from the work of others to be α6-integrin dependent (Colognato *et al.*, 2002). The molecular mechanisms underlying CXCL1-mediated OPC migration arrest are unknown, but may well entail chemokine-triggered integrin clustering or conformational change, as occurs in inflammatory leukocyte recruitment. Interestingly, CXCR2 is the chemokine receptor responsible for monocyte "arrest" under physiological flow conditions and is essential for recruitment of monocytes to sites of chronic inflammation.

1. *CXCL1 Synergizes with PDGF to Stimulate OPC Proliferation*

Astrocyte-conditioned medium (ACM) contains a factor that synergizes with PDGF to drive OPC proliferation (Robinson *et al.*, 1998). We identified CXCL1 as the necessary and sufficient ingredient in ACM to evoke this synergistic proliferative response. Furthermore, CXCL1 was produced in the ventral spinal cord around postnatal days 4–7 and in the dorsal spinal cord around days 8–12, precisely at the times and places where migrating OPCs arrest and proliferate in the developing mouse spinal cord (Robinson *et al.*, 1998; Tsai *et al.*, 2002). *Importantly, CXCL1 completely fails to affect OPC proliferation unless PDGF is also present.*

2. *Disordered Myelination in CXCR2$^{-/-}$ Mice*

In CXCR2$^{-/-}$ mice, OPCs travel past their usual point of migration arrest and accumulate at the pial surface, as predicted from our results *in vitro* (Tsai *et al.*, 2002). OPC proliferation and OPC programmed cell death are both significantly reduced in the developing spinal cords of CXCR2$^{-/-}$ mice (Tsai *et al.*, 2002). Reduced OPC proliferation and cell death summate in slightly lower numbers of mature oligodendrocytes in the spinal cords of adult CXCR2$^{-/-}$ mice (Padovani-Claudio *et al.*, 2006). In CXCR2$^{-/-}$ mice, the developmental timing of myelination is altered and the g ratio (axonal diameter/fiber diameter, an indicator of relative myelin thickness) is abnormal, with myelin thickness reduced particularly for larger fibers (Padovani-Claudio *et al.*, 2006). Myelin periodicity and paranodal components are normal in CXCR2$^{-/-}$ mice, arguing against dysfunction of mature oligodendrocytes (Padovani-Claudio *et al.*, 2006).

B. EAE in CXCR2$^{-/-}$ Mice: CXCR2 Deficiency Dramatically Reduces Susceptibility to EAE

CXCR2 induces migration arrest and (with PDGF) proliferation for OPCs during development, but the function of this receptor during myelin repair in the CNS has not been addressed. CXCR2$^{-/-}$ mice were reported in 1994 and were generated on a BALB/c background to study myelopoiesis (Cacalano *et al.*, 1994). We crossed $CXCR2^{-/-}$ mice for five generations to SWXJ (H2q/s), to conduct EAE experiments. We found that $CXCR2^{-/-}$ mice were relatively resistant to induction of EAE. In three separate experiments, we immunized a total of 24 $CXCR2^{-/-}$ mice, of which 9 died shortly after immunization and were removed from analysis. Only 5/15 surviving $CXCR2^{-/-}$ mice developed EAE, as compared with 20/28 of their $CXCR2^{+/+}$ littermates ($p < 0.05$). EAE onset, kinetics, mortality, and peak EAE severity in $CXCR2^{-/-}$ mice, which became ill, were all equivalent to that seen in $CXCR2^{+/+}$ littermates. Interestingly, recovery from the initial attack of EAE was faster and more complete in $CXCR2^{-/-}$ mice, and severity scores in the resolution phase were significantly lower than in $CXCR2^{+/+}$ mice. Interpreting these observations was challenging because the expression of CXCR2 on monocytes as well as OPCs could mediate a diversity of effects on lesion-associated tissue injury and repair.

1. *EAE in $CXCR2^{+/-} \rightarrow CXCR2^{-/-}$ Radiation Chimerae: Absence of CNS CXCR2 Improves Lesion Repair*

We proposed the hypothesis that CXCR2 might mediate pathogenic effects in EAE at two separate levels. First, CXCR2 acts as an "arrest receptor" for monocytes and could promote accumulation of inflammatory effectors in the CNS as reported for atherosclerosis models. Second, the CXCR2 ligand CXCL1 is expressed by reactive astrocytes at lesion edges, as we reported previously, from *in situ* hybridization studies (Glabinski *et al.*, 1997). Interestingly, the relevance of CXCL1 production in EAE lesions was strengthened by compatible findings in studies of MS tissue (Omari *et al.*, 2006). We speculated that OPCs might encounter CXCL1 at MS lesion borders, undergo migration arrest, and, possibly, proliferation, and would then be at a postmigratory stage in which they might fail to enter lesions, to carry out repair. In this regard, oligodendrocyte proliferation at MS lesion edges was reported in 1981 (Raine *et al.*, 1981). *Therefore, we speculated that EAE lesions might resolve more quickly or completely in $CXCR2^{-/-}$ mice.*

We addressed this issue by constructing radiation bone marrow chimerae that lacked CXCR2 only on OPCs, as well as appropriate controls. To do so, we lethally irradiated both $CXCR2^{+/+}$ and $CXCR2^{-/-}$ mice and transferred $CXCR2^{+/-}$ bone marrow, yielding $CXCR2^{+/-} \rightarrow CXCR2^{-/-}$ or $CXCR2^{+/-} \rightarrow CXCR2^{+/+}$ mice. The former lacked CXCR2 only on OPCs (and other nonradioresistant cells), while the latter retained CXCR2 on all cells; in particular, both chimerae expressed CXCR2 on hematopoietic cells. We found that engraftment could be conveniently verified by demonstrating normalization of myeloid cell populations in the circulation, as well as PCR assays of tail and PBDNA. We found strong evidence that absence of CXCR2 in the CNS was associated with improved lesion repair in EAE. $CXCR2^{+/-} \rightarrow CXCR2^{-/-}$ or $CXCR2^{+/-} \rightarrow CXCR2^{+/+}$ mice had equivalent incidence of EAE, indicating that loss of function of CXCR2 on leukocytes (probably monocytes) caused reduced incidence of EAE in CXCR2$^{-/-}$ mice. The day of onset, day of peak disease (reflecting kinetics of neurological impairment), and severity of EAE at the peak were all equivalent in both groups of mice. However, as seen previously in CXCR2$^{-/-}$ mice, the pace and extent of recovery were accelerated in mice lacking CXCR2 in the CNS.

This result suggested that absence of CXCR2 in the CNS led to improved repair of EAE lesions. This possibility was examined further by CNS tissue analysis. We initially speculated that remyelination might be more effective in mice lacking CXCR2 in the CNS and prepared axial toluidine blue-stained semithin spinal cord sections to address this possibility, with plans to identify remyelinated segments with thin myelin sheaths. We found that the developmental differences in myelin thickness in CXCR2$^{-/-}$ mice precluded this approach. Therefore, we quantified lesion area both at the peak of disease and at recovery. We found that lesion areas in CXCR2$^{+/-}$ → CXCR2$^{+/+}$ and CXCR2$^{+/-}$ → CXCR2$^{-/-}$ mice were indistinguishable at the peak of EAE, as predicted by the equivalent disease scores. However, at the recovery stage (normalized by assaying at day 10 after onset of neurological signs), we found significantly smaller lesion areas in CXCR2$^{+/-}$ → CXCR2$^{-/-}$ mice, consistent with the hypothesis that accelerated repair underlay their improved neurological recovery. *Overall, the data supported the hypothesis that deleting CXCR2 from the CNS led to more efficient or more complete repair of EAE lesions.*

In summary, we showed that mice lacking CXCR2 developed EAE at a much lower incidence than did wild-type littermate controls. CXCR2$^{-/-}$ mice with EAE recovered more quickly and more completely. We extended the analysis by constructing radiation bone marrow chimerae. Replacement of CXCR2 on hematopoietic cells rescued the low-incidence phenotype for EAE. CXCR2$^{+/-}$ → CXCR2$^{-/-}$ mice (which lacked CXCR2 on OPCs) showed quicker and more complete recovery from EAE attacks. Correlating with their EAE severity scores, CXCR2$^{+/-}$ → CXCR2$^{-/-}$ mice also showed reduced lesion area at the recovery phase of EAE, consistent with improved lesion repair. These data support our hypothesis regarding the functions of CXCR2 in EAE.

The medical implications of this research: EAE is a useful model for analyzing inflammatory demyelination of the CNS, beginning with breaking tolerance to myelin proteins, and culminating with a multifaceted highly focused attack on CNS myelin. The research described in this chapter carries significant translational potential. One important underlying point is that the components under study (fractalkine/CX3CR1; CXCL1/CXCR2) are closely homologous in humans and mice. For example, human and murine ligands for CXCR2 and CX3CR1 signal efficiently across species.

Our studies of EAE in CX3CR1$^{-/-}$ mice will help motivate research into the role of the human polymorphic variant may represent a major risk factor for accelerated neurodegeneration in MS and other diseases. The CX3CR1$^{I249/M280}$ polymorphic variant, present in about 30% of the US population, causes blunted signaling, even in heterozygous individuals

(McDermott *et al.*, 2003). The variant CX3CR1 receptor is functionally consequential, as shown by reduced atherosclerotic end point events in Framingham cohort participants bearing this allele (McDermott *et al.*, 2003). We propose, however, that variant CX3CR1 may be deficient for signaling to inhibit microglial neurotoxic effects in CNS inflammation. These questions cannot presently be addressed in mice due to structural characteristics of murine CX3CR1, and lead us to propose to "humanize" the mouse at the CX3CR1 locus. First, mice bear the equivalent of a mixed human genotype, with variant sequence I^{249} and reference sequence T^{280}. Second, mouse CX3CR1 signals differently from the human homologue due to replacement of a conserved S^{326} residue with proline in the murine receptor (Davis and Harrison, 2006). Importantly, individuals bearing the mutant form of CX3CR1 were shown to be at increased risk for age-related macular degeneration (AMD) (Tuo *et al.*, 2004), providing the first proof of principle that dysregulation of CNS myeloid cells through lack of CX3CR1 signaling might elevate the risk of neurotoxicity in some individuals.

We propose the hypothesis that blockade of CXCR2 might promote remyelination in MS lesions, as well as reducing inflammation. Chemokines and their receptors may be relevant targets for stimulating neuroprotection. Blocking CXCR2, for example, might confer both neuroprotection and anti-inflammatory effects. Leukocyte trafficking is a validated treatment target for MS, from studies using natalizumab and FTY-720 (Kappos *et al.*, 2006). However, a drug safer than natalizumab is urgently needed (Ransohoff, 2005). Chemokine receptor blockade may provide a safer, yet still effective, treatment approach.

References

Adler, M. W., and Rogers, T. J. (2005). Are chemokines the third major system in the brain? *J. Leukoc. Biol.* **78,** 1204–1209.

Cacalano, G., Lee, J., Kikly, K., Ryan, A. M., Pitts-Meek, S., Hultgren, B., Wood, W. I., and Moore, M. W. (1994). Neutrophil and B cell expansion in mice that lack the murine IL-8 receptor homolog. *Science* **265,** 682–684.

Cardona, A. E., Pioro, E. P., Sasse, M. E., Kostenko, V., Cardona, S. M., Dijkstra, I. M., Huang, D., Kidd, G., Dombrowski, S., Dutta, R., Lee, J. C., Cook, D. N., *et al.* (2006). Control of microglial neurotoxicity by the fractalkine receptor. *Nat. Neurosci.* **9,** 917–924.

Charo, I. F. (2001). Fractalkine and atherosclerosis: A new role for a curious chemokine. *Blood* **97,** 1905.

Charo, I. F., and Ransohoff, R. M. (2006). The many roles of chemokines and chemokine receptors in inflammation. *N. Engl. J. Med.* **354,** 610–621.

Colognato, H., Baron, W., Avellana-Adalid, V., Relvas, J. B., Baron-Van Evercooren, A., Georges-Labouesse, E., and Ffrench-Constant, C. (2002). CNS integrins switch growth factor signalling to promote target-dependent survival. *Nat. Cell Biol.* **4,** 833–841.

Cook, D. N., Chen, S. C., Sullivan, L. M., Manfra, D. J., Wiekowski, M. T., Prosser, D. M., Vassileva, G., and Lira, S. A. (2001). Generation and analysis of mice lacking the chemokine fractalkine. *Mol. Cell. Biol.* **21,** 3159–3165.

Davis, C. N., and Harrison, J. K. (2006). Proline 326 in the C terminus of murine CX3CR1 prevents G-protein and phosphatidylinositol 3-kinase-dependent stimulation of Akt and extracellular signal-regulated kinase in Chinese hamster ovary cells. *J. Pharmacol. Exp. Ther.* **316,** 356–363.

Fife, B. T., Huffnagle, G. B., Kuziel, W. A., and Karpus, W. J. (2000). CC chemokine receptor 2 is critical for induction of experimental autoimmune encephalomyelitis. *J. Exp. Med.* **192,** 899–906.

Garton, K. J., Gough, P. J., Blobel, C. P., Murphy, G., Greaves, D. R., Dempsey, P. J., and Raines, E. W. (2001). Tumor necrosis factor-alpha-converting enzyme (ADAM17) mediates the cleavage and shedding of fractalkine (CX3CL1). *J. Biol. Chem.* **276,** 37993–38001.

Gaupp, S., Pitt, D., Kuziel, W. A., Cannella, B., and Raine, C. S. (2003). Experimental autoimmune encephalomyelitis (EAE) in CCR2(−/−) mice: Susceptibility in multiple strains. *Am. J. Pathol.* **162,** 139–150.

Glabinski, A. R., Tani, M., Strieter, R. M., Tuohy, V. K., and Ransohoff, R. M. (1997). Synchronous synthesis of alpha- and beta-chemokines by cells of diverse lineage in the central nervous system of mice with relapses of chronic experimental autoimmune encephalomyelitis. *Am. J. Pathol.* **150,** 617–630.

Harrison, J. K. (2002). Constitutive roles of SDF-1/CXCR4 and fractalkine/CX3CR1 in the CNS. *In* "Universes in Delicate Balance: Chemokines and the Nervous System" (R. M. Ransohoff, K. Suzuki, A. E. I. Proudfoot, W. F. Hickey, and J. K. Harrison, Eds.), pp. 237–244. Elsevier, the Netherlands.

Haskell, C. A., Cleary, M. D., and Charo, I. F. (1999). Molecular uncoupling of fractalkine-mediated cell adhesion and signal transduction. Rapid flow arrest of CX3CR1-expressing cells is independent of G-protein activation. *J. Biol. Chem.* **274,** 10053–10058.

Haskell, C. A., Hancock, W. W., Salant, D. J., Gao, W., Csizmadia, V., Peters, W., Faia, K., Fituri, O., Rottman, J. B., and Charo, I. F. (2001). Targeted deletion of CX(3)CR1 reveals a role for fractalkine in cardiac allograft rejection. *J. Clin. Invest.* **108,** 679–688.

Huang, D., Shi, F. D., Jung, S., Pien, G. C., Wang, J., Salazar-Mather, T. P., He, T. T., Weaver, J. T., Ljunggren, H. G., Biron, C. A., Littman, D. R., and Ransohoff, R. M. (2006). The neuronal chemokine CX3CL1/fractalkine selectively recruits NK cells that modify experimental autoimmune encephalomyelitis within the central nervous system. *FASEB J.* **20,** 896–905.

Huang, D. R., Wang, J., Kivisakk, P., Rollins, B. J., and Ransohoff, R. M. (2001). Absence of monocyte chemoattractant protein 1 in mice leads to decreased local macrophage recruitment and antigen-specific T helper cell type 1 immune response in experimental autoimmune encephalomyelitis. *J. Exp. Med.* **193,** 713–726.

Huising, M. O., Stet, R. J., Kruiswijk, C. P., Savelkoul, H. F., and Lidy Verburg-van Kemenade, B. M. (2003). Molecular evolution of CXC chemokines: Extant CXC chemokines originate from the CNS. *Trends Immunol.* **24,** 307–313.

Hundhausen, C., Misztela, D., Berkhout, T. A., Broadway, N., Saftig, P., Reiss, K., Hartmann, D., Fahrenholz, F., Postina, R., Matthews, V., Kallen, K. J., Rose-John, S., *et al.* (2003). The disintegrin-like metalloproteinase ADAM10 is involved in constitutive cleavage of CX3CL1 (fractalkine) and regulates CX3CL1-mediated cell-cell adhesion. *Blood* **102,** 1186–1195.

Huo, Y., Weber, C., Forlow, S. B., Sperandio, M., Thatte, J., Mack, M., Jung, S., Littman, D. R., and Ley, K. (2001). The chemokine KC, but not monocyte chemoattractant protein-1, triggers monocyte arrest on early atherosclerotic endothelium. *J. Clin. Invest.* **108,** 1307–1314.

Izikson, L., Klein, R. S., Charo, I. F., Weiner, H. L., and Luster, A. D. (2000). Resistance to experimental autoimmune encephalomyelitis in mice lacking the CC chemokine receptor (CCR)2. *J. Exp. Med.* **192,** 1075–1080.

Jamieson, T., Cook, D. N., Nibbs, R. J., Rot, A., Nixon, C., McLean, P., Alcami, A., Lira, S. A., Wiekowski, M., and Graham, G. J. (2005). The chemokine receptor D6 limits the inflammatory response *in vivo*. *Nat. Immunol.* **6,** 403–411.

Jung, S., Aliberti, J., Graemmel, P., Sunshine, M. J., Kreutzberg, G. W., Sher, A., and Littman, D. R. (2000). Analysis of fractalkine receptor CX(3)CR1 function by targeted deletion and green fluorescent protein reporter gene insertion. *Mol. Cell. Biol.* **20,** 4106–4114.

Kappos, L., Antel, J., Comi, G., Montalban, X., O'Connor, P., Polman, C. H., Haas, T., Korn, A. A., Karlsson, G., and Radue, E. W. (2006). Oral fingolimod (FTY720) for relapsing multiple sclerosis. *N. Engl. J. Med.* **355,** 1124–1140.

Lazarini, F., Tham, T. N., Casanova, P., Arenzana-Seisdedos, F., and Dubois-Dalcq, M. (2003). Role of the alpha-chemokine stromal cell-derived factor (SDF-1) in the developing and mature central nervous system. *Glia* **42,** 139–148.

Liu, L., Graham, G. J., Damodaran, A., Hu, T., Lira, S. A., Sasse, M., Canasto-Chibuque, C., Cook, D. N., and Ransohoff, R. M. (2006). Cutting edge: The silent chemokine receptor d6 is required for generating T cell responses that mediate experimental autoimmune encephalomyelitis. *J. Immunol.* **177,** 17–21.

Mahad, D., Callahan, M. K., Williams, K. A., Ubogu, E. E., Kivisakk, P., Tucky, B., Kidd, G., Kingsbury, G. A., Chang, A., Fox, R. J., Mack, M., Sniderman, M. B., *et al.* (2006). Modulating CCR2 and CCL2 at the blood-brain barrier: Relevance for multiple sclerosis pathogenesis. *Brain* **129,** 212–223.

Mahad, D. J., and Ransohoff, R. M. (2003). The role of MCP-1 (CCL2) and CCR2 in multiple sclerosis and experimental autoimmune encephalomyelitis (EAE). *Semin. Immunol.* **15,** 23–32.

McDermott, D. H., Fong, A. M., Yang, Q., Sechler, J. M., Cupples, L. A., Merrell, M. N., Wilson, P. W., D'Agostino, R. B., O'Donnell, C. J., Patel, D. D., and Murphy, P. M. (2003). Chemokine receptor mutant CX3CR1-M280 has impaired adhesive function and correlates with protection from cardiovascular disease in humans. *J. Clin. Invest.* **111,** 1241–1250.

Miller, R. H. (2002). Regulation of oligodendrocyte development in the vertebrate CNS. *Prog. Neurobiol.* **67,** 451–467.

Nibbs, R., Graham, G., and Rot, A. (2003). Chemokines on the move: Control by the chemokine "interceptors" Duffy blood group antigen and D6. *Semin. Immunol.* **15,** 287–294.

Omari, K. M., John, G., Lango, R., and Raine, C. S. (2006). Role for CXCR2 and CXCL1 on glia in multiple sclerosis. *Glia* **53,** 24–31.

Padovani-Claudio, D. A., Liu, L., Ransohoff, R. M., and Miller, R. H. (2006). Alterations in the oligodendrocyte lineage, myelin, and white matter in adult mice lacking the chemokine receptor CXCR2. *Glia* **54,** 471–483.

Raine, C. S., Scheinberg, L., and Waltz, J. M. (1981). Multiple sclerosis. Oligodendrocyte survival and proliferation in an active established lesion. *Lab. Invest.* **45,** 534–546.

Ransohoff, R. M. (2005). Natalizumab and PML. *Nat. Neurosci.* **8,** 1275.

Rivest, S. (2003). Molecular insights on the cerebral innate immune system. *Brain Behav. Immunol.* **17,** 13–19.

Robinson, S., Tani, M., Strieter, R. M., Ransohoff, R. M., and Miller, R. H. (1998). The chemokine growth-regulated oncogene-alpha promotes spinal cord oligodendrocyte precursor proliferation. *J. Neurosci.* **18,** 10457–10463.

Rollins, B. J. (2001). Chemokines and atherosclerosis: What Adam Smith has to say about vascular disease. *J. Clin. Invest.* **108,** 1269–1271.

Rot, A., and von Andrian, U. H. (2004). Chemokines in innate and adaptive host defense: Basic chemokinese grammar for immune cells. *Annu. Rev. Immunol.* **22,** 891–928.

Rothwell, N. (2003). Interleukin-1 and neuronal injury: Mechanisms, modification, and therapeutic potential. *Brain Behav. Immunol.* **17,** 152–157.

Smith, D. F., Galkina, E., Ley, K., and Huo, Y. (2005). GRO family chemokines are specialized for monocyte arrest from flow. *Am. J. Physiol. Heart Circ. Physiol.* **289,** H1976–H1984.

Tran, P. B., and Miller, R. J. (2003). Chemokine receptors: Signposts to brain development and disease. *Nat. Rev. Neurosci.* **4,** 444–455.

Tsai, H. H., Frost, E., To, V., Robinson, S., Ffrench-Constant, C., Geertman, R., Ransohoff, R. M., and Miller, R. H. (2002). The chemokine receptor CXCR2 controls positioning of oligodendrocyte precursors in developing spinal cord by arresting their migration. *Cell* **110,** 373–383.

Tuo, J., Smith, B. C., Bojanowski, C. M., Meleth, A. D., Gery, I., Csaky, K. G., Chew, E. Y., and Chan, C. C. (2004). The involvement of sequence variation and expression of CX3CR1 in the pathogenesis of age-related macular degeneration. *FASEB J.* **18,** 1297–1299.

Zhang, B., Yamamura, T., Kondo, T., Fujiwara, M., and Tabira, T. (1997). Regulation of experimental autoimmune encephalomyelitis by natural killer (NK) cells. *J. Exp. Med.* **186,** 1677–1687.

SYSTEMIC AND ACQUIRED IMMUNE RESPONSES IN ALZHEIMER'S DISEASE

Markus Britschgi* and Tony Wyss-Coray*,†

*Department of Neurology and Neurological Sciences
Stanford University School of Medicine, Stanford, California 94305, USA
†Veterans Affairs Palo Alto Health Care System, Palo Alto, California 94304, USA

Alzheimer's disease (AD) is a neurodegenerative disorder characterized clinically by a progressive cognitive decline and dementia. AD brains are marked by amyloid plaques and neurofibrillary tangles, neuronal cell loss, and a prominent activation of glial cells, and innate immune responses. A growing number of studies in AD have also reported alterations in systemic immune responses including changes in lymphocyte and macrophage distribution and activation, the presence of autoantibodies, or abnormal cytokine production. Studies in animal models for AD support the notion that immune cells infiltrate the brain and may modulate the disease. Here we will review evidence for systemic alterations in immune responses and a role for acquired immunity in AD and discuss their potential contribution to the disease.

I. Alzheimer's Neuropathology

Alzheimer's disease (AD) is characterized by the accumulation of amyloid-β (Aβ) peptides into extracellular amyloid plaques and cerebrovascular deposits and by the aggregation of abnormally phosphorylated protein tau into intraneuronal

INTERNATIONAL REVIEW OF
NEUROBIOLOGY, VOL. 82
DOI: 10.1016/S0074-7742(07)82011-3

0074-7742/07 $35.00

tangles (Terry *et al.*, 1994). These protein deposits are accompanied by a dramatic loss of neurons in the cortex and hippocampus, and the past decade provided a wealth of information on the molecular mechanisms leading to these lesions (Blennow *et al.*, 2006).

Apart from this typical neuropathology, local immune responses involving glial cells and the complement system are activated in the AD brain as well (Akiyama *et al.*, 2000; Wyss-Coray, 2006). The inflammatory processes in the brain are part of a specialized inflammation that involves mostly effector mechanisms and cells of the innate immune response. This includes an increased local production of cytokines and acute phase proteins including C reactive protein, kinins, histamine, and arachidonate metabolites, and components of the complement cascade (Akiyama *et al.*, 2000). Microglia, which are the professional phagocytes of the brain and are derived from the myeloid lineage, are strongly activated and produce most of these mediators. Astrocytes are also capable of producing a wide array of immune molecules and they are invariably activated early in AD (Akiyama *et al.*, 2000; Eddleston and Mucke, 1993; Gasque *et al.*, 2000; Mennicken *et al.*, 1999).

In contrast to the activation of resident innate immune responses in the AD brain, acquired immune responses are much less prominent and systemic immune responses in AD remain poorly characterized. Nevertheless, blood-derived cells seem to accumulate in the AD brain (Rogers *et al.*, 1988; Togo *et al.*, 2002) or in mouse models of the disease (Malm *et al.*, 2005) at increased frequencies. Other reports show changes in the distribution or reactivity of immune cells in the blood (Fiala *et al.*, 2005; Monsonego *et al.*, 2003; Richartz-Salzburger *et al.*, 2007), the presence of CNS-specific or Aβ-specific antibodies in plasma (Du *et al.*, 2001; Moir *et al.*, 2005; Mruthinti *et al.*, 2004; Weksler *et al.*, 2002), or changes in individual cytokines in the plasma between those with or without AD (Teunissen *et al.*, 2003). Epidemiological studies showing roughly half the risk for developing AD in patients with arthritis and chronic use of nonsteroidal anti-inflammatory drugs (McGeer *et al.*, 1990) further suggest that immune and inflammatory mechanisms may be dysregulated in AD. Finally, the promise of Aβ vaccines in mice and humans (Weiner and Frenkel, 2006) has spurred the interest of immunologists in understanding immune responses in AD.

II. Cellular Immune Responses

Blood borne immune cells and mediators can enter most tissues but the brain limits their access and maintains a privileged or rather, specialized state. The concept of the "immune-privileged" brain arose because transplants into the brain survive much longer than those into peripheral organs (Barker and

Billingham, 1977) and was supported by the lack of a classical lymphatic drainage system and the presence of the blood–brain barrier (BBB), which consists of a layer of endothelial cells that lack fenestrae and astrocytic end-feet that encase the blood vessels (Ransohoff *et al.*, 2003). In addition, the CNS is largely devoid of classical antigen presenting cells, and large numbers of inflammatory cell infiltrates are only seen in few neurological diseases.

However, it is clear now that interactions between the brain and the immune system do occur on a continuing basis (Ransohoff *et al.*, 2003). At the cellular level, small numbers of lymphocytes and monocytes can enter the brain and CNS-specific T cells can be retained and cause disease (Engelhardt and Ransohoff, 2005; Ransohoff *et al.*, 2003). T cell trafficking to the CNS is increased after injury, and it is therefore conceivable that immune cells are attracted to the local CNS damage in the AD brain. In addition, it is possible that injury signals or peptides derived from the AD brain stimulate peripheral immune responses (Fig. 1).

In addition, certain cytokines and growth factors cross the BBB via specific transport systems (Table I). On the other hand, proteins and antigens can leave the CNS via specialized drainage systems that can transport molecules from the brain parenchyma to CSF or the lymphatics and result in efficient immune surveillance (Cserr and Knopf, 1992; Ransohoff *et al.*, 2003). This may involve

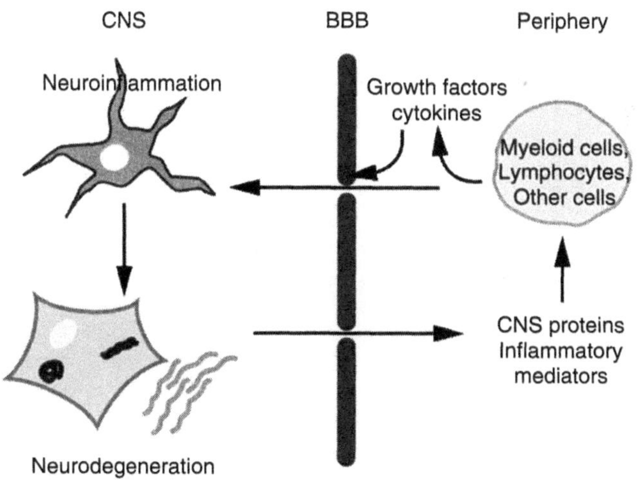

FIG. 1. Communication between central and systemic immune responses in neurodegeneration. Ongoing neurodegeneration and neuroinflammation can result in changes of efflux of CNS proteins (e.g., Aβ) or inflammatory mediators across the blood–brain barrier (BBB). These CNS-derived proteins and mediators may induce systemic immune reactions and/or recruit myeloid or lymphocytic cells into the CNS. Systemic immune or inflammatory processes may also affect CNS cells independently of local inflammatory reactions, modulating neuroinflammation and degeneration.

TABLE I

Selected Secreted Signaling Proteins with Reported Expression Changes in AD or with Potential Future Implications in AD

Factor *GeneID* OMIM[a]	Peripheral source[b]	CNS source[b]	Brain parenchyma		Cerebrospinal fluid		Plasma (Pl) or serum (Se)[c]		Crosses rodent BBB[d]
Cytokines, growth factors, and soluble receptors									
IL-1[e] *3553* 147720	Mostly MΦ, Mc, but also Ly, Fb, EndC	BEndC, Ac, Mg, ODC	↑ AD and DS	Griffin *et al.*, 1989	↔	Hasegawa *et al.*, 2000; Lanzrein *et al.*, 1998; Martinez *et al.*, 1993; Pirttila *et al.*, 1994; Tarkowski *et al.*, 1999	Ø Se	Hasegawa *et al.*, 2000	Some* Banks and Kastin, 1991; Banks *et al.*, 1989, 1991, 1994a, 2001
							↔ Se	Cacabelos *et al.*, 1991b (IL-1α,β)	
							↔ Se	Blum-Degen *et al.*, 1995; Lanzrein *et al.*, 1998; Pirttila *et al.*, 1994; Tarkowski *et al.*, 1999	
					↓	Richartz *et al.*, 2005			
					↑	Cacabelos *et al.*, 1991a; Blum-Degen *et al.*, 1995	↓ Se	Richartz *et al.*, 2005	
							↑ Pl	Licastro *et al.*, 2000	
IL-1ra *3557* 147679	Mostly MΦ, Mc, but also Ly, Fb, EndC	BEndC, Ac, Mg	↔ Cx	Wood *et al.*, 1993	↔	Lanzrein *et al.*, 1998	↔ Se	Hasegawa *et al.*, 2000; Lanzrein *et al.*, 1998	Yes* Gutierrez *et al.*, 1994
					↓	Tarkowski *et al.*, 2001			

IL-6 *3569* 147620	Mostly Ly, but also Mc, MΦ, bone marrow stromal cells, Fb, keratinocytes, EndC	Neu, Ac, Mg, BEndC	↑ Neuronal IR in diffuse and neuritic plaques ↑ ParCx ↓ OcCx	Strauss *et al.*, 1992 Hampel *et al.*, 2005	↔ ↑	Chao *et al.*, 1994; Engelborghs *et al.*, 1999; Hampel *et al.*, 1997; Kalman *et al.*, 1997; Lanzrein *et al.*, 1998; Marz *et al.*, 1997; Tarkowski *et al.*, 1999 Blum-Degen *et al.*, 1995; Martinez *et al.*, 2000	↔ Se ↔ Pl ↑ Se	Angelis *et al.*, 1998; Chao *et al.*, 1994; Lanzrein *et al.*, 1998; van Duijn *et al.*, 1990 Marz *et al.*, 1997 Bonaccorso *et al.*, 1998; Kalman *et al.*, 1997; Licastro *et al.*, 2000; Maes *et al.*, 1999; Richartz *et al.*, 2005; Tarkowski *et al.*, 1999	Yes* Banks *et al.*, 1994b
IL-6sR *3570* 47880	Ly, Mc, epithelial cells, Fb, hepatocytes	Sensory Neu, Ac, BEndC	↑ Fr and OcCx ↓ TCx and Cer	Hampel *et al.*, 2005	↔ ↓ ↑	Kordula *et al.*, 1998; Marz *et al.*, 1997; Hampel *et al.*, 1998 Hampel *et al.*, 1999; Richartz *et al.*, 2005 Bagli *et al.*, 2003	Ø Se ↔ Pl ↓ Se ↑ Pl	Hasegawa *et al.*, 2000 Marz *et al.*, 1997; Richartz *et al.*, 2005; Teunissen *et al.*, 2003 Bagli *et al.*, 2003	n/a

(*Continued*)

TABLE I (*Continued*)

Factor *GeneID* OMIM[a]	Peripheral source[b]	CNS source[b]	Brain parenchyma		Cerebrospinal fluid		Plasma (Pl) or serum (Se)[c]		Crosses rodent BBB[d]
M-CSF *1435* 120420	Ly, Mc, Fb, EndC, and multiple other	Neu, Ac, BEndC, Mg	↑ IR in Neu in proximity to Aβ deposits	Yan *et al.*, 1997	↑	Yan *et al.*, 1997			n/a
PDGF-BB *5155* 190040	platelets, megakaryocytes, MΦ, EndC, vasc smooth muscle cells	Mg, Ac, Neu	↓ Number of PDGF-BB IR pyramidal Neu in AD; IR in NFT in AD	Masliah *et al.*, 1995					No Kastin *et al.*, 2003a
TGF-β[f] *7040* 190180 190230 190220	Most nucleated cells including T-Ly, EndC; platelets (TGF-β1,β2)	Neu, Ac, Mg, BEndC	↔ TGF-β1,3 ↑ TGF-β2 ↑ TGF-β1 mRNA ↑ TGF-β in microvas TGF-β1,2,3 coloc with AD lesions	Flanders *et al.*, 1995 Wyss-Coray *et al.*, 1997 Grammas and Ovase, 2002 Peress and Perillo, 1995; van der Wal *et al.*, 1993	↑ ↑	Tarkowski *et al.*, 2002 (TGF-β) Rota *et al.*, 2006; Zetterberg *et al.*, 2004 (TGF-β1)	↔ Se ↓ Pl ↑ Se	Rota *et al.*, 2006 (TGF-β1) De Servi *et al.*, 2002; Mocali *et al.*, 2004 (TGF-β1) Chao *et al.*, 1994 (TGF-β1)	No Kastin *et al.*, 2003b

TNF-α *7124* 191160	Mainly by MΦ, but also Ly and Fb	Mg, Ac, BEndC	↓ FrCx, STGyr, Ent	Lanzrein *et al.*, 1998	↔	Hasegawa *et al.*, 2000; Lanzrein *et al.*, 1998	↔	Chao *et al.*, 1994; Lanzrein *et al.*, 1998; Maes *et al.*, 1999; Tarkowski *et al.*, 1999	Yes* Gutierrez *et al.*, 1994; Pan and Kastin, 2002
					↓	Richartz *et al.*, 2005			
					↑	Tarkowski *et al.*, 1999 (AD, VD)	↓	Cacabelos *et al.*, 1994	
					↑	Tarkowski *et al.*, 2003 (MCI)	↓ Se	Paganelli *et al.*, 2002 (mi and mod AD vs sev AD and VD)	
							↓ Se	Alvarez *et al.*, 1996 (early and late onset AD)	
							↑ Se	Galimberti *et al.*, 2003; Sun *et al.*, 2003	
							↑ Pl	Bruunsgaard *et al.*, 1999 (centenarians with AD)	
Chemokines									
CCL2/ MCP-1 *6347* 158105	Wide range of cells	Mg, perivasc MΦ, Ac, BEndC, ODC	In mature senile plaques and reactive Mg	Ishizuka *et al.*, 1997	↑	Galimberti *et al.*, 2003; Sun *et al.*, 2003	↑ Se	Galimberti *et al.*, 2006a (MCI, mi AD, not sev AD)	n/a

(Continued)

TABLE I (*Continued*)

Factor *GeneID* OMIM[a]	Peripheral source[b]	CNS source[b]	Brain parenchyma		Cerebrospinal fluid	Plasma (Pl) or serum (Se)[c]	Crosses rodent BBB[d]
CCL3/ MIP-1α *6348* 182283	Activated Ly, Langerhans dendritic cells, neutrophils, MΦ	BEndC, activated meningeal MΦ, Mg	↔	Xia and Hyman, 1999; Xia *et al.*, 1998			No Banks and Kastin, 1996
CCL4/ MIP-1β *6351* 182284	Ly, MΦ	BEndC, Ac	Mostly in a subpop of reactive Ac around Aβ plaques ↑ in AD	Xia *et al.*, 1998			No Banks and Kastin, 1996
CCL5/ RANTES *6352* 187011	T Ly, MΦ	Mg, Ac, Neu, ODC	↑ CCR3+, CCR5+ Mg around plaques, Ø CCL5 IR in AD brains	Xia *et al.*, 1998			n/a
			IR Neu of FrLb, Hp of AD	Fiala *et al.*, 2005			

CXCL8/ IL-8 *3576* 146930	Various cell types and tissues	Ac	↑ microvas	Grammas *et al.*, 2006 Reviewed by Xia and Hyman, 1999	↑	Galimberti *et al.*, 2006b (MCI, AD)	↔ Se	Bonaccorso *et al.*, 1998; Magaki *et al.*, 2007	Yes Pan *et al.*, 2003
Neurotrophic factors									
BDNF *627* 113505	T Ly, platelets	Neu, Ac, Mg	↑ Around plaques ↓ mRNA and protein levels	Ferrer *et al.*, 1999 Connor *et al.*, 1997; Ferrer *et al.*, 1999; Hock *et al.*, 2000; Holsinger *et al.*, 2000; Michalski and Fahnestock, 2003; Webster *et al.*, 2001	↔ Ø	Laske *et al.*, 2007 Laske *et al.*, 2006	↑ ↓ ↓ Se	Laske *et al.*, 2006 (early AD) Laske *et al.*, 2006 (during disease progress) Laske *et al.*, 2007	Yes* Pan *et al.*, 1998a; Poduslo and Curran, 1996

(*Continued*)

TABLE I (*Continued*)

Factor *GeneID* OMIM[a]	Peripheral source[b]	CNS source[b]	Brain parenchyma		Cerebrospinal fluid		Plasma (Pl) or serum (Se)[c]		Crosses rodent BBB[d]
NGF[g] *4803* 162030	Ly, Eos, prostate cells	Various cell types	↔ Hp, ParCx ↑ DGyr	Murase *et al.*, 1993 Narisawa-Saito *et al.*, 1996	↔	Murase *et al.*, 1993	↔ Se	Dicou *et al.*, 1997; Lorigados *et al.*, 1998; Murase *et al.*, 1993	Yes* Pan *et al.*, 1998b; Poduslo and Curran, 1996
NT-3 *4908* 162660	Skin, spleen, thymus, liver, muscle, lung, intestine, kidney, ovary	Probably various cell types	↔ mRNA Hp, ParCx ↔ ↓ Motor Cx	Murase *et al.*, 1994 Hock *et al.*, 2000 Narisawa-Saito *et al.*, 1996	↔	Murase *et al.*, 1994	↔ Se	Murase *et al.*, 1994	Yes* Pan *et al.*, 1998b; Poduslo and Curran, 1996
NT-4/5 *4909* 162662	Prostate, thymus, placenta, and skeletal muscle	Various CNS cells	↔ mRNA Par Cx, Cer ↓ Hp, Cer	Hock *et al.*, 1998 Hock *et al.*, 2000					Yes* Pan *et al.*, 1998b

[a]As online reference for each factor, the *Entrez GeneID* (http://www.ncbi.nlm.nih.gov/entrez/query.fcgi?db=gene) and the Online Mendelian Inheritance in Man (OMIM) ID number and link (www.ncbi.nlm.nih.gov/omim/) is given.

[b]Detected in humans and/or mice as reported in *The Cytokine FactsBook* (Fitzgerald *et al.*, 2001) and in *Cytokines and the CNS* (Ransohoff and Benveniste, 2006) and reviewed by the Neuroinflammation Working Group (Akiyama *et al.*, 2000).

[c]Expression levels of several of the stated factors differ between serum and plasma due to platelet degranulation or protein degradation in serum.

[d]Crossing of intact blood–brain barrier (BBB) was mostly shown in mice and a few factors were also studied in rats. BBB crossing by several of these factors may still be controversial, as the technique to analyze this mechanism has mainly been established by Banks, Pan, Kastin, and colleagues (for review see Pan and Kastin, 2004). Poduslo and Curran (1996) reproduced supporting evidence for the ability of neurotrophic factors to cross the BBB. Remarks refer to factor's ability to enter the brain parenchyma. The asterisk (*) indicates if the entry is linked to a saturable transport mechanism. n/a, not available.

[e]329 reports are found in PubMed with the search terms "alzheimer's interleukin-1alpha" but an extensive search of these articles failed to find evidence for changes in IL-1α. Instead, most articles report on Il-1β or IL-1.

[f]The multifunctional signaling protein transforming growth factor (TGF)-β would belong to the neurotrophic factors as well. In this table, it is listed with the cytokines.

[g]Data for NGF and its bioactive subunit β-NGF were combined in this table.

Changes in expression levels (mRNA or protein) or abnormal presence of the listed proteins in brain parenchyma, CSF, plasma, or serum, in human AD reported in PubMed articles before Oct. 31, 2006 (www.pubmed.gov). In the reported studies, AD samples were compared with samples from healthy age-matched controls if not mentioned different in the table. We did not list any reports of genetic associations between the listed factors and AD. Meta-analyses of multiple genetic studies have so far failed to produce any significant genetic effects of inflammatory genes (Wyss-Coray, 2006). Updates on systematic meta-analyses of genetic association studies in AD can be found on the Alzgene website http://www.alzgene.org (Bertram *et al.*, 2007).

Ac, astrocytes; BEndC, brain endothelial cells; Cer, cerebellum; coloc, colocalize; Cx, cortex; DGyr, dentate gyrus; DS, Down's syndrome; EndC, endothelial cells; Ent, entorhinal cortex; Eos, eosinophils; Fb, fibroblasts; Fr, frontal; Hp, hippocampus; IR, immunoreactive or immunoreactivity; Lb, lobe; Ly, Lymphocytes; MC, monocytes; MCI, mild cognitive impairment; Mg, microglia; mi, mild; microvas, microvasculature; mo, moderate; MΦ, macrophages; Neu, neurons; NFT, neurofibrillary tangles; Oc, occipital; ODC, oligodendrocytes; Par, parietal; sev, severe; STGyr, superior temporal gyrus; subpop, subpopulation; T, temporal; VD, vascular dementia; vasc, vascular; ↑ increased (in); ↓ decreased (in); ↔ no change (in); Ø not detectable (in); empty cells, no reports found.

perivascular drainage channels along penetrating arteries which connect to the subarachnoid space (Weller *et al.*, 1998), or specialized sites of exchange from the CSF into cervical lymph nodes (Cserr and Knopf, 1992). Perivascular drainage has been suggested to have a role in the clearance of Aβ in AD and the development of vascular amyloidosis in cerebral amyloid angiopathy (CAA) (Weller *et al.*, 1998).

A. Lymphocytes

Several reports describe increased numbers of T cells in AD brains (Itagaki *et al.*, 1988; Rogers *et al.*, 1988; Togo *et al.*, 2002) and in patients with CAA (Yamada *et al.*, 1996) where T cells remain mostly in the vicinity of blood vessels. In addition, several studies describe altered distribution of lymphocyte subsets in AD blood. Patients with AD had a general decrease in T and B cells, while the number of natural killer (NK) cells was not affected (Richartz-Salzburger *et al.*, 2007). Within the T cell population, CD8+ cells were decreased (Lombardi *et al.*, 1999; Pirttila *et al.*, 1992; Richartz-Salzburger *et al.*, 2007; Skias *et al.*, 1985) while the number of CD4+ cells was increased in AD (Lombardi *et al.*, 1999; Richartz-Salzburger *et al.*, 2007). However, no correlation was found between T cell subsets and cognitive decline (Richartz-Salzburger *et al.*, 2007). Speciale *et al.* (2007) analyzed lymphocyte subsets and NK cells in fresh blood from a large cohort of mild to moderate AD patients and healthy controls by flow cytometry. Consistent with the above reports, they found a highly significant reduction in B cells and no changes in NK cell numbers. However, they observed no changes in overall T cell number or CD4 and CD8 subsets but rather a selective increase in activated cells in the CD8 cell population expressing the surface markers CD71 and CD28 (Speciale *et al.*, 2007). These changes in lymphocyte distribution could indicate an impairment of peripheral immune function in AD.

At least some T cells in AD are reactive to Aβ (Monsonego *et al.*, 2003) and T cell activation by Aβ and other APP-derived peptides seemed to be reduced in AD patients (Trieb *et al.*, 1996). Similarly, T cells from APP transgenic mice also respond poorly to Aβ, but normally to other antigens, indicating the induction of relative T cell tolerance to Aβ (Monsonego *et al.*, 2001). However, if a more immunogenic Aβ peptide and more sensitive assays were used, patients with AD had increased T lymphocyte reactivity against Aβ (Monsonego *et al.*, 2003). The role of T cells in AD is unclear but activated T lymphocytes patrolling the CNS might secrete neurotrophins and be neuroprotective (Hohlfeld *et al.*, 2000; Schwartz *et al.*, 1999). Thus, brain-derived neurotrophic factor (BDNF)-producing T cells infiltrating the CNS may mediate the beneficial effect of the immune stimulant glatiramer acetate in the treatment of multiple sclerosis patients

(Aharoni *et al.*, 2003; Ziemssen *et al.*, 2005). Alternatively, T cells might exert cytotoxic effects and promote neuronal death in the CNS (Lewicki *et al.*, 2003; Medana *et al.*, 2001). That some T cells may also be beneficial in AD is supported by a study in which spleen cells isolated from wild-type mice immunized with Aβ were restimulated with Aβ *in vitro* and injected into APP/PS1 mice (Ethell *et al.*, 2006). Mice treated with these cells but not with CD4 T cell depleted fractions showed significant improvements in cognitive function, reduced levels of Aβ in the hippocampus, and reduced neuropathology (Ethell *et al.*, 2006).

B. Monocytes/Macrophages

Monocytes or macrophages may be dysfunctional in AD (Fiala *et al.*, 2005) and at least in mouse models of the disease, they seem to enter the brain at increased frequencies (Malm *et al.*, 2005; Simard *et al.*, 2006; Stalder *et al.*, 2005). Cells of the myeloid/monocytic lineage populate the developing brain to become microglia, the macrophages, and chief phagocytes of the brain (Ladeby *et al.*, 2005). In adulthood, small numbers of cells of hematopoietic origin continue to migrate to the CNS and develop into perivascular macrophages or microglia. These numbers can increase strongly after brain injury and infiltrating macrophages may participate in the clearance of dying cells or Aβ. Accordingly, in mouse models of AD, blood cells of the myeloid lineage infiltrate the brain and assume microglial properties (Malm *et al.*, 2005; Simard *et al.*, 2006; Stalder *et al.*, 2005) and at least some of these cells may limit Aβ accumulation in APP mice (Simard *et al.*, 2006). Recruitment was more pronounced if mice were transplanted before amyloid was deposited (Malm *et al.*, 2005; Stalder *et al.*, 2005) and further enhanced by CNS inflammation (Malm *et al.*, 2005). Many of these cells had ameboid morphology and were likely microglia, although a surprisingly high number of T lymphocytes was observed as well (Stalder *et al.*, 2005). One limitation of these three studies is that mice had to be lethally irradiated before bone marrow transplantation and this is known to result in vascular inflammation and increased infiltration of immune cells into the brain (Chiang *et al.*, 1993). In addition, the number of GFP positive cells in the brain was overall quite small and it remains to be seen whether these cells have an active role in limiting AD pathology in humans.

Monocytes and macrophages have also been studied in the blood of AD patients. On stimulation with the mitogen phytohemagglutinin (PHA) monocytes of AD patients seem to adopt a proinflammatory phenotype and secrete higher levels of interleukin (IL)-6 than cells from individuals without disease (Shalit *et al.*, 1994). On the other hand, freshly isolated macrophages from AD patients were less efficient in the phagocytosis of Aβ and monocytes showed limited potential for differentiation into macrophages (Fiala *et al.*, 2005). If monocytes do indeed migrate in significant numbers to the brain to limit Aβ accumulation, a dysfunction of these cells in AD may promote disease progression.

C. Molecular Profiles of Peripheral Immune Cells

An active communication between the CNS and the immune system in AD (Fig. 1) would be expected to result not only in changes in lymphocyte subset frequencies in the blood but also in changes in the molecular profile of peripheral immune cells and alterations in levels of immune mediators. Two groups have tested this hypothesis and found significant changes in gene expression in freshly isolated, unstimulated blood cells from AD and nondemented control patients (Kalman *et al.*, 2005; Maes *et al.*, 2006). A prominent dysregulation of gene expression in peripheral blood mononuclear cells (PBMCs) was observed in an analysis of over 6000 genes in 14 patients with mild AD and 14 age-matched nondemented controls (Maes *et al.*, 2006). Significant reductions in expression of genes associated with cytoskeletal integrity, cellular trafficking, cellular defenses, and DNA repair mechanisms was observed; remarkably, many of these changes were consistent with previous gene expression studies in AD brains or mouse models for AD (Maes *et al.*, 2006). Similarly, Kalman *et al.* (2005) confirm that gene expression in blood cells from AD patients are globally altered.

Possibly related to these findings at the transcriptional level, freshly isolated PBMC from AD patients produced higher levels of IL-1β, IL-6, and oncostatin M (OSM) than cells from nondemented controls and stimulation with PHA further accentuated these differences (Reale *et al.*, 2005). Interestingly, unstimulated PBMCs from the same patients treated for 1 month with acetylcholinesterase inhibitors, a group of approved medications for the treatment of AD, secreted lower levels of OSM but strongly increased levels of IL-1β and IL-6 (Reale *et al.*, 2005). Another study reported a remarkable deficit in vascular endothelial growth factor (VEGF) secretion in freshly isolated NK cells from patients with AD compared to those without (Solerte *et al.*, 2005). Although these studies again support a dysfunction of immune cells in AD their significance is unclear. In general, studies of unmanipulated, freshly isolated cells, including gene expression or cell surface marker analyses, will be less prone to artifacts.

III. Humoral Immune Responses in the Periphery

Key aspects of immune function are regulated by soluble factors, which make up the so-called humoral immune response. This response includes antibodies, the complement system, acute phase proteins, and cytokines and related proteins.

Apart from antibodies, most of these immune factors can be produced not only by hematopoietic or other peripheral cells but also by resident brain cells. Furthermore, immune factors may be secreted from the brain into the blood and at least some of them can enter the brain from the periphery via active transport across the BBB (Fig. 1 and Table I). Consequently, an equilibrium between blood and CNS levels may exist for specific factors. Indeed, studies in mice have shown that levels for BDNF, which is produced at high levels in the brain and blood, correlate tightly with plasma levels (Pan *et al.*, 1998a).

In AD, neurodegeneration and inflammation in the brain may result in the direct release of immune factors into the blood or alter their production in the periphery (Fig. 1). It is also likely that chemokines secreted from the brain are responsible for the above-mentioned attraction of myeloid and lymphoid cells into the brain in APP mice (Malm *et al.*, 2005; Simard *et al.*, 2006; Stalder *et al.*, 2005). Peripheral immune molecules in the blood, in turn, may have additional ways to reach the brain besides an active transport across the BBB, including a somewhat leaky BBB in AD or access via the circumventricular organs (Banks and Kastin, 1996; Gutierrez *et al.*, 1994). Cytokines might also activate vascular endothelial cells, perivascular macrophages, or parenchymal microglia in the brain.

For example, intracerebroventricular administration of IL-1β in rats and primates led to increases in serum IL-6 or IL-1 receptor antagonist (IL-1ra) levels, respectively (De Simoni *et al.*, 1993; Xiao *et al.*, 1999). Blockade of the IL-1β receptor with IL-1ra has been shown to prevent the onset of a sickness response to lipopolysaccharide (LPS) in mice (Dantzer and Kelley, 2007). In addition, intravenous injection of IL-1α reduced memory function significantly in mice (Banks *et al.*, 2001). It is tempting to speculate that these findings relate to the well-known transient impairment of cognition and attention in the elderly during and after systemic inflammation (Perry *et al.*, 2003) or the aggrevation of AD progression in twins and in AD patients with elevated IL-1β levels in serum (Holmes *et al.*, 2003; Perry *et al.*, 2003).

A. Antibodies

Autoreactive B lymphocytes and autoantibodies are present in healthy individuals at low levels but can increase with age and in pathological conditions. Aβ autoantibodies were found to be increased and to enhance Aβ toxicity in patients with AD compared to those without (Nath *et al.*, 2003) and to correlate inversely with disease severity (Du *et al.*, 2001; Weksler *et al.*, 2002). Others found the levels to be unchanged (Hyman *et al.*, 2001). Interestingly, plasma autoantibodies specific for oligomeric forms of Aβ, which is considered its most toxic aggregation state, were found to be lower in AD than in nondemented controls; in contrast, titers for autoantibodies recognizing monomeric Aβ were similar in the two groups (Moir *et al.*, 2005). In another report, autoantibodies against Aβ and the receptor for advanced glycation end products (RAGE), a protein implicated in AD, were several fold higher in AD patients and antibody titers correlated inversely with cognitive function (Mruthinti *et al.*, 2004). Autoantibodies entering the brain may facilitate the clearance of extracellular Aβ assemblies or they may mediate the removal of cellular debris and dying cells by recruiting phagocytes. Such clearance mechanisms could involve Fc receptors or complement and its receptors on phagocytes. Aβ fibrils bound to the cell surface, on the other hand, could trigger membrane attack complex (MAC) formation by activating complement without antibodies or as postulated by Aβ-specific autoantibodies that reach the CNS and activate the classical pathway of complement (D'Andrea, 2003). In addition, autoantibodies in the blood could form complexes with antigens such as Aβ and activate complement-mediated clearance by binding to erythrocytes (Rogers *et al.*, 2006; see below).

Artificial activation of autoimmune responses by active or passive immunization against Aβ reduced amyloid pathology and inflammation, and improved cognitive function in mouse models of AD (Schenk, 2002; Schenk *et al.*, 1999). Anti-Aβ antibodies can activate microglia to phagocytose Aβ (Bard *et al.*, 2000) or directly solubilize fibrillar amyloid (McLaurin *et al.*, 2002; Solomon *et al.*, 1996). Antibodies can also clear Aβ from blood without entering the brain, possibly by creating a peripheral "sink" (DeMattos *et al.*, 2001). Unfortunately, an Aβ vaccine for AD in humans was stopped when 6% of patients developed aseptic meningoencephalitis (Orgogozo *et al.*, 2003). This could be due to the development of cytotoxic T cells in this subset of patients or the generation of "toxic" autoantibodies that induced inflammation (Ferrer *et al.*, 2004; Nicoll *et al.*, 2003).

B. Complement

Because of its key role in initiating an inflammatory response (Carroll, 1998), activation of the complement system in the AD brain is of particular interest. Glial cells and neurons in the CNS can produce most components of this complex cascade, and their production is increased in AD (Emmerling *et al.*, 2000; McGeer and McGeer, 1999; Rogers *et al.*, 1996). Complement activation products, including the MAC, colocalize with amyloid plaques and tangle-bearing neurons in AD and Down's syndrome (Emmerling *et al.*, 2000; Fonseca *et al.*, 2004a; Head *et al.*, 2001; Itagaki *et al.*, 1994; Loeffler *et al.*, 2004; Shen *et al.*, 2001; Stoltzner *et al.*, 2000; Webster *et al.*, 1997). The function of the complement system in the CNS remains largely unknown but a number of studies in APP mouse models suggest that complement proteins may promote the clearance of Aβ and degenerating neurons (Wyss-Coray *et al.*, 2002). On the other hand, APP mice lacking complement C1q had less neuronal damage than complement sufficient mice suggesting a role for C1q in neuronal integrity (Fonseca *et al.*, 2004b).

More evidence for changes in the complement system in AD comes from proteomic studies of human serum or plasma. Unbiased quantitative 2D gel electrophoresis identified a number of complement factors including Factor H, Factor Bb, and fragments of C3 to be present at different levels in serum from AD patients compared to those with other neurodegenerative diseases or non-demented controls (Sheta *et al.*, 2006). In an independent unbiased study using the same technique with plasma from 50 AD patients and 50 elderly controls, Factor H precursor was strongly increased while C4 precursor was significantly decreased in AD (Hye *et al.*, 2006), indicating a potentially altered response of the complement system in the disease. Changes in Factor H are of particular interest as genetic variants in Factor H appear to be a major risk factor for the development of age-related macular degeneration (reviewed in Daiger, 2005). Although it is not clear yet how these mutations affect degeneration in the eye, it will be interesting to analyze Factor H polymorphisms in AD and explore potentially overlapping mechanisms of complement action in the two diseases. One way how complement may function in AD in plasma is by clearing Aβ from the circulation and thus creating a peripheral sink that has been proposed to reduce cerebral accumulation of Aβ based on mouse studies (DeMattos *et al.*, 2001). In support of this possibility Rogers *et al.* (2006) showed that Aβ is bound to complement receptor 1 on red blood cells and that AD patients at their earliest stages have significantly reduced levels of Aβ bound to these cells (Rogers *et al.*, 2006). Together, these reports suggest that complement activation may be altered not only in the AD brain but in the blood as well.

C. Cytokines and Related Proteins

Cytokines, chemokines, and neurotrophic and growth factors are part of a soluble network of communication factors between cells. Most of these factors are pleiotropic and regulate diverse cellular processes, including proliferation, survival, and differentiation, and they are important for the development and function of the hematopoietic and nervous systems. There is significant evidence that cytokine production is increased in the AD brain (Akiyama *et al.*, 2000) whereas neurotrophic factor signaling appears to be decreased (Connor and Dragunow, 1998; Murer *et al.*, 2001; Table I). These changes in the brain parenchyma are frequently accompanied by changes in protein levels in the CSF (Table I). The consequences of these changes on brain function and neurodegeneration relevant to AD are increasingly being studied in genetic mouse models (Wyss-Coray, 2006; Wyss-Coray and Mucke, 2002).

A number of immune factors have also been studied in the blood, although results from different studies are often inconsistent (Table I). Small sample sizes, variations in sample collection and processing, and differences or uncertainties in diagnoses are likely reasons for these discrepancies. In addition, serum is typically collected after blood coagulation at room temperature for an hour or more and is not suitable for most detailed protein studies, instead, plasma should be used. Because of their prominent roles as master regulators of immune function IL-1β, IL-6, transforming growth factor (TGF)-β1, and tumor necrosis factor (TNF)-α have been studied the most. While IL-6 seems to be unchanged or increased in AD serum or plasma, there is no consensus on IL-1β or TGF-β1 (Table I). TNF-α may be reduced early in AD but increases again in later stages (Bruunsgaard *et al.*, 1999; Paganelli *et al.*, 2002). These studies illustrate how difficult it will be to use individual cytokines as biomarkers of disease or clinical progression (Solfrizzi *et al.*, 2006; Teunissen *et al.*, 2002).

In addition to these classical cytokines, other immune factors of potential relevance for AD have been studied in the CNS or blood (Table I). Of interest is macrophage-colony stimulating factor (M-CSF) that has been implicated in AD and the potential clearance of Aβ in *op/op* mice. These mice, which lack M-CSF and as a consequence have fewer macrophages and only around 2/3 the number of microglia (Wegiel *et al.*, 1998), seem to develop Aβ immunoreactive deposits spontaneously in the brain parenchyma (Kaku *et al.*, 2003). M-CSF was also reported to be increased in CSF of AD patients (Yan *et al.*, 1997), which may indicate an attempt of the body to generate more monocytes. Likewise, chemokines which are important in attracting myeloid cells or lymphocytes to the brain (Rebenko-Moll *et al.*, 2006) may have altered levels of expression in AD. In this context, monocyte chemotactic protein-1 (CCL2/MCP-1) was reduced in early AD (Galimberti *et al.*, 2006a) and macrophage inflammatory peptide-1α (CCL3/MIP-1α) was increased in Down's syndrome with early AD-like dementia (Carta *et al.*, 2002).

Compelling arguments can also be made why to study neurotrophic factors in AD blood. Neurotrophins or neurotrophin signaling appears to be deficient in AD brains (Table I and Connor and Dragunow, 1998; Murer *et al.*, 2001). Additionally, neurotrophins may be produced by immune cells and cross the BBB efficiently (Table I). However, no consistent differences were observed between AD and nondemented control serum for these factors (Table I).

It is difficult at this point to draw conclusions from these studies of immune factors in AD blood, although it is possible that some factors are dysregulated. Maybe the most convincing evidence for a role of any of these factors in AD pathogenesis, short of a therapeutic effect in patients, would come from peripheral manipulations of individual proteins in mouse models for AD.

IV. Conclusion

As a whole, the studies summarized here suggest that AD is accompanied by changes in peripheral immune responses and by a limited activation of acquired immunity. Most convincing are the gene expression studies in freshly isolated blood cells or the proteomic studies in plasma, as well as the presence of Aβ-specific T cells and antibodies. It is however, difficult to focus on a particular molecule or cell type that consistently produced changes in different studies. Because the reviewed studies are descriptive in nature, it is also not possible to gauge the significance of any changes. For such assessments it will be necessary to manipulate peripheral immune responses in animal models for AD or reproduce such changes in these models. Successful therapeutic strategies targeting peripheral immune responses or manipulating acquired immune responses in AD patients would of course provide ultimate proof of their relevance.

Acknowledgments

This work was supported by the John Douglas French Alzheimer's Foundation, the National Institutes on Aging (AG20603, T.W.C.), and the Veterans Administration Geriatric Research, Education and Clinical Center.

References

Aharoni, R., Kayhan, B., Eilam, R., Sela, M., and Arnon, R. (2003). Glatiramer acetate-specific T cells in the brain express T helper 2/3 cytokines and brain-derived neurotrophic factor *in situ*. *Proc. Natl. Acad. Sci. USA* **100,** 14157–14162.

Akiyama, H., Barger, S., Barnum, S., Bradt, B., Bauer, J., Cole, G. M., Cooper, N. R., Eikelenboom, P., Emmerling, M., Fiebich, B. L., Finch, C. E., Frautschy, S., *et al.* (2000). Inflammation and Alzheimer's disease. Neuroinflammation Working Group. *Neurobiol. Aging* **21,** 383–421.

Alvarez, X. A., Franco, A., Fernandez-Novoa, L., and Cacabelos, R. (1996). Blood levels of histamine, IL-1 beta, and TNF-alpha in patients with mild to moderate Alzheimer disease. *Mol. Chem. Neuropathol.* **29,** 237–252.

Angelis, P., Scharf, S., Mander, A., Vajda, F., and Christophidis, N. (1998). Serum interleukin-6 and interleukin-6 soluble receptor in Alzheimer's disease. *Neurosci. Lett.* **244,** 106–108.

Bagli, M., Papassotiropoulos, A., Hampel, H., Becker, K., Jessen, F., Burger, K., Ptok, U., Rao, M. L., Moller, H. J., Maier, W., and Heun, R. (2003). Polymorphisms of the gene encoding the inflammatory cytokine interleukin-6 determine the magnitude of the increase in soluble interleukin-6 receptor levels in Alzheimer's disease. Results of a pilot study. *Eur. Arch. Psychiatry Clin. Neurosci.* **253,** 44–48.

Banks, W. A., and Kastin, A. J. (1991). Blood to brain transport of interleukin links the immune and central nervous systems. *Life Sci.* **48,** PL117–PL121.

Banks, W. A., and Kastin, A. J. (1996). Reversible association of the cytokines MIP-1 alpha and MIP-1 beta with the endothelia of the blood-brain barrier. *Neurosci. Lett.* **205,** 202–206.

Banks, W. A., Kastin, A. J., and Durham, D. A. (1989). Bidirectional transport of interleukin-1 alpha across the blood-brain barrier. *Brain Res. Bull.* **23,** 433–437.

Banks, W. A., Ortiz, L., Plotkin, S. R., and Kastin, A. J. (1991). Human interleukin (IL) 1 alpha, murine IL-1 alpha and murine IL-1 beta are transported from blood to brain in the mouse by a shared saturable mechanism. *J. Pharmacol. Exp. Ther.* **259,** 988–996.

Banks, W. A., Kastin, A. J., and Ehrensing, C. A. (1994a). Blood-borne interleukin-1 alpha is transported across the endothelial blood-spinal cord barrier of mice. *J. Physiol.* **479**(Pt. 2), 257–264.

Banks, W. A., Kastin, A. J., and Gutierrez, E. G. (1994b). Penetration of interleukin-6 across the murine blood-brain barrier. *Neurosci. Lett.* **179,** 53–56.

Banks, W. A., Farr, S. A., La Scola, M. E., and Morley, J. E. (2001). Intravenous human interleukin-1alpha impairs memory processing in mice: Dependence on blood-brain barrier transport into posterior division of the septum. *J. Pharmacol. Exp. Ther.* **299,** 536–541.

Bard, F., Cannon, C., Barbour, R., Burke, R., Games, D., Grajeda, H., Guido, T., Hu, K., Huang, J., Johnson-Wood, K., Khan, K., Kholodenko, D., *et al.* (2000). Peripherally administered antibodies against amyloid β-peptide enter the central nervous system and reduce pathology in a mouse model of Alzheimer disease. *Nat. Med.* **6,** 916–919.

Barker, C. F., and Billingham, R. E. (1977). Immunologically privileged sites. *Adv. Immunol.* **25,** 1–54.

Bertram, L., McQueen, M. B., Mullin, K., Blacker, D., and Tanzi, R. E. (2007). Systematic meta-analyses of Alzheimer disease genetic association studies: The AlzGene database. *Nat. Genet.* **39,** 17–23.

Blennow, K., de Leon, M. J., and Zetterberg, H. (2006). Alzheimer's disease. *Lancet* **368,** 387–403.

Blum-Degen, D., Muller, T., Kuhn, W., Gerlach, M., Przuntek, H., and Riederer, P. (1995). Interleukin-1 beta and interleukin-6 are elevated in the cerebrospinal fluid of Alzheimer's and *de novo* Parkinson's disease patients. *Neurosci. Lett.* **202,** 17–20.

Bonaccorso, S., Lin, A., Song, C., Verkerk, R., Kenis, G., Bosmans, E., Scharpe, S., Vandewoude, M., Dossche, A., and Maes, M. (1998). Serotonin-immune interactions in elderly volunteers and in patients with Alzheimer's disease (DAT): Lower plasma tryptophan availability to the brain in the elderly and increased serum interleukin-6 in DAT. *Aging (Milano)* **10,** 316–323.

Bruunsgaard, H., Andersen-Ranberg, K., Jeune, B., Pedersen, A. N., Skinhoj, P., and Pedersen, B. K. (1999). A high plasma concentration of TNF-alpha is associated with dementia in centenarians. *J. Gerontol. A Biol. Sci. Med. Sci.* **54,** M357–M364.

Cacabelos, R., Barquero, M., Garcia, P., Alvarez, X. A., and Varela de Seijas, E. (1991a). Cerebrospinal fluid interleukin-1 beta (IL-1 beta) in Alzheimer's disease and neurological disorders. *Methods Find. Exp. Clin. Pharmacol.* **13,** 455–458.

Cacabelos, R., Franco-Maside, A., and Alvarez, X. A. (1991b). Interleukin-1 in Alzheimer's disease and multi-infarct dementia: Neuropsychological correlations. *Methods Find. Exp. Clin. Pharmacol.* **13,** 703–708.

Cacabelos, R., Alvarez, X. A., Franco-Maside, A., Fernandez-Novoa, L., and Caamano, J. (1994). Serum tumor necrosis factor (TNF) in Alzheimer's disease and multi-infarct dementia. *Methods Find. Exp. Clin. Pharmacol.* **16,** 29–35.

Carroll, M. C. (1998). The role of complement and complement receptors in induction and regulation of immunity. *Annu. Rev. Immunol.* **16,** 545–568.

Carta, M. G., Serra, P., Ghiani, A., Manca, E., Hardoy, M. C., Del Giacco, G. S., Diaz, G., Carpiniello, B., and Manconi, P. E. (2002). Chemokines and pro-inflammatory cytokines in Down's syndrome: An early marker for Alzheimer-type dementia? *Psychother. Psychosom.* **71,** 233–236.

Chao, C. C., Ala, T. A., Hu, S., Crossley, K. B., Sherman, R. E., Peterson, P. K., and Frey, W. H., II. (1994). Serum cytokine levels in patients with Alzheimer's disease. *Clin. Diagn. Lab. Immunol.* **1,** 433–436.

Chiang, C. S., McBride, W. H., and Withers, H. R. (1993). Radiation-induced astrocytic and microglial responses in mouse brain. *Radiother. Oncol.* **29,** 60–68.

Connor, B., and Dragunow, M. (1998). The role of neuronal growth factors in neurodegenerative disorders of the human brain. *Brain Res. Brain Res. Rev.* **27,** 1–39.

Connor, B., Young, D., Yan, Q., Faull, R. L., Synek, B., and Dragunow, M. (1997). Brain-derived neurotrophic factor is reduced in Alzheimer's disease. *Brain Res. Mol. Brain Res.* **49,** 71–81.

Cserr, H. F., and Knopf, P. M. (1992). Cervical lymphatics, the blood-brain barrier and the immunoreactivity of the brain: A new view. *Immunol. Today* **13,** 507–512.

Daiger, S. P. (2005). Genetics. Was the human genome project worth the effort? *Science* **308,** 362–364.

D'Andrea, M. R. (2003). Evidence linking neuronal cell death to autoimmunity in Alzheimer's disease. *Brain Res.* **982,** 19–30.

Dantzer, R., and Kelley, K. W. (2007). Twenty years of research on cytokine-induced sickness behavior. *Brain Behav. Immun.* **21**(2), 153–160.

De Servi, B., La Porta, C. A., Bontempelli, M., and Comolli, R. (2002). Decrease of TGF-beta1 plasma levels and increase of nitric oxide synthase activity in leukocytes as potential biomarkers of Alzheimer's disease. *Exp. Gerontol.* **37,** 813–821.

De Simoni, M. G., De Luigi, A., Gemma, L., Sironi, M., Manfridi, A., and Ghezzi, P. (1993). Modulation of systemic interleukin-6 induction by central interleukin-1. *Am. J. Physiol.* **265,** R739–R742.

DeMattos, R. B., Bales, K. R., Cummins, D. J., Dodart, J.-C., Paul, S. M., and Holtzman, D. M. (2001). Peripheral anti-Aβ antibody alters CNS and plasma Aβ clearance and decreases brain Aβ burden in a mouse model of Alzheimer's disease. *Proc. Natl. Acad. Sci. USA* **98,** 8850–8855.

Dicou, E., Vermersch, P., Penisson-Besnier, I., Dubas, F., and Nerriere, V. (1997). Anti-NGF autoantibodies and NGF in sera of Alzheimer patients and in normal subjects in relation to age. *Autoimmunity* **26,** 189–194.

Du, Y., Dodel, R., Hampel, H., Buerger, K., Lin, S., Eastwood, B., Bales, K., Gao, F., Moeller, H.-J., Oertel, W., Farlow, M., and Paul, S. (2001). Reduced levels of amyloid β-peptide antibody in Alzheimer's disease. *Neurology* **57,** 801–805.

Eddleston, M. P., and Mucke, L. (1993). Molecular profile of reactive astrocytes—implications for their role in neurologic disease. *Neuroscience* **54,** 15–36.

Emmerling, M. R., Watson, M. D., Raby, C. A., and Spiegel, K. (2000). The role of complement in Alzheimer's disease pathology. *Biochim. Biophys. Acta* **1502,** 158–171.

Engelborghs, S., De Brabander, M., De Cree, J., D'Hooge, R., Geerts, H., Verhaegen, H., and De Deyn, P. P. (1999). Unchanged levels of interleukins, neopterin, interferon-gamma and tumor necrosis factor-alpha in cerebrospinal fluid of patients with dementia of the Alzheimer type. *Neurochem. Int.* **34,** 523–530.

Engelhardt, B., and Ransohoff, R. M. (2005). The ins and outs of T-lymphocyte trafficking to the CNS: Anatomical sites and molecular mechanisms. *Trends Immunol.* **26,** 485–495.

Ethell, D. W., Shippy, D., Cao, C., Cracchiolo, J. R., Runfeldt, M., Blake, B., and Arendash, G. W. (2006). Abeta-specific T-cells reverse cognitive decline and synaptic loss in Alzheimer's mice. *Neurobiol. Dis.* **23,** 351–361.

Ferrer, I., Marin, C., Rey, M. J., Ribalta, T., Goutan, E., Blanco, R., Tolosa, E., and Marti, E. (1999). BDNF and full-length and truncated TrkB expression in Alzheimer disease. Implications in therapeutic strategies. *J. Neuropathol. Exp. Neurol.* **58,** 729–739.

Ferrer, I., Boada Rovira, M., Sanchez Guerra, M. L., Rey, M. J., and Costa-Jussa, F. (2004). Neuropathology and pathogenesis of encephalitis following amyloid-beta immunization in Alzheimer's disease. *Brain Pathol.* **14,** 11–20.

Fiala, M., Lin, J., Ringman, J., Kermani-Arab, V., Tsao, G., Patel, A., Lossinsky, A. S., Graves, M. C., Gustavson, A., Sayre, J., Sofroni, E., Suarez, T., *et al.* (2005). Ineffective phagocytosis of amyloid-beta by macrophages of Alzheimer's disease patients. *J. Alzheimers Dis.* **7,** 221–232, discussion 255–262.

Fitzgerald, K. A., O'Neill, L. A. J., Gearing, A. J. H., and Callard, R. E. (Eds.) (2001). "The Cytokine FactsBook." Academic Press, San Diego.

Flanders, K. C., Lippa, C. F., Smith, T. W., Pollen, D. A., and Sporn, M. B. (1995). Altered expression of transforming growth factor-b in Alzheimer's disease. *Neurology* **45,** 1561–1569.

Fonseca, M. I., Kawas, C. H., Troncoso, J. C., and Tenner, A. J. (2004a). Neuronal localization of C1q in preclinical Alzheimer's disease. *Neurobiol. Dis.* **15,** 40–46.

Fonseca, M. I., Zhou, J., Botto, M., and Tenner, A. J. (2004b). Absence of C1q leads to less neuropathology in transgenic mouse models of Alzheimer's disease. *J. Neurosci.* **24,** 6457–6465.

Galimberti, D., Schoonenboom, N., Scarpini, E., and Scheltens, P. (2003). Chemokines in serum and cerebrospinal fluid of Alzheimer's disease patients. *Ann. Neurol.* **53,** 547–548.

Galimberti, D., Fenoglio, C., Lovati, C., Venturelli, E., Guidi, I., Corra, B., Scalabrini, D., Clerici, F., Mariani, C., Bresolin, N., and Scarpini, E. (2006a). Serum MCP-1 levels are increased in mild cognitive impairment and mild Alzheimer's disease. *Neurobiol. Aging* **27,** 1763–1768.

Galimberti, D., Schoonenboom, N., Scheltens, P., Fenoglio, C., Bouwman, F., Venturelli, E., Guidi, I., Blankenstein, M. A., Bresolin, N., and Scarpini, E. (2006b). Intrathecal chemokine synthesis in mild cognitive impairment and Alzheimer disease. *Arch. Neurol.* **63,** 538–543.

Gasque, P., Dean, Y. D., McGreal, E. P., Beek, J. V., and Morgan, B. P. (2000). Complement components of the innate immune system in health and disease in the CNS. *Immunopharmacology* **49,** 171–186.

Grammas, P., and Ovase, R. (2002). Cerebrovascular transforming growth factor-b contributes to inflammation in the Alzheimer's disease brain. *Am. J. Pathol.* **160,** 1583–1587.

Grammas, P., Samany, P. G., and Thirumangalakudi, L. (2006). Thrombin and inflammatory proteins are elevated in Alzheimer's disease microvessels: Implications for disease pathogenesis. *J. Alzheimers Dis.* **9,** 51–58.

Griffin, W. S., Stanley, L. C., Ling, C., White, L., MacLeod, V., Perrot, L. J., White, C. L., III, and Araoz, C. (1989). Brain interleukin 1 and S-100 immunoreactivity are elevated in Down syndrome and Alzheimer disease. *Proc. Natl. Acad. Sci. USA* **86,** 7611–7615.

Gutierrez, E. G., Banks, W. A., and Kastin, A. J. (1994). Blood-borne interleukin-1 receptor antagonist crosses the blood-brain barrier. *J. Neuroimmunol.* **55,** 153–160.

Hampel, H., Schoen, D., Schwarz, M. J., Kotter, H. U., Schneider, C., Sunderland, T., Dukoff, R., Levy, J., Padberg, F., Stubner, S., Buch, K., Muller, N., *et al.* (1997). Interleukin-6 is not altered in cerebrospinal fluid of first-degree relatives and patients with Alzheimer's disease. *Neurosci. Lett.* **228,** 143–146.

Hampel, H., Sunderland, T., Kotter, H. U., Schneider, C., Teipel, S. J., Padberg, F., Dukoff, R., Levy, J., and Moller, H. J. (1998). Decreased soluble interleukin-6 receptor in cerebrospinal fluid of patients with Alzheimer's disease. *Brain Res.* **780,** 356–359.

Hampel, H., Teipel, S. J., Padberg, F., Haslinger, A., Riemenschneider, M., Schwarz, M. J., Kotter, H. U., Scheloske, M., Buch, K., Stubner, S., Dukoff, R., Lasser, R., *et al.* (1999). Discriminant power of combined cerebrospinal fluid tau protein and of the soluble interleukin-6 receptor complex in the diagnosis of Alzheimer's disease. *Brain Res.* **823,** 104–112.

Hampel, H., Haslinger, A., Scheloske, M., Padberg, F., Fischer, P., Unger, J., Teipel, S. J., Neumann, M., Rosenberg, C., Oshida, R., Hulette, C., Pongratz, D., *et al.* (2005). Pattern of interleukin-6 receptor complex immunoreactivity between cortical regions of rapid autopsy normal and Alzheimer's disease brain. *Eur. Arch. Psychiatry Clin. Neurosci.* **255,** 269–278.

Hasegawa, Y., Sawada, M., Ozaki, N., Inagaki, T., and Suzumura, A. (2000). Increased soluble tumor necrosis factor receptor levels in the serum of elderly people. *Gerontology* **46,** 185–188.

Head, E., Azizeh, B. Y., Lott, I. T., Tenner, A. J., Cotman, C. W., and Cribbs, D. H. (2001). Complement association with neurons and β-amyloid deposition in the brains of aged individuals with Down syndrome. *Neurobiol. Dis.* **8,** 252–265.

Hock, C., Heese, K., Muller-Spahn, F., Hulette, C., Rosenberg, C., and Otten, U. (1998). Decreased trkA neurotrophin receptor expression in the parietal cortex of patients with Alzheimer's disease. *Neurosci. Lett.* **241,** 151–154.

Hock, C., Heese, K., Hulette, C., Rosenberg, C., and Otten, U. (2000). Region-specific neurotrophin imbalances in Alzheimer disease: Decreased levels of brain-derived neurotrophic factor and increased levels of nerve growth factor in hippocampus and cortical areas. *Arch. Neurol.* **57,** 846–851.

Hohlfeld, R., Kerschensteiner, M., Stadelmann, C., Lassmann, H., and Wekerle, H. (2000). The neuroprotective effect of inflammation: Implications for the therapy of multiple sclerosis. *J. Neuroimmunol.* **107,** 161–166.

Holmes, C., El-Okl, M., Williams, A. L., Cunningham, C., Wilcockson, D., and Perry, V. H. (2003). Systemic infection, interleukin 1beta, and cognitive decline in Alzheimer's disease. *J. Neurol. Neurosurg. Psychiatry* **74,** 788–789.

Holsinger, R. M., Schnarr, J., Henry, P., Castelo, V. T., and Fahnestock, M. (2000). Quantitation of BDNF mRNA in human parietal cortex by competitive reverse transcription-polymerase chain reaction: Decreased levels in Alzheimer's disease. *Brain Res. Mol. Brain Res.* **76,** 347–354.

Hye, A., Lynham, S., Thambisetty, M., Causevic, M., Campbell, J., Byers, H. L., Hooper, C., Rijsdijk, F., Tabrizi, S. J., Banner, S., Shaw, C. E., Foy, C., *et al.* (2006). Proteome-based plasma biomarkers for Alzheimer's disease. *Brain* **129,** 3042–3050.

Hyman, B. T., Smith, C., Buldyrev, I., Whelan, C., Brown, H., Tang, M. X., and Mayeux, R. (2001). Autoantibodies to amyloid-beta in Alzheimer's disease. *Ann. Neurol.* **49,** 808–810.

Ishizuka, K., Kimura, T., Igata-yi, R., Katsuragi, S., Takamatsu, J., and Miyakawa, T. (1997). Identification of monocyte chemoattractant protein-1 in senile plaques and reactive microglia of Alzheimer's disease. *Psychiatry Clin. Neurosci.* **51,** 135–138.

Itagaki, S., Akiyama, H., Saito, H., and McGeer, P. L. (1994). Ultrastructural localization of complement membrane attack complex (MAC)-like immunoreactivity in brains of patients with Alzheimer's disease. *Brain Res.* **645,** 78–84.

Itagaki, S., McGeer, P. L., and Akiyama, H. (1988). Presence of T-cytotoxic suppressor and leucocyte common antigen positive cells in Alzheimer's disease brain tissue. *Neurosci. Lett.* **91,** 259–264.

Kaku, M., Tsutsui, K., Motokawa, M., Kawata, T., Fujita, T., Kohno, S., Tohma, Y., Ohtani, J., Tenjoh, K., and Tanne, K. (2003). Amyloid beta protein deposition and neuron loss in osteopetrotic (op/op) mice. *Brain Res. Brain Res. Protoc.* **12,** 104–108.

Kalman, J., Juhasz, A., Laird, G., Dickens, P., Jardanhazy, T., Rimanoczy, A., Boncz, I., Parry-Jones, W. L., and Janka, Z. (1997). Serum interleukin-6 levels correlate with the severity of dementia in Down syndrome and in Alzheimer's disease. *Acta Neurol. Scand.* **96,** 236–240.

Kalman, J., Kitajka, K., Pakaski, M., Zvara, A., Juhasz, A., Vincze, G., Janka, Z., and Puskas, L. G. (2005). Gene expression profile analysis of lymphocytes from Alzheimer's patients. *Psychiatr. Genet.* **15,** 1–6.

Kastin, A. J., Akerstrom, V., Hackler, L., and Pan, W. (2003a). Different mechanisms influencing permeation of PDGF-AA and PDGF-BB across the blood-brain barrier. *J. Neurochem.* **87,** 7–12.

Kastin, A. J., Akerstrom, V., and Pan, W. (2003b). Circulating TGF-beta1 does not cross the intact blood-brain barrier. *J. Mol. Neurosci.* **21,** 43–48.

Kordula, T., Rydel, R. E., Brigham, E. F., Horn, F., Heinrich, P. C., and Travis, J. (1998). Oncostatin M and the interleukin-6 and soluble interleukin-6 receptor complex regulate alpha1-antichymotrypsin expression in human cortical astrocytes. *J. Biol. Chem.* **273,** 4112–4118.

Ladeby, R., Wirenfeldt, M., Garcia-Ovejero, D., Fenger, C., Dissing-Olesen, L., Dalmau, I., and Finsen, B. (2005). Microglial cell population dynamics in the injured adult central nervous system. *Brain Res. Brain Res. Rev.* **48,** 196–206.

Lanzrein, A. S., Johnston, C. M., Perry, V. H., Jobst, K. A., King, E. M., and Smith, A. D. (1998). Longitudinal study of inflammatory factors in serum, cerebrospinal fluid, and brain tissue in Alzheimer disease: Interleukin-1beta, interleukin-6, interleukin-1 receptor antagonist, tumor necrosis factor-alpha, the soluble tumor necrosis factor receptors I and II, and alpha1-antichymotrypsin. *Alzheimer Dis. Assoc. Disord.* **12,** 215–227.

Laske, C., Stransky, E., Leyhe, T., Eschweiler, G. W., Wittorf, A., Richartz, E., Bartels, M., Buchkremer, G., and Schott, K. (2006). Stage-dependent BDNF serum concentrations in Alzheimer's disease. *J. Neural. Transm.* **113,** 1217–1224.

Laske, C., Stransky, E., Leyhe, T., Eschweiler, G. W., Maetzler, W., Wittorf, A., Soekadar, S., Richartz, E., Koehler, N., Bartels, M., Buchkremer, G., and Schott, K. (2007). BDNF serum and CSF concentrations in Alzheimer's disease, normal pressure hydrocephalus and healthy controls. *J. Psychiatr. Res.* **41**(5), 387–394.

Lewicki, H., Tishon, A., Homann, D., Mazarguil, H., Laval, F., Asensio, V. C., Campbell, I. L., DeArmond, S., Coon, B., Teng, C., Gairin, J. E., and Oldstone, M. B. (2003). T cells infiltrate the brain in murine and human transmissible spongiform encephalopathies. *J. Virol.* **77,** 3799–3808.

Licastro, F., Pedrini, S., Caputo, L., Annoni, G., Davis, L. J., Ferri, C., Casadei, V., and Grimaldi, L. M. (2000). Increased plasma levels of interleukin-1, interleukin-6 and alpha-1-antichymotrypsin in patients with Alzheimer's disease: Peripheral inflammation or signals from the brain? *J. Neuroimmunol.* **103,** 97–102.

Loeffler, D. A., Camp, D. M., Schonberger, M. B., Singer, D. J., and LeWitt, P. A. (2004). Early complement activation increases in the brain in some aged normal subjects. *Neurobiol. Aging* **25,** 1001–1007.

Lombardi, V. R., Garcia, M., Rey, L., and Cacabelos, R. (1999). Characterization of cytokine production, screening of lymphocyte subset patterns and *in vitro* apoptosis in healthy and Alzheimer's Disease (AD) individuals. *J. Neuroimmunol.* **97,** 163–171.

Lorigados, L., Pavon, N., Serrano, T., and Robinson, M. A. (1998). [Nerve growth factor and neurological diseases]. *Rev. Neurol.* **26,** 744–748.

Maes, M., DeVos, N., Wauters, A., Demedts, P., Maurits, V. W., Neels, H., Bosmans, E., Altamura, C., Lin, A., Song, C., Vandenbroucke, M., and Scharpe, S. (1999). Inflammatory markers in younger vs elderly normal volunteers and in patients with Alzheimer's disease. *J. Psychiatr. Res.* **33,** 397–405.

Maes, O. C., Xu, S., Yu, B., Chertkow, H. M., Wang, E., and Schipper, H. M. (2006). Transcriptional profiling of Alzheimer blood mononuclear cells by microarray. *Neurobiol. Aging.* [Epub ahead of print: Sept 15, 2006; PMID: 16979800].

Magaki, S., Mueller, C., Dickson, C., and Kirsch, W. (2007). Increased production of inflammatory cytokines in mild cognitive impairment. *Exp. Gerontol.* **42**(3), 233–240.

Malm, T. M., Koistinaho, M., Parepalo, M., Vatanen, T., Ooka, A., Karlsson, S., and Koistinaho, J. (2005). Bone-marrow-derived cells contribute to the recruitment of microglial cells in response to beta-amyloid deposition in APP/PS1 double transgenic Alzheimer mice. *Neurobiol. Dis.* **18,** 134–142.

Martinez, M., Frank, A., and Hernanz, A. (1993). Relationship of interleukin-1 beta and beta 2-microglobulin with neuropeptides in cerebrospinal fluid of patients with dementia of the Alzheimer type. *J. Neuroimmunol.* **48,** 235–240.

Martinez, M., Fernandez-Vivancos, E., Frank, A., De la Fuente, M., and Hernanz, A. (2000). Increased cerebrospinal fluid fas (Apo-1) levels in Alzheimer's disease. Relationship with IL-6 concentrations. *Brain Res.* **869,** 216–219.

Marz, P., Heese, K., Hock, C., Golombowski, S., Muller-Spahn, F., Rose-John, S., and Otten, U. (1997). Interleukin-6 (IL-6) and soluble forms of IL-6 receptors are not altered in cerebrospinal fluid of Alzheimer's disease patients. *Neurosci. Lett.* **239,** 29–32.

Masliah, E., Mallory, M., Alford, M., DeTeresa, R., and Saitoh, T. (1995). PDGF is associated with neuronal and glial alterations of Alzheimer's disease. *Neurobiol. Aging* **16,** 549–556.

McGeer, P. L., and McGeer, E. G. (1999). Inflammation of the brain in Alzheimer's disease: Implications for therapy. *J. Leukoc. Biol.* **65,** 409–415.

McGeer, P. L., McGeer, E., Rogers, J., and Sibley, J. (1990). Anti-inflammatory drugs and Alzheimer disease. *Lancet* **335,** 1037.

McLaurin, J., Cecal, R., Kierstead, M. E., Tian, X., Phinney, A. L., Manea, M., French, J. E., Lambermon, M. H., Darabie, A. A., Brown, M. E., Janus, C., Chishti, M. A., *et al.* (2002). Therapeutically effective antibodies against amyloid-beta peptide target amyloid-beta residues 4–10 and inhibit cytotoxicity and fibrillogenesis. *Nat. Med.* **8,** 1263–1269.

Medana, I., Martinic, M. A., Wekerle, H., and Neumann, H. (2001). Transection of major histocompatibility complex class I-induced neurites by cytotoxic T lymphocytes. *Am. J. Pathol.* **159,** 809–815.

Mennicken, F., Maki, R., de Souza, E. B., and Quirion, R. (1999). Chemokines and chemokine receptors in the CNS: A possible role in neuroinflammation and patterning. *Trends Pharmacol. Sci.* **20,** 73–78.

Michalski, B., and Fahnestock, M. (2003). Pro-brain-derived neurotrophic factor is decreased in parietal cortex in Alzheimer's disease. *Brain Res. Mol. Brain Res.* **111,** 148–154.

Mocali, A., Cedrola, S., Della Malva, N., Bontempelli, M., Mitidieri, V. A., Bavazzano, A., Comolli, R., Paoletti, F., and La Porta, C. A. (2004). Increased plasma levels of soluble CD40, together with the decrease of TGF beta 1, as possible differential markers of Alzheimer disease. *Exp. Gerontol.* **39,** 1555–1561.

Moir, R. D., Tseitlin, K. A., Soscia, S., Hyman, B. T., Irizarry, M. C., and Tanzi, R. E. (2005). Autoantibodies to redox-modified oligomeric Abeta are attenuated in the plasma of Alzheimer's disease patients. *J. Biol. Chem.* **280,** 17458–17463.

Monsonego, A., Maron, R., Zota, V., Selkoe, D. J., and Weiner, H. L. (2001). Immune hyporesponsiveness to amyloid β-peptide in amyloid precursor protein transgenic mice: Implications for the pathogenesis and treatment of Alzheimer's disease. *Proc. Natl. Acad. Sci. USA* **98,** 10273–10278.

Monsonego, A., Zota, V., Karni, A., Krieger, J. I., Bar-Or, A., Bitan, G., Budson, A. E., Sperling, R., Selkoe, D. J., and Weiner, H. L. (2003). Increased T cell reactivity to amyloid beta protein in older humans and patients with Alzheimer disease. *J. Clin. Invest.* **112,** 415–422.

Mruthinti, S., Buccafusco, J. J., Hill, W. D., Waller, J. L., Jackson, T. W., Zamrini, E. Y., and Schade, R. F. (2004). Autoimmunity in Alzheimer's disease: Increased levels of circulating IgGs binding Abeta and RAGE peptides. *Neurobiol. Aging* **25,** 1023–1032.

Murase, K., Nabeshima, T., Robitaille, Y., Quirion, R., Ogawa, M., and Hayashi, K. (1993). NGF level of is not decreased in the serum, brain-spinal fluid, hippocampus, or parietal cortex of individuals with Alzheimer's disease. *Biochem. Biophys. Res. Commun.* **193,** 198–203.

Murase, K., Igarashi, K., and Hayashi, K. (1994). Neurotrophin-3 (NT-3) levels in the developing rat nervous system and in human samples. *Clin. Chim. Acta* **227,** 23–36.

Murer, M. G., Yan, Q., and Raisman-Vozari, R. (2001). Brain-derived neurotrophic factor in the control human brain, and in Alzheimer's disease and Parkinson's disease. *Prog. Neurobiol.* **63,** 71–124.

Narisawa-Saito, M., Wakabayashi, K., Tsuji, S., Takahashi, H., and Nawa, H. (1996). Regional specificity of alterations in NGF, BDNF and NT-3 levels in Alzheimer's disease. *Neuroreport* **7,** 2925–2928.

Nath, A., Hall, E., Tuzova, M., Dobbs, M., Jons, M., Anderson, C., Woodward, J., Guo, Z., Fu, W., Kryscio, R., Wekstein, D., Smith, C., *et al.* (2003). Autoantibodies to amyloid beta-peptide (Abeta) are increased in Alzheimer's disease patients and Abeta antibodies can enhance Abeta neurotoxicity: Implications for disease pathogenesis and vaccine development. *Neuromolecular Med.* **3,** 29–39.

Nicoll, J. A., Wilkinson, D., Holmes, C., Steart, P., Markham, H., and Weller, R. O. (2003). Neuropathology of human Alzheimer disease after immunization with amyloid-beta peptide: A case report. *Nat. Med.* **9,** 448–452.

Orgogozo, J. M., Gilman, S., Dartigues, J. F., Laurent, B., Puel, M., Kirby, L. C., Jouanny, P., Dubois, B., Eisner, L., Flitman, S., Michel, B. F., Boada, M., *et al.* (2003). Subacute meningoencephalitis in a subset of patients with AD after Abeta42 immunization. *Neurology* **61,** 46–54.

Paganelli, R., Di Iorio, A., Patricelli, L., Ripani, F., Sparvieri, E., Faricelli, R., Iarlori, C., Porreca, E., Di Gioacchino, M., and Abate, G. (2002). Proinflammatory cytokines in sera of elderly patients with dementia: Levels in vascular injury are higher than those of mild-moderate Alzheimer's disease patients. *Exp. Gerontol.* **37,** 257–263.

Pan, W., and Kastin, A. J. (2002). TNFalpha transport across the blood-brain barrier is abolished in receptor knockout mice. *Exp. Neurol.* **174,** 193–200.

Pan, W., and Kastin, A. J. (2003). Transport of cytokines and neurotrophins across the BBB and their regulation after spinal cord injury. *In* "Blood—Spinal Cord and Brain Barriers in Health and Disease" (H. Shanker and J. Westman, Eds.), pp. 395–407. Academic Press, Amsterdam.

Pan, W., and Kastin, A. J. (2004). Why study transport of peptides and proteins at the neurovascular interface. *Brain Res. Brain Res. Rev.* **46,** 32–43.

Pan, W., Banks, W. A., Fasold, M. B., Bluth, J., and Kastin, A. J. (1998a). Transport of brain-derived neurotrophic factor across the blood-brain barrier. *Neuropharmacology* **37,** 1553–1561.

Pan, W., Banks, W. A., and Kastin, A. J. (1998b). Permeability of the blood-brain barrier to neurotrophins. *Brain Res.* **788,** 87–94.

Peress, N. S., and Perillo, E. (1995). Differential expression of TGF-beta 1, 2 and 3 isotypes in Alzheimer's disease: A comparative immunohistochemical study with cerebral infarction, aged human and mouse control brains. *J. Neuropathol. Exp. Neurol.* **54,** 802–811.

Perry, V. H., Newman, T. A., and Cunningham, C. (2003). The impact of systemic infection on the progression of neurodegenerative disease. *Nat. Rev. Neurosci.* **4,** 103–112.

Pirttila, T., Mattinen, S., and Frey, H. (1992). The decrease of CD8-positive lymphocytes in Alzheimer's disease. *J. Neurol. Sci.* **107,** 160–165.

Pirttila, T., Mehta, P. D., Frey, H., and Wisniewski, H. M. (1994). Alpha 1-antichymotrypsin and IL-1 beta are not increased in CSF or serum in Alzheimer's disease. *Neurobiol. Aging* **15,** 313–317.

Poduslo, J. F., and Curran, G. L. (1996). Permeability at the blood-brain and blood-nerve barriers of the neurotrophic factors: NGF, CNTF, NT-3, BDNF. *Brain Res. Mol. Brain Res.* **36,** 280–286.

Ransohoff, R. M., and Benveniste, E. N. (2006). "Cytokines and the CNS." CRC Press Taylor & Francis Group, Boca Raton, London, New York.

Ransohoff, R. M., Kivisakk, P., and Kidd, G. (2003). Three or more routes for leukocyte migration into the central nervous system. *Nat. Rev. Immunol.* **3,** 569–581.

Reale, M., Iarlori, C., Gambi, F., Lucci, I., Salvatore, M., and Gambi, D. (2005). Acetylcholinesterase inhibitors effects on oncostatin-M, interleukin-1 beta and interleukin-6 release from lymphocytes of Alzheimer's disease patients. *Exp. Gerontol.* **40,** 165–171.

Rebenko-Moll, N. M., Liu, L., Cardona, A., and Ransohoff, R. M. (2006). Chemokines, mononuclear cells and the nervous system: Heaven (or hell) is in the details. *Curr. Opin. Immunol.* **18,** 683–689.

Richartz, E., Stransky, E., Batra, A., Simon, P., Lewczuk, P., Buchkremer, G., Bartels, M., and Schott, K. (2005). Decline of immune responsiveness: A pathogenetic factor in Alzheimer's disease? *J. Psychiatr. Res.* **39,** 535–543.

Richartz-Salzburger, E., Batra, A., Stransky, E., Laske, C., Kohler, N., Bartels, M., Buchkremer, G., and Schott, K. (2007). Altered lymphocyte distribution in Alzheimer's disease. *J. Psychiatr. Res.* **41,** 174–178.

Rogers, J., Luber-Narod, J., Styren, S. D., and Civin, W. H. (1988). Expression of immune system-associated antigens by cells of the human central nervous system: Relationship to the pathology of Alzheimer's disease. *Neurobiol. Aging* **9,** 339–349.

Rogers, J., Webster, S., Lue, L. F., Brachova, L., Civin, W. H., Emmerling, M., Shivers, B., Walker, D., and McGeer, P. (1996). Inflammation and Alzheimer's disease pathogenesis. *Neurobiol. Aging* **17,** 681–686.

Rogers, J., Li, R., Mastroeni, D., Grover, A., Leonard, B., Ahern, G., Cao, P., Kolody, H., Vedders, L., Kolb, W. P., and Sabbagh, M. (2006). Peripheral clearance of amyloid beta peptide by complement C3-dependent adherence to erythrocytes. *Neurobiol. Aging* **27,** 1733–1739.

Rota, E., Bellone, G., Rocca, P., Bergamasco, B., Emanuelli, G., and Ferrero, P. (2006). Increased intrathecal TGF-beta1, but not IL-12, IFN-gamma and IL-10 levels in Alzheimer's disease patients. *Neurol. Sci.* **27,** 33–39.

Schenk, D. (2002). Amyloid-beta immunotherapy for Alzheimer's disease: The end of the beginning. *Nat. Rev. Neurosci.* **3,** 824–828.

Schenk, D., Barbour, R., Dunn, W., Gordon, G., Grajeda, H., Guido, T., Hu, K., Huang, J., Johnson-Wood, K., Khan, K., Kholodenko, D., Lee, M., *et al.* (1999). Immunization with amyloid-β attenuates Alzheimer-disease-like pathology in the PDAPP mouse. *Nature* **400,** 173–177.

Schwartz, M., Cohen, I., Lazarov-Spiegler, O., Moalem, G., and Yoles, E. (1999). The remedy may lie in ourselves: Prospects for immune cell therapy in central nervous system protection and repair. *J. Mol. Med.* **77,** 713–717.

Shalit, F., Sredni, B., Stern, L., Kott, E., and Huberman, M. (1994). Elevated interleukin-6 secretion levels by mononuclear cells of Alzheimer's patients. *Neurosci. Lett.* **174,** 130–132.

Shen, Y., Lue, L.-F., Yang, L.-B., Roher, A., Kuo, Y.-M., Strohhmeyer, R., Goux, W. J., Lee, V., Johnson, G. V. W., Webster, S. D., Cooper, N. R., Bradt, B., *et al.* (2001). Complement activation by neurofibrillary tangles in Alzheimer's disease. *Neurosci. Lett.* **305,** 165–168.

Sheta, E. A., Appel, S. H., and Goldknopf, I. L. (2006). 2D gel blood serum biomarkers reveal differential clinical proteomics of the neurodegenerative diseases. *Expert Rev. Proteomics* **3,** 45–62.

Simard, A. R., Soulet, D., Gowing, G., Julien, J. P., and Rivest, S. (2006). Bone marrow-derived microglia play a critical role in restricting senile plaque formation in Alzheimer's disease. *Neuron* **49,** 489–502.

Skias, D., Bania, M., Reder, A. T., Luchins, D., and Antel, J. P. (1985). Senile dementia of Alzheimer's type (SDAT): Reduced T8+-cell-mediated suppressor activity. *Neurology* **35,** 1635–1638.

Solerte, S. B., Ferrari, E., Cuzzoni, G., Locatelli, E., Giustina, A., Zamboni, M., Schifino, N., Rondanelli, M., Gazzaruso, C., and Fioravanti, M. (2005). Decreased release of the angiogenic peptide vascular endothelial growth factor in Alzheimer's disease: Recovering effect with insulin and DHEA sulfate. *Dement. Geriatr. Cogn. Disord.* **19,** 1–10.

Solfrizzi, V., D'Introno, A., Colacicco, A. M., Capurso, C., Todarello, O., Pellicani, V., Capurso, S. A., Pietrarossa, G., Santamato, V., Capurso, A., and Panza, F. (2006). Circulating biomarkers of cognitive decline and dementia. *Clin. Chim. Acta* **364,** 91–112.

Solomon, B., Koppel, R., Hanan, E., and Katzav, T. (1996). Moncolonal antibodies inhibit *in vitro* fibrillar aggregation of the Alzheimer beta-amyloid peptide. *Proc. Natl. Acad. Sci. USA* **93,** 452–455.

Speciale, L., Calabrese, E., Saresella, M., Tinelli, C., Mariani, C., Sanvito, L., Longhi, R., and Ferrante, P. (2007). Lymphocyte subset patterns and cytokine production in Alzheimer's disease patients. *Neurobiol. Aging* **28,** 1163–1169. [Epub ahead of print: June 27, 2006; PMID: 16814429].

Stalder, A. K., Ermini, F., Bondolfi, L., Krenger, W., Burbach, G. J., Deller, T., Coomaraswamy, J., Staufenbiel, M., Landmann, R., and Jucker, M. (2005). Invasion of hematopoietic cells into the brain of amyloid precursor protein transgenic mice. *J. Neurosci.* **25,** 11125–11132.

Stoltzner, S. E., Grenfell, T. J., Mori, C., Wisniewski, K. E., Wisniewski, T. M., Selkoe, D. J., and Lemere, C. A. (2000). Temporal accrual of complement proteins in amyloid plaques in Down's syndrome with Alzheimer's disease. *Am. J. Pathol.* **156,** 489–499.

Strauss, S., Bauer, J., Ganter, U., Jonas, U., Berger, M., and Volk, B. (1992). Detection of interleukin-6 and alpha 2-macroglobulin immunoreactivity in cortex and hippocampus of Alzheimer's disease patients. *Lab. Invest.* **66,** 223–230.

Sun, Y. X., Minthon, L., Wallmark, A., Warkentin, S., Blennow, K., and Janciauskiene, S. (2003). Inflammatory markers in matched plasma and cerebrospinal fluid from patients with Alzheimer's disease. *Dement. Geriatr. Cogn. Disord.* **16,** 136–144.

Tarkowski, E., Blennow, K., Wallin, A., and Tarkowski, A. (1999). Intracerebral production of tumor necrosis factor-alpha, a local neuroprotective agent, in Alzheimer disease and vascular dementia. *J. Clin. Immunol.* **19,** 223–230.

Tarkowski, E., Liljeroth, A. M., Nilsson, Å., Minthon, L., and Blennow, K. (2001). Decreased levels of intrathecal interleukin 1 receptor antagonist in Alzheimer's disease. *Dement. Geriatr. Cogn. Disord.* **12,** 314–317.

Tarkowski, E., Issa, R., Sjogren, M., Wallin, A., Blennow, K., Tarkowski, A., and Kumar, P. (2002). Increased intrathecal levels of the angiogenic factors VEGF and TGF-beta in Alzheimer's disease and vascular dementia. *Neurobiol. Aging* **23,** 237–243.

Tarkowski, E., Andreasen, N., Tarkowski, A., and Blennow, K. (2003). Intrathecal inflammation precedes development of Alzheimer's disease. *J. Neurol. Neurosurg. Psychiatry* **74,** 1200–1205.

Terry, R. D., Masliah, E., and Hansen, L. A. (1994). Structural basis of the cognitive alterations in Alzheimer disease. *In* "Alzheimer Disease" (R. D. Terry, R. Katzman, and K. L. Bick, Ed.), pp. 179–196. Raven Press, New York.

Teunissen, C. E., de Vente, J., Steinbusch, H. W., and De Bruijn, C. (2002). Biochemical markers related to Alzheimer's dementia in serum and cerebrospinal fluid. *Neurobiol. Aging* **23,** 485–508.

Teunissen, C. E., Lutjohann, D., von Bergmann, K., Verhey, F., Vreeling, F., Wauters, A., Bosmans, E., Bosma, H., van Boxtel, M. P., Maes, M., Delanghe, J., Blom, H. J., *et al.* (2003). Combination of serum markers related to several mechanisms in Alzheimer's disease. *Neurobiol. Aging* **24,** 893–902.

Togo, T., Akiyama, H., Iseki, E., Kondo, H., Ikeda, K., Kato, M., Oda, T., Tsuchiya, K., and Kosaka, K. (2002). Occurrence of T cells in the brain of Alzheimer's disease and other neurological diseases. *J. Neuroimmunol.* **124,** 83–92.

Trieb, K., Ransmayr, G., Sgonc, R., Lassmann, H., and Grubeck-Loebenstein, B. (1996). APP peptides stimulate lymphocyte proliferation in normals, but not in patients with Alzheimer's disease. *Neurobiol. Aging* **17,** 541–547.

van der Wal, E. A., Gómez-Pinilla, F., and Cotman, C. W. (der Wal 1993). Transforming growth factor-b1 is in plaques in Alzheimer and Down pathologies. *Neuroreport* **4,** 69–72.

van Duijn, C. M., Hofman, A., and Nagelkerken, L. (Duijn 1990). Serum levels of interleukin-6 are not elevated in patients with Alzheimer's disease. *Neurosci. Lett.* **108,** 350–354.

Webster, S., Lue, L.-F., Brachova, L., Tenner, A. J., McGeer, P. L., Terai, K., Walker, D. G., Bradt, B., Cooper, N. R., and Rogers, J. (1997). Molecular and cellular characterization of the membrane attack complex, C5b-9, in Alzheimer's disease. *Neurobiol. Aging* **18,** 415–421.

Webster, S. D., Galvan, M. D., Ferran, E., Garzon-Rodriguez, W., Glabe, C. G., and Tenner, A. J. (2001). Antibody-mediated phagocytosis of the amyloid b-peptide in microglia is differentially modulated by C1q. *J. Immunol.* **166,** 7496–7503.

Wegiel, J., Wisniewski, H. M., Dziewiatkowski, J., Tarnawski, M., Kozielski, R., Trenkner, E., and Wiktor-Jedrzejczak, W. (1998). Reduced number and altered morphology of microglial cells in colony stimulating factor-1-deficient osteopetrotic op/op mice. *Brain Res.* **804,** 135–139.

Weiner, H. L., and Frenkel, D. (2006). Immunology and immunotherapy of Alzheimer's disease. *Nat. Rev. Immunol.* **6,** 404–416.

Weksler, M. E., Relkin, N., Turkenich, R., LaRusse, S., Zhou, L., and Szabo, P. (2002). Patients with Alzheimer disease have lower levels of serum anti-amyloid peptide antibodies than healthy elderly individuals. *Exp. Gerontol.* **37,** 943–948.

Weller, R. O., Massey, A., Newman, T. A., Hutchings, M., Kuo, Y. M., and Roher, A. E. (1998). Cerebral amyloid angiopathy. Amyloid β accumulates in putative interstitial fluid drainage pathways in Alzheimer's disease. *Am. J. Pathol.* **153,** 725–733.

Wood, J. A., Wood, P. L., Ryan, R., Graff-Radford, N. R., Pilapil, C., Robitaille, Y., and Quirion, R. (1993). Cytokine indices in Alzheimer's temporal cortex: No changes in mature IL-1 beta or IL-1RA but increases in the associated acute phase proteins IL-6, alpha 2-macroglobulin and C-reactive protein. *Brain Res.* **629,** 245–252.

Wyss-Coray, T. (2006). Inflammation in Alzheimer disease: Driving force, bystander or beneficial response? *Nat. Med.* **12,** 1005–1015.

Wyss-Coray, T., and Mucke, L. (2002). Inflammation in neurodegenerative disease: A double-edged sword. *Neuron* **35,** 419–432.

Wyss-Coray, T., Masliah, E., Mallory, M., McConlogue, L., Johnson-Wood, K., Lin, C., and Mucke, L. (1997). Amyloidogenic role of cytokine TGF-b1 in transgenic mice and Alzheimer's disease. *Nature* **389,** 603–606.

Wyss-Coray, T., Yan, F., Lin, A. H., Lambris, J. D., Alexander, J. J., Quigg, R. J., and Masliah, E. (2002). Prominent neurodegeneration and increased plaque formation in complement-inhibited Alzheimer's mice. *Proc. Natl. Acad. Sci. USA* **99,** 10837–10842.

Xia, M. Q., and Hyman, B. T. (1999). Chemokines/chemokine receptors in the central nervous system and Alzheimer's disease. *J. Neurovirol.* **5,** 32–41.

Xia, M. Q., Qin, S. X., Wu, L. J., Mackay, C. R., and Hyman, B. T. (1998). Immunohistochemical study of the beta-chemokine receptors CCR3 and CCR5 and their ligands in normal and Alzheimer's disease brains. *Am. J. Pathol.* **153,** 31–37.

Xiao, E., Xia, L., Ferin, M., and Wardlaw, S. L. (1999). Intracerebroventricular injection of interleukin-1 stimulates the release of high levels of interleukin-6 and interleukin-1 receptor antagonist into peripheral blood in the primate. *J. Neuroimmunol.* **97,** 70–76.

Yamada, M., Itoh, Y., Shintaku, M., Kawamura, J., Jensson, O., Thorsteinsson, L., Suematsu, N., Matsushita, M., and Otomo, E. (1996). Immune reactions associated with cerebral amyloid angiopathy. *Stroke* **27,** 1155–1162.

Yan, S. D., Zhu, H., Fu, J., Yan, S. F., Roher, A., Tourtellotte, W. W., Rajavashisth, T., Chen, X., Godman, G. C., Stern, D., and Schmidt, A. M. (1997). Amyloid-β peptide-receptor for advanced glycation endproduct interaction elicits neuronal expression of macrophage-colony stimulating factor: A proinflammatory pathway in Alzheimer disease. *Proc. Natl. Acad. Sci. USA* **94,** 5296–5301.

Zetterberg, H., Andreasen, N., and Blennow, K. (2004). Increased cerebrospinal fluid levels of transforming growth factor-beta1 in Alzheimer's disease. *Neurosci. Lett.* **367,** 194–196.

Ziemssen, T., Kumpfel, T., Schneider, H., Klinkert, W. E., Neuhaus, O., and Hohlfeld, R. (2005). Secretion of brain-derived neurotrophic factor by glatiramer acetate-reactive T-helper cell lines: Implications for multiple sclerosis therapy. *J. Neurol. Sci.* **233,** 109–112.

NEUROINFLAMMATION IN ALZHEIMER'S DISEASE AND PARKINSON'S DISEASE: ARE MICROGLIA PATHOGENIC IN EITHER DISORDER?

Joseph Rogers, Diego Mastroeni, Brian Leonard, Jeffrey Joyce, and Andrew Grover

The L. J. Roberts Center for Alzheimer's Research, Sun Health Research Institute
Sun City, Arizona 85351, USA; and
The Christopher Center for Parkinson's Research, Sun Health Research Institute
Sun City, Arizona 85351, USA

Microglial activation similar to that which occurs in peripheral macrophages during inflammatory attack was first demonstrated in the Alzheimer's disease (AD) brain two decades ago. Localization to pathologically vulnerable regions of AD cortex, localization to sites of specific AD pathology such as amyloid-β peptide (Aβ) deposits, and the ability of activated microglia to release toxic inflammatory factors suggested that the activation of microglia in AD might play a pathogenic role. However, proving this hypothesis in a disease in which so many profound pathologies occur (e.g., Aβ deposition, neurofibrillary tangle formation, inflammation, neuronal loss, neuritic loss, synaptic loss, neuronal dysfunction, vascular alterations) has proven difficult.

Although investigations of microglia in Parkinson's disease (PD) are more recent and therefore less extensive, demonstration of a pathogenic role for microglial activation may actually be much simpler in PD than AD because the root pathological event in PD, loss of dopamine (DA)-secreting substantia nigra neurons, is already well established. Indeed, indirect but converging evidence of a pathogenic

INTERNATIONAL REVIEW OF
NEUROBIOLOGY, VOL. 82
DOI: 10.1016/S0074-7742(07)82012-5

role for activated microglia in PD has already begun to emerge. The nigra reportedly has the highest density of microglia in brain, and, in PD, nigral microglia are not only highly activated but also highly clustered around dystrophic DA neurons. 6-OHDA and MPTP models of PD in rodents induce substantia nigra microglial activation. More cogent, injections of the classic microglial/macrophage activator lipopolysaccharide into or near the rodent nigra cause a specific loss of DA neurons there. Culture models with human microglia and human cellular targets replicate this phenomenon. Notably, nearly all the proposed etiologies of PD, including brain bacterial and viral exposure, pesticides, drug contaminants, and repeated head trauma, are known to cause brain inflammation.

A mechanism by which activated microglia might specifically target DA neurons remains a critical missing link in the proof of a pathogenic role for activated microglia in PD. If such a link could be established, however, clinical intervention trials with agents that dampen microglial activation might be warranted in PD.

I. Introduction

The history of neuroinflammatory mechanisms in Alzheimer's disease (AD) in large part began with our laboratory's discovery that microglia express the classic activation marker major histocompatibility complex type II (MHCII) cell surface glycoprotein in pathologically vulnerable regions of the AD brain (Luber-Narod and Rogers, 1988; Rogers *et al.*, 1986, 1988). Since that time, some two decades ago, a virtual textbook of upregulated inflammatory mediators has been found in association with AD pathology (reviewed in Akiyama *et al.*, 2000). However, then, as now, the pervasive question remained: was neuroinflammation causing pathology or was it simply a response to pathology? Was neuroinflammation directly involved in the formation of amyloid-β peptide (Aβ) deposits, neurofibrillary tangles, and the loss of neurons and neurites, or was it merely a more sophisticated means of detritus removal than we had previously imagined?

In many ways, the history of neuroinflammation in Parkinson's disease (PD) appears to be taking a similar course to that in AD. Initially, activated microglia were reported to cluster around dead or dystrophic, pigmented dopamine (DA) neurons in the PD substantia nigra (McGeer *et al.*, 1988). Elevated cytokines and other inflammatory mediators were subsequently found both in the nigra and striatum of PD patients (reviewed in Hunot and Hirsch, 2003). Were they cause or response?

II. Neuroinflammation in AD and PD

In addition to thousands of reports on the presence of various inflammatory mediators and mechanisms in AD brain (Akiyama *et al.*, 2000), animal studies and epidemiological surveys have been informative with respect to the actual role of neuroinflammation in AD neurodegeneration. Transgenic mice genetically engineered to overexpress the Aβ precursor protein show less pathology when they are chronically treated with anti-inflammatory agents (Heneka *et al.*, 2005; Jantzen *et al.*, 2002). Human subjects, in nearly two dozen epidemiological studies (reviewed in McGeer *et al.*, 2006), are less at risk for AD if they chronically take anti-inflammatory drugs. Nonetheless, these studies still do not tell us whether the effects of the anti-inflammatories are direct or indirect, and, indeed, AD treatment trials with anti-inflammatories have been equivocal at best (reviewed in van Gool *et al.*, 2003).

PD epidemiological studies have shown results similar to those of AD studies (Chen *et al.*, 2003, 2005), but PD animal experiments may more definitively demonstrate that neuroinflammation can play a pathogenic role. As in animal models of AD, anti-inflammatory drugs appear to be protective in animal models of PD. The prior and/or concurrent administration of anti-inflammatory drugs, for example, results in significantly less severe DA depletion and PD symptomology in rodents exposed to MPTP (Kurkowska-Jastrzebska *et al.*, 2004) or 6-OHDA (Carrasco and Werner, 2002; Sanchez-Pernaute *et al.*, 2004). In addition, the administration of inflammation-inducing agents, such as lipopolysaccharide (LPS), appears to directly replicate the primary deficit in PD: LPS administration into or near the nigra causes a selective, permanent depletion of nigral DA (Herrera *et al.*, 2000; Kim *et al.*, 2000).

III. PD May Provide a More Facile Model for Demonstrating a Pathogenic Role of Neuroinflammation

Although studies of neuroinflammatory mechanisms in PD do not have the history or burgeoning literature that is extant in AD, we believe that PD may actually provide advantages with respect to determining the pathogenic as opposed to pathological role of inflammation in neurological disorders. In particular, where AD entails a diverse set of potentially interactive toxic mechanisms, from plaques and tangles to vascular abnormalities, as well as a wide swath of neuroanatomical structures and cell types, from neocortical pyramidal cells to subcortical neurons of the nucleus basalis, PD is classically characterized by the loss of a single neuron type, the DA neuron, in a specific brain structure, the substantia nigra. Certainly, greater complexities for PD could be mentioned [e.g., striatal changes (Chase *et al.*, 2001), actions of growth factors (Sullivan and O'Keeffe, 2005), bacterial exposure (Carvey *et al.*, 2006)]; yet the apparent DA specificity of PD provides a singular target by which we may more directly assess whether activated microglia are killers, garbage collectors, or both.

IV. Advantages of Microglial Cell Cultures

To address the critical question of neuroinflammatory pathogenicity, our laboratory has made, over the years, a number of simplifying assumptions. The first is that although the variety of inflammatory mechanisms is wide ranging and complex, a good case can be made that the activation of microglia is pivotal in the initiation of neuroinflammation. To begin with, microglia are the only intrinsic cells in the central nervous system that have an inflammatory lineage. They are of mesodermal rather than ectodermal origin, and they are close cousins to, if not transformed examples of, the macrophage (Cossmann *et al.*, 1997). Like macrophages, microglia are capable of movement through tissue, including nervous tissue (Nimmerjahn *et al.*, 2005). Like macrophages, when exposed to various pathological conditions microglia become activated: that is, they upregulate the expression of molecules associated with the inflammatory response, including cytokines, chemokines, presentation molecules (e.g., MHCII), reactive oxygen and nitrogen species, complement, and inflammation-related receptors (reviewed in Akiyama *et al.*, 2000). Although astrocytes have a few of these properties (Akiyama *et al.*, 2000), and neurons may even express mRNA for several complement components (Shen *et al.*, 1997), no other cell type in brain has the relatively complete inflammatory repertoire of the microglia. Activation of microglia, moreover, appears to occur in virtually all neurological disorders in which neurodegeneration can be documented, including amyotrophic lateral sclerosis (Kawamata *et al.*, 1992), multiple sclerosis (Woodroofe *et al.*, 1986), prion disease (Sasaki *et al.*, 1993), stroke (Giulian and Vaca, 1993), progressive supranuclear palsy (Jellinger and Stadelmann, 2000), AD (Akiyama *et al.*, 2000; Luber-Narod and Rogers, 1988; Rogers *et al.*, 1988), and PD (McGeer *et al.*, 1988), among others.

Like macrophages, activated microglia can also be phagocytic (Croisier *et al.*, 2005), and it is this property, perhaps more than any other, that makes it difficult to discern whether microglia clustered at the site of an Aβ deposit are scavenging the Aβ, expressing cytotoxic factors that kill neighboring neurons and neurites, or both. Similarly, activated microglia that surround nigral DA neurons in PD could be attacking the neurons, scavenging them once dead, or both. In static tissue sections of AD and PD pathology, there is simply no way to tell.

For this reason, a second assumption we have made with respect to the pathogenicity of neuroinflammation is that cell culture models should have distinct experimental advantages (Lue *et al.*, 2001c). Because they permit dynamic assessment of experimental manipulations, cell cultures make it possible to determine specific cause and affect relationships in a much less equivocal manner. Although such models have been criticized for being removed from *in vivo* conditions, to the extent that they begin with experimental conditions that parallel those in animal models or human trials, and to the extent that they produce results consistent with animal models or human trials, culture studies can be appreciated for their ability to make accessible the molecular mechanisms that underlie basic pathogenic phenomena. Our laboratory has demonstrated this in AD (Lue *et al.*, 2001a,b,c; Strohmeyer *et al.*, 2005), and we believe our preliminary studies demonstrate it in PD as well.

Importantly, our experiments begin with cultures from rapidly autopsied human AD, PD, and normal elderly control subjects (Lue *et al.*, 1996). Microglia from human cell lines and from rodents may be obtained with relative ease, and have been used to advantage in many studies, but they differ in several critical respects from human elderly microglia. For example, THP-1 monocytes require priming to evince responses to Aβ **(cf. Lorton *et al.*, 2000), whereas human elderly microglia from rapid autopsies do not (cf. Lue *et al.*, 2001a).

V. Responses of Cultured Microglia to AD and PD Pathology

Cultured microglia from AD and normal elderly subjects rapidly migrate to 100-μl spots of aggregated Aβ dried down to the well floor. Within days, they cover the spot and begin to phagocytose it (Lue *et al.*, 2001c; Strohmeyer *et al.*, 2005; Fig. 1), confirming widespread speculation from static images of microglia and Aβ deposits in the AD brain. In the process, microglial exposure to Aβ results in a dose-dependent secretion of proinflammatory mediators, including chemokines, cytokines, growth factors, and elements of the respiratory burst (Lue *et al.*, 2001a; Fig. 2). These findings further confirm the dichotomous nature of neuroinflammatory mechanisms: the scavenging of Aβ, which is presumably beneficial, and the elaboration of potential cytotoxins, which is presumably detrimental. Moreover, confirming animal studies and human trials with Aβ immunization (Bacskai *et al.*, 2001; Masliah *et al.*, 2005), opsonizing the Aβ spots with an anti-Aβ antibody promotes Aβ removal and increases the secretion of toxic inflammatory mediators (Strohmeyer *et al.*, 2005; Fig. 3), suggesting that the application of anti-Aβ antibodies in human AD patients could have both beneficial effects and dangerous side effects, as has proven to be the case (Masliah *et al.*, 2005).

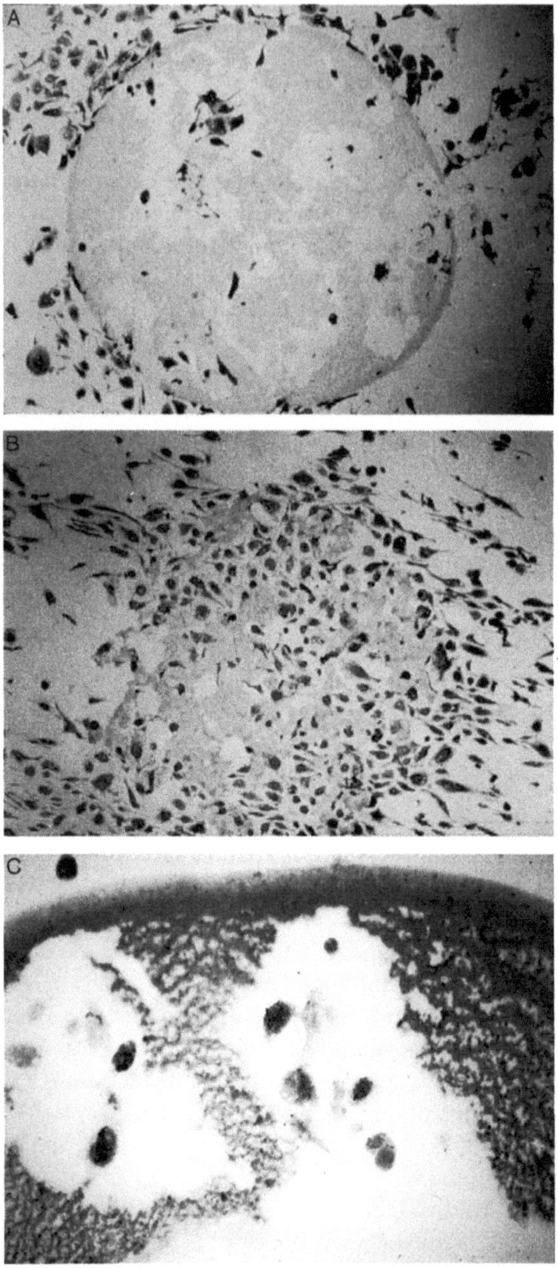

FIG. 1. Human elderly microglia exhibit chemotactic and phagocytic responses to Aβ *in vitro*. Here, a 100-μl spot of preaggregated Aβ42 was dried down to the well floor, followed by seeding with cultured human elderly microglia. Within days, there was pronounced migration to the spot (A), such

To our knowledge, however, no study—whether in human subjects, animal models, or cell culture—has ever directly demonstrated microglial killing of neurons under the conditions that apply in AD.

The same may not be true in PD. Our preliminary cell culture studies, replicating the conditions of animal models and PD itself, suggest that activated microglia preferentially migrate to and attack neurons with a DA phenotype. By cell counts or chemical measures of neurodegeneration, coculture of activated, human elderly microglia with undifferentiated SH-SY5Y cells, SH-SY5Y cells differentiated into a cholinergic phenotype, or SH-SY5Y cells differentiated into a DA phenotype result in significantly greater microglial accumulation around and neurotoxicity to the DA phenotype. Activation of microglia by LPS significantly enhances all the cytotoxic effects. The cell cultures also provide clues to the mechanisms that may be involved. In particular, release of DA by the DA phenotype cells increases toxicity by some 100-fold, and human elderly microglial chemotaxis in Boyden chambers is significantly stimulated by DA compared even to the classic chemokine macrophage chemoattractant protein-1.

These findings suggest that interactions of DA itself with activated microglia may drive the selective vulnerability of DA nigral neurons in PD. In this vein, it might be worth considering (1) that nearly all the proposed etiologies of PD, including environmental toxins (Sherer *et al.*, 2003), bacterial and viral exposure (Carvey *et al.*, 2006; Ling *et al.*, 2004), and repeated head trauma (Lees, 1997), induce microglial activation; (2) that, in their resting state, microglia subserve important supportive roles for neurons that are lost when the microglia become activated (Streit, 2002); and (3) that the substantia nigra contains the highest density of microglia in brain (Teismann *et al.*, 2003). As such, the activation of microglia by various insults should be especially damaging to nigral DA neurons compared to other brain structures and neuron types, and PD could be the most readily engendered manifestation of acute or chronic microglial activation.

VI. Conclusions

Neuroinflammation is not an epiphenomenon, nor a mere scavenger of other, putatively more important, pathogenic mechanisms (i.e., those studied by our reviewers). It is a complex set of processes that in brain may begin with microglial

---

that it was ultimately covered by microglia (B). Staining with anti-Aβ antibodies provided clear evidence of phagocytosis (C). The Aβ spot became progressively thinner, and in many areas the Aβ was completely removed. Concurrently, the microglia became immunoreactive for Aβ, whereas there was little to no evidence of such immunoreactivity in the microglia prior to Aβ exposure (Lue *et al.*, 2001c; Strohmeyer *et al.*, 2005). (See Color Insert.)

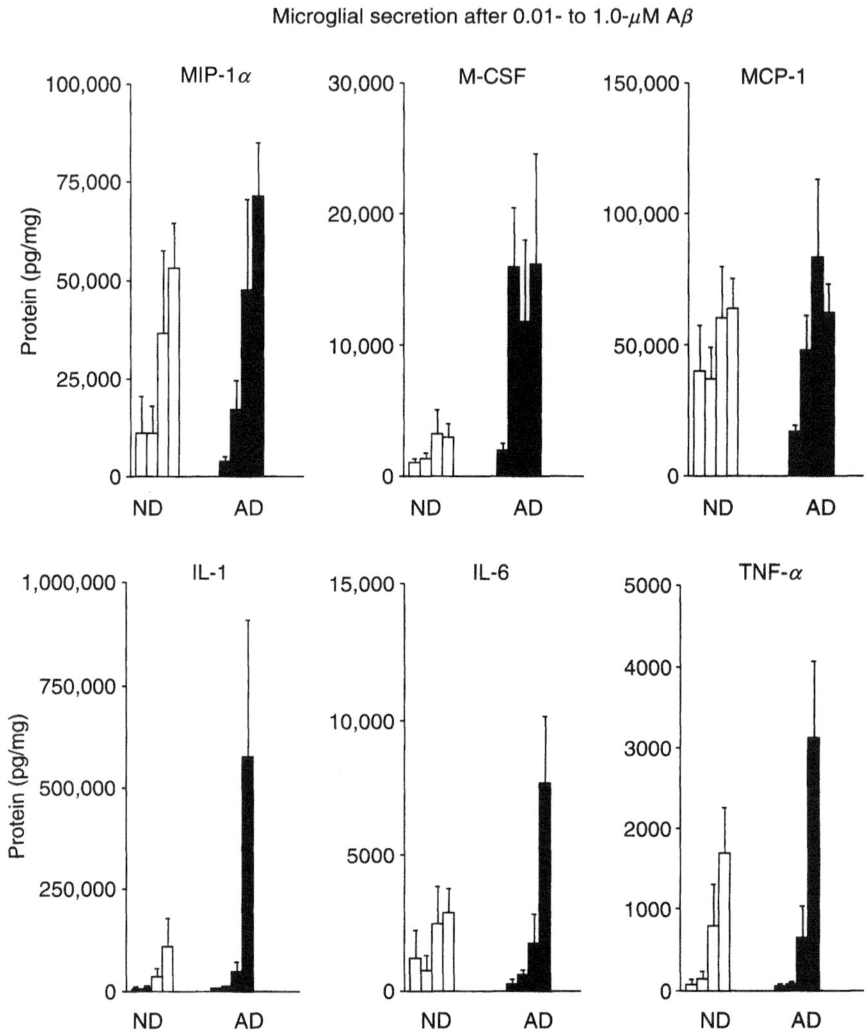

FIG. 2. Human elderly microglia secrete a wide range of inflammation-related activating and cytotoxic factors after exposure to graded, nanomolar doses of Aβ42 *in vitro* (Lue *et al.*, 2001a). Although microglia from AD patients typically exhibit heightened responses compared to microglia from elderly normal controls, we believe this may reflect increased exposure of AD microglia to a proinflammatory milieu prior to culture (i.e., in the AD cortex) rather than to some intrinsic difference in the cells from the two populations.

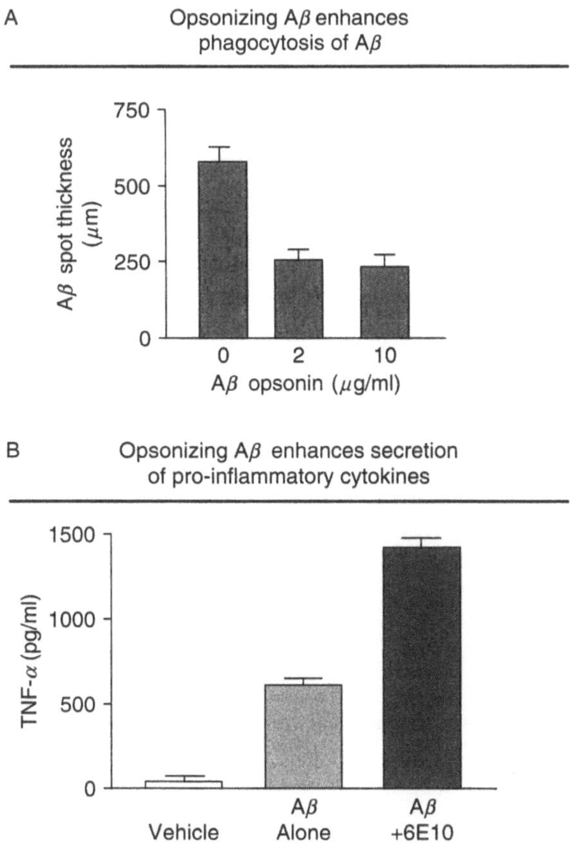

FIG. 3. Prior incubation of Aβ42 spots with anti-Aβ antibody enhances Aβ phagocytosis by microglia (A), but also enhances secretion of potential cytotoxins such as TNF-α (B).

activation and may end in both beneficial and harmful effects. The balance of these forces may yet prove, in AD, to be positive, and it is certainly true that without microglial removal the death of nigral DA neurons in PD would otherwise result in a putrid, floridly proinflammatory lesion. Nonetheless, our data on culture models of AD and PD do show very clearly that neuroinflammation in general and microglial activation in particular can not only clear pathology but kill neurons.

References

Akiyama, H., Barger, S., Barnum, S., Bradt, B., Bauer, J., Cole, G. M., Cooper, N. R., Eikelenboom, P., Emmerling, M., Fiebich, B. L., Finch, C. E., Frautschy, S., *et al.* (2000). Inflammation and Alzheimer's disease. *Neurobiol. Aging* **21,** 383–421.

Bacskai, B. J., Kajdasz, S. T., Christie, R. H., Carter, C., Games, D., Seubert, P., Schenk, D., and Hyman, B. T. (2001). Imaging of amyloid-beta deposits in brains of living mice permits direct observation of clearance of plaques with immunotherapy. *Nat. Med.* **7,** 369–372.

Carrasco, E., and Werner, P. (2002). Selective destruction of dopaminergic neurons by low concentrations of 6-OHDA and MPP+: Protection by acetylsalicylic acid aspirin. *Parkinsonism Relat. Disord.* **8,** 407–411.

Carvey, P. M., Punati, A., and Newman, M. B. (2006). Progressive dopamine neuron loss in Parkinson's disease: The multiple hit hypothesis. *Cell Transplant.* **15,** 239–250.

Chase, T. N., Konitsiotis, S., and Oh, J. D. (2001). Striatal molecular mechanisms and motor dysfunction in Parkinson's disease. *Adv. Neurol.* **86,** 355–360.

Chen, H., Zhang, S. M., Hernan, M. A., Schwarzschild, M. A., Willett, W. C., Colditz, G. A., Speizer, F. E., and Ascherio, A. (2003). Nonsteroidal anti-inflammatory drugs and the risk of Parkinson disease. *Arch. Neurol.* **60,** 1059–1064.

Chen, H., Jacobs, E., Schwarzschild, M. A., McCullough, M. L., Calle, E. E., Thun, M. J., and Ascherio, A. (2005). Nonsteroidal antiinflammatory drug use and the risk for Parkinson's disease. *Ann. Neurol.* **58,** 963–967.

Cossmann, P. H., Eggli, P. S., Christ, B., and Kurz, H. (1997). Mesoderm-derived cells proliferate in the embryonic central nervous system: Confocal microscopy and three-dimensional visualization. *Histochem. Cell Biol.* **107,** 205–213.

Croisier, E., Moran, L. B., Dexter, D. T., Pearce, R. K., and Graeber, M. B. (2005). Microglial inflammation in the parkinsonian substantia nigra: Relationship to alpha-synuclein deposition. *J. Neuroinflammation* **2,** 14.

Giulian, D., and Vaca, K. (1993). Inflammatory glia mediate delayed neuronal damage after ischemia in the central nervous system. *Stroke* **24,** I84–I90.

Heneka, M. T., Sastre, M., Dumitrescu-Ozimek, L., Hanke, A., Dewachter, I., Kuiperi, C., O'Banion, K., Klockgether, T., Van Leuven, F., and Landreth, G. E. (2005). Acute treatment with the PPARgamma agonist pioglitazone and ibuprofen reduces glial inflammation and Abeta1–42 levels in APPV717I transgenic mice. *Brain* **128,** 1442–1453.

Herrera, A. J., Castano, A., Venero, J. L., Cano, J., and Machado, A. (2000). The single intranigral injection of LPS as a new model for studying the selective effects of inflammatory reactions on dopaminergic system. *Neurobiol. Dis.* **7,** 429–447.

Hunot, S., and Hirsch, E. C. (2003). Neuroinflammatory processes in Parkinson's disease. *Ann. Neurol.* **53**(Suppl. 3), S49–S58. discussion S58–S60.

Jantzen, P. T., Connor, K. E., DiCarlo, G., Wenk, G. L., Wallace, J. L., Rojiani, A. M., Coppola, D., Morgan, D., and Gordon, M. N. (2002). Microglial activation and beta -amyloid deposit reduction caused by a nitric oxide-releasing nonsteroidal anti-inflammatory drug in amyloid precursor protein plus presenilin-1 transgenic mice. *J. Neurosci.* **22,** 2246–2254.

Jellinger, K. A., and Stadelmann, C. (2000). Mechanisms of cell death in neurodegenerative disorders. *J. Neural. Transm. Suppl.* **59,** 95–114.

Kawamata, T., Akiyama, H., Yamada, T., and McGeer, P. L. (1992). Immunologic reactions in amyotrophic lateral sclerosis brain and spinal cord tissue. *Am. J. Pathol.* **140,** 691–707.

Kim, W. G., Mohney, R. P., Wilson, B., Jeohn, G. H., Liu, B., and Hong, J. S. (2000). Regional difference in susceptibility to lipopolysaccharide-induced neurotoxicity in the rat brain: Role of microglia. *J. Neurosci.* **20,** 6309–6316.

Kurkowska-Jastrzebska, I., Litwin, T., Joniec, I., Ciesielska, A., Przybylkowski, A., Czlonkowski, A., and Czlonkowska, A. (2004). Dexamethasone protects against dopaminergic neurons damage in a mouse model of Parkinson's disease. *Int. Immunopharmacol.* **4,** 1307–1318.

Lees, A. J. (1997). Trauma and Parkinson disease. *Rev. Neurol. (Paris)* **153,** 541–546.

Ling, Z. D., Chang, Q., Lipton, J. W., Tong, C. W., Landers, T. M., and Carvey, P. M. (2004). Combined toxicity of prenatal bacterial endotoxin exposure and postnatal 6-hydroxydopamine in the adult rat midbrain. *Neuroscience* **124,** 619–628.

Lorton, D., Schaller, J., Lala, A., and De Nardin, E. (2000). Chemotactic-like receptors and Abeta peptide induced responses in Alzheimer's disease. *Neurobiol. Aging* **21,** 463–473.

Luber-Narod, J., and Rogers, J. (1988). Immune system associated antigens expressed by cells of the human central nervous system. *Neurosci. Lett.* **94,** 17–22.

Lue, L. F., Brachova, L., Walker, D. G., and Rogers, J. (1996). Characterization of glial cultures from rapid autopsies of Alzheimer's and control patients. *Neurobiol. Aging* **17,** 421–429.

Lue, L. F., Rydel, R., Brigham, E. F., Yang, L. B., Hampel, H., Murphy, G. M., Jr., Brachova, L., Yan, S. D., Walker, D. G., Shen, Y., and Rogers, J. (2001a). Inflammatory repertoire of Alzheimer's disease and nondemented elderly microglia *in vitro*. *Glia* **35,** 72–79.

Lue, L. F., Walker, D. G., Brachova, L., Beach, T. G., Rogers, J., Schmidt, A. M., Stern, D. M., and Yan, S. D. (2001b). Involvement of microglial receptor for advanced glycation endproducts (RAGE) in Alzheimer's disease: Identification of a cellular activation mechanism. *Exp. Neurol.* **171,** 29–45.

Lue, L. F., Walker, D. G., and Rogers, J. (2001c). Modeling microglial activation in Alzheimer's disease with human postmortem microglial cultures. *Neurobiol. Aging* **22,** 945–956.

Masliah, E., Hansen, L., Adame, A., Crews, L., Bard, F., Lee, C., Seubert, P., Games, D., Kirby, L., and Schenk, D. (2005). Abeta vaccination effects on plaque pathology in the absence of encephalitis in Alzheimer disease. *Neurology* **64,** 129–131.

McGeer, P. L., Itagaki, S., Boyes, B. E., and McGeer, E. G. (1988). Reactive microglia are positive for HLA-DR in the substantia nigra of Parkinson's and Alzheimer's disease brains. *Neurology* **38,** 1285–1291.

McGeer, P. L., Rogers, J., and McGeer, E. G. (2006). Inflammation, anti-inflammatory agents and Alzheimer disease: The last 12 years. *J. Alzheimers Dis.* **9,** 271–276.

Nimmerjahn, A., Kirchhoff, F., and Helmchen, F. (2005). Resting microglial cells are highly dynamic surveillants of brain parenchyma *in vivo*. *Science* **308,** 1314–1318.

Rogers, J., Luber-Narod, J., and Singer, R. (1986). Neurovirologic and neuroimmunological considerations in Alzheimer's disease. *Soc. Neurosci.* **12,** 125. (Abstract).

Rogers, J., Luber-Narod, J., Styren, S. D., and Civin, W. H. (1988). Expression of immune system-associated antigens by cells of the human central nervous system: Relationship to the pathology of Alzheimer's disease. *Neurobiol. Aging* **9,** 339–349.

Sanchez-Pernaute, R., Ferree, A., Cooper, O., Yu, M., Brownell, A. L., and Isacson, O. (2004). Selective COX-2 inhibition prevents progressive dopamine neuron degeneration in a rat model of Parkinson's disease. *J. Neuroinflammation* **1,** 6.

Sasaki, A., Hirato, J., and Nakazato, Y. (1993). Immunohistochemical study of microglia in the Creutzfeldt-Jakob diseased brain. *Acta Neuropathol. (Berl)* **86,** 337–344.

Shen, Y., Li, R., McGeer, E. G., and McGeer, P. L. (1997). Neuronal expression of mRNAs for complement proteins of the classical pathway in Alzheimer brain. *Brain Res.* **769,** 391–395.

Sherer, T. B., Betarbet, R., Kim, J. H., and Greenamyre, J. T. (2003). Selective microglial activation in the rat rotenone model of Parkinson's disease. *Neurosci. Lett.* **341,** 87–90.

Streit, W. J. (2002). Microglia as neuroprotective, immunocompetent cells of the CNS. *Glia* **40,** 133–139.

Strohmeyer, R., Kovelowski, C. J., Mastroeni, D., Leonard, B., Grover, A., and Rogers, J. (2005). Microglial responses to amyloid beta peptide opsonization and indomethacin treatment. *J. Neuroinflammation* **2,** 18.

Sullivan, A. M., and O'Keeffe, G. W. (2005). The role of growth/differentiation factor 5 (GDF5) in the induction and survival of midbrain dopaminergic neurones: Relevance to Parkinson's disease treatment. *J. Anat.* **207,** 219–226.

Teismann, P., Tieu, K., Cohen, O., Choi, D. K., Wu du, C., Marks, D., Vila, M., Jackson-Lewis, V., and Przedborski, S. (2003). Pathogenic role of glial cells in Parkinson's disease. *Mov. Disord.* **18,** 121–129.

van Gool, W. A., Aisen, P. S., and Eikelenboom, P. (2003). Anti-inflammatory therapy in Alzheimer's disease: Is hope still alive? *J. Neurol.* **250,** 788–792.

Woodroofe, M. N., Bellamy, A. S., Feldmann, M., Davison, A. N., and Cuzner, M. L. (1986). Immunocytochemical characterisation of the immune reaction in the central nervous system in multiple sclerosis. Possible role for microglia in lesion growth. *J. Neurol. Sci.* **74,** 135–152.

CYTOKINES AND NEURONAL ION CHANNELS IN HEALTH AND DISEASE

Barbara Viviani,* Fabrizio Gardoni,† and Marina Marinovich*

*Laboratory of Toxicology, Department of Pharmacological Sciences, University of Milan, Italy
†Department of Pharmacological Sciences, University of Milan, Italy

The biophysical properties and the spatial distribution of ion channels define the signaling characteristics of individual neurons. Function, number localization, and ratio of receptor and ion channels are dynamically modulated in response to diverse stimuli and undergo dynamic changes in both physiological and pathological conditions.

Increasing evidence indicates that cytokines may specifically interact with receptor and ion channels regulating neuronal excitability, synaptic plasticity, and injury. Interleukin (IL)-1β and tumor necrosis factor (TNF)-α, two proinflammatory cytokines implicated in various pathophysiological conditions of the CNS, have been particularly studied. Literature data indicate that these cytokines (1) directly and promptly modulate ion channel activity, (2) exert different (and often opposite) effects on the same channels, and (3) act on ion channels both at physiological and pathological concentrations. Consequently, cytokines are now regarded as novel neuromodulators, opening important perspectives in the current view of brain behavior.

I. Introduction

The term cytokines defines a broad range of multifunctional proteins, which act as humoral regulators in femto- to nanomolar concentrations and modulate the functional activities of individual cells and tissues under physiological,

DOI: 10.1016/S0074-7742(07)82013-7

0074-7742/07 $35.00

pathological, and toxicological conditions. Most of cytokines were initially described as immune cell mediators in the periphery. The discovery that cytokines along with the related "transducing machinery" are expressed within the nervous system and are involved in the modulation of several neurological functions and dysfunctions (Table I), opens up new perspective in the current view of brain behavior and new frontiers in therapeutic intervention.

In the central nervous system (CNS), a large number of cytokines and cytokine receptors are expressed in astrocytes, microglia, neurons, and oligodendrocytes either constitutively or by induction following brain damage. Due to the predominant role of cytokines in the inflammatory response at the "periphery" and because of the dramatic increase of their expression during damage, most of the emerging experimentation was directed to understand their role in promoting inflammation during pathological processes within the CNS. Initially, attention has been thus mainly devoted to the ability of cytokines in activating astrocytes and microglia (the "immune cells" resident within the CNS). Presently, we realize that these molecules may also exert a direct impact on neuronal function. Cytokines in fact recruit different responses in glial cells and neuron through cell-type-specific mediated signaling (Srinivasan *et al.*, 2004; Vitkovic *et al.*, 2000). This has been described for interleukin-1β (IL-1β) (Srinivasan *et al.*, 2004), a proinflammatory cytokine highly produced by CNS glia under conditions of damage, stress, and disease (Allan *et al.*, 2005). IL-1β seems to promote the p38 mitogen-activated protein kinase (MAPK) signaling pathway and CREB activation in hippocampal neurons. On the contrary, IL-1β activates NF-κB in hippocampal astrocytes (Srinivasan *et al.*, 2004), indicating a distinct and direct effect on both these two cell population, including neurons.

Up to now, the mechanisms governing the diverse action of cytokines in neurons are not well understood. In this chapter, we will focus our attention on ion channels, a particular target scarcely explored in relation to cytokines, but fundamental for neuronal activity.

Ion channels are sophisticated structures that can conduct ions with high degree of specificity, under tight regulation making possible rapid changes of membrane potential in nerve cells and controlling the influx of second messenger ions, such as Ca^{2+}, leading to activation of various enzymes and proteins and responsible for many metabolic processes. Thanks to these properties, ion channels play a key role in regulating synaptic transmission and activity (Bliss and Collingridge, 1993; Kennedy, 1989), neurotransmitter release (Augustine *et al.*, 1987), synaptogenesis (Mattson *et al.*, 1988), and gene expression (West *et al.*, 2001). Indeed, when their function goes awry, there can be serious consequences. Thus, certain neurological disorders, such as Lambert–Eaton Myasthenic syndrome, are thought to result from the action of specific antibodies interfering with channels function (Flink and Atchison, 2003). Excessive activation of ion channels may also contribute to neuronal degeneration in several disorders, including stroke (Choi, 1988), AIDS dementia

TABLE I

Impact of Cytokines on Brain Functions and Behavior in Health and Disease

Neurological diseases		
Progression and exacerbation of the damage		Allan and Rothwell, 2001
Alzheimer's disease	IL-1β, TNF-α, IL-8	Xia and Hyman, 1999
Traumatic brain injury	IL-β, TNF-α	
Epilepsy	IL-1β	Vezzani *et al.*, 2000
Parkinson's disease	IL-1β, TNF-α	
Ischemia	IL-1β, TNF- α, IL-6	Maeda *et al.*, 1994; Yamasaki *et al.*, 1995
AIDS dementia complex	IL-1β, IL-2, TNF-α	Bagetta *et al.*, 1999; Bezzi *et al.*, 2001
	IL-6, IL-8	Qiu and Gruol, 2003
Neuroprotection		
AIDS dementia complex	TGF-β1	Meucci and Miller, 1996
Traumatic brain injury	TNF-α	Scherbel *et al.*, 1999
Ischemia	IL-6	Loddick *et al.*, 1998
Alzheimer's disease	TNF-α	Barger *et al.*, 1995
Host defense responses to disease		
Febrile response		
Signalling of the peripheral immune activation to the CNS	IL-1β,TNF-α	Alheim and Bartfai, 1998; Rothwell, 1991; Saper and Breder, 1992
Hypophagia	IL-1β, TNF-α	
Sickness syndrome	IL-1β	
Neuroendocrine changes	IL-1β	Rothwell and Hopkins, 1995
Pain		
Create and maintain exaggerated pain responses	IL-1β, TNF-α, IL-6	Watkins *et al.*, 2001
block and/or reverse enhanced	IL-10	Milligan *et al.*, 2005

(*Continued*)

TABLE I (*Continued*)

Sleep		
Regulation of sleep-wake cycle	IL-1β, TNF-α	Krueger *et al.*, 1998
Nonrapid eye movements		
Synaptic plasticity		
LTP		
Inhibition	IL-1β	Cunningham *et al.*, 1996; Katsuki *et al.*, 1990; Schneider *et al.*, 1998
	TNF-α	Tancredi *et al.*, 1992
Induction	IL-1β	Ross *et al.*, 2003
Cognition		
Memory consolidation impairment	IL-1β TNF-α	Rachal Pugh *et al.*, 2001

complex (Kaul *et al.*, 2001), amyotrophic lateral sclerosis (Rothstein, 1995), and Alzheimer's disease (Ueda *et al.*, 1997). It is thus conceivable that modulation of neuronal ion channels response by cytokines explains part of the effects exerted on brain functions and behavior in health and disease (Table I). In this chapter, we will provide emerging evidence regarding IL-1β and tumor necrosis factor-α (TNF-α), the two most studied proinflammatory cytokines in the brain.

II. Properties of Ion Channels

To better appreciate the various ways cytokines interact with these targets and thus modulate neuronal functions, it is essential to first briefly describe the properties of ion channels. Ion channels are gated pores that permit the flow of ions down to their electrochemical gradients. They are formed by one (voltage-gated Ca^{2+} and Na^{+} channels) or more (four-glutamate receptors; five-Cys-loop receptors) subunits delimiting a pore, often associated with accessory subunits. Accessory subunits specify the location and abundance of ion channels in the plasma membrane, modulate their biophysical properties, and fine-tune their sensitivity to physiological ligands and pharmacological agents.

The cardinal properties of ion channels are ion selectivity and gating. Selectivity refers to the ability of some channels to discriminate between ion species and is achieved through a physical–chemical interaction between the ion and various amino acid residues lining the channel. Gating is the process of transition between the open and closed states; for most channels it involves a conformational change and is regulated by biological signals such as binding of intracellular or extracellular ligands (ligand-gated channels), changes in membrane potential (voltage-gated channels), and changes in mechanical stress. Channel activity may be further modified by different ions that act as blockers (i.e., Mg^{2+} for NMDA receptor, NMDAR) and/or by cellular metabolic reactions, including protein phosphorylation. Accessory proteins may also contribute to the modification of channel activity, for example, serving as kinases that phosphorylate the pore forming subunit changing channel's biochemical properties (i.e., MPS-1 for voltage-gated K^{+} channels) (Cai *et al.*, 2005)

III. Distribution and Targeting of Neuronal Ion Channels

The signaling characteristics of a neuron are not only determined by the biophysical properties of ion channels but also by their spatial distribution. The development of patch-clamp recording has allowed electrophysiological analyses of different subcellular compartments of neurons characterized by a variable distribution of ion channels in soma, dendrites, and axons. Different channels and channel isoforms are targeted to different parts of a neuron to allow for diverse properties of axonal and somatodendritic regions and to carry out specific functions (Lai and Jan, 2006). Furthermore, properties and functions of neurons can be modulated, thanks to the dynamical distribution of ion channels among different compartments on the neuronal plasma membrane (Perez-Otano and Ehlers, 2005; Triller and Choquet, 2005). An exhaustive example of this is provided by synaptic plasticity. Recent data on synaptic plasticity has shown that changes in synaptic strength derived partly from modification of postsynaptic receptor numbers (such as ion channels like $GABA_A$ receptors, AMPA receptors, and NMDARs). The presence of receptors at the active postsynaptic site can be modulated by different mechanisms such as:

- *New synthesis and delivery of receptors from the endoplasmic reticulum (ER).* Receptors are assembled in the ER, and then transported to the Golgi in vesicles. Targeting to the synapse or other sites occurs via post-Golgi transport. Numerous accessory proteins act like molecular chaperones modulating the trafficking of receptors subunits to the plasma membrane (Prybylowski and Wenthold, 2004)
- *Internalization to intracellular stores.* Removal of the receptors from the neuronal plasma membrane occurs through endocytosis (Perez-Otano and Ehlers, 2005)
- *Lateral diffusion from/to extrasynaptic sites on the plasma membrane.* In physiological conditions, only receptors at the postsynaptic surface immediately apposed to the presynaptic site contribute to the synaptic response. Extrasynaptic receptors have been limited to activation by spillover of neurotransmitter outside the synaptic cleft during massive release (Clark and Cull-Candy, 2002; Kullmann and Asztely, 1998) or glutamate release by neighboring glia (Newman, 2003). Lateral diffusion of receptors is now well accepted for AMPA (Triller and Choquet, 2005) and also for NMDARs (Groc *et al.*, 2006).

IV. Ion Channels Are Targeted by Proinflammatory Cytokines

In the 1990s, the first evidence of a cytokine able of modulating neuronal ion channel activity has been reported (Miller *et al.*, 1991; Plata-Salaman and Ffrench-Mullen, 1992) and dealt with IL-1β, voltage-dependent Ca^{2+}-channels (VDCCs), and $GABA_A$ receptors ($GABA_AR$). Part of these observations were correlated to behavioral studies indicating that the enhancement of $GABA_A$ receptor functions by IL-1β accounts for somnogenic and motor-depressant effects of this cytokine (Miller *et al.*, 1991). This observation provided evidence for the involvement of ion channel in at least part of the *in vivo* effects exerted IL-1β (Table I).

Due to the relevance of ion channels in the modulation of physiological and pathological functions, the possible existence of a relation between ion channels modulation by cytokines and behavioral effects appears quite reasonable. Nevertheless, we have to wait almost 10 years to collect new data on this subject. Now we know that other cytokines besides IL-1β modulate ion channels, as well as other channels than Ca^{2+}-VOC and $GABA_A$ are involved (Figs. 1 and 2). Therefore, just to bring few examples, IL-8 shares IL-1β inhibitory effect on L- and N-type Ca^{2+}-VOC (Puma *et al.*, 2001), TNF-α increases the mean frequency of AMPA-dependent miniature excitatory postsynaptic currents (mEPSCs) (Beattie *et al.*, 2002) and decreases $GABA_A$-dependent inhibitory synaptic strength (Stellwagen *et al.*, 2005), and IL-2 reduces kainate-activated currents (Ye *et al.*, 2005). On the other hand, short exposure (5 min) to IL-1β also inhibits voltage-gated Na^+ channels (Liu *et al.*, 2006), reduces the frequency of AMPA-dependent mEPSCs (Yang *et al.*, 2005) and enhances NMDAR functions (Viviani *et al.*, 2003; Yang *et al.*, 2005). Two questions

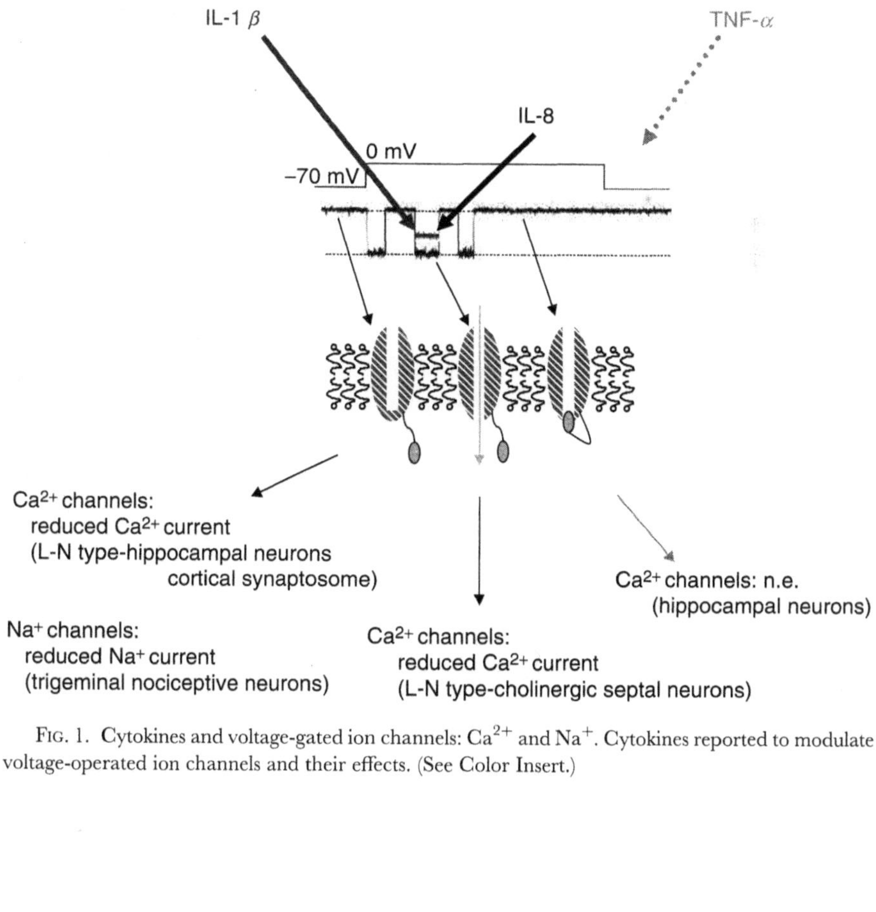

FIG. 1. Cytokines and voltage-gated ion channels: Ca^{2+} and Na^+. Cytokines reported to modulate voltage-operated ion channels and their effects. (See Color Insert.)

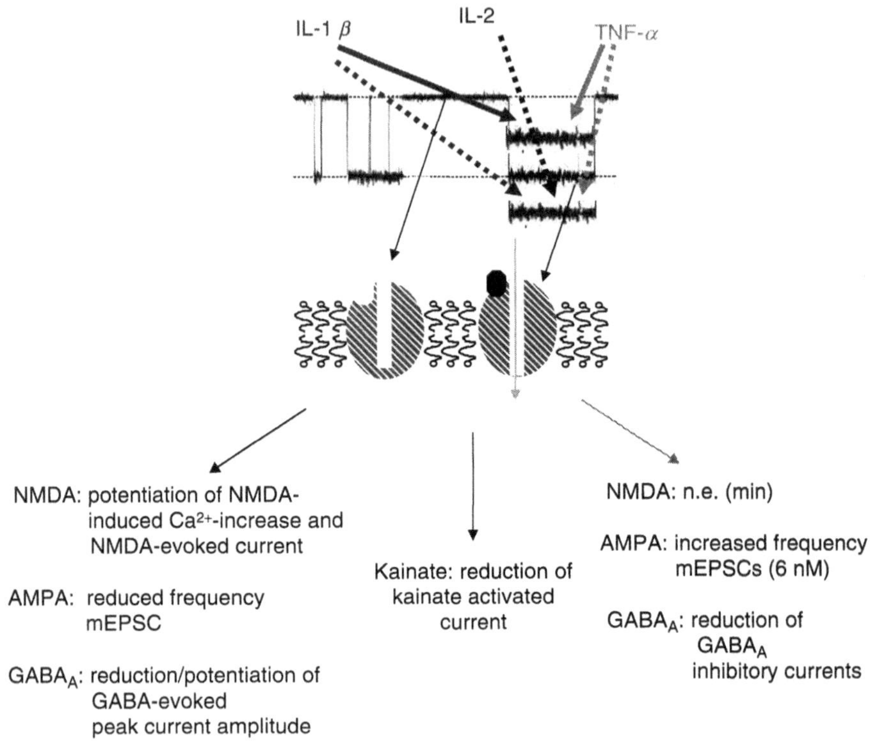

FIG. 2. Cytokines and ligand-gated ion channels. Cytokines reported to modulate ligand-gated ion channels and their effects. (See Color Insert.)

rise, which are the pathways recruited by cytokines to modulate ion channels and what is the impact on neuronal functions and brain behavior? In order to provide the answers we will focus on the following few examples.

V. IL-1β and Voltage-Dependent Ca^{2+} Channels

IL-1β has the proclivity to rapidly inhibit VDCCs activity promoted by a depolarizing stimuli or by amyloid-β (Aβ-(1–40)) (a peptide fragment derived from the proteolysis processing of Aβ precursor protein, which give rise to the senile plaques associated with Alzheimer's disease). This effect rises shortly after (few seconds) IL-1β application in acutely dissociated hippocampal CA1 neurons (Plata-Salaman and Ffrench-Mullen, 1992, 1994), in rat cortical synaptosome, in cultured cortical neurons (MacManus *et al.*, 2000), and in ventricular myocytes (Schreur and Liu, 1997). The lack of inhibition of neuronal VDCCs following exposure to other cytokines (IL-6, epidermal growth factor, basic fibroblast growth factor) or bacterial lipopolysaccharide indicates a specific action by IL-1β (Plata-Salaman and Ffrench-Mullen, 1994). In neurons, IL-1β inhibitory action is prevented by the classical IL-1 receptor antagonist (IL-1ra) and by the extracellular application of pertuxis toxin (PTX) and two different PKC inhibitors (H-7, staurosporine) (Plata-Salaman and Ffrench-Mullen, 1994). These observations indicate the recruitment of a PTX-sensitive G-protein and PKC following the activation of IL-1 receptor type I (IL-1RI). The involvement of PKC is confirmed by the observation that phorbol 12-myristate 13-acetate depresses voltage-dependent Ca^{2+} currents in a nonadditive manner with IL-1β (Plata-Salaman and Ffrench-Mullen, 1994).

IL-1β-induced inhibition of VDCCs and the resulting Ca^{2+} influx may impact on its ability to reduce neurotransmitter release (Murray *et al.*, 1997; Rada *et al.*, 1991), impair long-term potentiation in the hippocampus (Cunningham *et al.*, 1996; Katsuki *et al.*, 1990; Schneider *et al.*, 1998), and modulate synaptic transmission in the neocortex (D'Arcangelo *et al.*, 1997).

Concerning the pathological role of IL-1β, there is now extensive evidence to support the direct involvement of IL-1β in neuronal injury occurring in both acute and chronic neurodegenerative disorders (Allan *et al.*, 2005). Despite this, the inhibitory effect on VDCCs has been interpreted as a mechanism of neuroprotection since the consequent reduction in intracellular Ca^{2+} increases may limit neuronal damage by reducing presynaptic release of excitatory neurotransmitters and/or counteracting the activation of Ca^{2+}-dependent degenerative pathways. Nevertheless, the inhibitory effect of IL-1β on KCl-stimulated release of glutamate is evident with incubation time up to 60 s, while the effect is minimal after 5-min incubation (Murray *et al.*, 1997). This suggests a role for this pathway in the modulation of a physiological response characterized by transient variation of cytokines levels rather than a pathological condition. Furthermore, a general elevation in $[Ca^{2+}]_i$ does not necessarily predict neuronal death. According to the calcium source specificity hypothesis, Ca^{2+} toxicity occurs not simply as a function of increased Ca^{2+} concentration, but as a link to the route of Ca^{2+} entry and the distinct pathways that are activated as a result. Different studies have shown that specific neurotoxicants do not induce neuronal death despite the large elevation in Ca^{2+} (Dubinsky and Rothman, 1991; Ghosh and Greenberg, 1995). Finally, Ca^{2+} loads consequent to VDCCs are not harmful whereas similar $[Ca^{2+}]_i$ increases via NMDAR are toxic (Madden *et al.*, 1990; Tymianski *et al.*, 1993). Furthermore, Ca^{2+} flux through L-VDCCs has been tightly linked to gene expression and survival (West *et al.*, 2001). Thus, we can also hypothesize that an effect specifically directed to inhibit VDCCs activity may counteract a physiological or survival response resulting in final toxicity.

VI. IL-1β and NMDAR

Several *in vitro* and *in vivo* evidence suggest the existence of a functional interaction between IL-1β and NMDAR. Thus, IL-1β modulates physiological and pathological conditions depending on NMDAR such as control of synaptic activity (Cunningham *et al.*, 1996; Katsuki *et al.*, 1990; Ross *et al.*, 2003; Schneider *et al.*, 1998); regulation of sleep–wake cycle (Krueger *et al.*, 1998); seizure activity (Vezzani *et al.*, 1999); ischemic, traumatic, or excitotoxic brain damage (Loddick *et al.*, 1998; Yamasaki *et al.*, 1995). We observed that IL-1β increases the entry of Ca^{2+} through the NMDAR channel (Viviani *et al.*, 2003). Other authors, using an electrophysiological approach (Yang *et al.*, 2005), have later confirmed these data.

Potentiation of the NMDA response by IL-1β in purified hippocampal neurons requires a few minutes (Viviani *et al.*, 2003; Yang *et al.*, 2005); occurs in the absence of neurotransmitters release; is not affected by specific blockers of AMPA receptors, VDCCs, intracellular Ca^{2+}-store release; and is abolished by IL-1ra and specific inhibitors of the Src family of kinases (lavendustin A and PP2) (Viviani *et al.*, 2003). This suggests a direct effect of IL-1β on NMDAR through the activation of Src family of kinases downstream to IL-1RI. Actually, IL-1β induces phosphorylation of the Src-family-specific NR2B-Tyr-1472 site (Viviani *et al.*, 2003). All these effects are also recruited by endogenous IL-1β produced in a pathological condition and appear to be responsible of the stabilization of the NMDAR subunit NR2B at the synaptic sites (Viviani *et al.*, 2006).

Src is a lead member of a family of protein tyrosine kinases expressed in the CNS. By phosphorylating the NR2B subunit of the NMDAR, Src upregulates NMDAR function and increases its localization at the postsynaptic membrane (Collingridg *et al.*, 2004; Salter and Kalia, 2004). Electrophysiological recordings from neurons show that a balance between tyrosine phosphorylation and dephosphorylation governs NMDA currents (Kalia and Salter, 2003). Whether phosphorylation causes the increase in NMDAR gating still remains unclear (Salter and Kalia, 2004). On the other hand, tyrosine phosphorylation of NR2 subunits might also prevent the removal of signaling molecules from the NMDAR complex by protecting the subunits against degradation from the calcium-activated protease, calpain (Rong *et al.*, 2001). Furthermore, studies on recombinant NMDARs indicate that their association with the clathrin-mediated endocytosis machinery, a complex of proteins involved in the removal of receptors from the cell surface, is regulated by Src-mediated tyrosine phosphorylation of NMDAR subunits (Roche *et al.*, 2001; Vissel *et al.*, 2001). The NR2B-Tyr-1472 consensus domain is part of the internalization signal motif, a binding domain for the adaptor protein complex AP2, which associates with endocytic clathrin-coated vesicles. In addition, Src family tyrosine kinases have been shown to interact with NMDARs by binding to the scaffolding protein PSD-95 (Kalia and Salter, 2003). This interaction is strictly correlated to tyrosine phosphorylation of the NMDARs subunits (Collingridge *et al.*, 2004; Rong *et al.*, 2001; Song *et al.*, 2003). Finally, results show that stabilization of NR2B-containing receptors at the synapse is dynamically regulated by binding to a PDZ protein, such as PSD-95, and internalization through an interaction with AP-2 (Prybylowski *et al.*, 2005). With this view, NR2B–PDZ protein interaction may keep Tyr-1472 phosphorylated, and consequently unable to interact with AP-2 (Prybylowski *et al.*, 2005). The final result of this event is an increased localization of NR2B within the postsynaptic compartment, probably through prevention of endocytosis of NR2B-containing synaptic receptors (Collingridg *et al.*, 2004). Our results (Viviani *et al.*, 2003, 2006) indicate that all these mechanisms may be recruited by IL-1β, suggesting that the induced potentiation of NMDA response could be favored by both the enhancement of NMDAR single-channel gating and the permanence of the receptor at the postsynaptic membrane, where it can be stimulated. Such an effect persists for several minutes after removal of the cytokine (Viviani *et al.*, 2003) suggesting that even a transient exposure of neurons to this cytokine may favor long-lasting activation of the NMDAR functions. This may have functional relevance for either neuronal excitability or excitotoxicity. Hippocampal neurons exposed to IL-1β may be thus more susceptible to glutamatergic exitation through the NMDAR component. This may give insights into part of the molecular mechanisms underlying several of the *in vivo* effects exerted by this cytokine (Table I).

In our hands, the enhancement of the NMDAR function occurs at relatively low concentrations of IL-1β (0.01–0.1 ng/ml), whereas 1 ng/ml has a slight inhibitory effect on NMDA-mediated increases in $[Ca^{2+}]_i$ (Viviani *et al.*, 2003). An inhibitory effect was previously reported in the work of Coogan and O'Connor (1997) at IL-1β 1 ng/ml in rat dentate gyrus *in vitro*. Yang *et al.* (2005) also observed an increased NMDAR-mediated current in primary hippocampal neurons but at IL-1β 10 and 100 ng/ml. The limited overlapping between the concentrations of IL-1β and the response on the NMDAR observed in different work must not be surprising and is probably the consequence of several factors, including the functional state and the type of neurons, the duration of time that neurons are exposed to the cytokine, and, last but not least, the source and the batch of IL-1β used. These factors make it extremely difficult to compare results obtained by different groups. Nevertheless, all together these finding generally suggest a dual effect of IL-1β on NMDAR functions consistent with previous evidence showing that dose–response effects of IL-1β often display a bell-shaped curve (Allan and Rothwell, 2001; Rothwell and Hopkins, 1995).

A dual effect of IL-1β is also evident on another ion receptor, the $GABA_A$ receptor. Exogenous IL-1β was shown to inhibit GABAergic currents irreversibly in primary hippocampal neurons (Wang *et al.*, 2000) and to reduce synaptically mediated GABAergic inhibition in dentate gyrus and CA3 pyramidal neurons (Zeise *et al.*, 1997). In contrast, other authors have shown that IL-1β can also produce reversible and long-lasting increases in synaptically mediated GABAergic inhibition in the CA1 pyramidal neurons (Bellinger *et al.*, 1993).

VII. TNF-α: Few Final Considerations

As already introduced, a large number of cytokines are expressed in the CNS either constitutively or induced by brain damage. TNF-α is another cytokine particularly studied for its implication in brain pathophysiology. Together with IL-1β, this cytokine is central to the inflammatory response that occurs after injury and during prolonged CNS disease, and may contribute to the processes of neuronal cell death (Allan and Rothwell, 2001). TNF-α was found to be a new factor that enhances synaptic efficacy by increasing surface expression of AMPA receptors (Beattie *et al.*, 2002). This effect was observed after a short exposure (15 min) of primary hippocampal neurons to exogenous TNF-α and was reproduced following the application of astrocytes conditioned media (Beattie *et al.*, 2002). In the latter condition, AMPAR surface expression and mEPSCs were both prevented by TNFR1 and anti-TNF-α, suggesting that endogenous TNF-α influences AMPAR surface expression and synaptic strength (Beattie *et al.*, 2002). There were no changes in the amount of surface localization of the NMDAR (Beattie *et al.*, 2002) or amplitude of NMDA-induced currents at this time point (Furukawa and Mattson, 1998), indicating an effect specifically directed to AMPAR after a short time exposure to TNF-α. The increase in AMPAR surface expression is triggered by the specific TNFR1 receptor, requires phosphatidylinositol 3-kinase (PI3K) activity (Stellwagen *et al.*, 2005), and seems to be involved in synaptic scaling during activity blockade more than in the control of long-term potentiation or long-term depression (Stellwagen and Malenka, 2006). Synaptic scaling is a bidirectional phenomenon in which excitatory synapses scale-up in response to activity reduction, but scale-down in response to increases in activity (Turrigiano and Nelson, 2004). Scaling up of excitatory synapses requires TNF-α (Stellwagen and Malenka, 2006), showing a new neuronal function for constitutively released TNF-α. Furthermore, it has been hypothesized that TNF-α also may contribute to neuronal injury by increasing AMPAR surface expression. Indeed, coinjection of TNF-α with AMPAR agonists enhances necrotic cell death in the spinal cord (Hermann *et al.*, 2001).

VIII. Conclusions

Literature data underlines three important characteristics shared by cytokines concerning their action on ion channels:

- Cytokines action may occur within seconds or minutes, indicating the possibility of a prompt effect mainly due to posttranslational modification of ion channels subunits. In this chapter, we focused only on the fast (sec to min) action of cytokines.

Nevertheless, cytokines may also affect ion channels function after long time exposure (i.e., 24–48 h) (Furukawa and Mattson, 1998; Liu *et al.*, 2006). These temporal constraints are more consistent with a mechanism requiring altered gene expression rather than posttranslational modifications of channels proteins. In both cases, these actions modulate neuronal ion currents and the many physiological and pathological processes ruled by ion channels.

- Cytokines specifically affect ion channels function. In fact, different cytokines exert different effects on the same channel, as well as, the same cytokine exert different effects on different ion channels.
- Cytokine action is manifested mainly after the activation of the channel with an appropriate stimulus (i.e., depolarization step, KCl, specific agonists)

For all these reasons, cytokines are now regarded as novel neuromodulators.

With regard to the specific molecular pathway that mediates acute cytokines effect on ion channels, literature data provide evidence for the recruitment of different kinases (i.e., PKC, Src, PI3K). Kinases, by phosphorylating channels subunit, modify their biophysical properties and/or their stabilization at the postsynaptic membrane. Both these effects contribute to modulate neuronal responses to physiological and pathological stimuli.

In the present chapter, we reported just a few examples, representing the principal mode of action recruited by cytokines to affect ion channels. In our opinion, this evidence indicates ion channels as a relevant target to cytokine action at the CNS. Nevertheless, we are far from providing an exhaustive explanation of how cytokines can impact brain behavior through this pathway. The explanation has to be searched in the complexity of the real condition, where a sophisticated balance between more than one cytokine and channel at the same time rules physiological and pathological conditions. Efforts in defining such a complex network should be the future direction.

References

Alheim, K., and Bartfai, T. (1998). The interleukin-1 system: Receptors, ligands, and ICE in the brain and their involvement in the fever response. *Ann. NY Acad. Sci.* **840,** 51–58.

Allan, S. M., and Rothwell, N. J. (2001). Cytokines and acute neurodegeneration. *Nat. Rev. Neurosci.* **2,** 734–744.

Allan, S. M., Tyrrell, P. J., and Rothwell, N. J. (2005). Interleukin-1 and neuronal injury. *Nat. Rev. Immunol.* **5,** 629–640.

Augustine, G. J., Charlton, M. P., and Smith, S. J. (1987). Calcium action in synaptic transmitter release. *Annu. Rev. Neurosci.* **10,** 633–693.

Bagetta, G., Corasaniti, M. T., Berliocchi, L., Nistico, R., Giammarioli, A. M., Malorni, W., Aloe, L., and Finazzi-Agro, A. (1999). Involvement of interleukin-1beta in the mechanism of human immunodeficiency virus type 1 (HIV-1) recombinant protein gp120-induced apoptosis in the neocortex of rat. *Neuroscience* **89,** 1051–1066.

Barger, S. W., Van Eldik, L. J., and Mattson, M. P. (1995). S100 beta protects hippocampal neurons from damage induced by glucose deprivation. *Brain Res.* **677,** 167–170.

Beattie, E. C., Stellwagen, D., Morishita, W., Bresnahan, J. C., Ha, B. K., Von Zastrow, M., Beattie, M. S., and Malenka, R. C. (2002). Control of synaptic strength by glial TNFalpha. *Science* **295,** 2282–2285.

Bellinger, F. P., Madamba, S., and Siggins, G. R. (1993). Interleukin 1 beta inhibits synaptic strength and long-term potentiation in the rat CA1 hippocampus. *Brain Res.* **628,** 227–234.

Bezzi, P., Domercq, M., Brambilla, L., Galli, R., Schols, D., De Clercq, E., Vescovi, A., Bagetta, G., Kollias, G., Meldolesi, J., and Volterra, A. (2001). CXCR4-activated astrocyte glutamate release via TNFalpha: Amplification by microglia triggers neurotoxicity. *Nat. Neurosci.* **4,** 702–710.

Bliss, T. V., and Collingridge, G. L. (1993). A synaptic model of memory: Long-term potentiation in the hippocampus. *Nature* **361,** 31–39.

Cai, S. Q., Hernandez, L., Wang, Y., Park, K. H., and Sesti, F. (2005). MPS-1 is a K+ channel beta-subunit and a serine/threonine kinase. *Nat. Neurosci.* **8,** 1503–1509.

Choi, D. W. (1988). Glutamate neurotoxicity and diseases of the nervous system. *Neuron* **1,** 623–634.

Clark, B. A., and Cull-Candy, S. G. (2002). Activity-dependent recruitment of extrasynaptic NMDA receptor activation at an AMPA receptor-only synapse. *J. Neurosci.* **22,** 4428–4436.

Collingridge, G. L., Isaac, J. T., and Wang, Y. T. (2004). Receptor trafficking and synaptic plasticity. *Nat. Rev. Neurosci.* **5,** 952–962.

Coogan, A., and O'Connor, J. J. (1997). Inhibition of NMDA receptor-mediated synaptic transmission in the rat dentate gryrus *in vitro* by IL-1 beta. *Neuroreport* **8,** 2107–2110.

Cunningham, A. J., Murray, C. A., O'Neill, L. A., Lynch, M. A., and O'Connor, J. J. (1996). Interleukin-1 beta (IL-1 beta) and tumour necrosis factor (TNF) inhibit long-term potentiation in the rat dentate gyrus *in vitro*. *Neurosci. Lett.* **203,** 17–20.

D'Arcangelo, G., Dodt, H. U., and Zieglgansberger, W. (1997). Reduction of excitation by interleukin-1 beta in rat neocortical slices visualized using infrared-darkfield videomicroscopy. *Neuroreport* **8,** 2079–2083.

Dubinsky, J. M., and Rothman, S. M. (1991). Intracellular calcium concentrations during "chemical hypoxia" and excitotoxic neuronal injury. *J. Neurosci.* **11,** 2545–2551.

Flink, M. T., and Atchison, W. D. (2003). Ca2+ channels as targets of neurological disease: Lambert-Eaton Syndrome and other Ca2+ channelopathies. *J. Bioenerg. Biomembr.* **35,** 697–718.

Furukawa, K., and Mattson, M. P. (1998). The transcription factor NF-kappaB mediates increases in calcium currents and decreases in NMDA- and AMPA/kainate-induced currents induced by tumor necrosis factor-alpha in hippocampal neurons. *J. Neurochem.* **70,** 1876–1886.

Ghosh, A., and Greenberg, M. E. (1995). Calcium signaling in neurons: Molecular mechanisms and cellular consequences. *Science* **268,** 239–247.

Groc, L., Heine, M., Cousins, S. L., Stephenson, F. A., Lounis, B., Cognet, L., and Choquet, D. (2006). NMDA receptor surface mobility depends on NR2A-2B subunits. *Proc. Natl. Acad. Sci. USA* **103** (49), 18769–18774.

Hermann, G. E., Rogers, R. C., Bresnahan, J. C., and Beattie, M. S. (2001). Tumor necrosis factor-alpha induces cFOS and strongly potentiates glutamate-mediated cell death in the rat spinal cord. *Neurobiol. Dis.* **8,** 590–599.

Kalia, L. V., and Salter, M. W. (2003). Interactions between Src family protein tyrosine kinases and PSD-95. *Neuropharmacology* **45,** 720–728.

Katsuki, H., Nakai, S., Hirai, Y., Akaji, K., Kiso, Y., and Satoh, M. (1990). Interleukin-1 beta inhibits long-term potentiation in the CA3 region of mouse hippocampal slices. *Eur. J. Pharmacol.* **181,** 323–326.

Kaul, M., Garden, G. A., and Lipton, S. A. (2001). Pathways to neuronal injury and apoptosis in HIV-associated dementia. *Nature* **410,** 988–994.

Kennedy, M. B. (1989). Regulation of synaptic transmission in the central nervous system: Long-term potentiation. *Cell* **59,** 777–787.

Krueger, J. M., Fang, J., Taishi, P., Chen, Z., Kushikata, T., and Gardi, J. (1998). Sleep. A physiologic role for IL-1 beta and TNF-alpha. *Ann. N Y Acad. Sci.* **856,** 148–159.

Kullmann, D. M., and Asztely, F. (1998). Extrasynaptic glutamate spillover in the hippocampus: Evidence and implications. *Trends Neurosci.* **21,** 8–14.

Lai, H. C., and Jan, L. Y. (2006). The distribution and targeting of neuronal voltage-gated ion channels. *Nat. Rev. Neurosci.* **7,** 548–562.

Liu, L., Yang, T. M., Liedtke, W., and Simon, S. A. (2006). Chronic IL-1beta signaling potentiates voltage-dependent sodium currents in trigeminal nociceptive neurons. *J. Neurophysiol.* **95,** 1478–1490.

Loddick, S. A., Turnbull, A. V., and Rothwell, N. J. (1998). Cerebral interleukin-6 is neuroprotective during permanent focal cerebral ischemia in the rat. *J. Cereb. Blood Flow Metab.* **18,** 176–179.

MacManus, A., Ramsden, M., Murray, M., Henderson, Z., Pearson, H. A., and Campbell, V. A. (2000). Enhancement of (45)Ca(2+) influx and voltage-dependent Ca(2+) channel activity by beta-amyloid-(1–40) in rat cortical synaptosomes and cultured cortical neurons. Modulation by the proinflammatory cytokine interleukin-1beta. *J. Biol. Chem.* **275,** 4713–4718.

Madden, K. P., Clark, W. M., Marcoux, F. W., Probert, A. W., Jr., Weber, M. L., Rivier, J., and Zivin, J. A. (1990). Treatment with conotoxin, an 'N-type' calcium channel blocker, in neuronal hypoxic-ischemic injury. *Brain Res.* **537,** 256–262.

Maeda, Y., Matsumoto, M., Hori, O., Kuwabara, K., Ogawa, S., Yan, S. D., Ohtsuki, T., Kinoshita, T., Kamada, T., and Stern, D. M. (1994). Hypoxia/reoxygenation-mediated induction of astrocyte interleukin 6: A paracrine mechanism potentially enhancing neuron survival. *J. Exp. Med.* **180,** 2297–2308.

Mattson, M. P., Dou, P., and Kater, S. B. (1988). Outgrowth-regulating actions of glutamate in isolated hippocampal pyramidal neurons. *J. Neurosci.* **8,** 2087–2100.

Meucci, O., and Miller, R. J. (1996). gp120-induced neurotoxicity in hippocampal pyramidal neuron cultures: Protective action of TGF-beta1. *J. Neurosci.* **16,** 4080–4088.

Miller, L. G., Galpern, W. R., Dunlap, K., Dinarello, C. A., and Turner, T. J. (1991). Interleukin-1 augments gamma-aminobutyric acidA receptor function in brain. *Mol. Pharmacol.* **39,** 105–108.

Milligan, E. D., Sloane, E. M., Langer, S. J., Cruz, P. E., Chacur, M., Spataro, L., Wieseler-Frank, J., Hammack, S. E., Maier, S. F., Flotte, T. R., Forsayeth, J. R., Leinwand, L. A., *et al.* (2005). Controlling neuropathic pain by adeno-associated virus driven production of the anti-inflammatory cytokine, interleukin-10. *Mol. Pain* **1,** 9–22.

Murray, C. A., McGahon, B., McBennett, S., and Lynch, M. A. (1997). Interleukin-1 beta inhibits glutamate release in hippocampus of young, but not aged, rats. *Neurobiol. Aging* **18,** 343–348.

Newman, E. A. (2003). New roles for astrocytes: Regulation of synaptic transmission. *Trends Neurosci.* **26,** 536–542.

Perez-Otano, I., and Ehlers, M. D. (2005). Homeostatic plasticity and NMDA receptor trafficking. *Trends Neurosci.* **28,** 229–238.

Plata-Salaman, C. R., and Ffrench-Mullen, J. M. (1992). Interleukin-1 beta depresses calcium currents in CA1 hippocampal neurons at pathophysiological concentrations. *Brain Res. Bull.* **29,** 221–223.

Plata-Salaman, C. R., and Ffrench-Mullen, J. M. (1994). Interleukin-1 beta inhibits Ca2+ channel currents in hippocampal neurons through protein kinase C. *Eur. J. Pharmacol.* **266,** 1–10.

Prybylowski, K., and Wenthold, R. J. (2004). N-Methyl-D-aspartate receptors: Subunit assembly and trafficking to the synapse. *J. Biol. Chem.* **279,** 9673–9676.

Prybylowski, K., Chang, K., Sans, N., Kan, L., Vicini, S., and Wenthold, R. J. (2005). The synaptic localization of NR2B-containing NMDA receptors is controlled by interactions with PDZ proteins and AP-2. *Neuron* **47,** 845–857.

Puma, C., Danik, M., Quirion, R., Ramon, F., and Williams, S. (2001). The chemokine interleukin-8 acutely reduces Ca(2+) currents in identified cholinergic septal neurons expressing CXCR1 and CXCR2 receptor mRNAs. *J. Neurochem.* **78,** 960–971.

Qiu, Z., and Gruol, D. L. (2003). Interleukin-6, beta-amyloid peptide and NMDA interactions in rat cortical neurons. *J. Neuroimmunol.* **139,** 51–57.

Rachal Pugh, C., Fleshner, M., Watkins, L. R., Maier, S. F., and Rudy, J. W. (2001). The immune system and memory consolidation: A role for the cytokine IL-1beta. *Neurosci. Biobehav. Rev.* **25,** 29–41.

Rada, P., Mark, G. P., Vitek, M. P., Mangano, R. M., Blume, A. J., Beer, B., and Hoebel, B. G. (1991). Interleukin-1 beta decreases acetylcholine measured by microdialysis in the hippocampus of freely moving rats. *Brain Res.* **550,** 287–290.

Roche, K. W., Standley, S., McCallum, J., Dune Ly, C., Ehlers, M. D., and Wenthold, R. J. (2001). Molecular determinants of NMDA receptor internalization. *Nat. Neurosci.* **4,** 794–802.

Rong, Y., Lu, X., Bernard, A., Khrestchatisky, M., and Baudry, M. (2001). Tyrosine phosphorylation of ionotropic glutamate receptors by Fyn or Src differentially modulates their susceptibility to calpain and enhances their binding to spectrin and PSD-95. *J. Neurochem.* **79,** 382–390.

Ross, F. M., Allan, S. M., Rothwell, N. J., and Verkhratsky, A. (2003). A dual role for interleukin-1 in LTP in mouse hippocampal slices. *J. Neuroimmunol.* **144,** 61–67.

Rothstein, J. D. (1995). Excitotoxic mechanisms in the pathogenesis of amyotrophic lateral sclerosis. *Adv. Neurol.* **68,** 7–20; discussion 21–27.

Rothwell, N. J. (1991). Functions and mechanisms of interleukin 1 in the brain. *Trends Pharmacol. Sci.* **12,** 430–436.

Rothwell, N. J., and Hopkins, S. J. (1995). Cytokines and the nervous system II: Actions and mechanisms of action. *Trends Neurosci.* **18,** 130–136.

Salter, M. W., and Kalia, L. V. (2004). Src kinases: A hub for NMDA receptor regulation. *Nat. Rev. Neurosci.* **5,** 317–328.

Saper, C. B., and Breder, C. D. (1992). Endogenous pyrogens in the CNS: Role in the febrile response. *Prog. Brain Res.* **93,** 419–428; discussion 428–429.

Scherbel, U., Raghupathi, R., Nakamura, M., Saatman, K. E., Trojanowski, J. Q., Neugebauer, E., Marino, M. W., and McIntosh, T. K. (1999). Differential acute and chronic responses of tumor necrosis factor-deficient mice to experimental brain injury. *Proc. Natl. Acad. Sci. USA* **96,** 8721–8726.

Schneider, H., Pitossi, F., Balschun, D., Wagner, A., del Rey, A., and Besedovsky, H. O. (1998). A neuromodulatory role of interleukin-1beta in the hippocampus. *Proc. Natl. Acad. Sci. USA* **95,** 7778–7783.

Schreur, K. D., and Liu, S. (1997). Involvement of ceramide in inhibitory effect of IL-1 beta on L-type Ca2+ current in adult rat ventricular myocytes. *Am. J. Physiol.* **272,** H2591–H2598.

Song, C., Zhang, Y., Parsons, C. G., and Liu, Y. F. (2003). Expression of polyglutamine-expanded huntingtin induces tyrosine phosphorylation of N-methyl-D-aspartate receptors. *J. Biol. Chem.* **278,** 33364–33369.

Srinivasan, D., Yen, J. H., Joseph, D. J., and Friedman, W. (2004). Cell type-specific interleukin-1beta signaling in the CNS. *J. Neurosci.* **24,** 6482–6488.

Stellwagen, D., and Malenka, R. C. (2006). Synaptic scaling mediated by glial TNF-alpha. *Nature* **440,** 1054–1059.

Stellwagen, D., Beattie, E. C., Seo, J. Y., and Malenka, R. C. (2005). Differential regulation of AMPA receptor and GABA receptor trafficking by tumor necrosis factor-alpha. *J. Neurosci.* **25,** 3219–3228.

Tancredi, V., D'Arcangelo, G., Grassi, F., Tarroni, P., Palmieri, G., Santoni, A., and Eusebi, F. (1992). Tumor necrosis factor alters synaptic transmission in rat hippocampal slices. *Neurosci. Lett.* **146,** 176–178.

Triller, A., and Choquet, D. (2005). Surface trafficking of receptors between synaptic and extrasynaptic membranes: And yet they do move!. *Trends Neurosci.* **28,** 133–139.

Turrigiano, G. G., and Nelson, S. B. (2004). Homeostatic plasticity in the developing nervous system. *Nat. Rev. Neurosci.* **5,** 97–107.

Tymianski, M., Charlton, M. P., Carlen, P. L., and Tator, C. H. (1993). Source specificity of early calcium neurotoxicity in cultured embryonic spinal neurons. *J. Neurosci.* **13,** 2085–2104.

Ueda, K., Shinohara, S., Yagami, T., Asakura, K., and Kawasaki, K. (1997). Amyloid beta protein potentiates Ca2+ influx through L-type voltage-sensitive Ca2+ channels: A possible involvement of free radicals. *J. Neurochem.* **68,** 265–271.

Vezzani, A., Conti, M., De Luigi, A., Ravizza, T., Moneta, D., Marchesi, F., and De Simoni, M. G. (1999). Interleukin-1beta immunoreactivity and microglia are enhanced in the rat hippocampus by focal kainate application: Functional evidence for enhancement of electrographic seizures. *J. Neurosci.* **19,** 5054–5065.

Vezzani, A., Moneta, D., Conti, M., Richichi, C., Ravizza, T., De Luigi, A., De Simoni, A., De Simoni, M. G., Sperk, G., Andell-Jonsson, S., Lundkvist, J., Iverfeldt, K., *et al.* (2000). Powerful anticonvulsant action of IL-1 receptor antagonist on intracerebral injection and astrocytic overexpression in mice. *Proc. Natl. Acad. Sci. USA* **97,** 11534–11539.

Vissel, B., Krupp, J. J., Heinemann, S. F., and Westbrook, G. L. (2001). A use-dependent tyrosine dephosphorylation of NMDA receptors is independent of ion flux. *Nat. Neurosci.* **4,** 587–596.

Vitkovic, L., Bockaert, J., and Jacque, C. (2000). "Inflammatory" cytokines: Neuromodulators in normal brain? *J. Neurochem.* **74,** 457–471.

Viviani, B., Bartesaghi, S., Gardoni, F., Vezzani, A., Behrens, M. M., Bartfai, T., Binaglia, M., Corsini, E., Di Luca, M., Galli, C. L., and Marinovich, M. (2003). Interleukin-1beta enhances NMDA receptor-mediated intracellular calcium increase through activation of the Src family of kinases. *J. Neurosci.* **23,** 8692–8700.

Viviani, B., Gardoni, F., Bartesaghi, S., Corsini, E., Facchi, A., Galli, C. L., Di Luca, M., and Marinovich, M. (2006). Interleukin-1beta released by gp120 drives neural death through tyrosine phosphorylation and trafficking of NMDA receptors. *J. Biol. Chem.* **281,** 30212–30222.

Wang, S., Cheng, Q., Malik, S., and Yang, J. (2000). Interleukin-1beta inhibits gamma-aminobutyric acid type A (GABA(A)) receptor current in cultured hippocampal neurons. *J. Pharmacol. Exp. Ther.* **292,** 497–504.

Watkins, L. R., Milligan, E. D., and Maier, S. F. (2001). Glial activation: A driving force for pathological pain. *Trends Neurosci.* **24,** 450–455.

West, A. E., Chen, W. G., Dalva, M. B., Dolmetsch, R. E., Kornhauser, J. M., Shaywitz, A. J., Takasu, M. A., Tao, X., and Greenberg, M. E. (2001). Calcium regulation of neuronal gene expression. *Proc. Natl. Acad. Sci. USA* **98,** 11024–11031.

Xia, M. Q., and Hyman, B. T. (1999). Chemokines/chemokine receptors in the central nervous system and Alzheimer's disease. *J. Neurovirol.* **5,** 32–41.

Yamasaki, Y., Matsuura, N., Shozuhara, H., Onodera, H., Itoyama, Y., and Kogure, K. (1995). Interleukin-1 as a pathogenetic mediator of ischemic brain damage in rats. *Stroke* **26,** 676–680; discussion 681.

Yang, S., Liu, Z. W., Wen, L., Qiao, H. F., Zhou, W. X., and Zhang, Y. X. (2005). Interleukin-1beta enhances NMDA receptor-mediated current but inhibits excitatory synaptic transmission. *Brain Res.* **1034,** 172–179.

Ye, J. H., Zalcman, S. S., and Tao, L. (2005). Kainate-activated currents in the ventral tegmental area of neonatal rats are modulated by interleukin-2. *Brain Res.* **1049,** 227–233.

Zeise, M. L., Espinoza, J., Morales, P., and Nalli, A. (1997). Interleukin-1beta does not increase synaptic inhibition in hippocampal CA3 pyramidal and dentate gyrus granule cells of the rat *in vitro. Brain Res.* **768,** 341–344.

CYCLOOXYGENASE-2, PROSTAGLANDIN E_2, AND MICROGLIAL ACTIVATION IN PRION DISEASES

Luisa Minghetti and Maurizio Pocchiari

Department of Cell Biology and Neurosciences, Degenerative and Inflammatory Neurological Diseases Unit, Istituto Superiore di Sanità, Viale Regina Elena 299, 00161 Rome, Italy

Cyclooxygenase (COX) catalyzes the first committed step in the synthesis of prostaglandins (PGs) and is the main target of nonsteroidal anti-inflammatory drugs (NSAIDs). The enzyme exists as constitutive (COX-1) and inducible (COX-2) isoforms, being the latter a major player in inflammation. In the brain, COX-2 expression has been associated with inflammatory and neurodegenerative processes of several human neurological diseases. Prion diseases, or transmissible spongiform encephalopathies, are a heterogeneous group of fatal neurodegenerative disorders, characterized by deposition of the protease-resistant prion protein, astrocytosis, and spongiform degeneration. In addition, an extensive microglial activation supports the occurrence of local chronic inflammatory response. In experimental prion diseases, COX-2 immunoreactivity was found specifically localized to microglial cells and increased with the progression of disease, along with the number of activated microglia. The induction of COX-2 was paralleled by a substantial raise in the brain homogenate PGE_2 levels. In these models, only few scattered COX-1-positive microglia-like cells were detected, suggesting that COX-2 is the major form in prion diseases. In line with the animal models, elevated levels of PGE_2 were found in the cerebrospinal fluid of subjects affected by sporadic, genetic, or variant CJD. In sporadic CJD patients, the most numerous group of patients examined, higher CSF levels of PGE_2 were associated with shorter survival. Although the mechanisms leading to microglial COX-2 expression as well as its potential implication in prion disease pathogenesis remain to be

DOI: 10.1016/S0074-7742(07)82014-9

0074-7742/07 $35.00

established, PGE_2 levels in the cerebrospinal fluid might represent an important index to predict survival and disease severity.

I. Introduction

Cyclooxygenase (COX) catalyzes the first steps in the conversion of arachidonic acid into a large group of substances, collectively called prostanoids and including prostaglandins (PGs), prostacyclin, and thromboxanes. Prostanoids are short-lived lipid mediators, which exert diverse biological functions through specific membrane and nuclear receptors expressed by virtually all cell types (Funk, 2001). Since 1971, when it was demonstrated that nonsteroidal anti-inflammatory drugs (NSAIDs) exert their anti-inflammatory properties through the inhibition of COX activity, this enzyme has been considered an essential target for the development of anti-inflammatory treatments for a variety of pathologies, including neurodegenerative diseases.

COX, also known as prostaglandin H synthase or PGHS, exists in two major isoforms. The constitutive isoform (COX-1) is widely distributed and is mainly involved in physiological responses such as gastric cytoprotection and platelet aggregation. The second and inducible isoform, termed COX-2, is rapidly expressed in several cell types in response to growth factors, cytokines, and inflammatory stimuli, and is emerged as a major player in inflammatory reactions.

COX-1 and COX-2 are coded by two distinct genes located on human chromosome 9 and 1, respectively. At protein level, the two enzymes show over 60% homology in humans and rodents. While the functional sites are conserved, a few crucial substitutions cause important conformational variations in the active site pocket, which account for the different sensitivities of COX-1 and COX-2 to specific inhibitors. In addition, an insertion of 18 amino acids near the COX-2 C-terminus has allowed the production of specific antibodies.

The distribution of COX isoforms has been extensively studied in rat and human tissues. Although COX-1 is the isoform constitutively expressed in the majority of the tissues, in brain, testes, and kidney, both COX-1 and COX-2 are expressed under physiological conditions (Smith *et al.*, 1996). In human and rat brain, the two enzymes are present in several regions, although COX-2 is the prominent isoform in discrete neuronal populations of cerebral cortex and hippocampus (Breder *et al.*, 1995; Yasojima *et al.*, 1999).

A third variant of COX, named COX-3, has been identified from canine and human cerebral cortex cDNAs as an alternative product of the COX-1 gene, retaining intron 1 in its mRNAs (Chandrasekharan *et al.*, 2002). COX-3 expression appears tissue specific, with the highest levels in the brain, particularly in the cerebral cortex where it accounts for approximately 5% of COX-1. COX-3 exhibits enzymatic activity; however, its contribution to the overall PG production as well as its functional role in human brain remain, at present, uncertain (Kis *et al.*, 2005; Qin *et al.*, 2005).

The potential role of COX isoforms and PGs in brain diseases has been extensively reviewed in the past years (see for review Minghetti, 2004). Overexpression of COX-2 has been associated with neurotoxiticy in acute conditions, such as hypoxia/ischemia and seizures, as well as in chronic neurodegenerative diseases, including amyotrophic lateral sclerosis, Parkinson's disease, and Alzheimer's disease (AD). In spite of intensive research over the past years, evidence of a direct role of COX-2 in neurodegenerative events in chronic diseases is still controversial and further experimental and clinical studies are required before anti-inflammatory therapies aimed at inhibiting COX-2 are recommended (Firuzi and Pratico, 2006; Psaty and Furberg, 2005). Furthermore, the knowledge of COX-2 in brain physiology needs to be improved in order to fully exploit the potential benefits of COX-2 inhibition in neurological diseases.

II. COXs and PGs in Brain Functions

In normal brain, neuronal COX-2 expression is localized to excitatory glutamatergic neurons and is dependent on normal synaptic activity (Kaufmann *et al.*, 1996). Its expression is developmentally regulated and coincides with the critical period of activity-dependent cortical development. In rat brain, neuronal COX-2 mRNA and protein are detectable after the first postnatal week, reach a peak of expression during the third and the fourth week, and decrease to adult levels by 2 months. In the neocortex, COX-2-positive neurons show a typical laminar distribution that is heavily disrupted in subjects affected by Rett's syndrome, a neurodevelopmental disorder characterized by a defective development of cortical neurons (Kaufmann *et al.*, 1997).

Increasing evidence supports a role for COX-2 in behavioral and cognitive functions. Selective destruction of basal forebrain cholinergic neurons in rats at postnatal day 7 is associated with decreased hippocampal levels of COX-2 mRNA at adulthood. These animals show impaired social memory, suggesting that the early loss of hippocampal cholinergic input may impact on the expression of COX-2 in hippocampal neurons and on the functional role of PGs in synaptic activity (Ricceri *et al.*, 2004). Other studies have shown that systemic administration or hippocampal infusion of COX-2 selective inhibitors impair hippocampal-dependent memory and learning of a spatial task in rats (Rall *et al.*, 2003; Sharifzadeh *et al.*, 2005; Teather *et al.*, 2002).

Among COX-derived metabolites, PGE_2 appears to have a prominent role in mediating COX-2-dependent brain functions. First, PGE_2 is preferentially formed during the enzymatic activity of COX-2 rather than of COX-1, as suggested by observation in COX-1 and COX-2 knockout mice (Bosetti *et al.*, 2004). Second, COX-2 and PGE synthase, the enzyme converting the intermediated product PGH_2 into PGE_2, are coinduced by inflammatory stimuli; and third, they colocalize at intracellular level (Blais *et al.*, 2005; Ek *et al.*, 2001; Vazquez-Tello *et al.*, 2004). In addition, the exogenous application of PGE_2, but not PGD_2 or $PGF_{2\alpha}$, reversed the suppression of LTP induced by COX-2 inhibitor in hippocampal dentate granule neurons *in vitro* (Chen *et al.*, 2002). These findings, together with the localization of the enzymes synthesizing PGE_2 at postsynaptic levels and that of PGE_2 receptor subtype EP2 in the presynaptic dendritic spines, have led to propose PGE_2 as retrograde messenger (Sang *et al.*, 2005). Other mechanisms by which PGE_2 could indirectly contribute to synaptic plasticity include modulation of adrenergic, noradrenegic, and glutamatergic neurotransmission and regulation of membrane excitability (Bazan, 2003 and references therein).

III. Prion Diseases

Prion diseases, also known as transmissible spongiform encephalopathies, are a heterogeneous group of fatal disorders affecting both humans and animals. While in animals prion diseases are mainly acquired, human forms occur either as sporadic, genetic, or acquired disorders (Table I).

Sporadic CJD is the commonest form of human prion diseases and is characterized by rapidly progressive dementia and death within 4–6 months from the onset, with approximately only 5% of patients surviving over 1 year from clinical onset (Pocchiari *et al.*, 2004; Will *et al.*, 1998). The neuropathological hallmarks of the disease are the presence of amyloid plaques, generated by the deposition of the

TABLE I
Human Prion Diseases

Sporadic forms	Sporadic CJD, Sporadic fatal insomnia
Genetic forms	Genetic CJD, Gerstmann-Sträussler-Scheinker syndrome, fatal familial insomnia
Acquired forms	
Bovine source	Variant CJD
Human source	Iatrogenic CJD, Kuru, variant CJD[a]

[a]Transmitted through blood transfusion from asymptomatic subjects affected by variant CJD (Hewitt *et al.*, 2006).

pathological form of the prion protein (protease-resistant prion protein, PrP^{res} or scrapie prion protein, PrP^{sc}) and spongiform degeneration in the gray matter (Ironside *et al.*, 2005). In addition, as for other neurodegenerative diseases, astrocytosis and microglial activation are present in both animal and human forms of prion diseases, which supports the occurrence of a local chronic inflammatory response (Perry *et al.*, 2002).

Sporadic CJD is characterized by a diversity of neuropathological features, which are determined in part by the polymorphism at prion protein gene (*PRPN*) codon 129, encoding both methionine (M) or valine (V), and the type of PrP^{res} deposited in the brain, defined as type 1 or type 2 according to the size of the protease-resistant core fragment of 20 or 19 kDa, respectively. A classification in six major sCJD subtypes (Table II) has been proposed on the basis of the specific genotype of *PRPN* codon 129 and the presence of type 1 or type 2 PrP^{res} (Parchi *et al.*, 1999).

A considerable degree of phenotypic variability is also present among cases of genetic prion diseases, which are linked to an increasing number of pathogenic *PRPN* point mutations or insertions. The phenotypic variability is dependent on the site and the nature of mutation and on the polymorphism of codon 129, although other unknown factors are likely to be involved.

In contrast to the sporadic and genetic forms, variant CJD, the acquired human form associated with the consumption of contaminated bovine products (Will *et al.*, 1996), and, possibly, with blood transfusion (Hewitt *et al.*, 2006), shows a high degree of homogeneity. All cases of variant CJD so far identified were characterized by homozygosity at codon 129 of *PRPN* (M/M) and accumulation of type 2 PrP^{sc}, referred to as type 2B for the characteristic glycotype that distinguishes it from type 2 in sporadic CJD. Further distinctive traits of variant CJD are the more prolonged survival and the younger age at onset compared to other CJD forms.

TABLE II
Sporadic CJD Classification According to Parchi *et al.* (1999)

	PrP^{sc} Type 1	PrP^{sc} Type 2
codon 129M/M	MM1 (53%)	MM2
codon 129M/V	MV1	MV2
codon 129V/V	VV1	VV2 (13%)

Parentheses indicate frequency of genotypes for the two major subtypes, updated to 2004 (Pocchiari *et al.*, unpublished data).

IV. COXs in Human and Experimental Prion Diseases

The involvement of COX isoforms in experimental prion diseases was investigated using a mouse model in which C57BL/6J mice are intracerebrally injected with brain homogenate containing mouse-adapted ME7 strain of the ovine prion agent or scrapie. These mice develop the disease with a well-defined time course. Pathology and typical clinical signs are observed by 20–22 weeks postinjection (p.i.) and animals become terminally ill by 24 weeks p.i. (Betmouni *et al.*, 1996). Immunohistochemical analysis of brain tissues from ME7-injected mice revealed the presence of COX-2 immunoreactivity specifically associated with microglia-like cells, since very early stages of disease (16 weeks p.i.). COX-2-positive cells were particularly dense in areas of typical spongiform degeneration and their number increased with disease progression, along with that of activated microglia. In control brains from mice injected with normal brain homogenate, COX-2 immunoreactivity was found associated with rare cells showing small cell body and fine processes typical of parenchymal resting microglia (Walsh *et al.*, 2000). Consistent with COX-2 induction, the levels of PGE_2 in hippocampal homogenates at 24 weeks p.i. were higher in ME7-injected mice than in mice injected with normal brain homogenate (Minghetti *et al.*, 2000).

The observations in the ME7-injected mouse model suggest that COX-2 expression and activity may be specifically linked to the process of microglial activation in prion diseases. Since microglial activation occurs in considerable variation in CJD and in animal model of the disease (Baker *et al.*, 1999; Puoti *et al.*, 2005), we verified the induction of COX-2 in C3H mice infected with human brain homogenates from subjects who died by CJD. Transmission to C3H mice was set up from two cases of genetic CJD associated with V210I PrP mutation and two cases of sporadic CJD either homozygous for methionine (MM) or valine (VV) at the *PRPN* codon 129, using a combination of intracerebral and intraperitoneal injections. In these mice, clinical signs were insidious and often remitting. The disease frequently started with nervousness or steadiness, with tremor of the head, arched back, stiffness of tail, and running in a circle. Terminal stage was reached by 80–109 weeks p.i., when they were unable to walk (Minghetti *et al.*, 2005). Histological examination of brains revealed spongiosis in all mice. COX-2-positive cells with the typical microglial morphology were found in areas of intense spongiform degeneration. The intensity of COX-2 immunoreactivity was variable but always clearly significant compared to control mice, which showed very few COX-2-positive cells, with the typical morphology of resting microglia (Minghetti *et al.*, 2005). Only one case, inoculated with 129VV sporadic CJD, did not show a clearly distinguishable positivity for COX-2 compared to control.

The induction of COX-2 in microglia-like cells in two experimental prion diseases with distinctive features such as mouse strains (C57BL/6J and C3H), prion agent (mouse-adapted scrapie and CJD), and time of incubation necessary to develop the disease (22–24 and >90 weeks) supports the hypothesis that this event may be relevant to the pathogenesis of prion diseases rather than a characteristic of a specific prion agent or mouse strain. Furthermore, in both models, few scattered COX-1-positive microglia-like cells were found in control and infected brains (Minghetti, 2004), suggesting that COX-2 is the major COX isoform in experimental prion diseases.

The potential involvement of COX isoforms in human prion diseases was investigated by measuring the levels of PGE_2 in cerebrospinal fluid (CSF) samples from 52 subjects affected by sporadic CJD. The median levels of PGE_2 were higher in patients with sporadic CJD [51.8 (29.9) pg/ml, median (half interquartile range)] than in a control group of 14 subjects, undergoing subdural anesthesia or affected by noninflammatory neurological diseases [8.0 pg/ml (2.8)]. A comparable increase in CSF levels of PGE_2 was found in 10 patients affected by genetic CJD (Minghetti *et al.*, 2000), and, later, in 18 cases of variant CJD (Minghetti *et al.*, 2002).

In sporadic CJD patients, higher PGE_2 levels were associated with shorter patient survival. PGE_2 levels were not dependent on the time of CSF sampling during the course of the disease, suggesting that PGE_2 may be an index of disease severity rather than progression (Minghetti *et al.*, 2000). Interestingly, the median levels of PGE_2 measured in CJD patients were about fivefold higher than those reported by Combrinck *et al.* (2006) in probable AD patients, showing moderate cognitive impairment [mini mental state examination score 19.0 (3.2)]. In these subjects, in contrast to what observed in CJD patients, higher PGE_2 levels were associated with longer survival. It has been suggested that in AD, the concentration of PGE_2 in the CSF may reflect a greater survival of COX-2-positive neurons in hippocampus and cortex, two regions particularly vulnerable to degeneration in this disease. This interpretation is consistent with autopsy studies reporting that the number of neurons expressing COX-2 negatively correlates with the Braak score for $A\beta$ deposits and that the highest levels of neuronal COX-2 is observed in the early stages of AD pathology (Hoozemans *et al.*, 2004, 2005). Alternatively, early inflammatory processes, involving among other mechanisms the synthesis of PGs, may exert protective effects and impede the later progression of AD.

Increased levels of PGE_2 in the CSF of subjects affected by sporadic CJD or AD are likely to reflect distinct phenomena (microglial activation or neuronal survival), with opposite (negative or positive) predictive values for subject survival.

A collaborative study involving 10 national registries in Western Europe and over 2000 cases of sporadic CJD has shown that younger age at disease onset, female gender, heterozygosity at *PRPN* codon 129 and PrP^{Sc} of type 2 are associated with longer survival of patients (Pocchiari *et al.*, 2004). The inverse relation between CSF levels of PGE_2 and survival of sporadic CJD patients suggests that this COX metabolite may represent a promising additional diagnostic determinant that may help in understanding the marked clinical heterogeneity of sporadic CJD and in predicting the clinical progression of disease.

V. Roles of COX-2 and PGE_2 in Prion Diseases

The high expression of COX-2 in microglial cells in experimental prion diseases and the increased levels of PGE_2 in the CSF of sporadic, genetic, and variant CJD patients suggest that these molecules may be relevant to the pathogenesis of prion diseases. However, whether COX-2 overexpression in microglia is a cause or a consequence of neurodegeneration or whether the levels of PGE_2 in the CSF represent a surrogate of disease severity remain to be established.

The increases in COX-2 expression and PGE_2 synthesis could be related to mechanisms of host defense. In primary microglial cultures, COX-2 expression and PGE_2 synthesis are specifically promoted by the interaction with apoptotic, but not necrotic, neurons (De Simone *et al.*, 2004). In CJD, abundance of apoptotic neurons correlated with the presence of activated microglia (Gray *et al.*, 1999), supporting the hypothesis that the increased levels of PGE_2 in the CSF of CJD patients and the high expression of COX-2 in microglial cells in experimental prion diseases may be associated with the clearance of apoptotic neurons. Although considered a proinflammatory molecule, PGE_2 is immunosuppressive and downregulates the process of macrophage/microglial activation (Harris *et al.*, 2002; Levi *et al.*, 1999). As suggested for peripheral macrophages, PGE_2 could be instrumental for the safe clearance of apoptotic bodies by preventing an overt inflammatory reaction. However, an excessive or persistent synthesis of PGE_2 could result in a deficient host defense. In line with this hypothesis, genetic ablation of the PGE_2 receptor subtype EP2 significantly increased microglial phagocytosis of $A\beta$ peptides (Shie *et al.*, 2005), suggesting that phagocytosis, as well as many other functions associated with the process of microglial activation, are under the negative control of PGE_2 and that high levels of PGE_2 in brain parenchyma may have as consequences inefficient clearance of apoptotic neurons and amyloid removal.

Another possibility to consider is that PGE_2 could be associated with pathological prion protein accumulation and mediate neurotoxic processes. In neuroblastoma cells, exposure to PrP peptides increased PGE_2 levels culture media and the presence of COX-1 inhibitors protected against PrP toxicity (Bate *et al.*, 2002). Nonetheless, the nonselective COX inhibitor indomethacin had no significant effect on onset of clinical signs and survival time in experimental prion disease. In the same study (Manuelidis *et al.*, 1998), dapsone, an antibiotic used to treat leprosy and skin infections with a broad spectrum of anti-inflammatory activities, significantly delayed clinical signs and extended disease duration. The protective effect of dapsone was not confirmed in a later study (Guenther *et al.*, 2001).

Further studies on both human prion diseases and animal models will help elucidating the mechanisms leading to microglial COX-2 expression and PGE_2 synthesis in prion diseases and whether COX-2 and PGE_2 are targets and/or markers of potential use for treatment or follow up of prion diseases.

Acknowledgments

We thank Franco Cardone, Roberta Galeno, Anita Greco, Anna Ladogana, and Susanna Almonti for their continuous and valuable collaboration. We are also grateful to Maurizio Nunziati and Angela Valanzano for their excellent technical support.

References

Baker, C. A., Lu, Z. Y., Zaitsev, I., and Manuelidis, L. (1999). Microglial activation varies in different models of Creutzfeldt-Jakob disease. *J. Virol.* **73,** 5089–5097.

Bate, C., Rutherford, S., Gravenor, M., Reid, S., and Williams, A. (2002). Cyclo-oxygenase inhibitors protect against prion-induced neurotoxicity *in vitro*. *Neuroreport* **13,** 1933–1938.

Bazan, N. G. (2003). Synaptic lipid signaling: Significance of polyunsaturated fatty acids and platelet-activating factor. *J. Lipid Res.* **44,** 2221–2233.

Betmouni, S., Perry, V. H., and Gordon, J. L. (1996). Evidence for an early inflammatory response in the central nervous system of mice with scrapie. *Neuroscience* **74,** 1–5.

Blais, V., Turrin, N. P., and Rivest, S. (2005). Cyclooxygenase 2 (COX-2) inhibition increases the inflammatory response in the brain during systemic immune stimuli. *J. Neurochem.* **95,** 1563–1574.

Bosetti, F., Langenbach, R., and Weerasinghe, G. R. (2004). Prostaglandin E2 and microsomal prostaglandin E synthase-2 expression are decreased in the cyclooxygenase-2-deficient mouse brain despite compensatory induction of cyclooxygenase-1 and Ca^{2+}-dependent phospholipase A_2. *J. Neurochem.* **91,** 1389–1397.

Breder, C. D., Dewitt, D., and Kraig, R. P. (1995). Characterization of inducible cyclooxygenase in rat brain. *J. Comp. Neurol.* **355,** 296–315.

Chandrasekharan, N. V., Dai, H., Roos, K. L., Evanson, N. K., Tomsik, J., Elton, T. S., and Simmons, D. L. (2002). COX-3, a cycloxygenase-1 variant inhibited by acetaminophen and other analgesic/antipyretic drugs: Cloning, structure and expression. *Proc. Natl. Acad. Sci. USA* **99,** 13926–13931.

Chen, C., Magee, J. C., and Bazan, N. G. (2002). Cyclooxygenase-2 regulates prostaglandin E_2 signaling in hippocampal long-term synaptic activity. *J. Neurophysiol.* **87,** 2851–2857.

Combrinck, M., Williams, J., De Berardinis, M. A., Warden, D., Puopolo, M., Smith, A. D., and Minghetti, L. (2006). Levels of CSF prostaglandin E_2, cognitive decline, and survival in Alzheimer's disease. *J. Neurol. Neurosurg. Psychiatr.* **77,** 85–88.

De Simone, R., Ajmone-Cat, M. A., and Minghetti, L. (2004). Atypical antiinflammatory activation of microglia induced by apoptotic neurons: Possible role of phosphatidylserine-phosphatidylserine receptor interaction. *Mol. Neurobiol.* **29,** 197–212.

Ek, M., Engblom, D., Saha, S., Blomqvist, A., Jakobsson, P., and Ericsson-Dahlstrand, A. (2001). Inflammatory response: Pathway across the blood-brain barrier. *Nature* **410,** 430–431.

Firuzi, O., and Pratico, D. (2006). Coxibs and Alzheimer's disease: Should they stay or should they go? *Ann. Neurol.* **59,** 219–228.

Funk, C. D. (2001). Prostaglandins and leukotrienes: Advances in eicosanoid biology. *Science* **294,** 1871–1875.

Gray, F., Chretien, F., Adle-Biassette, H., Dorandeu, A., Ereau, T., Delisle, M. B., Kopp, N., Ironside, J. W., and Vital, C. (1999). Neuronal apoptosis in Creutzfeldt-Jakob disease. *J. Neuropathol. Exp. Neurol.* **58,** 321–328.

Guenther, K., Deacon, R. M., Perry, V. H., and Rawlins, J. N. (2001). Early behavioural changes in scrapie-affected mice and the influence of dapsone. *Eur. J. Neurosci.* **14,** 401–409.

Harris, S. G., Padilla, J., Koumas, L., Ray, D., and Phipps, R. P. (2002). Prostaglandins as modulators of immunity. *Trends Immunol.* **23,** 144–150.

Hewitt, P. E., Llewelyn, C. A., Mackenzie, J., and Will, R. G. (2006). Creutzfeldt–Jakob disease and blood transfusion: Results of the UK transfusion medicine epidemiological review study. *Vox Sang.* **91,** 221–230.

Hoozemans, J. J., Veerhuis, R., Rozemuller, A. J., Arendt, T., and Eikelenboom, P. (2004). Neuronal COX-2 expression and phosphorylation of pRb precede p38 MAPK activation and neurofibrillary changes in AD temporal cortex. *Neurobiol. Dis.* **15,** 492–499.

Hoozemans, J. J., van Haastert, E. S., Veerhuis, R., Arendt, T., Scheper, W., Eikelenboom, P., and Rozemuller, A. J. (2005). Maximal COX-2 and ppRb expression in neurons occurs during early Braak stages prior to the maximal activation of astrocytes and microglia in Alzheimer's disease. *J. Neuroinflammation* **2,** 27.

Ironside, J. W., Ritchie, D. L., and Head, M. W. (2005). Phenotypic variability in human prion diseases. *Neuropathol. Appl. Neurobiol.* **31,** 565–579.

Kaufmann, W. E., Worley, P. F., Pegg, J., Bremer, M., and Isakson, P. (1996). COX-2, a synaptically induced enzyme, is expressed by excitatory neurons at postsynaptic sites in rat cerebral cortex. *Proc. Natl. Acad. Sci. USA* **93,** 2317–2321.

Kaufmann, W. E., Worley, P. F., Taylor, C. V., Bremer, M., and Isakson, P. C. (1997). Cyclooxygenase-2 expression during rat neocortical development and in Rett syndrome. *Brain Dev.* **19,** 25–34.

Kis, B., Snipes, J. A., and Busija, D. W. (2005). Acetaminophen and the cyclooxygenase-3 puzzle: Sorting out facts, fictions, and uncertainties. *J. Pharmacol. Exp. Ther.* **315,** 1–7.

Levi, G., Minghetti, L., and Aloisi, F. (1999). Regulation of prostanoid synthesis in microglial cells and effects of prostaglandin E2 on microglial functions. *Biochimie* **80,** 899–904.

Manuelidis, L., Fritch, W., and Zaitsev, I. (1998). Dapsone to delay symptoms in Creutzfeldt-Jakob disease. *Lancet* **352,** 456.

Minghetti, L. (2004). Cyclooxygenase-2 (COX-2) in inflammatory and degenerative brain diseases. *J. Neuropathol. Exp. Neurol.* **63,** 901–910.

Minghetti, L., Greco, A., Cardone, F., Puopolo, M., Ladogana, A., Almonti, S., Cunningham, C., Perry, V. H., Pocchiari, M., and Levi, G. (2000). Increased brain synthesis of prostaglandin E_2 and F_2-isoprostane in human and experimental transmissible spongiform encephalopathies. *J. Neuropathol. Exp. Neurol.* **59,** 866–871.

Minghetti, L., Cardone, F., Greco, A., Puopolo, M., Levi, G., Green, A. J., Knight, R., and Pocchiari, M. (2002). Increased CSF levels of prostaglandin E(2) in variant Creutzfeldt-Jakob disease. *Neurology* **58,** 127–129.

Minghetti, L., Sbriccoli, M., Geloso, M. C., Ingrosso, L., Di Bari, M. A., Greco, A., Cardone, F., and Pocchiari, M. (2005). Cyclooxygenases and prostaglandin E2 in animal and human prion diseases. Proceedings VII European Meeting on glial cell functions, 2005 May 17–21, Amsterdam. Medimond International Proceeding Publishers.

Parchi, P., Giese, A., Capellari, S., Brown, P., Schulz-Schaeffer, W., Windl, O., Zerr, I., Budka, H., Kopp, N., Piccardo, P., Poser, S., Rojiani, A., *et al.* (1999). Classification of sporadic Creutzfeldt-Jakob disease based on molecular and phenotypic analysis of 300 subjects. *Ann. Neurol.* **46,** 224–233.

Perry, V. H., Cunningham, C., and Boche, D. (2002). Atypical inflammation in the central nervous system in prion disease. *Curr. Opin. Neurol.* **15,** 349–354.

Pocchiari, M., Puopolo, M., Croes, E. A., Budka, H., Gelpi, E., Collins, S., Lewis, V., Sutcliffe, T., Guilivi, A., Delasnerie-Laupretre, N., Brandel, J. P., Alperovitch, A., *et al.* (2004). Predictors of survival in sporadic Creutzfeldt-Jakob disease and other human transmissible spongiform encephalopathies. *Brain* **127,** 2348–2359.

Psaty, B. M., and Furberg, C. D. (2005). COX-2 inhibitors: Lessons in drug safety. *N. Engl. J. Med.* **352,** 1133–1135.

Puoti, G., Giaccone, G., Mangieri, M., Limido, L., Fociani, P., Zerbi, P., Suardi, S., Rossi, G., Iussich, S., Capobianco, R., Di Fede, G., Marcon, G., *et al.* (2005). Sporadic Creutzfeldt-Jakob disease: The extent of microglia activation is dependent on the biochemical type of PrPSc. *J. Neuropathol. Exp. Neurol.* **64,** 902–909.

Qin, N., Zhang, S. P., Reitz, T. L., Mei, J. M., and Flores, C. M. (2005). Cloning, expression, and functional characterization of human cyclooxygenase-1 splicing variants: Evidence for intron 1 retention. *J. Pharmacol. Exp. Ther.* **315,** 1298–1305.

Rall, J. M., Mach, S. A., and Dash, P. K. (2003). Intrahippocampal infusion of a cyclooxygenase-2 inhibitor attenuates memory acquisition in rats. *Brain Res.* **968,** 273–276.

Ricceri, L., Minghetti, L., Moles, A., Popoli, P., Confaloni, A., De Simone, R., Piscopo, P., Scattoni, M. L., di Luca, M., and Calamandrei, G. (2004). Cognitive and neurological deficits induced by early and prolonged basal forebrain cholinergic hypofunction in rats. *Exp. Neurol.* **189,** 162–172.

Sang, N., Zhang, J., Marcheselli, V., Bazan, N. G., and Chen, C. (2005). Postsynaptically synthesized prostaglandin E2 (PGE2) modulates hippocampal synaptic transmission via a presynaptic PGE2 EP2 receptor. *J. Neurosci.* **25,** 9858–9870.

Sharifzadeh, M., Naghdi, N., Khosrovani, S., Ostad, S. N., Sharifzadeh, K., and Roghani, A. (2005). Post-training intrahippocampal infusion of the COX-2 inhibitor celecoxib impaired spatial memory retention in rats. *Eur. J. Pharmacol.* **511,** 159–166.

Shie, F. S., Breyer, R. M., and Montine, T. J. (2005). Microglia lacking E prostanoid receptor subtype 2 have enhanced Abeta phagocytosis yet lack Abeta-activated neurotoxicity. *Am. J. Pathol.* **166,** 1163–1172.

Smith, W. L., Garavito, R. M., and DeWitt, D. L. (1996). Prostaglandin endoperoxide H synthases (cyclooxygenases)-1 and -2. *J. Biol. Chem.* **271,** 33157–33160.

Teather, L. A., Packard, M. G., and Bazan, N. G. (2002). Post-training cyclooxygenase-2 (COX-2) inhibition impairs memory consolidation. *Learn. Mem.* **9,** 41–47.

Vazquez-Tello, A., Fan, L., Hou, X., Joyal, J. S., Mancini, J. A., Quiniou, C., Clyman, R. I., Gobeil, F., Jr., Varma, D. R., and Chemtob, S. (2004). Intracellular-specific colocalization of prostaglandin E2 synthases and cyclooxygenases in the brain. *Am. J. Physiol. Regul. Integr. Comp. Physiol.* **287,** R1155–R1163.

Walsh, D. T., Perry, V. H., and Minghetti, L. (2000). Cyclooxygenase-2 is highly expressed in microglial-like cells in a murine model of prion disease. *Glia* **29,** 392–396.

Will, R. G., Ironside, J. W., Zeidler, M., Cousens, S. N., Esstibeiro, K., Alperovitch, A., Poser, S., Pocchiari, M., Hofman, A., and Smith, P. G. (1996). A new variant Creutzfeldt-Jakob disease in the UK. *Lancet* **347,** 921–925.

Will, R. G., Alperovitch, A., Poser, S., Pocchiari, M., Hofman, A., Mitrova, E., de Silva, R., D'Alessandro, M., Delasnerie-Laupretre, N., Zerr, I., and van Duijn, C. (1998). Descriptive epidemiology of Creutzfeldt-Jakob disease in six European countries, 1993–1995. EU Collaborative Study Group for CJD. *Ann. Neurol.* **43,** 763–767.

Yasojima, K., Schwab, C., McGeer, E. G., and McGeer, P. L. (1999). Distribution of cyclooxygenase-1 and cyclooxygenase-2 mRNAs and proteins in human brain and peripheral organs. *Brain Res.* **830,** 226–236.

GLIA PROINFLAMMATORY CYTOKINE UPREGULATION AS A THERAPEUTIC TARGET FOR NEURODEGENERATIVE DISEASES: FUNCTION-BASED AND TARGET-BASED DISCOVERY APPROACHES

Linda J. Van Eldik, Wendy L. Thompson, Hantamalala Ralay Ranaivo, Heather A. Behanna, and D. Martin Watterson

Center for Drug Discovery and Chemical Biology, Northwestern University
Chicago, Illinois 60611, USA

Inflammation is the body's defense mechanism against threats such as bacterial infection, undesirable substances, injury, or illness. The process is complex and involves a variety of specialized cells that mobilize to neutralize and dispose of the injurious material so that the body can heal. In the brain, a similar inflammation process occurs when glia, especially astrocytes and microglia, undergo activation in response to stimuli such as injury, illness, or infection. Like peripheral immune cells, glia in the central nervous system also increase production of inflammatory cytokines and neutralize the threat to the brain. This brain inflammation, or neuroinflammation, is generally beneficial and allows the brain to respond to changes in its environment and dispose of damaged tissue or undesirable substances. Unfortunately, this beneficial process sometimes gets out of balance and the neuroinflammatory process persists, even when the inflammation-provoking stimulus is eliminated. Uncontrolled chronic neuroinflammation is now known to play a key role in the progression of damage in a number of neurodegenerative diseases. Thus, overproduction of proinflammatory cytokines offers a pathophysiology progression mechanism that can be targeted in new therapeutic development for multiple neurodegenerative diseases. We summarize in this chapter the

INTERNATIONAL REVIEW OF
NEUROBIOLOGY, VOL. 82
DOI: 10.1016/S0074-7742(07)82015-0

evidence supporting proinflammatory cytokine upregulation as a therapeutic target for neurodegenerative disorders, with a focus on Alzheimer's disease. In addition, we discuss the drug discovery process and two approaches, function-driven and target-based, that show promise for development of neuroinflammation-targeted, disease-modifying therapeutics for multiple neurodegenerative disorders.

I. Neuroinflammation and Disease Progression

Neurodegenerative disorders such as Alzheimer's disease (AD), Parkinson's disease (PD), multiple sclerosis (MS), traumatic brain injury (TBI), and stroke have a major unmet need for therapies that alter disease progression. An increasing number of clinical and preclinical investigations have revealed similarities in the progression of pathophysiology among these diverse neurodegenerative disorders and raised the possibility for new therapeutic development by targeting these apparently common mechanisms of disease progression (Leker and Shohami, 2002; Perl *et al.*, 1998; Ringheim and Conant, 2004). Neuroinflammation is one of these disease progression mechanisms that is increasingly being investigated as a potential therapeutic target (Craft *et al.*, 2005a; Fahrig *et al.*, 2005; Gao *et al.*, 2003; Streit *et al.*, 2004). Neuroinflammation is a term used to describe central nervous system (CNS) inflammatory responses from activated glia, especially astrocytes and microglia. It is important to emphasize that under normal conditions, the innate immune responses of glia to a traumatic or injurious event, foreign pathogen, or activating stimulus generally lead to beneficial outcomes such as phagocytosis or production of reparative or protective factors. However, the delicate balance in this homeostasis can be disturbed, resulting in disease or exacerbation of initiating factors that result in disease.

Neuroinflammation was previously viewed as a "bystander effect" or epiphenomenon with inflammation occurring when damaged neurons elicit an activation response from glia. However, it is now appreciated that dysregulation of glia activation responses in the diseased brain can have a profound impact on the functionality of the neurons and influence the severity of neuronal dysfunction and the progression of neuropathology (Craft *et al.*, 2005a; Wyss-Coray, 2006). An increasing body of evidence suggests that the more intense or prolonged glial activation seen in neurodegenerative disorders becomes detrimental through chronic or aberrant expression of factors that may have cumulative or age-related consequences in vulnerable areas of the brain. It appears that the balance between the beneficial and detrimental effects of activated glia is disrupted, and that the scales are tipped toward neuropathological outcomes. This underscores the important concept that aberrant regulation of a normally beneficial process

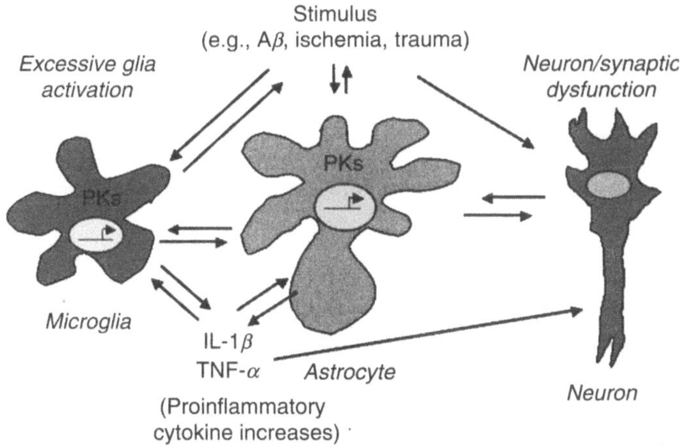

FIG. 1. The neuroinflammation cycle is a potential therapeutic intervention point for neurodegenerative disease drug discovery. The neuroinflammation cycle is characterized by a robust and sustained activation of astrocytes and microglia in response to a variety of disease-specific stimuli such as $A\beta$, ischemia, or trauma. These activated glial cells upregulate the production of proinflammatory cytokines, such as IL-1β, and TNF-α, which can serve to propagate the inflammatory responses and lead to neuronal and synaptic dysfunction. Many disease-relevant stimuli can also directly cause neuronal damage. Neuronal dysfunction or injury induces further glial activation, which exacerbates the neuropathological sequelae. Thus, excessive production of proinflammatory cytokines can be a driving force for pathophysiology progression in several CNS disorders. Targeting glial activation mechanisms such as pathways involving gene-regulating protein kinases (PKs) that lead to upregulation of proinflammatory cytokine production offers a focus for new therapeutic development strategies with potential for disease modification in multiple CNS disorders. (See Color Insert.)

may instead result in pathology. A diagrammatic outline of the neuroinflammatory cycle is shown in Fig. 1.

Chronic glial activation and overproduction of neuroinflammatory molecules such as proinflammatory cytokines are seen in a number of neurodegenerative disorders. Excessive production of proinflammatory cytokines can be a driving force for neuropathological progression of disease (Griffin and Mrak, 2002; Griffin *et al.*, 1998; Mrak and Griffin, 2005) and has been postulated to be involved in the neurological sequelae of several CNS disorders (Craft *et al.*, 2005a; Fahrig *et al.*, 2005; Gao *et al.*, 2003). It is clear that diverse diseases can sometimes have common mechanistic features that contribute to disease progression, even in complex disorders with multiple contributing causes. Therefore, innovative approaches that target common areas of disease progression have the potential to address major unmet needs in multiple disease areas. Upregulation of proinflammatory cytokine production offers a pathophysiology progression mechanism that can be targeted in new therapeutic development strategies with potential for disease modification in multiple diseases and clinical presentations.

In this chapter, we will focus primarily on AD, but the concepts and strategies are likely generalizable to multiple disorders where excessive activation of proinflammatory cytokines is a component of disease progression.

II. CNS Proinflammatory Cytokine Production as a Therapeutic Target for AD

There is extensive evidence from both clinical studies and preclinical animal models of disease that suggest the key role of proinflammatory cytokine overproduction as a potential driving force for pathology progression in AD. In principle, if proinflammatory cytokine upregulation is a driver of pathology rather than merely a response to the disease process, there are several testable predictions. First, increases in proinflammatory cytokines should occur early in the disease process (although increases later in the disease course could also accelerate disease progression). Second, manipulations that lead to overproduction of cytokines should lead to worsening of disease outcomes. Third, selective suppression of proinflammatory cytokine overproduction should lead to a reduction in disease-relevant end points. As discussed below, each of these predictions has been validated for AD by experimental and clinical data.

First, preclinical studies in diverse animal models relevant to AD showed a temporal correlation between early increases in the hippocampal levels of proinflammatory cytokines, such as interleukin (IL)-1β and tumor necrosis factor (TNF)-α, and the loss of surrogate biochemical markers of synaptic function in the hippocampus, with an associated worsening of hippocampal-dependent behavior (Craft *et al.*, 2004a,b, 2006; Mrak and Griffin, 2005; Tuppo and Arias, 2005). Second, evidence of a functional mechanistic link between neuroinflammation and neurodegeneration was provided by the finding that transgenic mice with upregulation of proinflammatory cytokine production and activity are more susceptible to loss of hippocampus synaptic markers when subjected to an AD-relevant injury stimulus, chronic exposure to toxic forms of human amyloid-β (Aβ1–42) (Craft *et al.*, 2005b,c). Third, further evidence for this causal link was provided by integrative chemical biology studies using bioavailable inhibitors of glial activation which showed that attenuation of the Aβ-induced increase in proinflammatory cytokine production resulted in improvement in the levels of hippocampus synaptic markers and hippocampal-dependent behavioral deficits (Craft *et al.*, 2004a,b). The identification of an orally bioavailable, brain-penetrant, nontoxic, lead compound by use of a *de novo* discovery paradigm focused on suppression of glial activation endpoints provides a further link between neuroinflammation, especially upregulated production of proinflammatory cytokines, and progression of disease-relevant pathophysiology (Ralay Ranaivo *et al.*, 2006). In addition, an open-label pilot study (Tobinick *et al.*, 2006) of 15 patients with mild-to-severe AD tested the effects of 6 months of treatment with CNS-delivered etanercept, a recombinant protein that interferes with TNF-α activity. The outcome was significant improvement of the patients in three different cognitive tests. These pilot results provide an important clinical precedent that inhibition of proinflammatory cytokines may hold promise as a therapeutic strategy for AD.

Taken in its entirety, the evidence is consistent with the hypothesis that proinflammatory cytokine upregulation is a comparatively early event in the progression of pathophysiology that is causally linked to synaptic dysfunction and hippocampal-dependent behavior deficits. Increases in proinflammatory cytokine production by glia as a contributor to neuronal loss is consistent with prevailing neuronal injury models in which a variety of glial products can contribute to neuronal dysfunction and death through glial stimulation of neuronal apoptosis pathways. The targeting of glial upregulation of proinflammatory cytokine production appears to be a viable strategy for the discovery of new classes of orally bioavailable compounds that attenuate human Aβ-induced neurodegeneration. However, approved cytokine-targeted drugs are macromolecules (Braddock and Quinn, 2004; Palladino *et al.*, 2003), and using macromolecules as a therapeutic approach has a number of disadvantages for clinical use in chronic CNS disorders such as high cost and inconvenient dosing regimens. Thus, there is a critical unmet need for small molecule, orally active, brain-penetrant, anticytokine compounds as new classes of CNS disease-modifying drugs.

A. Protein Phosphorylation Pathways as Regulators of Proinflammatory Cytokine Production

Over the last several years, significant advances have been made in our understanding of signal transduction pathways involved in regulation of expression of inflammatory proteins. Phosphorylation and activation of intracellular protein kinase cascades, followed by specific gene transcription appears to be a common mechanism for controlling the induction of many different inflammatory proteins, including proinflammatory cytokines, in response to extracellular stimuli. Many protein kinase pathways can regulate proinflammatory cytokine production in response to specific stimuli, but a particularly important set of pathways involve mitogen-activated protein kinase (MAPK) cascades (Adams *et al.*, 2001; O'Neill, 2006). MAPKs have been found to play critical roles in integrating and processing cellular responses to a number of diverse extracellular signals that lead to inflammatory responses (for general review, see Kyriakis and Avruch, 2001). Although a number of different MAPK pathways have been identified (ERK, JNK, and p38 MAPK being the most well studied), one of the major MAPK pathways mediating inflammation-related responses involves the p38 MAPK family (Kaminska, 2005;

Kumar *et al.*, 2003; Saklatvala, 2004). The p38 MAPKs are activated by a number of different extracellular signals, including toxins, environmental stress inducers, cell injury, apoptosis inducers, cytokines, inflammatory signals, endotoxin, and Aβ to name just a few. Like other MAPKs, the p38 MAPKs are activated by upstream kinases called MAPK kinases (MAPKK or MKKs), which are themselves phosphorylated and activated by MKK kinases (MAPKKK). On activation, p38 MAPKs can phosphorylate a variety of substrates, usually transcription factors or other protein kinases that culminate in specific biological responses. This three-tiered phosphorylation/activation cascade (MAPKKK to MAPKK to MAPK) and multiple kinases at each level of the cascade allow for exquisite control and selectivity of the cellular responses to different extracellular stimuli. There are also other levels of control of MAPK signaling that include cross talk among different MAPK pathways and between MAPK and non-MAPK signaling pathways, overlapping substrate specificities of the MAPKs, interactions with scaffolding proteins, specific localizations within cells, and different enzyme isoforms which can have distinct functions (Ashwell, 2006). Overall, the final biological responses of a cell to an extracellular signal are highly context dependent.

In terms of proinflammatory responses, the p38 MAPK pathway has been found to be a major point of convergence for diverse extracellular stimuli that induce inflammation. The p38 MAPK family consists of at least four isoforms, p38α, β, γ, δ, which are encoded by separate genes, expressed in different tissues, and are functionally distinct. Most evidence to date suggests that it is the p38α isoform that is primarily responsible for regulation of inflammatory signaling, although the other isoforms may contribute under specific sets of stimuli (Guo *et al.*, 2003). For the purposes of communication, we will use the term p38 MAPK generically throughout this chapter.

Related to p38 MAPK–dependent upregulation of cytokines, activation of p38 MAPK has been shown to regulate gene expression and lead to increased production of proinflammatory cytokines by a number of different mechanisms (reviewed in Schieven, 2005). First, p38 MAPK can directly phosphorylate and activate transcription factors, some of which can increase transcription of inflammatory cytokine genes. Second, p38 MAPK regulates IL-1β and TNF-α mRNA stability and translation, apparently by phosphorylation and activation of downstream kinases such as MAPK-activated protein kinase 2 (MAPKAP K2). The mRNAs for these inducible cytokines are short lived and contain AU-rich elements in the 3′ untranslated region of the mRNA that associate with AU-binding proteins. Under normal conditions, the AU-rich regions of the mRNA transcripts are occupied by AU-binding proteins, and so are not efficiently translated or are rapidly turned over. Following activation of p38 MAPK and MAPKAP K2, these AU-binding proteins get phosphorylated and dissociate from the mRNA, allowing mRNA stabilization and translation of the cytokines. Third, through its downstream kinase substrates MSK1 and MSK2, p38 MAPK can regulate the phosphorylation of histone H3 in chromatin at NF-κB binding sites of certain genes, which can influence transcriptional activation of certain cytokines. Thus, p38 MAPK can regulate a number of different signaling events that can converge on proinflammatory cytokine upregulation. In addition, because proinflammatory cytokine signaling can itself be dependent on p38 MAPK activation, the use of p38 MAPK inhibitors may not only lead to suppression of cytokine synthesis but also block their activity on a target cell.

Because of the large body of literature implicating p38 MAPK as central to the regulation of cytokine synthesis and signaling, the development of p38 MAPK inhibitors emerged as an attractive avenue of exploration for new therapies against inflammatory diseases. Another advantage of p38 MAPK as a drug discovery target is the availability of a number of X-ray crystal structures of the protein or the protein in complex with an inhibitor, thus facilitating structure-assisted design approaches (Dominguez *et al.*, 2005; Goldstein and Gabriel, 2005; Lee and Dominguez, 2005; Wrobleski and Doweyko, 2005). There has been a concerted effort by many pharmaceutical companies to develop p38 MAPK inhibitors as potential treatments for inflammatory diseases, especially peripheral inflammatory diseases like rheumatoid arthritis. The advancement of early p38 MAPK inhibitors into clinical trials was problematic because of safety issues in both the preclinical and clinical development stages (see Dominguez *et al.*, 2005; Goldstein and Gabriel, 2005 for review). There were a number of side effects reported for p38 MAPK inhibitors, including hepatotoxicity, cardiotoxicity, light-headedness, gastrointestinal effects, and infections (Dambach, 2005; Dominguez *et al.*, 2005; Lee and Dominguez, 2005). However, because of the remarkable diversity of chemotypes that can inhibit p38 MAPK (Diller *et al.*, 2005; Hynes and Leftheri, 2005; Lee and Dominguez, 2005), there were continued efforts to develop second-generation p38 MAPK inhibitors with a better therapeutic index. An increasing number of p38 MAPK inhibitors have been developed from multiple scaffolds, and different chemotypes have now advanced into clinical trials for peripheral disorders ranging from rheumatoid arthritis to dental pain and multiple myeloma. Based on the increased safety with these new-generation p38 MAPK inhibitors, it appears that the side effects encountered with the early inhibitors were chemotype-based or off-target effects rather than mechanism based. To date, there has been no evidence of unexpected mechanism-based safety problems in humans. Clinical side effects such as diarrhea, dizziness, gastrointestinal symptoms, and liver enzyme elevations have been compound specific. Based on these findings and the recent approval of anticytokine macromolecular therapy for multiple indications, it can be anticipated that small molecule p38 MAPK inhibitors will find an increased number of therapeutic indications.

B. The p38 MAPK as a Therapeutic Target for AD

The observations that many of the p38 MAPK inhibitors used in clinical trials have shown cytokine suppression in humans following oral administration of the compounds provide a precedent for exploring whether p38 MAPK inhibitors could be effective in CNS diseases where proinflammatory cytokine overproduction by glia is a component of disease progression. However, because there are a number of different signaling pathways that can modulate the levels of proinflammatory cytokines, it is critical to know if p38 MAPK is potentially involved in the upregulated cytokine production in specific neurodegenerative disorders we wish to target.

Extensive supporting evidence has been generated over the last several years that the p38 MAPK signaling cascade contributes to the cytokine overproduction and neurodegenerative sequelae seen in AD (reviewed in Johnson and Bailey, 2003), and that it may be a good therapeutic target for this disease (reviewed in Dalrymple, 2002). First, the p38 MAPK pathway is activated in the brain of AD patients in the early stages of the disease (Ferrer, 2004; Hensley *et al.*, 1999; Pei *et al.*, 2001; Sun *et al.*, 2003; Zhu *et al.*, 2000, 2001), consistent with a potential mechanistic linkage to disease progression. Second, p38 MAPK activation and a later induction of proinflammatory cytokines are stimulated in glial cell cultures by Aβ1–42 (Kim *et al.*, 2004; Koistinaho and Koistinaho, 2002; McDonald *et al.*, 1998). Third, activation of p38 MAPK is seen in AD-relevant animal models, including different APP transgenic mice strains (Ferrer, 2004; Giovannini *et al.*, 2002; Savage *et al.*, 2002). Fourth, inhibition of p38 MAPK by using a dominant negative approach or the SB203580 small molecule inhibitor blocked Aβ-induced neuronal death (Zhu *et al.*, 2005), suggesting that p38 MAPK inhibitors may also have beneficial effects on direct Aβ-induced neurotoxicity. Fifth, transgenic mice overexpressing APP751, which have increased neuroinflammation and are more vulnerable than wild-type mice to ischemic insult, were protected against the ischemic injury after treatment with a p38 MAPK inhibitor (Koistinaho *et al.*, 2002). All the available evidence to date suggests that p38 MAPK represents an important convergent point for integration of extracellular stimuli that induce CNS cytokine production, thus making it a compelling therapeutic target for treatment of neurodegenerative diseases like AD that have elevated levels of proinflammatory cytokines as a component of disease progression.

If p38 MAPK is such a potentially attractive drug discovery target, why are not the currently available p38 MAPK inhibitors being tested in neurodegenerative disorders? There are a number of reasons why p38 MAPK inhibitors have not been explored yet or are not optimal for use in CNS disorders. One of the main limitations of the current p38 MAPK compounds is that they were developed for peripheral inflammatory diseases, and in many cases were specifically designed to NOT cross the blood–brain barrier. Part of the rationale for this was historic in that there was a bias in the field that suppressing p38 MAPK activity in the CNS would be detrimental to the organism. Some of this bias likely arose from adverse effects seen with some of the early p38 MAPK inhibitors (Dambach, 2005). For example, the first p38 MAPK inhibitor from Vertex, VX-745, was discontinued due to the development of an unspecified CNS toxicity in dogs during the preclinical development studies, and lightheadedness was seen in humans in a phase 1 trial of the SCIOS 469 compound (the lightheadedness could be overcome by predosage feeding suggestive of susceptibility to a major P450 activity). However, many newer generation p38 MAPK inhibitors are based on chemotypes, such as pyridazinones (Natarajan *et al.*, 2006) and pyridazines (McIntyre *et al.*, 2002; Tamayo *et al.*, 2005), that are found in CNS-active drugs. This suggests that safe, CNS-active p38 MAPK inhibitors can be developed. In addition, the base of knowledge about targeting kinases for therapeutic development is rapidly increasing, and kinases represent a very sizable part of the druggable genome (Bhagwat *et al.*, 1999; Hopkins and Groom, 2002; Russ and Lampel, 2005). Finally, it is becoming clear that early concerns about potential side effects related to prolonged treatment with cytokine suppressors leading to impaired normal immune responses have not materialized in the few cases that have been examined. Nevertheless, because of the persistent perception that p38 MAPK inhibitors may not be safe for the CNS, and because CNS disorders are complex and more difficult and expensive for therapeutic development programs, the pharmaceutical industry has moved p38 MAPK inhibitors forward first for peripheral inflammatory diseases. However, it is very likely, based on the recent successes with macromolecule therapeutics to block cytokine production and signaling, that future emphasis will be placed on development of small molecule p38 MAPK inhibitors to target CNS disorders.

III. *De Novo* Lead Compound Discovery and the Recent Major Changes in Translational Research at the Chemistry–Biology Interface

A new chapter in the evolution of drug discovery and development began in 1997 with the seminal work by Lipinski *et al.* (2001) in which the trend toward higher failure rates with the wholesale adoption of high-throughput screening (HTS) technologies was found to correlate with a shift in molecular properties of lead compounds. Molecular properties of a chemical compound that are related to its drug-likeness include physical features such as molecular weight (MW), lipophilicity (log P), aqueous solubility (log S), polar surface area (PSA), and number of rotatable bonds (Amidon *et al.*, 1995; Lipinski, 2003; Proudfoot, 2005; Veber *et al.*, 2002; Vieth *et al.*, 2004; Wu and Benet, 2005). Many lead compounds coming out of HTS were characterized by higher MW and lipophilicity. Given the medicinal chemistry standard of adding 150–200 to the MW and increasing lipophilicity during medicinal chemistry refinement, the majority of compounds from the major pharmaceutical industries going into clinical development programs during the post-HTS era were characterized by a much higher MW and lipophilicity and poorer solubility than the profile of lead compounds pre-HTS. As discussed extensively in various drug discovery forums and reviews (Bleicher *et al.*, 2003; Kola and Landis, 2004; Posner, 2005; Sams-Dodd, 2006, 2007), HTS and combinatorial chemistry have made significant contributions to the drug discovery and drug development pipeline while exacerbating the problem of increasingly high failure rate that has characterized modern drug discovery over recent decades.

In parallel with this insight coming out of physical organic chemistry and computational biology, pharmacologists were observing that higher MW and lipophilicity were correlated with increased potential of compounds to serve as substrates for the major class of first pass metabolism enzymes, thereby making them more susceptible to metabolic instability and drug–drug or drug–food interactions. I and II addition, they noted the trend that increasing lipophilicity increased the probability that a compound could become a ligand for certain ion channels (Aronov, 2005, 2006), resulting in cardiotoxicity evidenced as prolongation of the cardiac QTc interval. Chemical engineers and clinicians involved in formulation and phase I and II clinical trials were beginning to observe a correlation with these same molecular properties and late-stage problems in clinical studies. Scientists in the pharmaceutical industry immediately realized that undesirable molecular properties in new molecular entities were a discrete contributing factor to the spiraling costs of late-stage drug development failures.

The increased appreciation of the primary importance of molecular properties has also altered basic science activities that form the foundation of translational research. For example, it is now realized that increasing lipophilicity of a lead compound may improve affinity for a molecular target via hydrophobic interactions, but does not necessarily improve *in vivo* efficacy. Increasing lipophilicity can result in the compound becoming a better substrate for key cytochrome P450 enzymes (de Groot and Ekins, 2002; Lewis *et al.*, 2004; Lombardo *et al.*, 2000), thereby decreasing the compound's metabolic stability. Further, increased lipophilicity is associated with nonspecific binding to off-target molecules and serum proteins, thereby decreasing the free drug concentration in blood and tissue (Liu and Chen, 2005). The realization that improving affinity for a molecular target does not necessarily result in improved *in vivo* efficacy and safety, a basic tenet of rational drug design, provides insight into why the wholesale shift to single molecular target drug discovery and development was correlated with a three decade long decline in the success rate of new chemical entities. This recent evolution in the medicinal chemistry and pharmacology activities at early stage drug discovery and in translational research at the chemistry–biology interface has now laid the foundation for a revamped drug discovery and development process that is already having an impact on reducing late-stage drug development failures, especially those failures directly or indirectly related to molecular properties. Rather than focusing solely on developing compounds that are optimized as ligands for affinity and selectivity against isolated targets, early drug discovery efforts must incorporate considerations of toxicology, pharmacokinetics, bioavailability, and efficacy. In fact, many drugs in clinical use today and the highest success rates in late-stage drug development are still those entering the development pipeline as compounds with early demonstration of safety, bioavailability, and some indication of *in vivo* efficacy, regardless of molecular target identification (Pritchard *et al.*, 2003).

The current trend of moving considerations of metabolic stability, toxicology, and bioavailability to earlier stages of drug discovery research places a demand on the chemistry platform to minimize the number of compounds made and subjected to such testing. The scientific and technical goals must be balanced in order to generate fewer, more diverse compounds that provide some insight into structure–activity relationships (SAR) when initially tested, and which have a higher probability for bioavailability and safety. This demand is being increasingly met by analysis of physical properties of novel compounds, using statistical analyses of what constitutes a good drug–like property, is not associated with late-stage problems, and is found in key molecular property measurements of clinical candidates. In summary, a range of MW, log P, log S, and PSA values must be considered in the refinement strategy such that a new chemical entity has an enhanced potential for success in drug development based on bioavailability, solubility, and metabolic stability.

A very attractive and promising, but still unproven, approach to contemporary drug discovery and development that has arisen from the study of molecular properties and in-parallel synthetic chemistries is the concept of *de novo* lead compound discovery. *De novo* lead compound discovery is an approach used by a new type of medicinal chemist-pharmacologist in which the prior art, discussed above, is leveraged in a "build-from-scratch" paradigm using the rapidly accumulating knowledge base about what makes a small molecule chemical a safe and bioavailable lead compound (see Wing *et al.*, 2006 for review of the approach). The hypothesis is that one can build focused chemical libraries of compounds with a high potential for pharmacological efficacy and safety by starting from small, generally inactive chemotypes or scaffolds and synthesizing a small number of compounds by using the pharmacoinformatics and chemoinformatics alluded to above. The front-end design of potential compounds uses decision filters driven by consideration of multiple molecular properties of a proposed compound and a desire to make chemically diverse compounds that sample a restricted chemical space as a starting point. For example, initial syntheses focus on products that maintain a minimal MW, restrict the number of rotatable bonds introduced into the initial expansion of the small and inactive molecular fragment, and sample a percentage of available chemical space at different points of molecular diversification. Additional pragmatic considerations that facilitate rapid progress and cost effectiveness are also used. For example, synthetic feasibility, simplicity, and scalability are important considerations (e.g., using in-parallel synthetic schemes where possible and minimizing the introduction of stereocenters). In addition, one must consider the potential for metabolic and chemical stability of the proposed product or its potential to undergo biologically facilitated reactions (e.g., covalent attachment of the product to proteins via Michael additions).

The *de novo* lead compound discovery approach works most effectively when it is highly integrated with a biological or biophysical screening platform to allow rapid, recursive feedback on which initial compounds in the syntheses have activity that can be improved by medicinal chemistry refinement. As discussed below, we used a *de novo* lead compound discovery approach that involved the screening of a highly focused chemical library based on diversification of the 3-amino-6-phenylpyridazine scaffold to discover a hit for neuroinflammation that was found to have the desired *in vivo* properties and efficacy in an AD animal model. Therefore, medicinal chemistry refinement was highly focused and centered on improvement of key molecular properties associated with later clinical success and desired by the FDA at the time of investigational new drug (IND) filing and phase 1 clinical trials.

IV. Development of Minozac: A Function-Driven Approach to Develop Small Molecule Compounds That Target Proinflammatory Cytokine Upregulation

The common theme of most FDA-approved drugs currently being examined for potential anti-neuroinflammatory action in neurodegenerative disease is that they were developed for other disease applications and have since been found to have potential actions in the brain. The straightforward approach of exploring alternative use of existing drugs is understandable based on the cost and time required to bring a new drug through drug development and FDA approval. However, there is no *a priori* reason to expect that a drug approved by the FDA for one use will work in another disease area.

In contrast to approaches that try to adapt existing compounds, we described a functional approach for CNS drug discovery that exploited the established success of modulating proinflammatory cytokine pathways in anti-inflammatory drug discovery for peripheral tissues and combined it with a CNS focus by using glia as the primary target for discovery (Hu *et al.*, 2005; Ralay Ranaivo *et al.*, 2006; Wing *et al.*, 2006). A *de novo* lead compound discovery approach (Wing *et al.*, 2006) was used that was based on the pyridazine chemotype (Fig. 2). We started with an

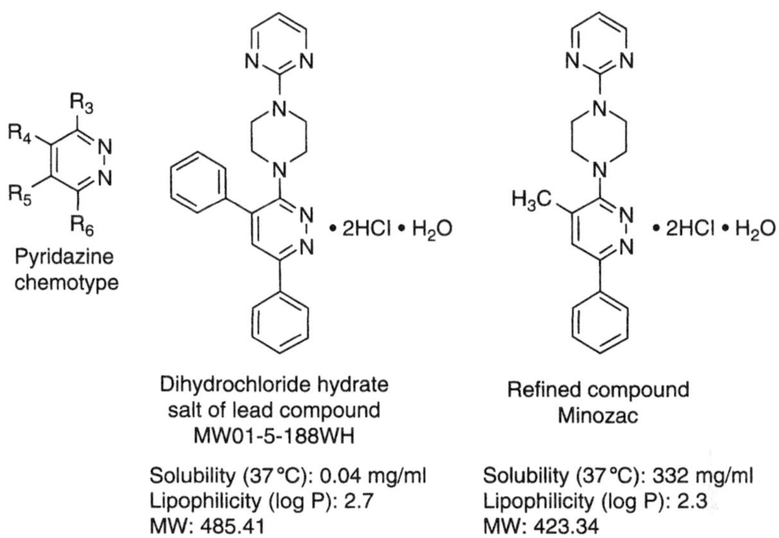

FIG. 2. Structures of the pyridazine chemotype, the lead compound MW01-5-188WH, and the refined compound Minozac. The pyridazine chemotype has several positions (R3–R6) available for chemical diversifications. The *de novo* lead compound discovery approach we used explored diversifications at R3 and R4 to make a highly focused chemical library from which the lead compound MW01-5-188WH emerged. Medicinal chemistry refinement of MW01-5-188WH to improve molecular properties yielded Minozac. Comparison of the molecular properties of the dihydrochloride hydrate salt of MW01-5-188WH and Minozac shows that Minozac has better molecular properties, including increased aqueous solubility, decreased lipophilicity, and lower molecular weight.

inactive phenylpyridazine fragment subjected to chemical diversifications based on a series of decision gates and used established chemistries amenable to in-parallel strategies for synthetic diversifications in a stepwise expansion of the inactive fragment and restricted sampling of chemical space to produce a focused synthetic chemical library of novel compounds. Synthetic chemistry design included the prioritization of synthetic efforts around compounds with potentially good molecular properties to enhance the potential for oral bioavailability and brain uptake while minimizing nonspecific serum and tissue protein binding. Synthetic feasibility is also considered, especially the potential in early syntheses to generate a focused set of molecules that sample part of a medicinal chemistry SAR. Synthetic candidates are also biased toward molecules that minimize stereocenters. Chirality can increase synthetic complexity and the individual enantiomers frequently differ in both their pharmacodynamic and pharmacokinetic profiles, often lowering the efficacy and safety of clinical candidates (Andersson, 2004; Baker and Prior, 2002; Hutt and Tan, 1996). The ability to make several compounds in parallel is facilitated by synthesis of intermediates that can be diversified using identical or highly similar sets of reaction conditions at the later steps. These considerations form the foundation of a synthetic chemistry platform that can rapidly produce active compounds and that has an inherent capability for rapid medicinal chemistry refinement.

The chemistry platform was integrated with a hierarchal biological platform focused on selective inhibition of proinflammatory cytokine production by activated glia, along with a series of screens to assess oral bioavailability, metabolic stability, brain uptake, and initial toxicity parameters. A lead compound, MW01-5-188WH (Fig. 2), emerged from this process (Ralay Ranaivo *et al.*, 2006). The compound is a selective suppressor of excessive glia proinflammatory cytokine production and attenuates synaptic dysfunction and hippocampal-dependent behavioral deficits in an AD animal model (Ralay Ranaivo *et al.*, 2006). MW01-5-188WH has attractive efficacy and safety screening profiles, but its aqueous solubility is not ideal for clinical development. As discussed above, a compound's molecular properties can have a significant influence on pharmacokinetics, toxicity, and efficacy, with success generally associated with low MW and good aqueous solubility. However, increasing the aqueous solubility of a compound while decreasing MW and retaining *in vivo* biological function is a nontrivial goal often considered paradoxical, especially for CNS lead compounds. Analysis of marketed drugs has led to an emerging consensus that multiproperty refinement is critical to improvement of the success rate in moving compounds through development. Solubility is a key molecular property that should be addressed early in the development process. Because the lead compound had the desired *in vivo* functions for a clinical candidate, we decided to test the hypothesis that a molecular property-driven medicinal chemistry refinement could improve aqueous solubility without increasing molecular weight, yet retain the highly attractive *in vivo* efficacy and safety in animal models.

Medicinal chemistry refinement (Hu *et al.*, 2007) yielded Minozac (Fig. 2), a hydrochloride hydrate salt that is a water-soluble, efficacious, safe, novel drug candidate that retains the *in vivo* functions of MW01-5-188WH but has improved molecular properties. Minozac has a promising safety and bioavailability profile, and is efficacious in animal models of AD (Hu *et al.*, 2007). A synthetic scheme amenable to large-scale production under good manufacturing protocol (GMP) conditions was also developed and transferred to a contract manufacturing organization (CMO) for production of clinical product. A clinical development campaign by an industry licensee is in progress. Follow-on compounds in the Minozac class are currently being developed by the function-driven approach that was successfully used for Minozac development (Fig. 3).

During screening of the same library that identified the lead compound MW01-5-188WH that was refined and developed into Minozac, we discovered a new class of pyridazine-based compounds that suppress glial activation and are

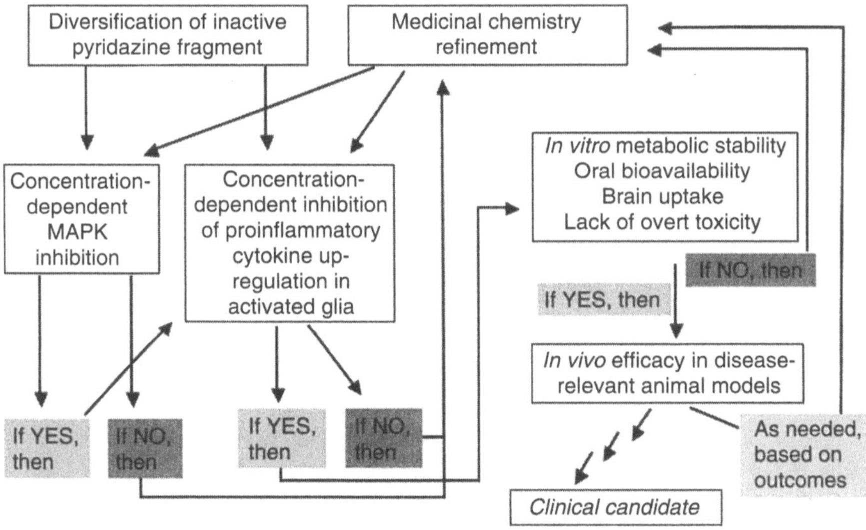

FIG. 3. Hierarchal strategy for development of glia inhibitors. For the function-driven discovery approach that yielded Minozac and that is currently being used for development of follow-on compounds, synthesized compounds are evaluated for selective suppression of the production of proinflammatory cytokines in activated glia cultures. For the target-based p38 MAPK inhibitor development project, synthesized compounds are first evaluated for concentration-dependent, selective inhibition of p38 MAPK enzyme activity. Positive compounds are then tested for concentration-dependent and selective suppression of the production of proinflammatory cytokines in activated glia cultures. Non-GLP tests of *in vitro* metabolic stability, oral bioavailability, brain uptake, and toxicity are done to restrict compounds to safe, orally bioavailable, brain-penetrant leads. Medicinal chemistry refinement is done to improve activities and properties as needed. Compounds passing these assays are then screened for efficacy to suppress disease-relevant endpoints in animal models. Refined compounds are taken recursively through the platform, and the clinical candidates are selected based on final biology and chemistry criteria of functional compounds.

p38 MAPK inhibitors (Minozac is not a p38 MAPK inhibitor). Minozac and its analogues represent diversifications of the 3-amino-6-phenylpyridazine scaffold at the R3 and R4 positions (Fig. 2). The new class of glial cytokine suppressors that are p38 inhibitors represents R5 diversifications. One of these p38 inhibitors has also been found to have efficacy *in vivo* in an AD animal model (Van Eldik and Watterson, unpublished observations). Efforts are underway to refine and optimize this new class of pyridazine-based compounds by using approaches that involve structure-assisted design in the context of molecular properties and other analyses as decision filters for the synthesis of a highly focused, small chemical library containing compounds with diverse lipophilicity values, a range of potential aqueous solubilities, and other attractive physical properties related to oral bioavailability and safety. The immediate insight into SAR by initial results demonstrates the power inherent with in-parallel synthesis of a compound family from common intermediates. Figure 3 shows an outline of the hierarchal screening platform for both the function-based development program (Minozac follow-on compounds) and the target-based development program (p38 MAPK-targeted compounds).

In summary, targeting key intracellular signaling pathways that regulate the overproduction of proinflammatory cytokines in activated glia provides an opportunity for the development of innovative, potentially disease-modifying therapeutics for neurodegenerative disorders. Our results support the feasibility of using either a function-based drug discovery strategy or a single molecular target-based strategy for AD drug discovery. In addition, the broad-ranging detrimental effects of excessive production of proinflammatory cytokines and the observation that unregulated neuroinflammation appears to be a component of disease progression in a variety of neurological disorders suggest that development of selective neuroinflammation-reducing drugs has the potential to address major unmet needs in multiple disease areas.

Acknowledgments

This work was supported in part by NIH grants U01 AG028561 (D.M.W., L.V.E), R01 NS047586 (D.M.W.), R37 AG013939 (L.V.E.), T32 AG000260 (H.A.B., W.L.T.), and F31 NS055471 (W.L.T.).

References

Adams, J. L., Badger, A. M., Kumar, S., and Lee, J. C. (2001). P38 MAP kinase: Molecular target for the inhibition of pro-inflammatory cytokines. *Prog. Med. Chem.* **38,** 1–60.

Amidon, G. L., Lennernas, H., Shah, V. P., and Crison, J. R. (1995). A theoretical basis for a biopharmaceutic drug classification: The correlation of *in vitro* drug product dissolution and *in vivo* bioavailability. *Pharm. Res.* **12,** 413–420.

Andersson, T. (2004). Single-isomer drugs: True therapeutic advances. *Clin. Pharmacokinet.* **43,** 279–285.

Aronov, A. M. (2005). Predictive in silico modeling for hERG channel blockers. *Drug Discov. Today* **10,** 149–155.

Aronov, A. M. (2006). Common pharmacophores for human uncharged human ether-a-go-go-related gene (hERG) blockers. *J. Med. Chem.* **49,** 6917–6921.

Ashwell, J. D. (2006). The many paths to p38 mitogen-activated protein kinase activation in the immune system. *Nat. Rev. Immunol.* **6,** 532–540.

Baker, G. B., and Prior, T. I. (2002). Stereochemistry and drug efficacy and development: Relevance of chirality to antidepressant and antipsychotic drugs. *Ann. Med.* **34,** 537–543.

Bhagwat, S. S., Manning, A. M., Hoekstra, M. F., and Lewis, A. (1999). Gene-regulating protein kinases as important anti-inflammatory targets. *Drug Discov. Today* **4,** 472–479.

Bleicher, K. H., Bohm, H. J., Muller, K., and Alanine, A. I. (2003). Hit and lead generation: Beyond high-throughput screening. *Nat. Rev. Drug Discov.* **2,** 369–378.

Braddock, M., and Quinn, A. (2004). Targeting IL-1 in inflammatory disease: New opportunities for therapeutic intervention. *Nat. Rev. Drug Discov.* **3,** 330–339.

Craft, J. M., Watterson, D. M., Frautschy, S. A., and Van Eldik, L. J. (2004a). Aminopyridazines inhibit beta-amyloid induced glial activation and neuronal damage *in vivo*. *Neurobiol. Aging* **25,** 1283–1292.

Craft, J. M., Van Eldik, L. J., Zasadzki, M., Hu, W., and Watterson, D. M. (2004b). Aminopyridazines attenuate hippocampus-dependent behavioral deficits induced by human beta-amyloid in a murine model of neuroinflammation. *J. Mol. Neurosci.* **24,** 115–122.

Craft, J. M., Watterson, D. M., and Van Eldik, L. J. (2005a). Neuroinflammation: A potential therapeutic target. *Expert Opin. Ther. Targets* **9,** 887–900.

Craft, J. M., Watterson, D. M., Marks, A., and Van Eldik, L. J. (2005b). Enhanced susceptibility of S-100B transgenic mice to neuroinflammation and neuronal dysfunction induced by intracerebroventricular infusion of human beta-amyloid. *Glia* **51,** 209–216.

Craft, J. M., Watterson, D. M., Hirsch, E., and Van Eldik, L. J. (2005c). Interleukin 1 receptor antagonist knockout mice show enhanced microglial activation and neuronal damage induced by intracerebroventricular infusion of human beta-amyloid. *J. Neuroinflammation* **2,** 15–23.

Craft, J. M., Watterson, D. M., and Van Eldik, L. J. (2006). Human amyloid beta-induced neuroinflammation is an early event in neurodegeneration. *Glia* **53,** 484–490.

Dalrymple, S. A. (2002). P38 mitogen activated protein kinase as a therapeutic target for Alzheimer's disease. *J. Mol. Neurosci.* **19,** 295–299.

Dambach, D. M. (2005). Potential adverse effects associated with inhibition of p38alpha/beta MAP kinases. *Curr. Top. Med. Chem.* **5,** 929–939.

de Groot, M. J., and Ekins, S. (2002). Pharmacophore modeling of cytochromes P450. *Adv. Drug Deliv. Rev.* **54,** 367–383.

Diller, D. J., Lin, T. H., and Metzger, A. (2005). The discovery of novel chemotypes of p38 kinase inhibitors. *Curr. Top. Med. Chem.* **5,** 953–965.

Dominguez, C., Powers, D. A., and Tamayo, N. (2005). p38 MAP kinase inhibitors: Many are made, but few are chosen. *Curr. Opin. Drug Discov. Devel.* **8,** 421–430.

Fahrig, T., Gerlach, I., and Horvath, E. (2005). A synthetic derivative of the natural product rocaglaol is a potent inhibitor of cytokine-mediated signaling and shows neuroprotective activity *in vitro* in animal models of Parkinson's disease and traumatic brain injury. *Mol. Pharmacol.* **67,** 1544–1555.

Ferrer, I. (2004). Stress kinases involved in tau phosphorylation in Alzheimer's disease, tauopathies and APP transgenic mice. *Neurotox. Res.* **6,** 469–475.

Gao, H. M., Liu, B., Zhang, W., and Hong, J. S. (2003). Novel anti-inflammatory therapy for Parkinson's disease. *Trends Pharmacol. Sci.* **24,** 395–401.

Giovannini, M. G., Scali, C., Prosperi, C., Bellucci, A., Vannucchi, M. G., Rosi, S., Pepeu, G., and Casamenti, F. (2002). Beta-amyloid-induced inflammation and cholinergic hypofunction in the rat brain *in vivo*: Involvement of the p38 MAPK pathway. *Neurobiol. Dis.* **11,** 257–274.

Goldstein, D. M., and Gabriel, T. (2005). Pathway to the clinic: Inhibition of p38 MAP kinase. A review of ten chemotypes selected for development. *Curr. Top. Med. Chem.* **5,** 1017–1029.

Griffin, W. S., and Mrak, R. E. (2002). Interleukin-1 in the genesis and progression of and risk for development of neuronal degeneration in Alzheimer's disease. *J. Leukoc. Biol.* **72,** 233–238.

Griffin, W. S., Sheng, J. G., Royston, M. C., Gentleman, S. M., McKenzie, J. E., Graham, D. I., Roberts, G. W., and Mrak, R. E. (1998). Glial-neuronal interactions in Alzheimer's disease: The potential role of a "cytokine cycle" in disease progression. *Brain Pathol.* **8,** 65–72.

Guo, X., Gerl, R. E., and Schrader, J. W. (2003). Defining the involvement of p38α MAPK in the production of anti- and proinflammatory cytokines using an SB 203580-resistant form of the kinase. *J. Biol. Chem.* **278,** 22237–22242.

Hensley, K., Floyd, R. A., Zheng, N. Y., Nael, R., Robinson, K. A., Nguyen, X., Pye, Q. N., Stewart, C. A., Geddes, J., Markesbery, W. R., Patel, E., Johnson, G. V., *et al.* (1999). P38 kinase is activated in the Alzheimer's disease brain. *J. Neurochem.* **72,** 2053–2058.

Hopkins, A. L., and Groom, C. R. (2002). The druggable genome. *Nat. Rev. Drug Discov.* **1,** 727–730.

Hu, W., Ralay Ranaivo, H., Craft, J. M., Van Eldik, L. J., and Watterson, D. M. (2005). Validation of the neuroinflammation cycle as a drug discovery target using integrative chemical biology and lead compound development with an Alzheimer's disease-related mouse model. *Curr. Alzheimer Res.* **2,** 197–205.

Hu, W., Ralay Ranaivo, H., Roy, S. M., Behanna, H. A., Wing, L. K., Munoz, L., Guo, L., Van Eldik, L. J., and Watterson, D. M. (2007). Development of a novel therapeutic suppressor of brain proinflammatory cytokine up-regulation that attenuates synaptic dysfunction and behavioral deficits. *Bioorg. Med. Chem. Lett.* **17,** 414–418.

Hutt, A. J., and Tan, S. C. (1996). Drug chirality and its clinical significance. *Drugs* **52,** 1–12.

Hynes, J., and Leftheri, K. (2005). Small molecule p38 inhibitors: Novel structural features and advances from 2002–2005. *Curr. Top. Med. Chem.* **5,** 967–985.

Johnson, G. V., and Bailey, C. D. (2003). The p38 MAP kinase signaling pathway in Alzheimer's disease. *Exp. Neurol.* **183,** 263–268.

Kaminska, B. (2005). MAPK signaling pathways as molecular targets for anti-inflammatory therapy—from molecular mechanisms to therapeutic benefits. *Biochim. Biophys. Acta* **1754,** 253–262.

Kim, S. H., Smith, C. J., and Van Eldik, L. J. (2004). Importance of MAPK pathways for microglial pro-inflammatory cytokine IL-1beta production. *Neurobiol. Aging* **25,** 431–439.

Koistinaho, M., and Koistinaho, J. (2002). Role of p38 and p44/42 mitogen-activated protein kinases in microglia. *Glia* **40,** 175–183.

Koistinaho, M., Kettunen, M. I., Goldsteins, G., Keinanen, R., Salminen, A., Ort, M., Bures, J., Liu, D., Kauppinen, R. A., Higgins, L. S., and Koistinaho, J. (2002). Beta-amyloid precursor protein transgenic mice that harbor diffuse Abeta deposits but do not form plaques show increased ischemic vulnerability: Role of inflammation. *Proc. Natl. Acad. Sci. USA* **99,** 1610–1615.

Kola, I., and Landis, J. (2004). Can the pharmaceutical industry reduce attrition rates? *Nat. Rev. Drug Discov.* **3,** 711–715.

Kumar, S., Boehm, J., and Lee, J. C. (2003). p38 MAP kinases: Key signaling molecules as therapeutic targets for inflammatory diseases. *Nat. Rev. Drug Discov.* **2,** 717–726.

Kyriakis, J. M., and Avruch, J. (2001). Mammalian mitogen-activated protein kinase signal transduction pathways activated by stress and inflammation. *Physiol. Rev.* **81,** 808–869.

Lee, M. R., and Dominguez, C. (2005). MAP kinase p38 inhibitors: Clinical results and an intimate look at their interactions with p38alpha protein. *Curr. Med. Chem.* **12,** 2979–2994.

Leker, R. R., and Shohami, E. (2002). Cerebral ischemia and trauma-different etiologies yet similar mechanisms: Neuroprotective opportunities. *Brain Res. Brain Res. Rev.* **39,** 55–73.

Lewis, D. F., Jacobs, M. N., and Dickins, M. (2004). Compound lipophilicity for substrate binding to human P450s in drug metabolism. *Drug Discov. Today* **9,** 530–537.

Lipinski, C. A. (2003). Compound properties and drug quality. *In* "The Practice of Medicinal Chemistry" (C. G. Wermuth, Ed.), 2nd. Academic Press, San Diego.

Lipinski, C. A., Lombardo, F., Dominy, B. W., and Feeney, P. J. (2001). Experimental and computational approaches to estimate solubility and permeability in drug discovery and development settings. *Adv. Drug Deliv. Rev* **46,** 3–26.

Liu, X., and Chen, C. (2005). Strategies to optimize brain penetration in drug discovery. *Curr. Opin. Drug Discov. Devel.* **8,** 505–512.

Lombardo, F., Shalaeva, M. Y., Tupper, K. A., Gao, F., and Abraham, M. H. (2000). ElogPoct: A tool for lipophilicity determination in drug discovery. *J. Med. Chem.* **43,** 2922–2928.

McDonald, D. R., Bamberger, M. E., Combs, C. K., and Landreth, G. E. (1998). Beta-amyloid fibrils activate parallel mitogen-activated protein kinase pathways in microglia and THP1 monocytes. *J. Neurosci.* **18,** 4451–4460.

McIntyre, C. J., Ponticello, G. S., Liverton, N. J., O'Keefe, S. J., O'Neill, E. A., Pang, M., Schwartz, C. D., and Claremon, D. A. (2002). Pyridazine based inhibitors of p38 MAPK. *Bioorg. Med. Chem. Lett.* **12,** 689–692.

Mrak, R. E., and Griffin, W. S. (2005). Glia and their cytokines in progression of neurodegeneration. *Neurobiol. Aging* **26,** 349–354.

Natarajan, S. R., Heller, S. T., Nam, K., Singh, S. B., Scapin, G., Patel, S., Thompson, J. E., Fitzgerald, C. E., and O'Keefe, S. J. (2006). p38 MAP kinase inhibitors. Part 6: 2-Arylpyridazin-3-ones as templates for inhibitor design. *Bioorg. Med. Chem. Lett.* **16,** 5809–5813.

O'Neill, L. A. (2006). Targeting signal transduction as a strategy to treat inflammatory diseases. *Nat. Rev. Drug Discov.* **5,** 549–563.

Palladino, M. A., Bahjat, F. R., Theodorakis, E. A., and Moldawer, L. L. (2003). Anti-TNF-alpha therapies: The next generation. *Nat. Rev. Drug Discov.* **2,** 736–746.

Pei, J. J., Braak, E., Braak, H., Grundke-Iqbal, I., Iqbal, K., Winblad, B., and Cowburn, R. F. (2001). Localization of active forms of C-jun kinase (JNK) and p38 kinase in Alzheimer's disease brains at different stages of neurofibrillary degeneration. *J. Alzheimers Dis.* **3,** 41–48.

Perl, D. P., Olanow, C. W., and Calne, D. (1998). Alzheimer's disease and Parkinson's disease: Distinct entities or extremes of a spectrum of neurodegeneration? *Ann. Neurol.* **44,** S19–S31.

Posner, B. A. (2005). High-throughput screening-driven lead discovery: Meeting the challenges of finding new therapeutics. *Curr. Opin. Drug Discov. Devel.* **8,** 487–494.

Pritchard, J. F., Jurima-Romet, M., Reimer, M. L., Mortimer, E., Rolfe, B., and Cayen, M. N. (2003). Making better drugs: Decision gates in non-clinical drug development. *Nat. Rev. Drug Discov.* **2,** 542–553.

Proudfoot, J. R. (2005). The evolution of synthetic oral drug properties. *Bioorg. Med. Chem. Lett.* **15,** 1087–1090.

Ralay Ranaivo, H., Craft, J. M., Hu, W., Guo, L., Wing, L. K., Van Eldik, L. J., and Watterson, D. M. (2006). Glia as a therapeutic target: Selective suppression of human amyloid-beta-induced upregulation of brain proinflammatory cytokine production attenuates neurodegeneration. *J. Neurosci.* **26,** 662–670.

Ringheim, G. E., and Conant, K. (2004). Neurodegenerative disease and the neuroimmune axis (Alzheimer's and Parkinson's disease, and viral infections). *J. Neuroimmunol.* **147,** 43–49.

Russ, A. P., and Lampel, S. (2005). The druggable genome: An update. *Drug Discov. Today* **10,** 1607–1610.

Saklatvala, J. (2004). The p38 MAP kinase pathway as a therapeutic target in inflammatory disease. *Curr. Opin. Pharmacol.* **4,** 372–377.

Sams-Dodd, F. (2006). Drug discovery: Selecting the optimal approach. *Drug Discov. Today* **11,** 465–472.

Sams-Dodd, F. (2007). Research & market stategy: How choice of drug discovery approach can affect market position. *Drug Discov. Today* **12,** 314–318.

Savage, M. J., Lin, Y. G., Ciallella, J. R., Flood, D. G., and Scott, R. W. (2002). Activation of c-Jun N-terminal kinase and p38 in an Alzheimer's disease model is associated with amyloid deposition. *J. Neurosci.* **22,** 3376–3385.

Schieven, G. L. (2005). The biology of p38 kinase: A central role in inflammation. *Curr. Top. Med. Chem.* **5,** 921–928.

Streit, W. J., Mrak, R. E., and Griffin, W. S. (2004). Microglia and neuroinflammation: A pathological perspective. *J. Neuroinflammation* **1,** 14–17.

Sun, A., Liu, M., Nguyen, X. V., and Bing, G. (2003). p38 MAP kinase is activated at early stages in Alzheimer's disease brain. *Exp. Neurol.* **183,** 394–405.

Tamayo, N., Liao, L., Goldberg, M., Powers, D., Tudor, Y. Y., Yu, V., Wong, L. M., Henkle, B., Middleton, S., Syed, R., Harvey, T., Jang, G., *et al.* (2005). Design and synthesis of potent pyridazine inhibitors of p38 MAP kinase. *Bioorg. Med. Chem. Lett.* **15,** 2409–2413.

Tobinick, E., Gross, H., Weinberger, A., and Cohen, H. (2006). TNF-alpha modulation for treatment of Alzheimer's disease: A 6-month pilot study. *Med. Gen. Med.* **8,** 25–36.

Tuppo, E. E., and Arias, H. R. (2005). The role of inflammation in Alzheimer's disease. *Int. J. Biochem. Cell Biol.* **37,** 289–305.

Veber, D. F., Johnson, S. R., Cheng, H. Y., Smith, B. R., Ward, K. W., and Kopple, K. D. (2002). Molecular properties that influence the oral bioavailability of drug candidates. *J. Med. Chem.* **45,** 2615–2623.

Vieth, M., Siegel, M. G., Higgs, R. E., Watson, I. A., Robertson, D. H., Savin, K. A., Durst, G. L., and Hipskind, P. A. (2004). Characteristic physical properties and structural fragments of marketed oral drugs. *J. Med. Chem.* **47,** 224–232.

Wing, L. K., Behanna, H. A., Van Eldik, L. J., Watterson, D. M., and Ralay Ranaivo, H. (2006). *De novo* and molecular target-independent discovery of orally bioavailable lead compounds for neurological disorders. *Curr. Alzheimer Res.* **3,** 205–214.

Wrobleski, S. T., and Doweyko, A. M. (2005). Structural comparison of p38 inhibitor–protein complexes: A review of recent p38 inhibitors having unique binding interactions. *Curr. Top. Med. Chem.* **5,** 1001–1016.

Wu, C. Y., and Benet, L. Z. (2005). Predicting drug disposition via application of BCS: Transport/absorption/elimination interplay and development of a biopharmaceutics drug disposition classification system. *Pharm. Res.* **22,** 11–23.

Wyss-Coray, T. (2006). Inflammation in Alzheimer disease: Driving force, bystander or beneficial response? *Nat. Med.* **12,** 1005–1015.

Zhu, X., Rottkamp, C. A., Boux, H., Takeda, A., Perry, G., and Smith, M. A. (2000). Activation of p38 kinase links tau phosphorylation, oxidative stress, and cell cycle-related events in Alzheimer disease. *J. Neuropathol. Exp. Neurol.* **59,** 880–888.

Zhu, X., Rottkamp, C. A., Hartzler, A., Sun, Z., Takeda, A., Boux, H., Shimohama, S., Perry, G., and Smith, M. A. (2001). Activation of MKK6, an upstream activator of p38, in Alzheimer's disease. *J. Neurochem.* **79,** 311–318.

Zhu, X., Mei, M., Lee, H. G., Wang, Y., Han, J., Perry, G., and Smith, M. A. (2005). P38 activation mediates amyloid-beta cytotoxicity. *Neurochem. Res.* **30,** 791–796.

OXIDATIVE STRESS AND THE PATHOGENESIS OF NEURODEGENERATIVE DISORDERS

Ashley Reynolds, Chad Laurie, R. Lee Mosley, and Howard E. Gendelman

Department of Pharmacology and Experimental Neuroscience
Center for Neurovirology and Neurodegenerative Disorders, University of Nebraska Medical Center, Omaha, Nebraska 68198, USA

Microglia-derived inflammatory neurotoxins play a principal role in the pathogenesis of neurodegenerative disorders including Alzheimer's disease, Parkinson's disease, amyotrophic lateral sclerosis, and HIV-associated dementia; chief among these is reactive oxygen species. The detrimental effects of oxidative stress in the brain and nervous system are primarily a result of the diminished capacity of the central nervous system to prevent ongoing oxidative damage. A spectrum of environmental cues, mitochondrial dysfunction, accumulation of aberrant misfolded proteins, inflammation, and defects in protein clearance are known to evolve and form as a result of disease progression. These factors likely affect glial function serving to accelerate the tempo of disease. Understanding the relationships between disease progression, free radical formation, neuroinflammation, and neurotoxicity is critical to elucidating disease mechanisms and the development of therapeutic modalities to combat disease processes. In an era where populations continue to age, the prevalence and incidence of age-related neurodegenerative diseases are on the rise; therefore, the need for novel therapeutic strategies that attenuate neuroinflammation and protect neurons against oxidative stress is ever more immediate.

INTERNATIONAL REVIEW OF
NEUROBIOLOGY, VOL. 82
DOI: 10.1016/S0074-7742(07)82016-2

0074-7742/07 $35.00

I. Introduction: Free Radicals, Immunity, and the Nervous System

Cells within the brain, notably neurons, are highly vulnerable to the detrimental effects of reactive oxygen species (ROS). This occurs because of their high metabolic rate, rich composition of fatty acids prone to peroxidation, high intracellular concentrations of transition metals capable of catalyzing the formation of reactive hydroxyl radicals, low levels of antioxidants, and reduced capability to regenerate. Glial cells (microglia and astrocytes) are the key support cells of the nervous system. However, during neuroinflammatory processes, they produce free radicals early and often, and serve as key pathogenic elements in neurodegenerative diseases [Alzheimer's and Parkinson's disease (AD and PD), amyotrophic lateral sclerosis (ALS), and HIV-1-associated dementia (HAD)]. Free radicals have the capacity to attack proteins, polysaccharides, lipid bilayers, and DNA, causing cellular oxidative damage. Nucleic acid oxidation occurs in neurons during disease and is detected as elevated levels of 8-hydroxyl-2-dexoyguanosine in DNA and 8-hydroxyguanosine in RNA. Hydroxyl radical-mediated DNA damage often results in strand breaks, DNA-protein cross-linking, and base modifications (Nunomura *et al.*, 1999). All of these events can lead to neuronal injury. Oxygen free radicals are also found in brain regions affected by neurodegenerative diseases. These include the hippocampus in AD patients, the substantia nigra and caudate putamen in PD patients, and spinal fluids in ALS and HAD patients. Therapeutic strategies that inhibit free radical formation and prevent downstream formation of hydroxyl radicals can slow disease and remain an active area of investigation.

The innate and adaptive immune responses modulate much of the free radical formation within the central nervous system (CNS) in both health and disease (Carlson *et al.*, 2002). As to the former, innate immunity is defined as a nonspecific defense response that is activated after an antigen (microbe, aberrant protein) emerges. Although nonspecific mechanisms also include physical barriers in the CNS, innate immunity is best characterized by the production and release of a plethora of secretory products by antigen-exposed immunocytes that are affected or regulated by redox pathways. The principal immune cells in the nervous system that react in this manner include microglia and astrocytes. Microglia are highly mobile cells with numerous roles in protecting the nervous system (acting as scavengers and chemical secretors, and to present antigen to induce an immune response). They also affect the pathobiology of neurodegenerative disorders by eliciting inflammatory reactions as a consequence of infection or disease. Activated microglia participate in inflammatory processes linked to neurodegeneration by producing neurotoxic factors including quinolinic acid, superoxide anions, matrix metalloproteinases, nitric oxide (NO), arachidonic acid and its metabolites, chemokines, proinflammatory cytokines, and excitotoxins including glutamate. Astrocytes are a major brain cell that provide physical support to neurons as well as assisting the microglia during cleanup of debris. Astrocytes also provide direct support to neurons by secreting growth factors required for proper function of the nervous system and control of fluid composition surrounding neurons. Neurons also play a role in innate immune function, but to a significantly lesser degree than microglia or astrocytes. The cellular machinery of microglia, astrocytes, and neurons and their production and regulation of toxic free radicals are illustrated in Figs. 1 and 2. Briefly, nervous system cells contain the nicotinamide adenine dinucleotide phosphate (NADPH) oxidase complex, which when assembled and activated, produces free radicals in abundance that can lead to tissue damage (McGeer and McGeer, 2002a). Another source of these destructive oxidants is credited to mitochondrial metabolism, and is estimated that up to 1% of the mitochondrial electron flow leads to formation of superoxide radicals (Lass *et al.*, 1997). Albeit the production of free radicals and ROS facilitates the function of the innate immune system in host defense against invading organisms, the secondary consequence of such reactions, if not properly controlled (as occurs in neurodegenerative diseases), is tissue injury.

Adaptive immunity refers to antigen-specific immune responses, and is significantly more complex. Antigen first must be processed and recognized by antigen-presenting cells such as dendritic cells, macrophages, and microglia. Once an antigen has been recognized, the adaptive immune system expands large numbers of immune cells (commonly T cells) specifically designed to attack and destroy the foreign antigen. The role of free radicals produced and regulated by T cells in the context of neurodegenerative diseases is incompletely understood. However, for neurodegenerative disorders, the presence within the brain of major T cell subsets in ratios exceeding those typically found in the periphery suggests a more profound role in disease than merely performing a surveillance function. How these T cells are activated, whether they are antigen-specific, or migrating in response to microglial inflammation has yet to be determined.

Oxidative abnormalities and damage are a major pathogenic feature of AD, PD, ALS, and HAD (Andersen, 2004; Tabner *et al.*, 2001). Importantly, compensatory mechanisms are operative and neurons have the capacity to upregulate antioxidant defenses, which suggest a balance for oxidant damage in disease (Carri *et al.*, 2003). However, when pro-oxidants surpass the endogenous controls or antioxidants, a formula for oxidative stress arises and affects disease. Oxidative stress has been linked to protein misfolding and aggregation, and ultimately to microglial activation, thus ascribing a pathogenic trigger to the chronic inflammatory response and neurodegeneration. The processes involved in oxidative stress and protein aggregation are not mutually exclusive. The exact mechanisms by which protein aggregates formed in AD, PD, and ALS mediate neuronal cell death remain incomplete, has been linked to ROS generation, either directly or through microglial activation that ultimately affects cell demise.

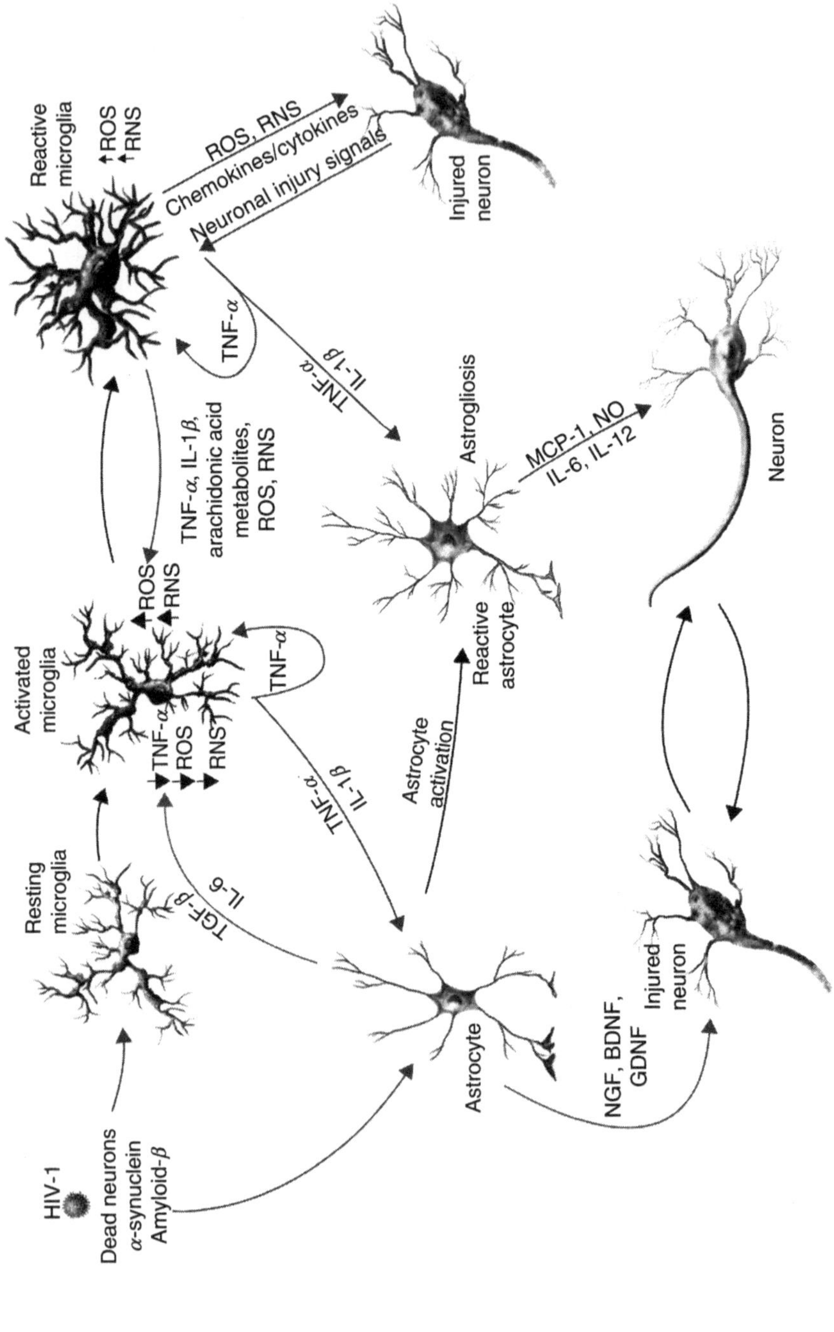
HIV-1
Dead neurons
α-synuclein
Amyloid-β
Resting microglia
Activated microglia
TNF-α
ROS
RNS
Reactive microglia
Injured neuron
ROS, RNS
Chemokines/cytokines
Neuronal injury signals
TNF-α, IL-1β, arachidonic acid metabolites, ROS, RNS
TGF-β
IL-6
IL-1β
Astrocyte activation
Astrocyte
Reactive astrocyte
Astrogliosis
MCP-1, NO
IL-6, IL-12
Neuron
NGF, BDNF, GDNF

A novel mechanism that may initiate the production of ROS is the effect of hydrogen peroxide on protein aggregation, which in the presence of transition metals, such as iron, transforms into hydroxyl radicals (Fig. 2) and initiates oxidative damage before antioxidant defenses can intervene. Genetic mutations found within AD, PD, and ALS disorders lead to irregular processing of misfolded proteins, promoting the deposition of protein aggregates. This occurs in the cytosol, extracellular spaces, and/or nucleus, leading to CNS amyloidosis. In this scenario, soluble proteins are altered into insoluble, filamentous polymers that begin to develop cross-linked β-pleated sheet structures (Sipe and Cohen, 2000). β-Sheets accumulate, form fibrillar amyloid deposits, and aid in the establishment of neurodegeneration (Forman *et al.*, 2004; Zhang *et al.*, 2005). Amyloid-β (Aβ), α-synuclein, and superoxide dismutase 1 (SOD1) are the primary proteins found within the protein deposits in tissues of patients with AD, PD, and ALS, respectively. Senile plaques along with neuronal and glial inclusions are typical of protein aggregates found in histological preparations from AD patients, while Lewy bodies and hyaline inclusions represent protein aggregates typically found in PD and ALS (Tabner *et al.*, 2005). Protein aggregates may also bind and attach to cellular components, or be phagocytosed by resident microglia and incite ROS production leading to oxidative damage (Thomas *et al.*, 2007; Zhang *et al.*, 2005). Microglial activation as a result of interactions with aggregated proteins encompasses a common mechanism by which aggregated proteins can facilitate ROS production and perpetuate the neurodegenerative process.

II. Neuropathogenesis of Neurodegeneration

AD is the most common neurodegenerative disease. Sporadic AD is extremely rare in individuals younger than 60 years of age (representing less than 5% of all reported cases), but the incidence of AD dramatically increases to 40% after the age of 85 (Forman *et al.*, 2004). Neuronal cell loss in the hippocampus and amygdala, the brain subregions responsible for learning and memory, underlies

---

FIG. 1. Neuroinflammatory interactions between microglia, astrocytes, and neurons during neurodegenerative diseases. Microglia and astrocytes communicate through numerous cell-signaling pathways, resulting in reactive cell phenotypes and hastening the neuroinflammatory process. Activation of resting microglia results in amplification of secretions and the production of ROS, RNS, chemokines, and proinflammatory cytokines initiating a generalized inflammatory response and ultimately leading to neuronal injury or death. Astrocyte activation (astrogliosis) develops from reactive microglia that alter the astrocytic phenotype from neurotrophic to neurotoxic. (Green arrow = protective response, red arrow = destructive response, and black arrow = spontaneous.) (See Color Insert.)

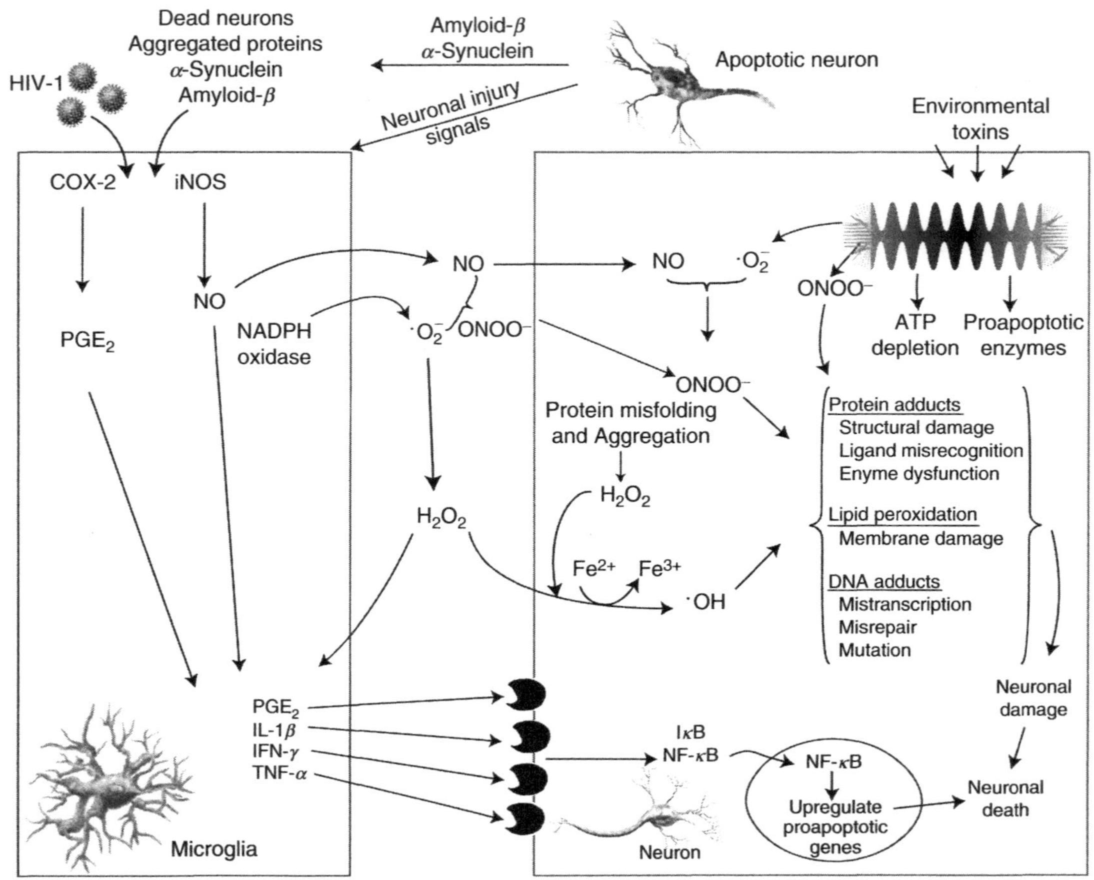
HIV-1
Dead neurons
Aggregated proteins
α-Synuclein
Amyloid-β
Amyloid-β
α-Synuclein
Neuronal injury signals
Apoptotic neuron
Environmental toxins
COX-2
iNOS
NO
PGE$_2$
NADPH oxidase
NO
$\cdot O_2^-$
ONOO⁻
H_2O_2
NO
$\cdot O_2^-$
ONOO⁻
ONOO⁻
ATP depletion
Proapoptotic enzymes
Protein misfolding and Aggregation
H_2O_2
Fe^{2+}
Fe^{3+}
$\cdot OH$
Protein adducts
Structural damage
Ligand misrecognition
Enyme dysfunction
Lipid peroxidation
Membrane damage
DNA adducts
Mistranscription
Misrepair
Mutation
Neuronal damage
Neuronal death
PGE$_2$
IL-1β
IFN-γ
TNF-α
Microglia
IκB
NF-κB
NF-κB
Upregulate proapoptotic genes
Neuron

AD pathobiology (Markesbery and Mira, 1996). Two types of lesions characterize AD neuropathology, senile plaques and neurofibrillary tangles (NFT) composed of Aβ. Aβ is a cleavage product of the amyloid precursor protein (APP), which is phosphorylated tau functioning as a microtubule stabilizer. When tau is hyperphosphorylated, it enhances the formation of fibrillar aggregates that leads to NFT. Although plaques and fibrils appear to be the most prevailing features of AD pathology, they alone are not sufficient to generate the significant and profound neuronal loss that is characteristic of disease. Neuronal damage caused by neurotoxic factors initiated from inflammatory responses by immune-activated glial cells appears to be the best link to cognitive deterioration. Aβ plaques prime microglia, and an inflammatory cascade is created and supported by secondary factors including proinflammatory cytokines, chemokines, or by T cells trafficking in and out of the nervous system (reviewed in Carlson *et al.*, 2002). These factors can also contribute to the breakdown of the blood–brain barrier (BBB), allowing leukocytes entry into the brain and propagating the CNS inflammatory cascade. As a consequence of BBB breakdown and local immune activation, the brain's resident macrophages or microglia release a plethora of neurotoxins such as proteases, glutamate, arachidonic acid and its metabolites, including platelet-activating factor and ROS; of which all affect synaptic transmission and can lead to neuronal damage. Epidemiological studies for AD have shown that nonsteroidal anti-inflammatory drugs (NSAIDs), which are known to reduce microglia responses, also reduce the risk of AD (McGeer and McGeer, 2002b).

PD is the second most common age-related neurodegenerative disease and is grouped among motor system disorders. PD is predominantly defined by the loss of dopaminergic neurons (dopamine-producing neurons) of the substantia nigra pars compacta (SNpc) and their axons projecting to the caudate-putamen or striatum. The etiology of PD remains unknown; however, patterns of familial inheritance and several animal models suggest a possible connection involving abnormal protein processing and accumulation. Indeed, a pathological feature of PD is the formation of inclusion bodies or Lewy bodies that are primarily composed of ubiquitin and α-synuclein aggregates. The majority of PD cases are sporadic and may be due, in part, to mitochondrial defects at complex I (Dauer and Przedborski, 2003). Complex I inhibitors, such as 1-methyl-4-phenylpyridine and rotenone, can recapitulate pathological features of PD in humans and animal

---

FIG. 2. Oxidative stress and neuronal damage. ROS can arise in several distinct ways such as glial cell activation, mitochondrial dysfunction, and protein aggregation. When ROS tips the balance outweighing antioxidants, oxidative stress is generated and neuronal cell injury or death ensues. Oxidative stress destroys several organic structures of the cell, including proteins, lipids, and DNA, causing irreversible and detrimental damage. This figure was adapted from F. Gao and colleagues (Du *et al.*, 2001). (See Color Insert.)

models of disease (Dauer and Przedborski, 2003; Dawson and Dawson, 2003). Reactive microglial responses, thought to play a central role in dopaminergic neuronal death, predominate the site of dopaminergic neuronal injury, and may be amplified through paracrine and autocrine processes (Dawson and Dawson, 2003). In PD, a clear connection exists between inflammation and neurodegeneration, even more so than aberrant protein processing and accumulation. There is an increase in activated microglia in the SNpc and striatum of patients with idiopathic PD (Kohutnicka *et al.*, 1998; Kurkowska-Jastrzebska *et al.*, 1999; Wu *et al.*, 2002). Indeed, attenuation of microglial activation in PD models can protect greater than 90% of the dopaminergic neurons otherwise destined to die (Choi *et al.*, 2005; Du *et al.*, 2001; Kurkowska-Jastrzebska *et al.*, 1999; Teismann and Ferger, 2001; Teismann *et al.*, 2003; Vijitruth *et al.*, 2006; Wu *et al.*, 2002). Furthermore, epidemiological data has shown that the use of NSAIDs decreases the risk for the PD as it does for AD (Chen *et al.*, 2003, 2005).

HAD is another neurodegenerative disorder that is secondary to HIV-1 infection. A common disturbance associated with HIV infection is the development of neurological disorders. Approximately 60% of HIV-1-infected patients display some form of neurological dysfunction, most likely due to early entry of HIV-1 into the CNS. The virus infects the brain through CD4+ T lymphocytes (Haase, 1999) and mononuclear phagocytes (MPs: dendritic cells, monocytes, and macrophages) (Tardieu and Boutet, 2002). The brain lacks sufficient viral control methods and adaptive immunity; therefore, active viral replication is virtually unrestricted and the most detrimental form of HIV-1-associated tissue damage can occur within the CNS. Antiretroviral therapy (ART) was introduced to help extend the amount of time a patient might have before the development of CNS disease and its associated immune suppression takeover (d'Arminio Monforte *et al.*, 2000; Yong *et al.*, 2001). Patients that responded poorly to therapy, or those that did not receive ART, display signs of damage to their immune systems and a more rapid progression of the disease (Ho *et al.*, 1995; Krishnakumar, 2005; Wei *et al.*, 1995). Even though HAD can be observed in up to 40% of infected individuals, ART has considerably reduced both disease incidence and its destruction. HIV-1-associated cognitive impairment is composed of a wide spectrum of conditions from the mild HIV-1 motor cognitive-motor disorder to the severe and debilitating HAD. The progression of HAD is extremely variable and may depend on several issues including genetic factors of the infected host, systemic and brain HIV burden, BBB integrity, the speed of CD4+ T-lymphocyte decline, and the genotype of the viral strains that gain access to the brain or induce neurovirulent activities (Anderson *et al.*, 2002). The pathology of HAD revolves around the formation of multinucleated giant cells of the MP lineage, which are also the main reservoirs for the virus in the brain. MPs can secrete neurotoxins and induce neuronal injury leading to neurocognitive impairments and ultimately to HAD (Aquaro *et al.*, 2000; Elbim *et al.*, 2001; Kaul *et al.*, 2001; Luo *et al.*, 2003; Xiong *et al.*, 2000). Under normal conditions, MPs secrete neurotrophins and eradicate foreign material to maintain CNS homeostasis (Gras *et al.*, 2003). However, during HAD, MP-mediated control over the neurotrophic factors becomes altered and infected MPs aberrantly affect the development of a metabolic encephalopathy. In addition, infected MPs secrete proinflammatory cytokines, chemokines, eicosanoids, excitatory amino acids, TNF-related apoptosis inducing ligand, viral proteins, reactive nitrogen species (RNS) and ROS (Ensoli *et al.*, 2000; Floyd *et al.*, 1999; Gendelman *et al.*, 1998; Raber *et al.*, 1998). The actual severity of the dementia is strongly associated with the numbers of activated macrophages and microglia rather than the actual viral load in the CNS (Adle-Biasette, 1999). Therefore, neuronal damage observed during HAD is linked more to macrophage activation than to HIV-1 itself.

ALS, also known as Lou Gehrig's disease, is a devastating motorneuron disorder. Muscle weakness is the hallmark sign of ALS, occurring in approximately 60% of patients. Unlike the previous diseases, progression of ALS is rapid. Individuals with ALS lose function of motorneurons within 3–5 years (Weydt and Moller, 2005) terminating with complete neuromuscular failure and death (typically caused by compromised respiratory function). ALS affects about 5–7 in 100,000 adults throughout the world (Rowland and Shneider, 2001). The majority of ALS cases are sporadic (90–95%) (Cleveland and Rothstein, 2001), while the remaining cases are attributable to familial etilogy (Cleveland and Rothstein, 2001; Valentine, 2002). The pathology of ALS is characterized by neuronal degeneration and atrophy confined almost entirely to the upper and lower motorneurons (Weydt and Moller, 2005). The hypothesized mechanisms that lead to neurodegeneration, such as glutamate toxicity, exogenous factors, neurofilament accumulation, neuroinflammation, and oxidative stress (Bruijn *et al.*, 2004; Rowland and Shneider, 2001; Strong, 2003), could be independent factors or could cooperate to cause motorneuron loss. A key discovery was the identification of missense mutations in the gene on chromosome 21 encoding for a Cu/Zn-binding protein called SOD1 (Weydt *et al.*, 2002). SOD1 is a key antioxidant in the front line of defense against oxidative stress; it detoxifies oxidative agents by converting superoxide to hydrogen peroxide and dioxygen, thus decreasing levels of superoxide within its proximity. This enzyme is located predominately in the nucleus, cytosol, and mitochondrial intermembrane space (Lyons *et al.*, 1999). Presently, over 100 SOD1 mutations have been identified in familial ALS patients (Andersen, 2001; Gaudette *et al.*, 2000; Guegan and Przedborski, 2003). The implication of SOD1 mutations is still under intense investigation; however, it is now hypothesized that the ALS phenotype is caused by a gain of a novel, unknown toxic property of the SOD1 mutant enzyme rather than by diminished SOD1 activity (Valentine, 2002).

III. Free Radicals and Neurodegenerative Disorders

Inflammatory responses induced by reactive microglia, macrophages, and proinflammatory T cells provide a primary source of free radicals, ROS (O_2^-, H_2O_2, -OH, HOCl, ferryl, peroxyl, and alkoxyl) and RNS [NO, peroxynitrite ($ONOO^-$), and peroxynitrous acid (ONOOH)] with the capacity to modify proteins, lipids, and nucleic acids. Copious amounts of ROS production, known as respiratory burst, have detrimental effects on delicate neuronal networks in the CNS. ROS include superoxide, hydrogen peroxide, and hydroxyl free radicals as well as nitrogen intermediates (NO and peroxynitrite) and can cause damage to neurons if produced in excess as occurs during prolonged neuroinflammatory processes. Microglial-derived ROS such as superoxide cannot efficiently traverse cellular membranes, and therefore unlikely to gain access to neurons and trigger intraneuronal toxic events (Beckman and Crow, 1993). However, superoxide can rapidly react with NO in the extracellular space to form peroxynitrite (Beckman and Koppenol, 1996), which can readily cross cell membranes and damage intracellular components. NO is produced in many tissues including neurons, astrocytes, and microglia, and can be secreted by glial cells on activation with proinflammatory cytokines including, but not limited to IL-1β and TNF-α (Mollace and Nistico, 1995). NO alone has a relatively low neurotoxic capacity; however, in conjunction with superoxide, it additively contributes to neuronal destruction. Nitrated species have been associated with the disruption of mitochondrial electron transport chain, lipid peroxidation, DNA damage, and protein nitration (Beckman, 1996). Therefore, superoxide production by microglia, by contributing to peroxynitrite formation, provides a significant contribution to the pathogenesis of neurodegenerative disorders.

The notion that oxidative stress is involved in AD stems from the free radical hypothesis of aging, which simply states that, as one grows older, accumulation of more ROS, coupled with diminished antioxidant capacity, results in more destruction to major cellular components (Beal, 1995; Pratico and Delanty, 2000; Reiter, 1995). ROS in AD is caused by several factors, including active microglia, redox-active metals, advanced glycation, and Aβ peptide. Near neuritic plaques and throughout the entire brain, activated microglia are heavily populated, releasing high levels of ROS. Both the NFT and Aβ deposits contain the redox-active transition metal iron (Smith *et al.*, 1997). Within affected neurons, iron is found in the cytoplasm and within lipofuscin granules. Additionally, alterations of iron homeostasis occur in AD and are supported by finding of elevated serum levels of the iron-binding protein p97 in diseased patients. This coupled with the presence of redox-available iron in AD brain regions and activation of the receptor for advanced glycation end products leading to oxygen radicals and cell injury supports the importance of oxidative stress in AD pathogenesis.

Free radicals that are considered to cause neuronal loss in AD are believed to be produced as a result of the deposition of aggregated Aβ peptide. However, whether Aβ functions as a producer of ROS or a modulator of redox reactions is not clear. Accumulated evidence of surrogate markers for oxidant injury from postmortem tissues indicates a strong association between oxidative stress and the pathology of AD. These markers show increased lipid peroxidation, protein oxidation, 8-OHdG levels, and a marked decline in oxidative-sensitive enzymes (Castellani *et al.*, 2001; Markesbery and Carney, 1999; Nunomura *et al.*, 1999, 2000, 2001; Zhu *et al.*, 2004).

Oxidative stress contributes to the cascade leading to dopaminergic neurodegeneration in PD (Tabner *et al.*, 2001). Brain regions that are rich in catecholamines, such as, adrenaline, noradrenaline, and dopamine, are exceptionally vulnerable to free radical generation. Catecholamines can spontaneously break down to free radicals, or be metabolized to free radicals by endogenous enzymes such as monoamine oxidases. Activated microglia also contribute to the degeneration of dopaminergic neurons by releasing neurotoxic factors such as NADPH oxidase-derived superoxide and cytokines. Peroxynitrite production contributes to mitochondrial dysfunction and nitration of tyrosine residues in cellular proteins and enzymes such as α-synuclein. Indeed, soluble nitrated α-synuclein is able to activate microglia to produce copious amounts of ROS through modulation of specific ion channels (Thomas *et al.*, 2007). Nitration of α-synuclein can significantly enhance fibril formation *in vitro*, similar to the biophysical properties of α-synuclein isolated form PD brains (Norris *et al.*, 2003). Aberrant protein conformations of modified α-synuclein can also potentially overload the cellular proteasome, and by doing so, may increase cellular stress associated with the accumulation of misfolded proteins in affected neurons (Vila and Przedborski, 2004). Hydrogen peroxide alone or with help from downstream ROS products can also facilitate toxic events in dopaminergic neurons by either intensifying other cytotoxic factors or by elevating the generation of neurotoxic factors in microglia (Andersen, 2004).

Oxidative stress plays a critical role in the neuropathogenesis of HIV-1. As stated earlier, the process of HIV infection leads to the generation of inflammatory products, thus in turn giving rise to an excessive amount of ROS. In HAD, production of excess-free radicals accompanied by dysregulation of antioxidants (e.g., SOD, glutathione, and catalase) diminishes the protective potential to establish a pro-oxidative stress environment (Mollace *et al.*, 2001). Superoxide anions, peroxynitrite, and NO are produced in this scenario. Peroxynitrite is a potent oxidant that can nitrate tyrosine residues of structural proteins. Neurofilament, a structural protein that provides stability to neurons, is one of the target proteins for peroxynitrite (Beckman, 1996). Besides affecting CNS injury, free radicals produced during HAD including ceramide, sphingomyelin, and 4-hydroxynonenal can also adversely affect the disease course including the permeability of the BBB (Mollace *et al.*, 2001), which may increase the tempo of HIV-associated nervous system dysfunction and lead to progressive cognitive deficits and death.

A primary role for oxidative stress in ALS is evident. The discovery of mutations in the *SOD1* gene encoding the Cu/Zn SOD in familial ALS patients was responsible for associating this disease to ROS metabolism (Deng *et al.*, 1993; Olanow, 1993; Rosen *et al.*, 1993). Mutant forms of SOD proteins have altered enzyme activity and can result in increased superoxide radicals such that the formation of peroxynitrite from superoxide and NO would be favored. A decline in the activity of SOD1 measured in patient's tissues (Deng *et al.*, 1993) suggests that higher levels of superoxide activity may form and increase the potential of developing hydroxyl radicals and eventually neuronal damage (de Belleroche *et al.*, 1996). Interestingly, knockout SOD1 mice show limited motor symptoms by 6 months of age (Shefner *et al.*, 1999). These mice do not develop an ALS phenotype, whereas mutants of the Cu/Zn SOD1 proteins are inevitably lethal. The human mutant (G93A) *SOD1* gene, introduced to a murine model, resulted in an ALS phenotype characterized by elevated levels of lipid peroxidation, protein oxidation, and DNA oxidation. This human G93A SOD1 mutation confers a toxic gain of function that destroys motorneurons (Cudkowicz *et al.*, 1997). This toxic gain of function was shown to come from the ability of familial ALS-mutant Cu/Zn SOD1 proteins to catalyze oxidation reactions of hydrogen peroxide or peroxynitrite to hydroxyl radicals (Valentine, 2002). These results support the notion that expressed mutations in the *SOD1* gene lead to ALS pathogenesis by increasing the amount of hydroxyl radicals and generating vast amounts of oxidative stress capable of inducing motorneuron degeneration (Barzilai *et al.*, 2002).

IV. Glutathione System, Glutamate–Glutamine Cycle, and the CNS

Production of ROS and NO in neurons is buffered primarily by the glutathione system. Glutathione is the major thiol present in brain tissue, and the most important redox buffer in cells. This antioxidant molecule cycles between reduced glutathione (GSH) and oxidized glutathione disulfide (GSSG), and serves as a vital sink for control of ROS levels in cells. GSH reacts with oxygen- and nitrogen-free radicals resulting in the reduction of peroxides (Dringen *et al.*, 2000). Although varying in different regions of the brain, all GSH levels diminish by about 30% in the elderly (Chen *et al.*, 1989), suggesting a possible link with the age-associated risk factor of AD and PD. Depletion of GSH may render cells more sensitive to toxic effects of oxidative stress and potentiate the toxic effects of reactive microglia (Dringen *et al.*, 2000; Sian *et al.*, 1994; Winterbourn and Metodiewa, 1994).

A common mechanism for neuronal demise may underlie most neurodegenerative diseases. The excitotoxicity metabolic pathway that modulates neuronal injury has remained a focus of neuroscience research efforts for the past three decades. In excitotoxicity, glutamate release overstimulates neuronal ionotropic glutamate receptors resulting in significant calcium influx and neuronal cell death stroke and neurodegenerative diseases. Indeed in this pathway, the major excitatory amino acid neurotransmitter is glutamate as well as aspartate, while γ-aminobutyric acid, glycine, and taurine are inhibitory. Glutamate is, without question, the principal excitatory neurotransmitter and is the key factor implicated in affecting neuronal death. Increased extracellular glutamate also leads to oxidative stress and oxidative glutamate toxicity. During steady state, neuronal glutamate acts as a neurotransmitter. Its extracellular concentration is kept to very low levels through transporters that are present in glia and primarily in astrocytes. Moreover, extracellular concentrations of glutamate are modulated through the glutamate–cystine antiporter that serves to regulate excitotoxicity and neuronal injury. Increased levels of extracellular glutamate deplete cells of cystine by blocking the gradient-driven glutamate–cystine antiporter, a dimeric transport system composed of a specific subunit, xCT, and a 4F2 heavy chain. Cystine is required for the synthesis of GSH, thus excess depletion renders cells incapable of removing ROS. Due to the involvement of increased oxidative stress in the pathogenesis of PD, it is thought that diminished levels of GSH in the substantia nigra precede neuronal degeneration. Indeed, when GSH levels are diminished, the uptake of cystine is induced leading to increased glutamate efflux and neurotoxic outcomes (Jiang *et al.*, 2001; Simantov, 1989) (Fig. 3).

Both *in vivo* and *in vitro* studies have demonstrated that activated microglia express the transporters and enzymes of the glutamate–glutamine cycle (reviewed in Gras *et al.*, 2006), suggesting that these cells may exhibit neuroprotective properties in addition to their neurotoxic properties and can partially compensate for the deleterious reactive state within the context of neurodegeneration. This is underscored by the observation that, unlike astrocytes, microglia do not constitutively express glutamate transporters, but are only expressed on activation. Therefore, the microglia constitute a local, inducible cell population that can mediate glutamate clearance and metabolism. This property may be critical to neuronal survival and disease outcome in CNS inflammation.

V. Modulators of Microglial Activation

Critical to understanding the role of oxidative stress and inflammation in neurodegeneration is an understanding of the disease process in and of itself. A common pathological feature to most neurodegenerative disorders is the

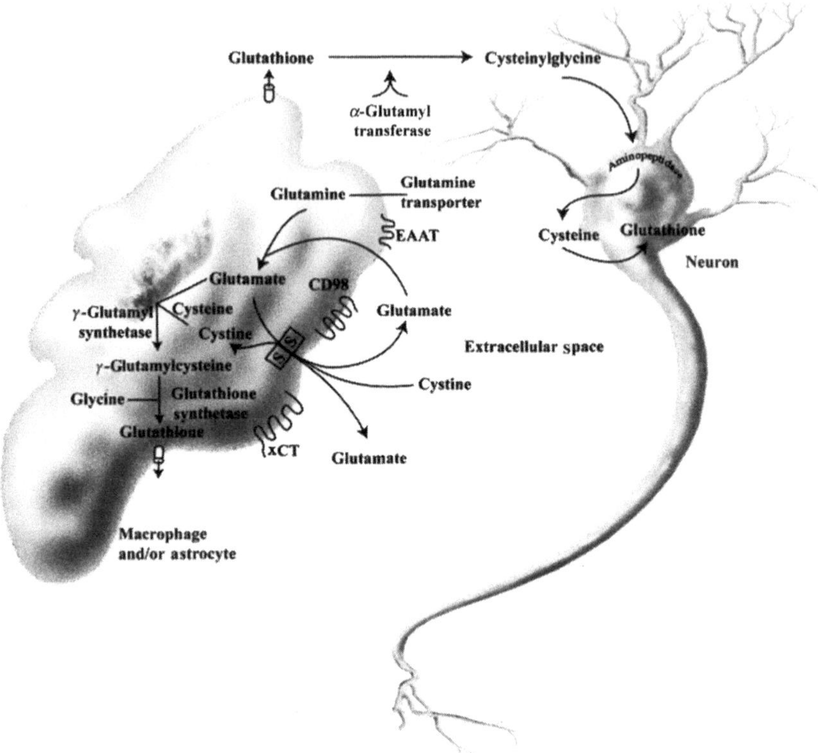

FIG. 3. Glutamate–glutamine transporters for glutamate, cysteine, and cystine in microglia (or astrocytes) and neurons. Communication links for redox pathways in the nervous system are outlined. A family of excitatory amino acid transporters (EAAT) and other transporter proteins (receptors) regulate extracellular concentrations of glutamate. When intracellular glutamate is higher than the plasma concentration, transport of glutamate across the luminal membrane occurs. The transporters on the ablumenal membrane surface of astrocytes or microglia provide a mechanism to increase intracellular glutamate concentration and removal of glutamate from cells. In microglia (or macrophages), cystine is taken up through the CD98/xCT cystine–glutamate antiporter. The cystine–glutamate antiporter then exchanges extracellular cystine for intracellular glutamate. (See Color Insert.)

activated microglial cell. However, mechanisms underlying microglia activation to incite a neuroinflammatory cascade in disease are incompletely understood.

The pathogenic process in AD involves deposition of insoluble aggregates of Aβ, oxidative stress, and activation of inflammatory cytokine cascades involving microglia. Activated microglia congregate in and around amyloid plaques, where they can produce cytokines, ROS, and excitotoxins that can kill or injure neurons. Oxidative stress in AD is more attributed to the free radical generation by affected neurons more so than activated microglia. This can occur as a result of impaired mitochondrial oxidative metabolism (Mattson *et al.*, 1997). Alternatively, a major source of free radicals in the AD brain is likely to include microglia, which have the potential to produce large amounts of ROS. Free radical generation by activated microglia has been demonstrated on direct interaction with Aβ peptides and activation of the NADPH oxidase complex to produce superoxide radicals (Bianca *et al.*, 1999; Klegeris *et al.*, 1997a,b; Van Muiswinkel *et al.*, 1999). Moreover, Aβ can potentiate the production of free radicals by phagocytic cells treated with other stimulatory agents such as IFN-γ, lipopolysaccharide (LPS), and TNF-α (Goodwin *et al.*, 1995; McDonald *et al.*, 1997; Meda *et al.*, 1996). In addition to oxygen-free radicals, large amounts of NO can be produced by activated microglia through induction of iNOS (Goodwin *et al.*, 1995). Activated microglia also release the excitotoxins glutamate (Piani *et al.*, 1992) and quinolinic acid, and on activation by amyloid plaques also evoke fulminant excitotoxicity activity (Giulian *et al.*, 1995). All of which can contribute to inflammatory mechanisms.

In PD, reactive microglia predominate within the SNpc of PD brains at autopsy (Croisier *et al.*, 2005; McGeer *et al.*, 1988; Yamada *et al.*, 1992). A significant increase in the number of reactive microglia shown phagocytosing dopaminergic neurons (McGeer *et al.*, 1988) correlated with the deposition of α-synuclein (Croisier *et al.*, 2005). In addition, microglia in the vicinity of dopaminergic neurons in disease appear to have an upregulated capacity for ROS production due to increased expression of NADPH oxidase. How microglia are activated in PD and affect disease are incompletely understood, but the release of aggregated and nitrated α-synuclein from dying or damaged dopaminergic neurons in the SN is thought to contribute, in part, to their activation (Thomas *et al.*, 2007; Zhang *et al.*, 2005). Several lines of evidence support this contention. First, evidence of α-synuclein linkage to familial PD is derived from the discovery of three missense mutations (A53T, A30P, and E46K) in the gene encoding α-synuclein (Kruger *et al.*, 1998; Polymeropoulos *et al.*, 1997; Spira *et al.*, 2001; Zarranz *et al.*, 2004) as well as duplication and triplication of the gene (Singleton *et al.*, 2003). Second, oxidation of α-synuclein leads to formation of aggregates and filaments found to be a major component of LB (Giasson *et al.*, 2000; Souza *et al.*, 2000). Third, α-synuclein itself can activate microglia, causing release of ROS and neurotoxicity (Thomas *et al.*, 2007; Zhang *et al.*, 2005). Fourth, oxidized and aggregated α-synuclein, when released from dying neurons, may stimulate scavenger receptors on microglia resulting in their sustained activation and subsequent dopaminergic neurodegeneration (Croisier *et al.*, 2005; Wersinger and Sidhu, 2006; Zhang *et al.*, 2005). Taken together, several lines of research provide corroborating evidence that implicate aggregation and oxidative modification of α-synuclein as key components that facilitate much of the neuroinflammation and degeneration in sporadic and some familial forms of PD.

Activated MPs are the primary perpetrators of neuronal injury in HIV-1-associated CNS disease. It is widely accepted that these MPs act to induce neuronal injury primarily through indirect mechanisms. These indirect mechanisms are alterations in secretory function of chemokines, cytokines, arachidonic acid derivatives, and platelet-activating factor, as well as NO, free radicals, and excitatory amino acids (Kadiu *et al.*, 2005). Direct mechanisms also bring about neurotoxicity, but probably play a lesser role. These mechanisms consist of soluble viral proteins and glycoproteins that work through neuronal receptors (D'Aversa *et al.*, 2005). MP activation during HIV-1 infection may occur by several mechanisms. Proinflammatory cytokines, such as IFN-γ and TNF-α, are potent MP activators. TNF-α allows astrocytes and microglia to amplify immune activation, resulting in coactivation of other microglia. Axonal injury also results in activation of MPs in the CNS. Chemokines are yet another method of macrophage activation. Macrophage inflammatory protein (MIP)-1β and RANTES (regulated on activation, normal T cell expressed and secreted) act through CCR5 (Cocchi *et al.*, 1995) on the cell surface, and stromal cell–derived factor (SDF)-1α through CXCR4 (Oberlin *et al.*, 1996). Fractalkine, a brain chemokine expressed by neurons, astrocytes, and endothelial cells, binds CX3CR1 to mediate macrophage recruitment and activation (Tong *et al.*, 2000). T cells also activate MPs by cytokines as well as direct contact. Activated T cells will enter the blast phase and the CNS. As they migrate through the parenchyma, they secrete the cytokines that serve as a source for microglia activation (Diesing *et al.*, 2002; Lawrence and Major, 2002). As these T cells die in the brain, the debris is removed by the brain macrophages, also contributing to macrophage activation. Viral proteins such as gp120 (Brenneman *et al.*, 1988), gp41 (Adamson *et al.*, 1996), and the nonstructural proteins Tat (New *et al.*, 1997; Price *et al.*, 2005), Nef, Vpr, and Rev secreted by infected MP can (Price *et al.*, 2006) directly disrupt glial and neuronal function through alteration of calcium homeostasis (Lannuzel *et al.*, 1995), induction of ROS and RNS (Mollace *et al.*, 1993), induction of apoptosis, or enhanced secretion of proinflammatory cytokines such as TNF-α and IL-1β as well as arachidonic acid metabolites that are implicated in HIV-1 neuropathogenesis (Jana and Pahan, 2004; Nath, 2002; Song *et al.*, 2003).

The interplay between motorneurons and glia cells is important in the pathological progression of motorneuron diseases, and release of ROS and RNS or cytokines form microglia could contribute to the demise of motorneurons as seen in ALS (reviewed in Agar and Durham, 2003). The deposition of specific proteins into intracellular hyaline inclusions in motorneurons and astrocytes is a characteristic neuropathological finding in familial ALS as well as in sporadic ALS (Kato *et al.*, 2000; Shibata *et al.*, 2000). These inclusions are strongly immunopositive for both mutant and wild-type SOD1 in patients and transgenic mice expressing mutated forms of human SOD1. Interestingly, mutant SOD1 is more susceptible to oxidation-induced aggregation *in vitro* than the wild-type enzyme (Rakhit *et al.*, 2002). Furthermore, studies on mice-expressing mutated human SOD1 revealed that elevated markers for activated microglia are present at times of neuronal loss suggesting that activated microglia may contribute to the oxidative insult. In addition to exogenous activators of microglia, microglia in ALS may have intrinsic cytotoxic potential as a result of a defect in SOD1 function that becomes apparent following activation (Weydt *et al.*, 2004). Therefore, in this context, mutant SOD1 renders microglia more susceptible to other toxic agents that generate ROS (Liu *et al.*, 1998).

VI. Growth Factors, Antioxidants, and Anti-Inflammatory Drug Therapies

Many diverse mechanisms, factors, and pathways are involved in neurodegenerative disorders; thus, several different therapeutic methods have been developed to target a specific factor or a whole intricate pathway with the intent of ameliorating, preventing, or reversing neuronal cell damage (Tabner *et al.*, 2001). Inflammation and oxidative stress are both time and site specific, thus in order to treat the effects of these components on neurodegenerative disorders, knowing the temporal and spatial activities involved in those processes is critical for development of efficacious therapeutics. Because inflammation and oxidative stress are linked, attenuation of inflammatory responses could slow the neurodegenerative process. Therapeutic paths could include neurotrophic factor enhancement (brain-derived neurotrophic factor, glia cell line-derived neurotrophic factor, and nerve growth factor), upregulation of anti-inflammatory cytokines (IL-4, IL-10, and TGF-β), inhibition of enzymatic activities that encourage neurotoxicity (GSK-3β, γ-secretase), Ca^{2+} and glutamate excitotoxicity blockers that inhibit NMDA receptor function, suppression of neuronal cytotoxicity (memantine, lithium, sodium valproate), attenuation of inflammation by NSAIDs, minocycline, dextromethorphan, and sequestering misfolded proteins with antibodies (Fig. 4A). Due in part to late diagnosis of the disease, treatment is initiated well after neurodegeneration has started and its detrimental effects become symptomatic. The failure in late- or end-stage clinical trials of promising therapeutic modalities emphasizes the need for presymptomatic treatment.

Growing evidence suggests that, in disease, microglia may contribute to redox stress by producing ROS during phagocytosis of debris from dying and degenerating neurons (Green *et al.*, 2001; Qin *et al.*, 2004; Zhang *et al.*, 2005). As compensatory ionic fluxes are required to sustain ROS generation, characterization of the relative roles of plasma membrane ion currents in the generation of ROS in response to pathological stimuli is important, with the long-term goal of abrogating neuronal damage by modulating these fluxes during disease-associated microglial activation. Depending on the neuroinflammatory context,

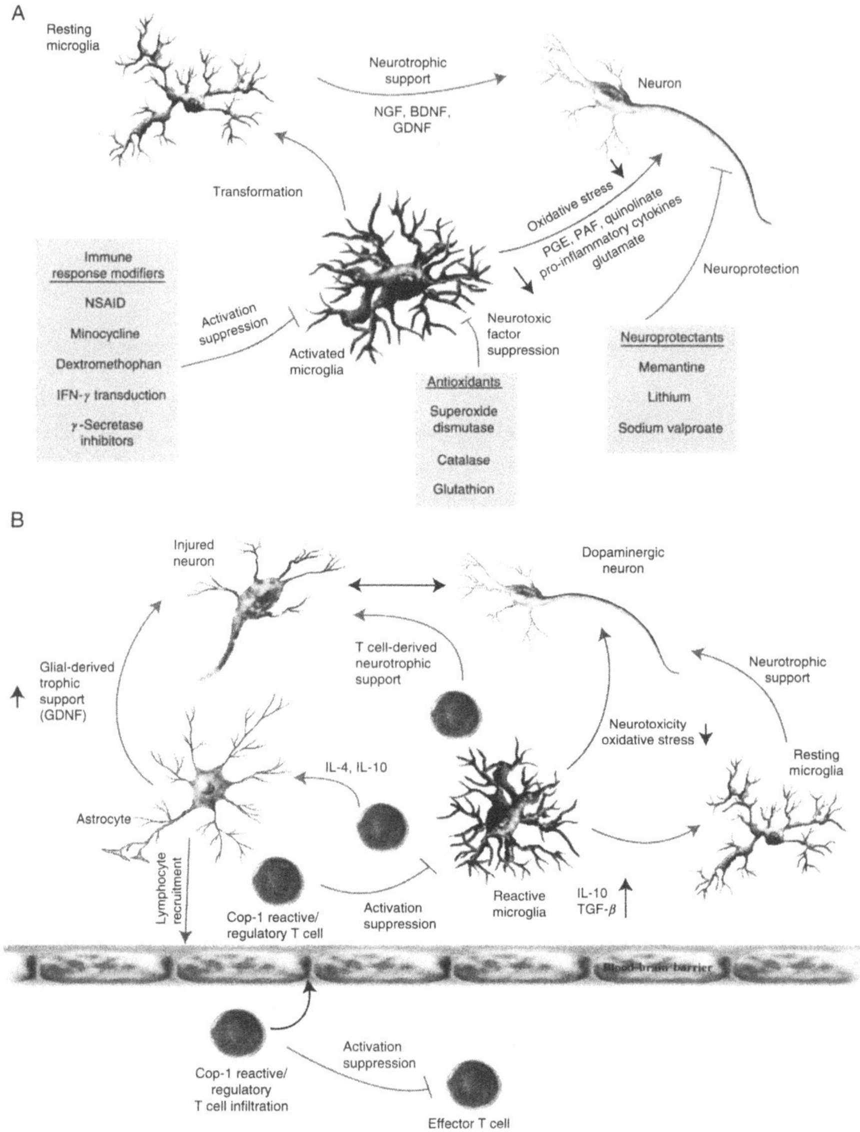

FIG. 4. Putative therapeutic anti-inflammatory and antioxidant interventions for neurodegenerative diseases. (A) Induced neuronal inflammation can be attenuated by reduction or alterations in neurodegenerative environmental cues (Aβ, α-synuclein, HIV, SOD-1) and from deactivation of activated microglia via anti-inflammatory agents and reduction in neurotoxic ROS activities through endogenous enzymes and exogenous vitamins responsible for stimulating antioxidants. (B) Immunomodulation using T cell–mediated therapies, such as immunization with Cop-1, has proven efficacious in a variety of neurological disorders. Whether Cop-1-mediated effects are through modulation of Th1/Th2 and regulatory T cell responses, or innate immunity to confer neuroprotection, the underlying etiology of disease may dictate the cellular response. (See Color Insert.)

different ionic conductances may play divergent roles that depend on unique signaling pathways. Indeed, while direct activation of microglial NADPH oxidase is mediated primarily through chloride currents, ROS production following microglial activation with aggregated α-synuclein was predominately mediated through voltage-activated K^+ currents or proton currents (Thomas *et al.*, 2007). Therefore, identification and blockade of specific ion channels may attenuate redox-related stress and may slow disease progression in a variety of neuroinflammatory conditions.

VII. Therapeutic Immunomodulation

T cell-mediated immune responses are another potential therapeutic avenue for neurodegeneration. While naive T cells are precluded from CNS entry, neuroinflammation aggressively recruits activated components of the adaptive immune system to sites of active neurodegeneration. Investigations provide evidence that a well-controlled response of activated T cells is neuroprotective (Angelov *et al.*, 2003; Benner *et al.*, 2004; Butovsky *et al.*, 2006b; Kipnis and Schwartz, 2002; Kipnis *et al.*, 2000). A generalized efficacy for immunization with Copolymer-1 (Cop-1, glatiramer acetate) in divergent models of human neurological disorders including multiple sclerosis, spinal cord injury, glaucoma, PD, AD, HAD, and ALS substantiates this observation (Arnon and Sela, 2003; Benner *et al.*, 2004; Butovsky *et al.*, 2006a; Haenggeli *et al.*, 2006; Kipnis *et al.*, 2000; Laurie *et al.*, 2007). To date, the exact mechanism of action of Cop-1 is unknown. Possible mechanisms include preferential induction of a Th2 response, competition between Cop-1 and myelin basic protein for binding sites on MHC class II molecules, or bystander suppression by a yet undefined mechanism. Cop-1 is a potent inducer of Th2 regulatory cells, which secrete IL-4, IL-5, IL-10, and TGF-β through induction of a cytokine shift (Aharoni *et al.*, 2000). Cop-1-induced Th2 adaptive immune responses can affect microglial responses and lead to neuroprotection in models of metabolic and traumatic disorders. T cells reactive to Cop-1 could also be a source of brain-derived neurotrophic factor and other neurotrophic factors (Arnon and Sela, 2003) or can induce production of neurotrophins by microglia or astrocytes (Benner *et al.*, 2004). Investigations in our laboratory and others have extended these observations in deciphering mechanisms of Cop-1-mediated neuroprotection in animal models of PD, AD, HAD, and ALS (Fig. 4B) (Benner *et al.*, 2004; Butovsky *et al.*, 2006a; Haenggeli *et al.*, 2006; Laurie *et al.*, 2007). Parallel investigations suggest that Cop-1 may induce the conversion of CD4+CD25− effector T cells to CD4+ CD25+ regulatory T cells (Hong *et al.*, 2005). Indeed, an increased regulatory T cell population in Cop-1-immunized animals demonstrates a possible role for regulatory T cells in neuroprotection.

Recent data generated in our laboratories demonstrated that regulatory T cells can suppress microglial-mediated ROS production to near prestimulatory levels in response to inflammatory activities including proinflammatory cytokines (e.g., TNF-α and LPS) (Fig. 5). These observations also support the notion that therapeutic strategies that induce regulatory T cell responses can be used to attenuate inflammation and promote neuronal survival for neurodegenerative disorders. A therapeutic vaccine approach using Cop-1 in conjunction with specific adjuvants that elicit specific regulatory T cell responses represents a potential interdictory modality for slowing the progression of neuroinflammation

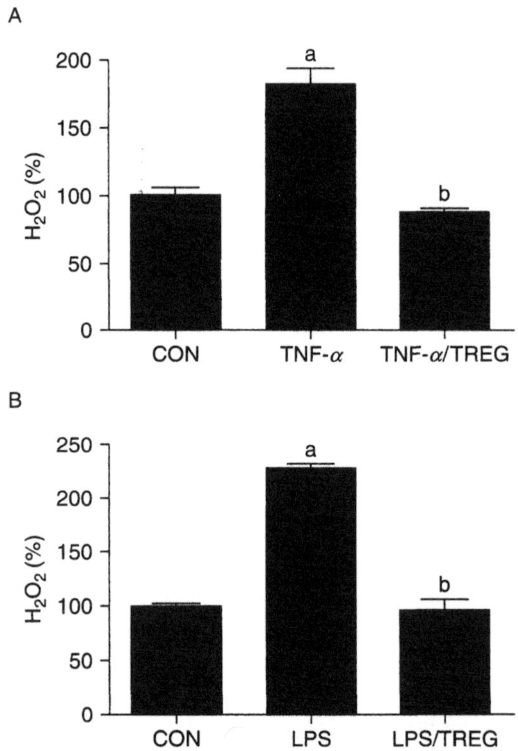

Fig. 5. CD4+CD25+ regulatory T cells attenuate microglial ROS production. ROS production by microglia in response to inflammatory stimuli was measured with Amplex Red (Molecular Probes, Carlsbad, CA). The total levels of hydrogen peroxide (H_2O_2) were determined based on the conversion of superoxide to H_2O_2 by horseradish peroxidase. Microglia produce copious amounts of ROS in response to inflammatory stimuli such as TNF-α (A) and LPS (B) ($p < 0.01$). Cocultivation of microglia with regulatory T cells with inflammatory stimuli reduced the ROS response.

and secondary neurodegeneration. This may be considered in conjunction with other anti-inflammatory or antioxidant therapies as part of a broad based therapeutic approach.

VIII. Summary

Thus, neurodegenerative disorders are strongly associated with inflammation and oxidative stress. Ironically, although essential for life, ubiquitous oxygen also is, by its products of metabolism, toxic to cells. Consequently, a strong association exists between production of free radicals, aging, and neurodegenerative diseases. Oxidative stress has been correlated with the development of cellular injuries leading to neuropathology in these various disease states. However, whether oxidative stress is the primary force driving neurodegenerative disorders is still unclear. A complete comprehension of the cellular and molecular mechanisms, and of the specificities of oxidative damage in these neurological disorders, may lead to the development of therapeutic strategies to prevent or slow the progression of disease.

Acknowledgments

The authors wish to thank Ms. Robin Taylor for excellent graphic and administrative assistance. The National Institutes of Health (NIH) grants that supported this work included P01 NS31492, R01 NS34239, P01 NS043985, and R37 NS36136 and P01 MH64570-03 (to H.E.G.) and R21 NS049264 (to R.L.M.).

References

Adamson, D. C., Wildmann, B., Sasaki, M., Glass, J. D., McArthur, J. C., Christov, V. I., Dawson, T. M., and Dawson, V. L. (1996). Immunologic NO synthase: Elevation in severe AIDS and indection by HIV-1 gp41. *Science* **274,** 1917–1921.

Adle-Biasette, H. (1999). Neuronal apoptosis does not correlate with dementia in HIV infection but is related to microglial activation and axonal damage. *Neuropathol. Appl. Neurobiol.* **25,** 123–133.

Agar, J., and Durham, H. (2003). Relevance of oxidative injury in the pathogenesis of motor neuron diseases. *Amyotroph. Lateral. Scler. Other Motor Neuron Disord.* **4,** 232–242.

Aharoni, R., Teitelbaum, D., Leitner, O., Meshorer, A., Sela, M., and Arnon, R. (2000). Specific Th2 cells accumulate in the central nervous system of mice protected against experimental autoimmune encephalomyelitis by copolymer 1. *Proc. Natl. Acad. Sci. USA* **97,** 11472–11477.

Andersen, J. K. (2004). Oxidative stress in neurodegeneration: Cause or consequence? *Nat. Med.* **10**(Suppl.), S18–S25.

Andersen, P. M. (2001). Genetics of sporadic ALS. *Amyotroph. Lateral. Scler. Other Motor Neuron Disord.* **2**(Suppl. 1), S37–S41.

Anderson, E., Zink, W., Xiong, H., and Gendelman, H. E. (2002). HIV-1-associated dementia: A metabolic encephalopathy perpetrated by virus-infected and immune-competent mononuclear phagocytes. *J. Acquir. Immune Defic. Syndr.* **31**(Suppl. 2), S43–S54.

Angelov, D. N., Waibel, S., Guntinas-Lichius, O., Lenzen, M., Neiss, W. F., Tomov, T. L., Yoles, E., Kipnis, J., Schori, H., Reuter, A., Ludolph, A., and Schwartz, M. (2003). Therapeutic vaccine for acute and chronic motor neuron diseases: Implications for amyotrophic lateral sclerosis. *Proc. Natl. Acad. Sci. USA* **100,** 4790–4795.

Aquaro, S., Panti, S., Caroleo, M. C., Balestra, E., Cenci, A., Forbici, F., Ippolito, G., Mastino, A., Testi, R., Mollace, V., Calio, R., and Perno, C. F. (2000). Primary macrophages infected by human immunodeficiency virus trigger CD95-mediated apoptosis of uninfected astrocytes. *J. Leukoc. Biol.* **68,** 429–435.

Arnon, R., and Sela, M. (2003). Immunomodulation by the copolymer glatiramer acetate. *J. Mol. Recognit.* **16,** 412–421.

Barzilai, A., Rotman, G., and Shiloh, Y. (2002). ATM deficiency and oxidative stress: A new dimension of defective response to DNA damage. *DNA Repair (Amst.)* **1,** 3–25.

Beal, M. F. (1995). Aging, energy, and oxidative stress in neurodegenerative diseases. *Ann. Neurol.* **38,** 357–366.

Beckman, J. S. (1996). Oxidative damage and tyrosine nitration from peroxynitrite. *Chem. Res. Toxicol.* **9,** 836–844.

Beckman, J. S., and Crow, J. P. (1993). Pathological implications of nitric oxide, superoxide and peroxynitrite formation. *Biochem. Soc. Trans.* **21,** 330–334.

Beckman, J. S., and Koppenol, W. H. (1996). Nitric oxide, superoxide, and peroxynitrite: The good, the bad, and ugly. *Am. J. Physiol.* **271,** C1424–C1437.

Benner, E. J., Mosley, R. L., Destache, C. J., Lewis, T. B., Jackson-Lewis, V., Gorantla, S., Nemachek, C., Green, S. R., Przedborski, S., and Gendelman, H. E. (2004). Therapeutic immunization protects dopaminergic neurons in a mouse model of Parkinson's disease. *Proc. Natl. Acad. Sci. USA* **101,** 9435–9440.

Bianca, V. D., Dusi, S., Bianchini, E., Dal Pra, I., and Rossi, F. (1999). Beta-amyloid activates the O-2 forming NADPH oxidase in microglia, monocytes, and neutrophils. A possible inflammatory mechanism of neuronal damage in Alzheimer's disease. *J. Biol. Chem.* **274,** 15493–15499.

Brenneman, D., Westbrook, G. L., Fitzgerald, S. P., Ennist, D. L., Elkins, K. L., Ruff, M. R., and Pert, C. B. (1988). Neuronal cell killing by the envelope protein of HIV and its prevention by vasoactive intestinal peptide. *Nature* **335,** 639–642.

Bruijn, L. I., Miller, T. M., and Cleveland, D. W. (2004). Unraveling the mechanisms involved in motor neuron degeneration in ALS. *Annu. Rev. Neurosci.* **27,** 723–749.

Butovsky, O., Koronyo-Hamaoui, M., Kunis, G., Ophir, E., Landa, G., Cohen, H., and Schwartz, M. (2006a). Glatiramer acetate fights against Alzheimer's disease by inducing dendritic-like microglia expressing insulin-like growth factor 1. *Proc. Natl. Acad. Sci. USA* **103,** 11784–11789.

Butovsky, O., Ziv, Y., Schwartz, A., Landa, G., Talpalar, A. E., Pluchino, S., Martino, G., and Schwartz, M. (2006b). Microglia activated by IL-4 or IFN-gamma differentially induce neurogenesis and oligodendrogenesis from adult stem/progenitor cells. *Mol. Cell. Neurosci.* **31,** 149–160.

Carlson, K. A., Cotter, R., Williams, C. E., Branecki, C. E., Zheng, J., and Gendelman, H. E. (2002). Macrophages and Alzheimer's disease. *In* "The Macrophage" (B. Burke and C. Lewis, Eds.), pp. 305–332. Oxford University Press, New York.

Carri, M. T., Ferri, A., Cozzolino, M., Calabrese, L., and Rotilio, G. (2003). Neurodegeneration in amyotrophic lateral sclerosis: The role of oxidative stress and altered homeostasis of metals. *Brain Res. Bull.* **61,** 365–374.

Castellani, R. J., Harris, P. L., Sayre, L. M., Fujii, J., Taniguchi, N., Vitek, M. P., Founds, H., Atwood, C. S., Perry, G., and Smith, M. A. (2001). Active glycation in neurofibrillary pathology of Alzheimer disease: N(epsilon)-(carboxymethyl) lysine and hexitol-lysine. *Free Radic. Biol. Med.* **31,** 175–180.

Chen, H., Zhang, S. M., Hernan, M. A., Schwarzschild, M. A., Willett, W. C., Colditz, G. A., Speizer, F. E., and Ascherio, A. (2003). Nonsteroidal anti-inflammatory drugs and the risk of Parkinson disease. *Arch. Neurol.* **60,** 1059–1064.

Chen, H., Jacobs, E., Schwarzschild, M. A., McCullough, M. L., Calle, E. E., Thun, M. J., and Ascherio, A. (2005). Nonsteroidal antiinflammatory drug use and the risk for Parkinson's disease. *Ann. Neurol.* **58,** 963–967.

Chen, T. S., Richie, J. P., Jr., and Lang, C. A. (1989). The effect of aging on glutathione and cysteine levels in different regions of the mouse brain. *Proc. Soc. Exp. Biol. Med.* **190,** 399–402.

Choi, D. K., Pennathur, S., Perier, C., Tieu, K., Teismann, P., Wu, D. C., Jackson-Lewis, V., Vila, M., Vonsattel, J. P., Heinecke, J. W., and Przedborski, S. (2005). Ablation of the inflammatory enzyme myeloperoxidase mitigates features of Parkinson's disease in mice. *J. Neurosci.* **25,** 6594–6660.

Cleveland, D. W., and Rothstein, J. D. (2001). From Charcot to Lou Gehrig: Deciphering selective motor neuron death in ALS. *Nat. Rev. Neurosci.* **2,** 806–819.

Cocchi, F., DeVico, A. L., Garzino-Demo, A., Arya, S. K., Gallo, R. C., and Lusso, P. (1995). Identification of RANTES, MIP-1 alpha, and MIP-1 beta as the major HIV-suppressive factors produced by CD8+ T cells. *Science* **270,** 1811–1815.

Croisier, E., Moran, L. B., Dexter, D. T., Pearce, R. K., and Graeber, M. B. (2005). Microglial inflammation in the parkinsonian substantia nigra: Relationship to alpha-synuclein deposition. *J. Neuroinflammation* **2,** 14.

Cudkowicz, M. E., McKenna-Yasek, D., Sapp, P. E., Chin, W., Geller, B., Hayden, D. L., Schoenfeld, D. A., Hosler, B. A., Horvitz, H. R., and Brown, R. H. (1997). Epidemiology of mutations in superoxide dismutase in amyotrophic lateral sclerosis. *Ann. Neurol.* **41,** 210–221.

d'Arminio Monforte, A., Duca, P. G., Vago, L., Grassi, M. P., and Moroni, M. (2000). Decreasing incidence of CNS AIDS-defining events associated with antiretroviral therapy. *Neurology* **54,** 1856–1859.

D'Aversa, T. G., Eugenin, E. A., and Berman, J. W. (2005). NeuroAIDS: Contributions of the human immunodeficiency virus-1 proteins Tat and gp120 as well as CD40 to microglial activation. *J. Neurosci. Res.* **81,** 436–446.

Dauer, W., and Przedborski, S. (2003). Parkinson's disease: Mechanisms and models. *Neuron* **39,** 889–909.

Dawson, T. M., and Dawson, V. L. (2003). Molecular pathways of neurodegeneration in Parkinson's disease. *Science* **302,** 819–822.

de Belleroche, J., Orrell, R. W., and Virgo, L. (1996). Amyotrophic lateral sclerosis: Recent advances in understanding disease mechanisms. *J. Neuropathol. Exp. Neurol.* **55,** 747–757.

Deng, H. X., Hentati, A., Tainer, J. A., Iqbal, Z., Cayabyab, A., Hung, W. Y., Getzoff, E. D., Hu, P., Herzfeldt, B., Roos, R. P., *et al.* (1993). Amyotrophic lateral sclerosis and structural defects in Cu, Zn superoxide dismutase. *Science* **261,** 1047–1051.

Diesing, T. S., Swindells, S., Gelbard, H., and Gendelman, H. E. (2002). HIV-1-associated dementia: A basic science and clinical perspective. *AIDS Read.* **12,** 358–368.

Dringen, R., Gutterer, J. M., and Hirrlinger, J. (2000). Glutathione metabolism in brain metabolic interaction between astrocytes and neurons in the defense against reactive oxygen species. *Eur. J. Biochem.* **267,** 4912–4916.

Du, Y., Ma, Z., Lin, S., Dodel, R. C., Gao, F., Bales, K. R., Triarhou, L. C., Chernet, E., Perry, K. W., Nelson, D. L., Luecke, S., Phebus, L. A., *et al.* (2001). Minocycline prevents nigrostriatal dopaminergic neurodegeneration in the MPTP model of Parkinson's disease. *Proc. Natl. Acad. Sci. USA* **98,** 14669–14674.

Elbim, C., Pillet, S., Prevost, M. H., Preira, A., Girard, P. M., Rogine, N., Hakim, J., Israel, N., and Gougerot-Pocidalo, M. A. (2001). The role of phagocytes in HIV-related oxidative stress. *J. Clin. Virol.* **20,** 99–109.

Ensoli, F., Fiorelli, V., De Cristofaro, M., Muratori, D. S., Novi, A., Isgro, A., and Aiuti, F. (2000). Role of immune-derived diffusible mediators in AIDS-associated neurological disorders. *Arch. Immunol. Ther. Exp. (Warsz.)* **48,** 259–266.

Floyd, R. A., Hensley, K., Jaffery, F., Maidt, L., Robinson, K., Pye, Q., and Stewart, C. (1999). Increased oxidative stress brought on by pro-inflammatory cytokines in neurodegenerative processes and the protective role of nitrone-based free radical traps. *Life Sci.* **65,** 1893–1899.

Forman, M. S., Trojanowski, J. Q., and Lee, V. M. (2004). Neurodegenerative diseases: A decade of discoveries paves the way for therapeutic breakthroughs. *Nat. Med.* **10,** 1055–1063.

Gaudette, M., Hirano, M., and Siddique, T. (2000). Current status of SOD1 mutations in familial amyotrophic lateral sclerosis. *Amyotroph. Lateral. Scler. Other Motor Neuron Disord.* **1,** 83–99.

Gendelman, H. E., Zheng, J., Coulter, C. L., Ghorpade, A., Che, M., Thylin, M., Rubocki, R., Persidsky, Y., Hahn, F., Reinhard, J., Jr., and Swindells, S. (1998). Suppression of inflammatory neurotoxins by highly active antiretroviral therapy in human immunodeficiency virus-associated dementia. *J. Infect. Dis.* **178,** 1000–1007.

Giasson, B. I., Duda, J. E., Murray, I. V., Chen, Q., Souza, J. M., Hurtig, H. I., Ischiropoulos, H., Trojanowski, J. Q., and Lee, V. M. (2000). Oxidative damage linked to neurodegeneration by selective alpha-synuclein nitration in synucleinopathy lesions. *Science* **290,** 985–989.

Giulian, D., Haverkamp, L. J., Li, J., Karshin, W. L., Yu, J., Tom, D., Li, X., and Kirkpatrick, J. B. (1995). Senile plaques stimulate microglia to release a neurotoxin found in Alzheimer brain. *Neurochem. Int.* **27,** 119–137.

Goodwin, J. L., Uemura, E., and Cunnick, J. E. (1995). Microglial release of nitric oxide by the synergistic action of beta-amyloid and IFN-gamma. *Brain Res.* **692,** 207–214.

Gras, G., Chretien, F., Vallat-Decouvelaere, A. V., Le Pavec, G., Porcheray, F., Bossuet, C., Leone, C., Mialocq, P., Dereuddre-Bosquet, N., Clayette, P., Le Grand, R., Creminon, C., *et al.* (2003). Regulated expression of sodium-dependent glutamate transporters and synthetase: A neuroprotective role for activated microglia and macrophages in HIV infection? *Brain Pathol.* **13,** 211–222.

Gras, G., Porcheray, F., Samah, B., and Leone, C. (2006). The glutamate-glutamine cycle as an inducible, protective face of macrophage activation. *J. Leukoc. Biol.* **80,** 1067–1075.

Green, S. P., Cairns, B., Rae, J., Errett-Baroncini, C., Hongo, J. A., Erickson, R. W., and Curnutte, J. T. (2001). Induction of gp91-phox, a component of the phagocyte NADPH oxidase, in microglial cells during central nervous system inflammation. *J. Cereb. Blood Flow Metab.* **21,** 374–384.

Guegan, C., and Przedborski, S. (2003). Programmed cell death in amyotrophic lateral sclerosis. *J. Clin. Invest.* **111,** 153–161.

Haase, A. T. (1999). Population biology of HIV-1 infection: Viral and CD4+ T cell demographics and dynamics in lymphatic tissues. *Annu. Rev. Immunol.* **17,** 625–656.

Haenggeli, C., Julien, J. P., Lee Mosley, R., Perez, N., Dhar, A., Gendelman, H. E., and Rothstein, J. D. (2006). Therapeutic immunization with a glatiramer acetate derivative does not alter survival in G93A and G37R SOD1 mouse models of familial ALS. *Neurobiol. Dis.* **26,** 146–152.

Ho, D. D., Neumann, A. U., Perelson, A. S., Chen, W., Leonard, J. M., and Markowitz, M. (1995). Rapid turnover of plasma virions and CD4 lymphocytes in HIV-1 infection. *Nature* **373,** 123–126.

Hong, J., Li, N., Zhang, X., Zheng, B., and Zhang, J. Z. (2005). Induction of CD4+CD25+ regulatory T cells by copolymer-I through activation of transcription factor Foxp3. *Proc. Natl. Acad. Sci. USA* **102,** 6449–6454.

Jana, A., and Pahan, K. (2004). Human immunodeficiency virus type 1 gp120 induces apoptosis in human primary neurons through redox-regulated activation of neutral sphingomyelinase. *J. Neurosci.* **24,** 9531–9540.

Jiang, Z. G., Piggee, C., Heyes, M. P., Murphy, C., Quearry, B., Bauer, M., Zheng, J., Gendelman, H. E., and Markey, S. P. (2001). Glutamate is a mediator of neurotoxicity in secretions of activated HIV-1-infected macrophages. *J. Neuroimmunol.* **117,** 97–107.

Kadiu, I., Glanzer, J. G., Kipnis, J., Gendelman, H. E., and Thomas, M. P. (2005). Mononuclear phagocytes in the pathogenesis of neurodegenerative diseases. *Neurotox. Res.* **8,** 25–50.

Kato, S., Takikawa, M., Nakashima, K., Hirano, A., Cleveland, D. W., Kusaka, H., Shibata, N., Kato, M., Nakano, I., and Ohama, E. (2000). New consensus research on neuropathological aspects of familial amyotrophic lateral sclerosis with superoxide dismutase 1 (SOD1) gene mutations: Inclusions containing SOD1 in neurons and astrocytes. *Amyotroph. Lateral. Scler. Other Motor Neuron Disord.* **1,** 163–184.

Kaul, M., Garden, G. A., and Lipton, S. A. (2001). Pathways to neuronal injury and apoptosis in HIV-associated dementia. *Nature* **410,** 988–994.

Kipnis, J., and Schwartz, M. (2002). Dual action of glatiramer acetate (Cop-1) in the treatment of CNS autoimmune and neurodegenerative disorders. *Trends Mol. Med.* **8,** 319–323.

Kipnis, J., Yoles, E., Porat, Z., Cohen, A., Mor, F., Sela, M., Cohen, I. R., and Schwartz, M. (2000). T cell immunity to copolymer 1 confers neuroprotection on the damaged optic nerve: Possible therapy for optic neuropathies. *Proc. Natl. Acad. Sci. USA* **97,** 7446–7451.

Klegeris, A., Walker, D. G., and McGeer, P. L. (1997a). Interaction of Alzheimer beta-amyloid peptide with the human monocytic cell line THP-1 results in a protein kinase C-dependent secretion of tumor necrosis factor-alpha. *Brain Res.* **747,** 114–121.

Klegeris, A., Walker, D. G., and McGeer, P. L. (1997b). Regulation of glutamate in cultures of human monocytic THP-1 and astrocytoma U-373 MG cells. *J. Neuroimmunol.* **78,** 152–161.

Kohutnicka, M., Lewandowska, E., Kurkowska-Jastrzebska, I., Czlonkowski, A., and Czlonkowska, A. (1998). Microglial and astrocytic involvement in a murine model of Parkinson's disease induced by 1-methyl-4-phenyl-1,2,3,6-tetrahydropyridine (MPTP). *Immunopharmacology* **39,** 167–180.

Krishnakumar, D. (2005). "Immunology of HIV-1." Oxford University Press, New York.

Kruger, R., Kuhn, W., Muller, T., Woitalla, D., Graeber, M., Kosel, S., Przuntek, H., Epplen, J. T., Schols, L., and Riess, O. (1998). Ala30Pro mutation in the gene encoding alpha-synuclein in Parkinson's disease. *Nat. Genet.* **18,** 106–108.

Kurkowska-Jastrzebska, I., Wronska, A., Kohutnicka, M., Czlonkowski, A., and Czlonkowska, A. (1999). The inflammatory reaction following 1-methyl-4-phenyl-1,2,3,6-tetrahydropyridine intoxication in mouse. *Exp. Neurol.* **156,** 50–61.

Lannuzel, A., Lledo, P. M., Lamghitnia, H. O., Vincent, J. D., and Tardieu, M. (1995). HIV-1 envelope proteins gp120 and gp160 potentiate NMDA-induced [Ca2+]i increase, alter [Ca2+]i homeostasis and induce neurotoxicity in human embryonic neurons. *Eur. J. Neurosci.* **7,** 2285–2293.

Lass, A., Agarwal, S., and Sohal, R. S. (1997). Mitochondrial ubiquinone homologues, superoxide radical generation, and longevity in different mammalian species. *J. Biol. Chem.* **272,** 19199–19204.

Laurie, C., Reynolds, A., Cuskun, O., Bowman, E., Gendelman, H. E., and Mosley, R. L. (2007). CD4 + T cells from Copolymer-1 immunized mice protect dopaminergic neurons in the 1-methyl-4-phenyl-1,2,3,6-tetrahydropyridine model of Parkinson's disease. *J. Neuroimmunol.* **183,** 60–68.

Lawrence, D. M., and Major, E. O. (2002). HIV-1 and the brain: Connections between HIV-1-associated dementia, neuropathology and neuroimmunology. *Microbes Infect.* **4,** 301–308.

Liu, Y., Brooks, B. R., Taniguchi, N., and Hartmann, H. A. (1998). CuZnSOD and MnSOD immunoreactivity in brain stem motor neurons from amyotrophic lateral sclerosis patients. *Acta Neuropathol. (Berl.)* **95,** 63–70.

Luo, X., Carlson, K. A., Wojna, V., Mayo, R., Biskup, T. M., Stoner, J., Anderson, J., Gendelman, H. E., and Melendez, L. M. (2003). Macrophage proteomic fingerprinting predicts HIV-1-associated cognitive impairment. *Neurology* **60,** 1931–1937.

Lyons, T. J., Gralla, E. B., and Valentine, J. S. (1999). Biological chemistry of copper-zinc superoxide dismutase and its link to amyotrophic lateral sclerosis. *Met. Ions Biol. Syst.* **36,** 125–177.

Markesbery, W. R., and Carney, J. M. (1999). Oxidative alterations in Alzheimer's disease. *Brain Pathol.* **9,** 133–146.

Markesbery, W. R., and Mira, S. (1996). "Alzheimer's Disease-Cause(s), Diagnosis, Treatment, and Care." CRC Press, New York.

Mattson, M. P., Barger, S. W., Furukawa, K., Bruce, A. J., Wyss-Coray, T., Mark, R. J., and Mucke, L. (1997). Cellular signaling roles of TGF beta, TNF alpha and beta APP in brain injury responses and Alzheimer's disease. *Brain Res. Brain Res. Rev.* **23,** 47–61.

McDonald, D. R., Brunden, K. R., and Landreth, G. E. (1997). Amyloid fibrils activate tyrosine kinase-dependent signaling and superoxide production in microglia. *J. Neurosci.* **17,** 2284–2294.

McGeer, P. L., and McGeer, E. G. (2002a). Inflammatory processes in amyotrophic lateral sclerosis. *Muscle Nerve* **26,** 459–470.

McGeer, P. L., and McGeer, E. G. (2002b). Local neuroinflammation and the progression of Alzheimer's disease. *J. Neurovirol.* **8,** 529–538.

McGeer, P. L., Itagaki, S., Boyes, B. E., and McGeer, E. G. (1988). Reactive microglia are positive for HLA-DR in the substantia nigra of Parkinson's and Alzheimer's disease brains. *Neurology* **38,** 1285–1291.

Meda, L., Bernasconi, S., Bonaiuto, C., Sozzani, S., Zhou, D., Otvos, L., Jr., Mantovani, A., Rossi, F., and Cassatella, M. A. (1996). Beta-amyloid (25–35) peptide and IFN-gamma synergistically induce the production of the chemotactic cytokine MCP-1/JE in monocytes and microglial cells. *J. Immunol.* **157,** 1213–1218.

Mollace, V., and Nistico, G. (1995). Release of nitric oxide from astroglial cells: A key mechanism in neuroimmune disorders. *Adv. Neuroimmunol.* **5,** 421–430.

Mollace, V., Colasanti, M., Persichini, T., Bagetta, G., Lauro, G. M., and Nistico, G. (1993). HIV gp120 glycoprotein stimulates the inducible isoform of no synthase in human cultured astrocytoma cells. *Biochem. Biophys. Res. Commun.* **194,** 439–445.

Mollace, V., Nottet, H. S., Clayette, P., Turco, M. C., Muscoli, C., Salvemini, D., and Perno, C. F. (2001). Oxidative stress and neuroAIDS: Triggers, modulators and novel antioxidants. *Trends Neurosci.* **24,** 411–416.

Nath, A. (2002). Human immunodeficiency virus (HIV) proteins in neuropathogenesis of HIV dementia. *J. Infect. Dis.* **186**(Suppl. 2), S193–S198.

New, D., Ma, M., Epstein, L. G., Nath, A., and Gelbard, H. A. (1997). Human immunodeficiency virus type-1 tat protein induces death by apoptosis in primary human neuron cultures. *J. Neurovirol.* **3,** 168–173.

Norris, E. H., Giasson, B. I., Ischiropoulos, H., and Lee, V. M. (2003). Effects of oxidative and nitrative challenges on alpha-synuclein fibrillogenesis involve distinct mechanisms of protein modifications. *J. Biol. Chem.* **278,** 27230–27240.

Nunomura, A., Perry, G., Pappolla, M. A., Wade, R., Hirai, K., Chiba, S., and Smith, M. A. (1999). RNA oxidation is a prominent feature of vulnerable neurons in Alzheimer's disease. *J. Neurosci.* **19,** 1959–1964.

Nunomura, A., Perry, G., Pappolla, M. A., Friedland, R. P., Hirai, K., Chiba, S., and Smith, M. A. (2000). Neuronal oxidative stress precedes amyloid-beta deposition in Down syndrome. *J. Neuropathol. Exp. Neurol.* **59,** 1011–1017.

Nunomura, A., Perry, G., Aliev, G., Hirai, K., Takeda, A., Balraj, E. K., Jones, P. K., Ghanbari, H., Wataya, T., Shimohama, S., Chiba, S., Atwood, C. S., *et al.* (2001). Oxidative damage is the earliest event in Alzheimer disease. *J. Neuropathol. Exp. Neurol.* **60,** 759–767.

Oberlin, E., Amara, A., Bachelerie, F., Bessia, C., Virelizier, J. L., Arenzana-Seisdedos, F., Schwartz, O., Heard, J. M., Clark-Lewis, I., Legler, D. F., Loetscher, M., Baggiolini, M., *et al.* (1996). The CXC chemokine SDF-1 is the ligand for LESTR/fusin and prevents infection by T-cell-line-adapted HIV-1. *Nature* **382,** 833–835.

Olanow, C. W. (1993). A radical hypothesis for neurodegeneration. *Trends Neurosci.* **16,** 439–444.

Piani, D., Spranger, M., Frei, K., Schaffner, A., and Fontana, A. (1992). Macrophage-induced cytotoxicity of N-methyl-D-aspartate receptor positive neurons involves excitatory amino acids rather than reactive oxygen intermediates and cytokines. *Eur. J. Immunol.* **22,** 2429–2436.

Polymeropoulos, M. H., Lavedan, C., Leroy, E., Ide, S. E., Dehejia, A., Dutra, A., Pike, B., Root, H., Rubenstein, J., Boyer, R., Stenroos, E. S., Chandrasekharappa, S., *et al.* (1997). Mutation in the alpha-synuclein gene identified in families with Parkinson's disease. *Science* **276,** 2045–2047.

Pratico, D., and Delanty, N. (2000). Oxidative injury in diseases of the central nervous system: Focus on Alzheimer's disease. *Am. J. Med.* **109,** 577–585.

Price, T. O., Ercal, N., Nakaoke, R., and Banks, W. A. (2005). HIV-1 viral proteins gp120 and Tat induce oxidative stress in brain endothelial cells. *Brain Res.* **1045,** 57–63.

Price, T. O., Uras, F., Banks, W. A., and Ercal, N. (2006). A novel antioxidant N-acetylcysteine amide prevents gp120- and Tat-induced oxidative stress in brain endothelial cells. *Exp. Neurol.* **201,** 193–202.

Qin, L., Liu, Y., Wang, T., Wei, S. J., Block, M. L., Wilson, B., Liu, B., and Hong, J. S. (2004). NADPH oxidase mediates lipopolysaccharide-induced neurotoxicity and proinflammatory gene expression in activated microglia. *J. Biol. Chem.* **279,** 1415–1421.

Raber, J., Sorg, O., Horn, T. F., Yu, N., Koob, G. F., Campbell, I. L., and Bloom, F. E. (1998). Inflammatory cytokines: Putative regulators of neuronal and neuro-endocrine function. *Brain Res. Brain Res. Rev.* **26,** 320–326.

Rakhit, R., Cunningham, P., Furtos-Matei, A., Dahan, S., Qi, X. F., Crow, J. P., Cashman, N. R., Kondejewski, L. H., and Chakrabartty, A. (2002). Oxidation-induced misfolding and aggregation of superoxide dismutase and its implications for amyotrophic lateral sclerosis. *J. Biol. Chem.* **277,** 47551–47556.

Reiter, R. J. (1995). Oxidative processes and antioxidative defense mechanisms in the aging brain. *FASEB J.* **9,** 526–533.

Rosen, D. R., Siddique, T., Patterson, D., Figlewicz, D. A., Sapp, P., Hentati, A., Donaldson, D., Goto, J., O'Regan, J. P., Deng, H. X., Rahmani, Z., Krizus, A., *et al.* (1993). Mutations in Cu/Zn superoxide dismutase gene are associated with familial amyotrophic lateral sclerosis. *Nature* **362,** 59–62.

Rowland, L. P., and Shneider, N. A. (2001). Amyotrophic lateral sclerosis. *N. Engl. J. Med.* **344,** 1688–1700.

Shefner, J. M., Reaume, A. G., Flood, D. G., Scott, R. W., Kowall, N. W., Ferrante, R. J., Siwek, D. F., Upton-Rice, M., and Brown, R. H., Jr. (1999). Mice lacking cytosolic copper/zinc superoxide dismutase display a distinctive motor axonopathy. *Neurology* **53,** 1239–1246.

Shibata, N., Hirano, A., Yamamoto, T., Kato, Y., and Kobayashi, M. (2000). Superoxide dismutase-1 mutation-related neurotoxicity in familial amyotrophic lateral sclerosis. *Amyotroph. Lateral. Scler. Other Motor Neuron Disord.* **1,** 143–161.

Sian, J., Dexter, D. T., Lees, A. J., Daniel, S., Agid, Y., Javoy-Agid, F., Jenner, P., and Marsden, C. D. (1994). Alterations in glutathione levels in Parkinson's disease and other neurodegenerative disorders affecting basal ganglia. *Ann. Neurol.* **36,** 348–355.

Simantov, R. (1989). Glutamate neurotoxicity in culture depends on the presence of glutamine: Implications for the role of glial cells in normal and pathological brain development. *J. Neurochem.* **52,** 1694–1699.

Singleton, A. B., Farrer, M., Johnson, J., Singleton, A., Hague, S., Kachergus, J., Hulihan, M., Peuralinna, T., Dutra, A., Nussbaum, R., Lincoln, S., Crawley, A., *et al.* (2003). Alpha-Synuclein locus triplication causes Parkinson's disease. *Science* **302,** 841.

Sipe, J. D., and Cohen, A. S. (2000). Review: History of the amyloid fibril. *J. Struct. Biol.* **130,** 88–98.

Smith, M. A., Richey Harris, P. L., Sayre, L. M., Beckman, J. S., and Perry, G. (1997). Widespread peroxynitrite-mediated damage in Alzheimer's disease. *J. Neurosci.* **17,** 2653–2657.

Song, L., Nath, A., Geiger, J. D., Moore, A., and Hochman, S. (2003). Human immunodeficiency virus type 1 Tat protein directly activates neuronal N-methyl-D-aspartate receptors at an allosteric zinc-sensitive site. *J. Neurovirol.* **9,** 399–403.

Souza, J. M., Giasson, B. I., Chen, Q., Lee, V. M., and Ischiropoulos, H. (2000). Dityrosine cross-linking promotes formation of stable alpha -synuclein polymers. Implication of nitrative and oxidative stress in the pathogenesis of neurodegenerative synucleinopathies. *J. Biol. Chem.* **275,** 18344–18349.

Spira, P. J., Sharpe, D. M., Halliday, G., Cavanagh, J., and Nicholson, G. A. (2001). Clinical and pathological features of a Parkinsonian syndrome in a family with an Ala53Thr alpha-synuclein mutation. *Ann. Neurol.* **49,** 313–319.

Strong, M. J. (2003). The basic aspects of therapeutics in amyotrophic lateral sclerosis. *Pharmacol. Ther.* **98,** 379–414.

Tabner, B. J., Turnbull, S., El-Agnaf, O., and Allsop, D. (2001). Production of reactive oxygen species from aggregating proteins implicated in Alzheimer's disease, Parkinson's disease and other neurodegenerative diseases. *Curr. Top. Med. Chem.* **1,** 507–517.

Tabner, B. J., El-Agnaf, O. M., German, M. J., Fullwood, N. J., and Allsop, D. (2005). Protein aggregation, metals and oxidative stress in neurodegenerative diseases. *Biochem. Soc. Trans.* **33,** 1082–1086.

Tardieu, M., and Boutet, A. (2002). HIV-1 and the central nervous system. *Curr. Top. Microbiol. Immunol.* **265,** 183–195.

Teismann, P., and Ferger, B. (2001). Inhibition of the cyclooxygenase isoenzymes COX-1 and COX-2 provide neuroprotection in the MPTP-mouse model of Parkinson's disease. *Synapse* **39,** 167–174.

Teismann, P., Tieu, K., Choi, D. K., Wu, D. C., Naini, A., Hunot, S., Vila, M., Jackson-Lewis, V., and Przedborski, S. (2003). Cyclooxygenase-2 is instrumental in Parkinson's disease neurodegeneration. *Proc. Natl. Acad. Sci USA* **100,** 5473–5478.

Thomas, M. P., Chartrand, K., Reynolds, A., Vitvitsky, V., Banerjee, R., and Gendelman, H. E. (2007). Ion channel blockade attenuates aggregated alpha synuclein induction of microglial reactive oxygen species: Relevance for the pathogenesis of Parkinson's disease. *J. Neurochem.* **100,** 503–519.

Tong, N., Perry, S. W., Zhang, Q., James, H. J., Guo, H., Brooks, A., Bal, H., Kinnear, S. A., Fine, S., Epstein, L. G., Dairaghi, D., Schall, T. J., *et al.* (2000). Neuronal fractalkine expression in HIV-1 encephalitis: Roles for macrophage recruitment and neuroprotection in the central nervous system. *J. Immunol.* **164,** 1333–1339.

Valentine, J. S. (2002). Do oxidatively modified proteins cause ALS? *Free Radic. Biol. Med.* **33,** 1314–1320.

Van Muiswinkel, F. L., Raupp, S. F., de Vos, N. M., Smits, H. A., Verhoef, J., Eikelenboom, P., and Nottet, H. S. (1999). The amino-terminus of the amyloid-beta protein is critical for the cellular binding and consequent activation of the respiratory burst of human macrophages. *J. Neuroimmunol.* **96,** 121–130.

Vijitruth, R., Liu, M., Choi, D. Y., Nguyen, X. V., Hunter, R. L., and Bing, G. (2006). Cyclooxygenase-2 mediates microglial activation and secondary dopaminergic cell death in the mouse MPTP model of Parkinson's disease. *J. Neuroinflammation* **3,** 6–28.

Vila, M., and Przedborski, S. (2004). Genetic clues to the pathogenesis of Parkinson's disease. *Nat. Med.* **10**(Suppl.), S58–S62.

Wei, X., Ghosh, S. K., Taylor, M. E., Johnson, V. A., Emini, E. A., Deutsch, P., Lifson, J. D., Bonhoeffer, S., Nowak, M. A., Hahn, B. H., Saag, M. S., and Shaw, G. M. (1995). Viral dynamics in human immunodeficiency virus type 1 infection. *Nature* **373,** 117–122.

Wersinger, C., and Sidhu, A. (2006). An inflammatory pathomechanism for Parkinson's disease? *Curr. Med. Chem.* **13,** 591–602.

Weydt, P., and Moller, T. (2005). Neuroinflammation in the pathogenesis of amyotrophic lateral sclerosis. *Neuroreport* **16,** 527–531.

Weydt, P., Weiss, M. D., Moller, T., and Carter, G. T. (2002). Neuro-inflammation as a therapeutic target in amyotrophic lateral sclerosis. *Curr. Opin. Investig. Drugs* **3,** 1720–1724.

Weydt, P., Yuen, E. C., Ransom, B. R., and Moller, T. (2004). Increased cytotoxic potential of microglia from ALS-transgenic mice. *Glia* **48,** 179–182.

Winterbourn, C. C., and Metodiewa, D. (1994). The reaction of superoxide with reduced glutathione. *Arch. Biochem. Biophys.* **314,** 284–290.

Wu, D. C., Jackson-Lewis, V., Vila, M., Tieu, K., Teismann, P., Vadseth, C., Choi, D. K., Ischiropoulos, H., and Przedborski, S. (2002). Blockade of microglial activation is neuroprotective in the 1-methyl-4-phenyl-1,2,3,6-tetrahydropyridine mouse model of Parkinson disease. *J. Neurosci.* **22,** 1763–1771.

Xiong, H., Zeng, Y. C., Lewis, T., Zheng, J., Persidsky, Y., and Gendelman, H. E. (2000). HIV-1 infected mononuclear phagocyte secretory products affect neuronal physiology leading to cellular demise: Relevance for HIV-1-associated dementia. *J. Neurovirol.* **6**(Suppl. 1), S14–S23.

Yamada, T., McGeer, P. L., and McGeer, E. G. (1992). Appearance of paired nucleated, Tau-positive glia in patients with progressive supranuclear palsy brain tissue. *Neurosci. Lett.* **135,** 99–102.

Yong, V. W., Power, C., Forsyth, P., and Edwards, D. R. (2001). Metalloproteinases in biology and pathology of the nervous system. *Nat. Rev. Neurosci.* **2,** 502–511.

Zarranz, J. J., Alegre, J., Gomez-Esteban, J. C., Lezcano, E., Ros, R., Ampuero, I., Vidal, L., Hoenicka, J., Rodriguez, O., Atares, B., Llorens, V., Gomez Tortosa, E., *et al.* (2004). The new mutation, E46K, of alpha-synuclein causes Parkinson and Lewy body dementia. *Ann. Neurol.* **55,** 164–173.

Zhang, W., Wang, T., Pei, Z., Miller, D. S., Wu, X., Block, M. L., Wilson, B., Zhang, W., Zhou, Y., Hong, J. S., and Zhang, J. (2005). Aggregated alpha-synuclein activates microglia: A process leading to disease progression in Parkinson's disease. *FASEB J.* **19,** 533–542.

Zhu, X., Raina, A. K., Lee, H. G., Casadesus, G., Smith, M. A., and Perry, G. (2004). Oxidative stress signalling in Alzheimer's disease. *Brain Res.* **1000,** 32–39.

DIFFERENTIAL MODULATION OF TYPE 1 AND TYPE 2 CANNABINOID RECEPTORS ALONG THE NEUROIMMUNE AXIS

Sergio Oddi,*,† Paola Spagnuolo,‡,§ Monica Bari,§
Antonella D'Agostino,* and Mauro Maccarrone*,†

*Department of Biomedical Sciences, University of Teramo, Teramo 64100, Italy
†European Center for Brain Research (CERC)/IRCCS S. Lucia Foundation, Rome 00143, Italy
‡Department of Pharmacobiology and University Centre for the Study of Adaptive Disorder and Headache (UCADH), Section of Neuropharmacology of Normal and Pathological Neuronal Plasticity, University of Calabria, Rende (CS) 87036, Italy
§Department of Experimental Medicine and Biochemical Sciences
University of Rome Tor Vergata, Rome 00133, Italy

Endocannabinoid-signaling chains have been implicated in a variety of pathophysiological functions, including memory, coordination, vasoregulation, reproduction, neurodegeneration, and inflammation. These activities were thought to be mediated by the activation of two G-protein–coupled receptors (GPCRs), type 1 and type 2 cannabinoid receptors (CB_1R and CB_2R). These two CBR subtypes share common agonists and trigger similar signaling pathways, yet they present several important differences in structure and cell distribution. In particular, recent research has shown that the CB_1R and CB_2R are differentially linked to lipid rafts, specialized microdomains of the plasma membrane involved in the signaling of many other GPCRs. We present an overview of the current literature on the effects that lipid raft perturbation have on CBRs activities, and provide a mechanistic model to interpret these data in terms of structural and functional aspects. These findings may also have important implications for the development of new therapeutic approaches, including lipid raft perturbing drugs, aimed to selectively modulate CB_1R signaling in a variety of pathological conditions.

INTERNATIONAL REVIEW OF
NEUROBIOLOGY, VOL. 82
DOI: 10.1016/S0074-7742(07)82017-4

0074-7742/07 $35.00

I. Introduction

The endogenous cannabinoid system is an evolutionarily ancient lipid-signaling system with important regulatory functions, both in central nervous system and peripheral tissues. In the last two decades, the molecular components of this novel signaling pathway have been identified and characterized. The main endocannabinoids are small molecules derived from arachidonic acid, that is *N*-arachidonoylethanolamine (anandamide, AEA) (Devane *et al.*, 1992) and 2-arachidonoylglycerol (2-AG) (Mechoulam *et al.*, 1995; Sugiura *et al.*, 1995). They bind to a family of G-protein–coupled receptors (GPCRs), of which cannabinoid receptor type 1 (CB_1R) is densely distributed in areas of the brain (Egertova *et al.*, 2003) related to motor control, cognition, emotional responses, motivated behavior, and homeostasis (Fride, 2002). Outside the brain, the functions of endocannabinoids are mainly mediated by CB_2Rs (Howlett *et al.*, 2004), which modulated several functions of the immune system (Sancho *et al.*, 2004). Endocannabinoids are released on demand from lipid precursors in a receptor-dependent manner (Di Marzo *et al.*, 1994), transported outside and inside the cells by an uptake system not yet well characterized (Battista *et al.*, 2005; Glaser *et al.*, 2005; Hillard and Jarrahian, 2003) and degraded by two enzymes, the fatty acid amide hydrolase (Cravatt *et al.*, 1996) and the monoacylglycerol lipase (Dinh *et al.*, 2002), specific for the hydrolysis of AEA and 2-AG, respectively.

In the last few years, mounting evidence has accumulated that lipid rafts, specialized membrane microdomains, are involved in the trafficking and function of CBRs (McFarland and Barker, 2005). This chapter discusses the available data on the differential modulation that lipid rafts seem to exert on the activities of the two subtypes of CBR. In line with this, we propose a model that could account for this different sensitivity, potentially opening the way to the design of specific drugs to be exploited for the treatment of CB_1-related neuroinflammatory diseases (Bari *et al.*, 2006a).

II. Lipid Rafts and Cannabinoid Receptors

Lipid rafts are membrane domains biochemically defined by the insolubility of their components in cold nonionic detergents (like Triton X-100 or Brij-98) (Brown and Rose, 1992; Simons and Ikonen, 1997). They are enriched in specific lipids with saturated fatty acid chains, like sphingomyelin and sphingolipids, and in cholesterol. Caveolae are a subclass of lipid rafts that occur as small invaginations of the plasma membrane and contain a family of integral membrane proteins called caveolins (Pike *et al.*, 2002; Razani *et al.*, 2002). Because of their ability of forming more liquid ordered domains within the membrane bilayers, lipid rafts also recruit different proteic machineries by which they are assumed to play important roles in cholesterol transport, endocytosis, and signal transduction (Michel *et al.*, 1997; Simons and Toomre, 2000). In particular, there is mounting evidence that several components of different GPCR signal transduction chains interact with and/or are localized within lipid rafts. Indeed, heterotrimeric GTP-binding proteins (G-proteins), regulators of G-proteins, and GPCRs themselves have been reported to be associated to rafts and caveolae (Huang *et al.*, 1997, 1999; Morales *et al.*, 1998; Song *et al.*, 1997; Wedegaertner *et al.*, 1995). Frequently, raft localization is a dynamic process, as it has been found that the β2-adrenergic receptor leaves caveolae on ligand binding, whereas muscarinic acetylcholine receptor and B2 bradykinin receptor that are uniformly distributed throughout the plasma membrane localize to caveolae only in the presence of their ligands (de Weerd and Leeb-Lundberg, 1997).

Lipid rafts are thought to be involved in the regulation of signal transduction through several different mechanisms (Barnett-Norris *et al.*, 2005). For example, rafts may contain incomplete signaling pathways that are activated when a receptor or other required molecules is recruited into the raft (Pike, 2003). These membrane microdomains may also play a role in signal termination by mediating the internalization of specific components of the signaling chain, such as GPCRs and G-proteins (Anderson, 1998; Le *et al.*, 2002).

CBRs are G-protein–coupled seven-transmembrane receptors coupled to $G_{i/o}$-proteins, which are activated by common agonists (like AEA or 2-AG), and trigger common signaling pathways, including the inhibition of adenylate cyclase (AC), the regulation of ionic currents (inhibition of voltage-gated L-, N-, and P/Q-type Ca^{2+} channels, activation of K^+ channels), the activation of focal adhesion kinase, of mitogen-activated protein kinase (MAPK), of cytosolic phospholipase A2, and of nitric oxide synthetase (reviewed by Bari *et al.*, 2006a; Piomelli, 2003).

In recent years, a functional link has been established between CB_1R signaling and lipid rafts integrity. In particular, it has been demonstrated that lipid raft disruption by methyl-β-cyclodextrin (MCD) (Kunzelmann-Marche *et al.*, 2002; Xiang *et al.*, 2002), a membrane cholesterol depletor (Ohvo and Slotte, 1996), doubles G-protein–dependent signaling of CB_1R via AC and MAPK in neuronal cells (Bari *et al.*, 2005a). In keeping with these data, it has also been shown that the ability of MCD to block AEA-induced apoptosis (Biswas *et al.*, 2003; Sarker and Maruyama, 2003) is dependent on its effect on the CB_1R activation (Bari *et al.*, 2005a) (Fig. 1). Consistently, it has been also demonstrated that CB_1R signal activity is significantly reduced by cholesterol enrichment (Bari *et al.*, 2005b), which is known to enforce the stability and function of lipid rafts (Mitter *et al.*, 2003; Simons and Ehehalt, 2002). Indeed, membrane cholesterol enrichment in nerve cells reduces by half the binding to CB_1R and subsequent G-protein–dependent signaling through AC and MAPK (Bari *et al.*, 2005b) (Fig. 2). Therefore, it is conceivable

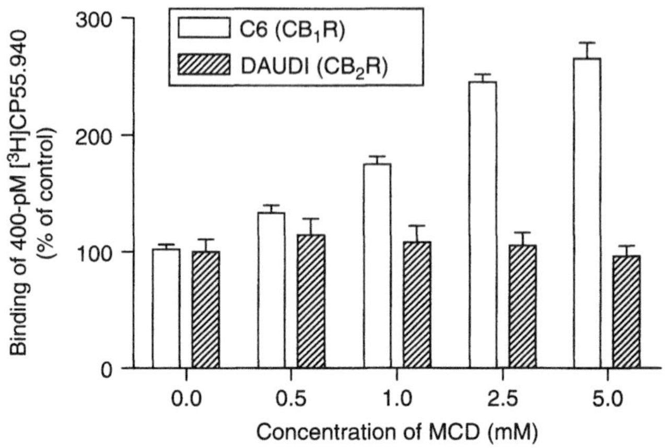

FIG. 1. Effect of MCD on CB_1R and CB_2R in neuronal C6 cells and immune DAUDI cells. Binding of 400-pM [^{3}H]CP55.940 to CB_2R in DAUDI cells, or to CB_1R in C6 cells. Both cells were pretreated for 30 min with various amounts of methyl-β-cyclodextrin (MCD). Values were expressed as percentage of the untreated control (100% = 170 $\pm$ 20 fmol/mg protein for DAUDI, 600 $\pm$ 70 fmol/mg protein for C6).

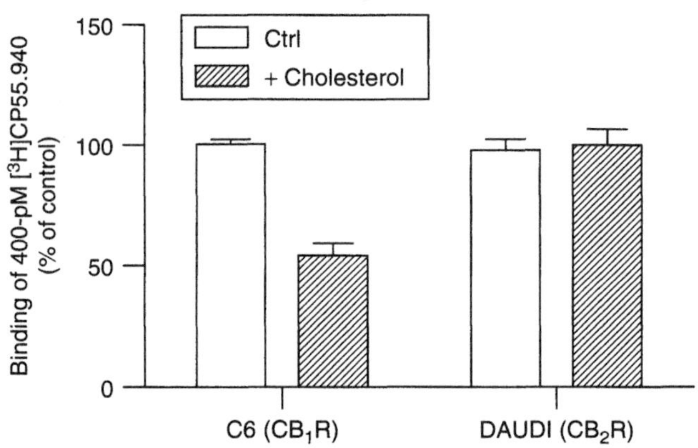

FIG. 2. Effect of cholesterol on CB_1R and CB_2R in neuronal C6 cells and immune DAUDI cells. Binding of 400-pM [^{3}H]CP55.940 to CB_2R in DAUDI cells, or to CB_1R in C6 cells. Both cells were untreated (CTR) or pretreated with cholesterol (CHOL). Values were expressed as percentage of the untreated control (100% = 160 $\pm$ 15 fmol/mg protein for DAUDI, 650 $\pm$ 50 fmol/mg protein for C6).

that lipid raft integrity is important for the CB_1R signal activity. In the same line, a paper demonstrated that CB_1R is physically associated with lipid rafts/caveolae in a breast cancer cell line, suggesting that they might represent a cellular device for its intracellular trafficking, as well as a favorable platform to regulate CB_1R signaling

(Sarnataro *et al.*, 2005). In particular, it has been found that in the absence of AEA, a significant portion of CB_1R resides within lipid rafts and that after agonist binding, the cell surface receptor was endocytosed from cell surface and travels to lysosomal compartment. Interestingly, the lysosomal localization has been found to be impaired by cholesterol depletion, suggesting that the lipid raft integrity is necessary for the trafficking of CB_1R in the lysosomes (Sarnataro *et al.*, 2005). The study of the subcellular distribution of CB_1R has shown that, due its natural constitutive activity, CB_1R permanently and constitutively cycles between plasma membrane and endosomes, leading to a predominantly intracellular localization. This complex process of endocytosis and recycling is mediated by the small GTPases Rab5 and Rab4, respectively (Leterrier *et al.*, 2004).

Unlike CB_1R in nerve cells, CB_2R in leukemic cells has not been found to be regulated by lipid rafts perturbation. In fact, neither MCD nor cholesterol enrichment did affect the binding and signaling of endocannabinoids through CB_2R (Bari *et al.*, 2006b) (Figs. 1 and 2). This conclusion was further supported by preliminary fractionation studies showing that CB_1R in C6 cells, but not CB_2R in DAUDI cells, colocalizes with the raft marker caveolin-1 (Bari *et al.*, 2006b and unpublished data).

III. Discussion

The molecular basis of the different sensitivity of CB_1R and CB_2R to raft integrity might be very complex and need a more detailed analysis of the three-dimensional (3D) structures of the two receptors and their interactions in the context of the membrane bilayers. The two CBR subtypes are encoded by different genes, exhibit 44% amino acid identity throughout the whole protein, and have been classified into the class A rhodopsin-like family of GPCR (Di Marzo *et al.*, 2002; Howlett *et al.*, 2004; Pertwee, 1997). The most important differences between CB_1R and CB_2R are located in the N-terminal extracellular loop II (EL2), in the C-terminus of transmembrane helix VII (TMH7), and in the C-terminus (Montero *et al.*, 2005) (Fig. 3). Moreover, a study using combined high resolution NMR and computer modeling has shown that CB_1R and CB_2R have conformational properties and salt bridge differences in the so-called juxtamembrane segment (or helix 8), that is critical for their activity and regulation and, more notably, is under the influence of the surrounding chemical environment (Xie and Chen, 2005). On this basis, it is tempting to speculate that lipid rafts might regulate CB_1Rs by interacting with the specific regions of its 3D dimensional structure, like helices 3 and 6 (Tian *et al.*, 2005), or helix 8 (Xie and Chen, 2005).

An interesting structural feature of CB_1R that could account for its specific interaction with lipid rafts is the presence in the C-terminal domain of a specific cysteine [C7.11(414)] that is constitutively palmitoylated

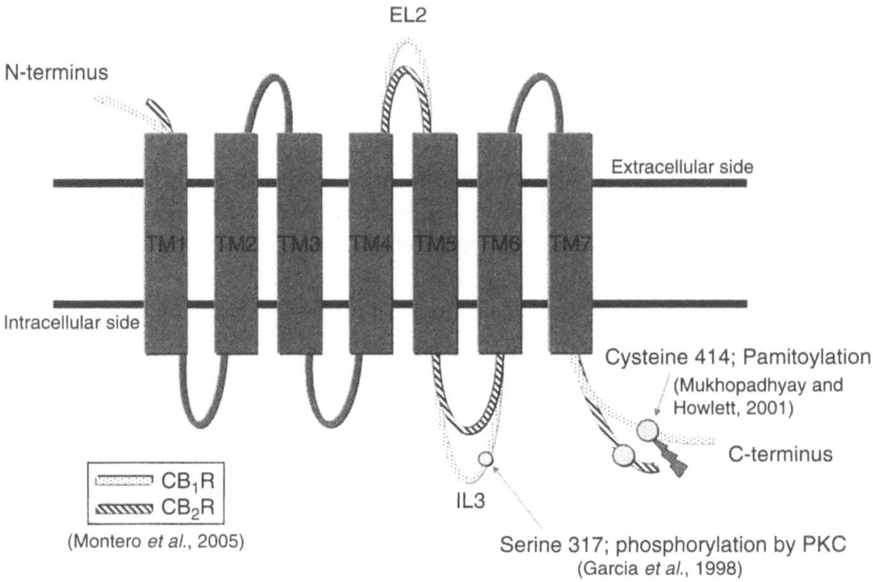

FIG. 3. 2D model of CB_1R and CB_2R. The differences in CBRs sequences are shown as different filled lines. TM, transmembrane; EL, extracellular; IL, intracellular.

(Mukhopadhyay and Howlett, 2001). Importantly, the palmitoylation state of CB_1R has been reported to influence the $G\alpha$ protein subtype with which CB_1R interacts (Mukhopadhyay *et al.*, 2002). In addition, the C-terminal domain of CB_1R between TMH7 and the palmitoylated cysteine (401–417) has been shown to interact with $G\alpha 0$ and $G\alpha i3$ (Mukhopadhyay and Howlett, 2001). Coimmunoprecipitation experiments in rat brain and N18TG2 neuronal cells that were chemically depalmitoylated with hydroxylamine showed that depalmitoylation irreversibly disrupted $CB_1/G\alpha 0$ and $CB_1/G\alpha i3$ association (Mukhopadhyay *et al.*, 2002). Since it has been shown that GPCRs can be targeted to lipid rafts via palmitoylation of specific cysteine residues, it is possible that this covalent modification is necessary to target CB_1R to specific subdomains in the lipid bilayer (Mukhopadhyay *et al.*, 2002).

Another important difference between the two receptors is the presence of a serine residue (at position 317) in the intracellular loop III (IL-3) of CB_1R, but not of CB_2R. Remarkably, the same serine residue has been found to be phosphorylated by PKC (Garcia *et al.*, 1998). Desensitization of GPCRs is often associated with phosphorylation of specific serines in this domain, followed by binding of β-arrestin and a reduction in affinity for G-proteins (Krupnick and Benovic, 1998; Zhang *et al.*, 1997), as well as the internalization of the GPCR (Moro *et al.*, 1993). This sequence of events effectively attenuates signaling by the GPCR and its ligand. Previous studies have shown that CB_1Rs undergo agonist-induced internalization in transfected CHO cells (Rinaldi-Carmona *et al.*, 1998), AtT20 cells (Hsieh *et al.*, 1999), and HEK-293 cells (Shapira *et al.*, 2003), as well as in cultured hippocampal neurons and F-11 cells, that naturally express CB1Rs (Coutts *et al.*, 2001).

The documented effects of MCD on CB_1R binding and signaling could be explained by means of a simplified model, where CB_1R unlike CB_2R interacts with lipid rafts (Fig. 4). It is still unclear whether CB_1R is constitutively present in lipid rafts or it is translocated there after agonist binding as part of the events involved in desensitization and internalization of the receptors. At any rate, it is known that the termination of the signaling activity of CB_1R, like many G-protein–coupled seven-transmembrane receptors, requires agonist induced (Garland *et al.*, 1996; Roth *et al.*, 1997) or constitutive endocytosis (Leterrier *et al.*, 2004), and hence cycling

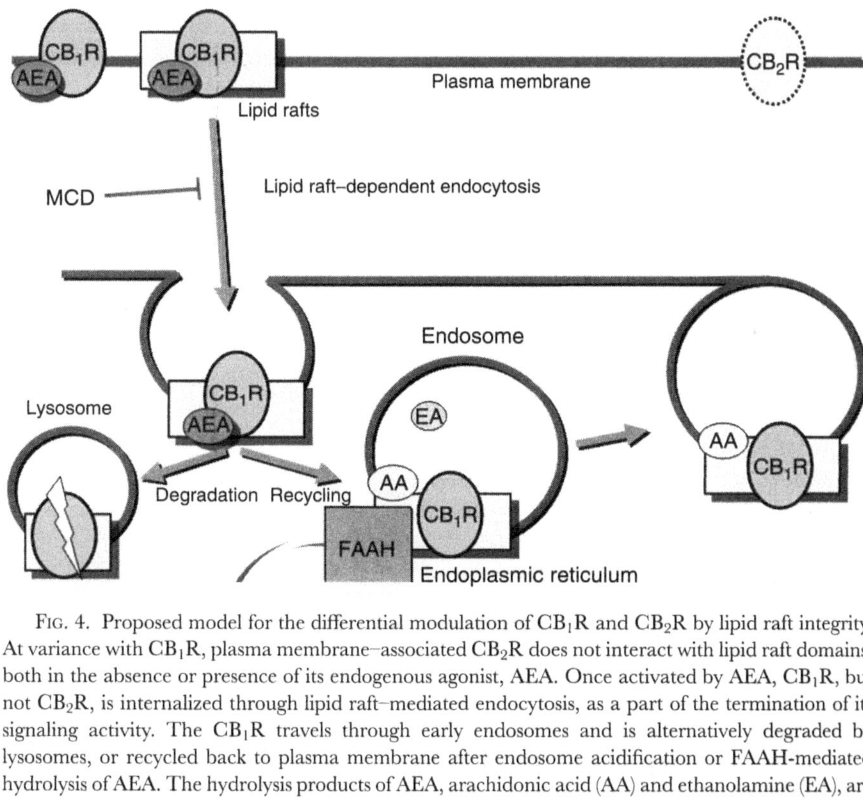

FIG. 4. Proposed model for the differential modulation of CB_1R and CB_2R by lipid raft integrity. At variance with CB_1R, plasma membrane–associated CB_2R does not interact with lipid raft domains, both in the absence or presence of its endogenous agonist, AEA. Once activated by AEA, CB_1R, but not CB_2R, is internalized through lipid raft–mediated endocytosis, as a part of the termination of its signaling activity. The CB_1R travels through early endosomes and is alternatively degraded by lysosomes, or recycled back to plasma membrane after endosome acidification or FAAH-mediated hydrolysis of AEA. The hydrolysis products of AEA, arachidonic acid (AA) and ethanolamine (EA), are in turn recycled back to plasma membrane via a caveolae-dependent mechanism (McFarland *et al.*, 2004). The hyperstimulation of the CB_1R-signaling activity mediated by MCD treatment may be interpreted as a consequence of the impairment of the lipid raft–dependent mechanism(s) involved in the interruption of CB_1R signaling.

between the plasma membrane and endosomes or lysosomes. It has been described that this process occurs by two different endocytotic pathways: one involving clathrin-coated pits and another one, more rapid, involving lipid rafts/caveolae (Keren and Sarne, 2003). Once internalized, the receptor can be degraded by the lysosome pathway (Sarnataro *et al.*, 2005), or it can be recycled to the cell surface after agonist removal, for example, by FAAH activity or by endosome acidification (Leterrier *et al.*, 2004). It has been also described that the products of AEA hydrolysis, arachidonic acid and ethanolamine, are in turn recycled back to plasma membranes via a caveolae-dependent mechanism (McFarland *et al.*, 2004).

Therefore, MCD could enhance the binding and signaling of CB_1R, but not of CB_2R, by interfering with lipid raft–dependent endocytosis and, in general, with those mechanisms responsible for the termination of CB_1R signaling. It should be stressed that the concept that the membrane environment might represent a key factor for the selective regulation of CB_1Rs could be exploited for the development of next generation therapeutics. These novel drugs might be used for the treatment of CB_1R-dependent human pathologies, including neuroinflammatory disorders like Alzheimer's disease or multiple sclerosis (Bari *et al.*, 2006a).

Acknowledgments

This investigation was supported by Ministero dell'Istruzione, dell'Università e della Ricerca (FIRB 2006), by Agenzia Spaziale Italiana (DCMC and MoMa projects 2005), and by Fondazione TERCAS (Research Programs 2004 and 2005) to M.M.

References

Anderson, R. G. W. (1998). The caveolae membrane system. *Annu. Rev. Biochem.* **67,** 199–225.

Bari, M., Battista, N., Fezza, F., Finazzi-Agrò, A., and Maccarrone, M. (2005a). Lipid rafts control signaling of type-1 cannabinoid receptors in neuronal cells. Implications for anandamide-induced apoptosis. *J. Biol. Chem.* **280,** 12212–12220.

Bari, M., Paradisi, A., Pasquariello, N., and Maccarrone, M. (2005b). Cholesterol-dependent modulation of type 1 cannabinoid receptors in nerve cells. *J. Neurosci. Res.* **81,** 275–283.

Bari, M., Battista, N., Fezza, F., Gasperi, V., and Maccarrone, M. (2006a). New insights into endocannabinoid degradation and its therapeutic potential. *Mini Rev. Med. Chem.* **6,** 257–268.

Bari, M., Spagnuolo, P., Fezza, F., Oddi, S., Pasquariello, N., Finazzi-Agrò, A., and Maccarrone, M. (2006b). Effect of lipid rafts on cb2 receptor signaling and 2-arachidonoyl-glycerol metabolism in human immune cells. *J. Immunol.* **177,** 4971–4980.

Barnett-Norris, J., Lynch, D., and Reggio, P. H. (2005). Lipids, lipid rafts and caveolae: Their importance for GPCR signaling and their centrality to the endocannabinoid system. *Life Sci.* **77,** 1625–1639.

Battista, N., Gasperi, V., Fezza, F., and Maccarrone, M. (2005). The anandamide membrane transporter and the therapeutic implication of its inhibition. *Therapy* **2,** 141–146.

Biswas, K. K., Sarker, K. P., Abeyama, K., Kawahara, K., Iino, S., Otsubo, Y., Saigo, K., Izumi, H., Hashiguchi, T., Yamakuchi, M., Yamaji, K., Endo, R., *et al.* (2003). Membrane cholesterol but not putative receptors mediates anandamide-induced hepatocyte apoptosis. *Hepatology* **38,** 1167–1177.

Brown, D. A., and Rose, J. K. (1992). Sorting of GPI-anchored proteins to glycolipid-enriched membrane subdomains during transport to the apical cell surface. *Cell* **68,** 533–544.

Coutts, A. A., Anavi-Goffer, S., Ross, R. A., MacEwan, D. J., Mackie, K., Pertwee, R. G., and Irving, A. J. (2001). Agonist-induced internalization and trafficking of cannabinoid CB1 receptors in hippocampal neurons. *J. Neurosci.* **21,** 2425–2433.

Cravatt, B. F., Giang, D. K., Mayfield, S. P., Boger, D. L., Lerner, R. A., and Gilula, N. B. (1996). Molecular characterization of an enzyme that degrades neuromodulatory fatty-acid amides. *Nature* **384,** 83–87.

Devane, W. A., Hanus, L., Breuer, A., Pertwee, R. G., Stevenson, L. A., Griffin, G., Gibson, D., Mandelbaum, A., Etinger, A., and Mechoulam, R. (1992). Isolation and structure of a brain constituent that binds to the cannabinoid receptor. *Science* **258,** 1946–1949.

de Weerd, W. F., and Leeb-Lundberg, L. M. (1997). Bradykinin sequesters B2 bradykinin receptors and the receptor-coupled Galpha subunits Galphaq and Galphai in caveolae in DDT1 MF-2 smooth muscle cells. *J. Biol. Chem.* **272,** 17858–17866.

Di Marzo, V., Fontana, A., Cadas, H., Schinelli, S., Cimino, G., Schwartz, J. C., and Pomelli, D. (1994). Formation and inactivation of endogenous cannabinoid anandamide in central neurons. *Nature* **372,** 686–691.

Di Marzo, V., De Petrocellis, L., Fezza, F., Ligresti, A., and Bisogno, T. (2002). Anandamide receptors. *Prostaglandins Leukot. Essent. Fatty Acids* **66,** 377–391.

Dinh, T. P., Carpenter, D., Leslie, F. M., Freund, T. F., Katona, I., Sensi, S. L., Kathuria, S., and Piomelli, D. (2002). Brain monoglyceride lipase participating in endocannabinoid inactivation. *Proc. Natl. Acad. Sci. USA* **99,** 10819–10824.

Egertova, M., Cravatt, B. F., and Elphick, M. R. (2003). Comparative analysis of fatty acid amide hydrolase and cb(1) cannabinoid receptor expression in the mouse brain: Evidence of a widespread role for fatty acid amide hydrolase in regulation of endocannabinoid signaling. *Neuroscience* **119,** 481–496.

Fride, E. (2002). Endocannabinoids in the central nervous system-an overview. *Prostaglandins Leukot. Essent. Fatty Acids* **66,** 221–233.

Garcia, D. E., Brown, S., Hille, B., and Mackie, K. (1998). Protein kinase C disrupts cannabinoid actions by phosphorylation of the CB1 cannabinoid receptor. *J. Neurosci.* **18,** 2834–2841.

Garland, A. M., Grady, E., Lovett, M., Vigna, S. R., Frucht, M. M., Krause, J. E., and Bunnett, N. W. (1996). Mechanisms of desensitization and resensitization of G protein-coupled neurokinin1 and neurokinin2 receptors. *Mol. Pharmacol.* **49,** 438–446.

Glaser, S. T., Kaczocha, M., and Deutsch, D. G. (2005). Anandamide transport: A critical review. *Life Sci.* **77,** 1584–1604.

Hillard, C. J., and Jarrahian, A. (2003). Cellular accumulation of anandamide: Consensus and controversy. *Br. J. Pharmacol.* **140,** 802–808.

Howlett, A. C., Breivogel, C. S., Childers, S. R., Deadwyler, S. A., Hampson, R. E., and Porrino, L. J. (2004). Cannabinoid physiology and pharmacology: 30 years of progress. *Neuropharmacology* **47,** 345–358.

Hsieh, C., Brown, S., Derleth, C., and Mackie, K. (1999). Internalization and recycling of the CB1 cannabinoid receptor. *J. Neurochem.* **73,** 493–501.

Huang, C., Hepler, J. R., Chen, L. T., Gilman, A. G., Anderson, R. G., and Mumby, S. M. (1997). Organization of G proteins and adenylyl cyclase at the plasma membrane. *Mol. Biol. Cell.* **8,** 2365–2378.

Huang, C., Duncan, J. A., Gilman, A. G., and Mumby, S. M. (1999). Persistent membrane association of activated and depalmitoylated G protein alpha subunits. *Proc. Natl. Acad. Sci. USA* **96,** 412–417.

Keren, O., and Sarne, Y. (2003). Multiple mechanisms of CB1 cannabinoid receptors regulation. *Brain Res.* **980,** 197–205.

Krupnick, J. G., and Benovic, J. L. (1998). The role of receptor kinases and arrestins in G protein-coupled receptor regulation. *Annu. Rev. Pharmacol. Toxicol.* **38,** 289–319.

Kunzelmann-Marche, C., Freyssinet, J. M., and Martinez, M. C. (2002). Loss of plasma membrane phospholipid asymmetry requires raft integrity. Role of transient receptor potential channels and ERK pathway. *J. Biol. Chem.* **277,** 19876–19881.

Le, P. U., Guay, G., Altschuler, Y., and Nabi, I. R. (2002). Caveolin-1 is a negative regulator of caveolae-mediated endocytosis to the endoplasmic reticulum. *J. Biol. Chem.* **277,** 3371–3379.

Leterrier, C., Bonnard, D., Carrel, D., Rossier, J., and Lenkei, Z. (2004). Constitutive endocytic cycle of the CB1 cannabinoid receptor. *J. Biol. Chem.* **279,** 36013–36021.

McFarland, M. J., and Barker, E. L. (2005). Lipid rafts: A nexus for endocannabinoid signaling? *Life Sci.* **77,** 1640–1650.

McFarland, M. J., Porter, A. C., Rakhshan, F. R., Rawat, D. S., Gibbs, R. A., and Barker, E. L. (2004). A role for caveolae/lipid rafts in the uptake and recycling of the endogenous cannabinoid anandamide. *J. Biol. Chem.* **279,** 41991–41997.

Mechoulam, R., Ben-Shabata, S., Hanusa, L., Ligumskya, M., Kaminskib, N. E., Schatzd, A. R., Gopherc, A., Almogc, S., Martind, B. R., Comptond, D. R., Pertweee, R. G., Griffine, G., *et al.* (1995). Identification of an endogenous 2-monoglyceride, present in canine gut, that binds to cannabinoid receptors. *Biochem. Pharmacol.* **50,** 83–90.

Michel, J. B., Feron, O., Sacks, D., and Michel, T. (1997). Reciprocal regulation of endothelial nitric-oxide synthase by Ca^{2+}-calmodulin and caveolin. *J. Biol. Chem.* **272,** 15583–15586.

Mitter, D., Reisinger, C., Hinz, B., Hollmann, S., Yelamanchili, S. V., Treiber-Held, S., Ohm, T. G., Herrmann, A., and Ahnert-Hilger, G. (2003). The synaptophysin/synaptobrevin interaction critically depends on the cholesterol content. *J. Neurochem.* **84,** 35–42.

Montero, C., Campillo, N. E., Goya, P., and Paez, J. A. (2005). Homology models of the cannabinoid CB1 and CB2 receptors. A docking analysis study. *Eur. J. Med. Chem.* **40,** 75–83.

Morales, J., Fishburn, C. S., Wilson, P. T., and Bourne, H. R. (1998). Plasma membrane localization of G alpha z requires two signals. *Mol. Biol. Cell* **9,** 1–14.

Moro, O., Lameh, J., and Sadee, W. (1993). Serine- and threonine-rich domain regulates internalization of muscarinic cholinergic receptors. *J. Biol. Chem.* **268,** 6862–6865.

Mukhopadhyay, S., and Howlett, A. C. (2001). CB1 receptor-G protein association. Subtype selectivity is determined by distinct intracellular domains. *Eur. J. Biochem.* **268,** 499–505.

Mukhopadhyay, S., Sandiford, S., Howlett, A. C., and Chambers, J. L. (2002). "Palmitoylation Regulates CB1 Cannabinoid Receptors-G Protein Selectivity" p. A133. International Union of Pharmacology, San Francisco, CA.

Ohvo, H., and Slotte, J. P. (1996). Cyclodextrin-mediated removal of sterols from monolayers: Effects of sterol structure and phospholipids on desorption rate. *Biochemistry* **35,** 8018–8024.

Pertwee, R. G. (1997). Pharmacology of cannabinoid CB1 and CB2 receptors. *Pharmacol. Ther.* **74,** 129–180.

Pike, L. J. (2003). Lipid rafts: Bringing order to chaos. *J. Lipid. Res.* **44,** 655–667.

Pike, L. J., Han, X., Chung, K. N., and Gross, R. W. (2002). Lipid rafts are enriched in arachidonic acid and plasmenylethanolamine and their composition is independent of caveolin-1 expression: A quantitative electrospray ionization/mass spectrometric analysis. *Biochemistry* **41,** 2075–2088.

Piomelli, D. (2003). The molecular logic of endocannabinoid signalling. *Nat. Rev. Neurosci.* **4,** 873–884.

Razani, B., Woodman, S. E., and Lisanti, M. P. (2002). Caveolae: From cell biology to animal physiology. *Pharmacol. Rev.* **54,** 431–467.

Rinaldi-Carmona, M., Le Duigou, A., Oustric, D., Barth, F., Bouaboula, M., Carayon, P., Casellas, P., and Le Fur, G. (1998). Modulation of CB1 cannabinoid receptor functions after a long-term exposure to agonist or inverse agonist in the Chinese hamster ovary cell expression system. *J. Pharmacol. Exp. Ther.* **287,** 1038–1047.

Roth, A., Kreienkamp, H. J., Meyerhof, W., and Richter, D. (1997). Phosphorylation of four amino acid residues in the carboxyl terminus of the rat somatostatin receptor subtype 3 is crucial for its desensitization and internalization. *J. Biol. Chem.* **272,** 23769–23774.

Sancho, R., Macho, A., de La Vega, L., Calzado, M. A., Fiebich, B. L., Appendino, G., and Munoz, E. (2004). Immunosuppressive activity of endovanilloids: N-arachidonoyl-dopamine inhibits activation of the NF-kappa B, NFAT, and activator protein 1 signaling pathways. *J. Immunol.* **172,** 2341–2351.

Sarker, K. P., and Maruyama, I. (2003). Anandamide induces cell death independently of cannabinoid receptors or vanilloid receptor 1: Possible involvement of lipid rafts. *Cell. Mol. Life Sci.* **60,** 1200–1208.

Sarnataro, D., Grimaldi, C., Pisanti, S., Gazzero, P., Laezza, C., Zurzolo, C., and Bifulco, M. (2005). Plasma membrane and lysosomal localization of CB1 cannabinoid receptor are dependent on lipid rafts and regulated by anandamide in human breast cancer cells. *FEBS Lett.* **579,** 6343–6349.

Shapira, M., Gafni, M., and Sarne, Y. (2003). Long-term interactions between opioid and cannabinoid agonists at the cellular level: Cross-desensitization and downregulation. *Brain Res.* **960,** 190–200.

Simons, K., and Ehehalt, R. (2002). Cholesterol, lipid rafts, and disease. *J. Clin. Invest.* **110,** 597–603.

Simons, K., and Ikonen, E. (1997). Functional rafts in cell membranes. *Nature* **387,** 569–572.

Simons, K., and Toomre, D. (2000). Lipid rafts and signal transduction. *Nat. Rev. Mol. Cell Biol.* **1,** 31–39.

Song, K. S., Sargiacomo, M., Galbiati, F., Parenti, M., and Lisanti, M. P. (1997). Targeting of a G alpha subunit (Gi1 alpha) and c-Src tyrosine kinase to caveolae membranes: Clarifying the role of N-myristoylation. *Cell. Mol. Biol. (Noisy-le-grand)* **43,** 293–303.

Sugiura, T., Kondo, S., Sukagawa, A., Nakane, S., Shinoda, A., Itoh, K., Yamashita, A., and Waku, K. (1995). 2-Arachidonoylglycerol: A possible endogenous cannabinoid receptor ligand in brain. *Biochem. Biophys. Res. Commun.* **215,** 89–97.

Tian, X., Guo, J., Yao, F., Yang, D. P., and Makriyannis, A. (2005). The conformation, location, and dynamic properties of the endocannabinoid ligand anandamide in a membrane bilayer. *J. Biol. Chem.* **280,** 29788–29795.

Wedegaertner, P. B., Wilson, P. T., and Bourne, H. R. (1995). Lipid modifications of trimeric G proteins. *J. Biol. Chem.* **270,** 503–506.

Xiang, Y., Rybin, V. O., Steinberg, S. F., and Kobilka, B. (2002). Caveolar localization dictates physiologic signaling of beta 2-adrenoceptors in neonatal cardiac myocytes. *J. Biol. Chem.* **277,** 34280–34286.

Xie, X. Q., and Chen, J. Z. (2005). NMR structural comparison of the cytoplasmic juxtamembrane domains of G-protein-coupled CB1 and CB2 receptors in membrane mimetic dodecylphosphocholine micelles. *J. Biol. Chem.* **280,** 3605–3612.

Zhang, J., Barak, L. S., Winkler, K. E., Caron, M. G., and Ferguson, S. S. (1997). A central role for beta-arrestins and clathrin-coated vesicle-mediated endocytosis in beta2-adrenergic receptor resensitization. Differential regulation of receptor resensitization in two distinct cell types. *J. Biol. Chem.* **272,** 27005–27014.

EFFECTS OF THE HIV-1 VIRAL PROTEIN TAT ON CENTRAL NEUROTRANSMISSION: ROLE OF GROUP I METABOTROPIC GLUTAMATE RECEPTORS

Elisa Neri, Veronica Musante, and Anna Pittaluga

Department of Experimental Medicine, Pharmacology and Toxicology Section
University of Genova, Genova, Italy

Human immunodeficiency virus 1 (HIV-1)-associated dementia (HAD) represents a common complication of HIV-1 infection. Antiretroviral therapy has diminished its incidence, but it is insufficient to eradicate the problem. HAD depends on the presence of the virus in central nervous system (CNS), but the molecular mechanisms involved are not completely understood. It is widely accepted that proteins shed by the virus, such as the envelope glycoprotein gp120 and the nonstructural viral protein Tat, may themselves cause alterations to CNS. By one side, viral proteins are toxic to neurons because of their ability (1) to act as excitotoxins and (2) to evoke the release of endogenous neurotoxins and/or proinflammatory cytokines. By the other side, evidences are emerging that viral components can alter neuronal functions either by modifying the release of neurotransmitters or by influencing the functions of classical receptors controlling central neurotransmission. We here review some results concerning the effects of Tat on cholinergic and noradrenergic neurotransmission in human and rat cortex. The protein can induce the release of acetylcholine from both human and rat cortical cholinergic nerve terminals in a specie-specific manner. In human cholinergic terminals, Tat-mediated releasing effect depends on activation of receptors belonging to I group of metabotropic glutamate receptors (mGluRs), while in rat terminals Tat-induced effect involves the activation of a so far unknown receptor. The protein, unable on its own to release noradrenaline from

INTERNATIONAL REVIEW OF
NEUROBIOLOGY, VOL. 82
DOI: 10.1016/S0074-7742(07)82018-6

0074-7742/07 $35.00

human and rat cortical noradrenergic nerve endings, potentiates the release of amine induced by presynaptic NMDA receptors. Also in this case, Tat effect involves activation of a receptor belonging to the group I mGluRs, in particular of the mGluR1 subtype.

The finding that group I mGluRs may represent a preferential target of the protein in CNS may be relevant to the proposal of new therapeutic approaches for the cure of HAD.

I. Neurological Complications of HIV-1 Infection

Acquired immunodeficiency syndrome (AIDS) is often accompanied by neuropsychiatric symptoms known as *Human immunodeficiency virus 1* (HIV-1)-associated dementia (HAD); (Diesing *et al.*, 2002; Lawrence and Major, 2002; Navia *et al.*, 1986), the origins of which are poorly understood. About one-third of adult infected patients, but as many as 60% of children and adolescents, develop this clinical manifestation so that HAD is considered one of the leading cause of dementia in people younger than 60 years of age (McArthur *et al.*, 1993). Longitudinal studies indicates that half of the HIV^+ patients receiving highly active antiretroviral therapy (HAART) develop neurocognitive impairments (McArthur, 2004) and taking into account that reducing the viral load by antiretroviral therapy does not impede HAD development (Lawrence and Major, 2002; McArthur *et al.*, 2005), neuronal manifestations are expected to represent a major aspect of HIV-1 infection in the future.

HAD is a consequence of HIV-1 infection of the brain rather than the consequence of opportunistic infections. In HIV-1-infected patients, the virus enters the central nervous system (CNS) at a relatively early stage of infection, presumably carried by circulating infected macrophages; during seroconversion, however, neurological dysfunctions are generally transient and not life threatening, but they become evident and persistent at the late stage of the disease.

Symptoms of HAD include attention and concentration impairments, loss of coordination, motor and mental retardation, language and comprehension problems, alterations in personality, and enhanced difficulty with daily activities and lifestyle (Ungvarski and Trzcianowska, 2000).

Neuroinflammation and neuronal death (collectively termed HIV-1 encephalopathy) are common and have also been implicated in the pathogenesis of HAD (Everall *et al.*, 1999; Masliah *et al.*, 1992). The occurrence of these neurotoxic events is sustained by histological observations made in autoptic cerebral tissue from HIV-1-positive patients (Masliah *et al.*, 1992, 1997; Navia *et al.*, 1986), where loss of neurons was observed. The hallmarks of HIV-1 encephalopathy include infiltration of macrophages, the formation of microglial nodules and multinucleated giant cells, astrogliosis, and loss of specific neuronal subpopulations. The hippocampus, the cortex, and the striatum are particularly susceptible to HIV-1-mediated injury (Masliah *et al.*, 1996).

Considering that HIV-1 encephalopathy is a gradual process, one may presume that dramatic neurodegenerative events may not be necessary to explain the occurrence of neuropsychiatric deficits. Impairments of memory and movement control, on the contrary, may probably depend on subtle changes in central neurotransmission induced by HIV-1 itself or by viral proteins. Taking into account that the virus cannot infect directly neurons, it is reasonable to hypothesize that changes in central neurotransmission may be induced either *directly*, by viral proteins shed by the virus (i.e., the capsidic protein gp120 or the constitutive protein Tat), or *indirectly*, by endogenous components (cytokines, chemokines, glutamate, quinolinate, NO) released by infected cells.

II. The HIV-1 Viral Protein Tat

The HIV-1 transactivator of transcription (Tat) belongs to the group of the nonstructural, regulatory proteins that are essential to productive virus infection. Tat is a potent activator of the HIV-LTR promoter (Green and Lowenstein, 1988), and it controls viral gene expression and replication (Kelly *et al.*, 1998).

Besides its transcriptional activity, Tat also acts as injurious molecule contributing to HAD. Tat mRNA levels are elevated in the serum (Westendorp *et al.*, 1995) and in the brain of patients with HIV-1 dementia (Wessenligh *et al.*, 1993) or HIV encephalitis (Wiley *et al.*, 1996). Tat is actively released from HIV-infected cells (Ensoli *et al.*, 1993), and then it is taken up by neighboring cells, where it activates a variety of cellular genes (Minghetti *et al.*, 2004); finally it interferes with neuronal and glial functions (Minghetti *et al.*, 2004; Nath and Geiger, 1998; Peruzzi *et al.*, 2005).

The protein comprises 101 amino acids encoded by two exons. The first exon (exon 1) forms the first 72-amino acid sequence that contains the transactivating domain. Peptides derived from the first exon were found to bind to and to excite neuronal membranes (Nath and Geiger, 1998; Sabatier *et al.*, 1991) The second exon has variable structure depending on the nature of HIV-1 strains; the C-terminal sequence encoded by this exon varies in length between 14–30 amino acids. This sequence is responsive of the binding of Tat to integrin receptors. (Brake *et al.*, 1990), as well as to other proteins (Peruzzi *et al.*, 2005).

If subtle changes in neurotransmission, independent of or preceding neurotoxicity, underlie the development of HAD, it is worthwhile to investigate if and how HIV-1 proteins alter neurotransmitter systems potentially involved in central functions. In this short chapter, we summarize some of recent results concerning the effects of Tat on the release of neurotransmitters from human and rat isolated nerve endings (synaptosomes in superfusion).

III. About the Experimental Approach

This work was performed by using a methodology which has been in use in various laboratories for several years. Nonetheless, some readers may appreciate a few explanatory considerations related to the experimental approach used here.

Synaptosomes are resealed pinched off nerve terminals, isolated by homogenizing the tissue in phosphate-buffered sucrose. These particles retain the physiological functions characteristic of nerve terminals, so that they can uptake, synthesize, and release neurotransmitters. They are endowed with native receptors that can mediate neurotransmitter release; they also possess the functional intraterminal machinery needed to allow neurotransmitter release and receptor-receptor cross talk.

After selective labeling with radioactive tracer, identical portions of synaptosomal suspension are filtered through microporous filters placed at the bottom of several (up to 24) parallel superfusion chambers thermostated at 37 °C and then up–down superfused at 0.5 ml/min (Raiteri *et al.*, 1974). Under these experimental conditions, endogenous compounds just released are immediately removed by the superfusion solution so that the concentration of endogenous compounds at nerve terminals is minimized. Additionally, the continuous superfusion avoids indirect effects due to endogenous substances since compounds released by one particle could not act at adjacent isolated terminals nor they can feedback and control the functions of the releasing particles.

If an exogenous compound (i.e., the viral protein Tat), added to the superfusion medium, modifies the amount of radioactive tracers released, it could be hypothesized that, in some way, the protein interferes with the molecular mechanism controlling the release of neurotransmitter. Inasmuch, if the releasing effect induced by the viral protein is impeded/prevented by contemporary administration of an antagonist having a well-known pharmacological profile, it can be proposed that the receptor selectively targeted by the antagonist may represent a preferential target of the viral protein.

To conclude, the system appears particularly convenient to investigate in general functional interactions among exogenous compounds (such as the viral proteins) with native receptors or other proteins colocalized on a well-identified neuronal system.

IV. Effects of Tat on the Release of Neurotransmitters in CNS

In the early 1990, iodinated Tat (the full-length protein and the 38–68 amino acid sequence) was shown to bind with moderate affinity to rat brain synaptosomal membranes in a saturable and competitive manner. Inasmuch, Tat 38–86 was found to induce a rapid and large membrane depolarization in the cercal-afferent giant interneuron synapses in the CNS of the cockroach *Periplaneta americana* (Sabatier *et al.*, 1991). The notion that Tat can excite neurons was further supported by findings showing that (1) Tat depolarized rat CA1 hippocampal neurons (Magnuson *et al.*, 1995) and cultured human fetal neurons (Cheng *et al.*, 1998; Nath *et al.*, 1996), and (2) Tat caused elevation of cytosolic Ca^{2+}, possibly by mobilizing the ion from intraterminal stores (Cheng *et al.*, 1998; Haughey *et al.*, 1999; Nath *et al.*, 1996).

All together, these observations may suggest that Tat can elicit a Ca^{2+}-dependent release of neurotransmitters from synaptic terminals. In an attempt to evaluate this possibility, and taking into account the excitotoxic nature of Tat-induced neuronal death (Cheng *et al.*, 1998; Nath and Geiger, 1998; Sabatier *et al.*, 1991; Wang *et al.*, 1999), attention first focused on the effects of the protein on the release of glutamate in CNS.

Despite the expectation, however, Tat was unable to affect the release of endogenous glutamate or preloaded [^{3}H]-D-aspartate ([^{3}H]-D-Asp, a nonmetabolized tracer used as marker for neuronal glutamate) from human and rat synaptosomes. Indeed, the spontaneous release of endogenous glutamate and of [^{3}H]-D-Asp (Feligioni *et al.*, 2003) as well as the release of endogenous glutamate induced by depolarizing (30 mM) stimuli (Haughey *et al.*, 2001) were unaffected by the protein.

The protein also failed to modify the spontaneous release of other endogenous neurotransmitters (Feligioni *et al.*, 2003), including aspartate, γ-amino butyric acid, glycine, and D/L-serine as well as the spontaneous release of noradrenaline (measured as release of preloaded [^{3}H]NA), but it strongly potentiated the release of [^{3}H]ACh from both human and rat cortical preparations (Feligioni *et al.*, 2003).

Tat induced the release of [^{3}H]ACh from human and rat cortical synaptosomes in a concentration-dependent manner (0.1–10 nM), reaching a maximal effect when applied at 1 nM, a concentration probably compatible with the one released in the CNS during virus infection (Westendorp *et al.*, 1995; but see also Nath and Geiger, 1998). Tat-evoked releasing effect depended on the presence of the protein since heat-shock denaturation or immunoprecipitation of Tat prevented the releasing effect (Feligioni *et al.*, 2003).

Removal of Ca^{2+} from the superfusion medium halved the Tat-induced release of acetylcholine from human synaptosomes, while left unchanged that from rat cortical synaptosomes (Feligioni *et al.*, 2003) to suggest that the protein can evoke the release of the neurotransmitter in a specie-specific manner. However, in both cases (human and rat), the Tat-induced release was almost completely prevented when the membrane-impermeant Ca^{2+} chelator BAPTA was entrapped into synaptosomes, suggesting that Ca^{2+} release from internal stores participates to the induced releasing effect induced by Tat in both human and rat cortical nerve endings (Feligioni *et al.*, 2003).

V. Effects of Tat on Presynaptic AMPA/Kainate Receptors

In cultured human fetal neurons, membrane depolarization and increase of Ca^{2+} ions caused by Tat were abolished by the selective α-amino-3-hydroxy-5-methyl-4-isoxazole propionate (AMPA)/kainate receptor antagonist NBQX to suggest that the Tat-mediated effect can rely on activation of non-*N*-methyl-D-aspartate (NMDA) receptors (Magnuson *et al.*, 1995; Nath *et al.*, 1996). In view of the possibility that AMPA receptors (AMPARs) may represent selective neuronal targets for the viral protein, we investigated whether and to what extent Tat can modulate the functions of AMPARs controlling the release of neurotransmitters in CNS.

The mammalian AMPARs are heterotetrameric complexes composed of four subunits, termed GluR1 to GluR4 (Dingledine *et al.*, 1999; Rosenmund *et al.*, 1998). AMPARs exist at the presynaptic level on cholinergic and noradrenergic terminals, where they mediate neurotransmitter release in a Ca^{2+}-dependent, exocytotic-like fashion (Ghersi *et al.*, 2003). These presynaptic AMPARs represent pharmacologically distinct subtypes based on sensitivity to antagonists and to the allosteric modulator cyclothiazide (Ghersi *et al.*, 2003; Pittaluga *et al.*, 1997).

As shown in Fig. 1, exposure of human and rat cortical cholinergic synaptosomes to 100-μM AMPA elicits the release of [^{3}H]ACh, an effect totally prevented by the broad spectrum AMPA/kainate antagonist NBQX. On the contrary, NBQX failed to inhibit the release of tritium caused by 1-nM Tat to suggest that the protein does not bind to the presynaptic AMPA/kainate receptors localized on cholinergic terminals. Moreover, the releasing effects caused by 1-nM Tat and 100-μM AMPA, respectively, were addictive in both synaptosomal preparations to further suggest that the two events occur through different mechanisms.

Human and rat cortical noradrenergic terminals are also endowed with presynaptic AMPARs, whose activation elicits the release of noradrenaline in an exocytotic manner; the AMPA-evoked release of [^{3}H]NA from both synaptosomal preparations was inhibited by NBQX (Pittaluga *et al.*, 2005) (Fig. 2).

Tat unaffected the functions of the presynaptic AMPARs located on noradrenergic terminals. As already stated, Tat was unable to mimic AMPA at the glutamate-binding site located on AMPARs since it failed to evoke on its own the release of preloaded [^{3}H]NA from both human and rat cortical noradrenergic nerve endings. Moreover, the protein failed to modify the release of amine caused by 100-μM AMPA (Fig. 2).

VI. Effects of Tat on Presynaptic NMDA Receptors

Functional native NMDA receptors (NMDARs) are voltage- and ligand-gated ion channels, the stimulation of which results in Na^+ and Ca^{2+} entry. They are heteromeric assembly of four/five subunits comprising NR1 and NR2 or NR3

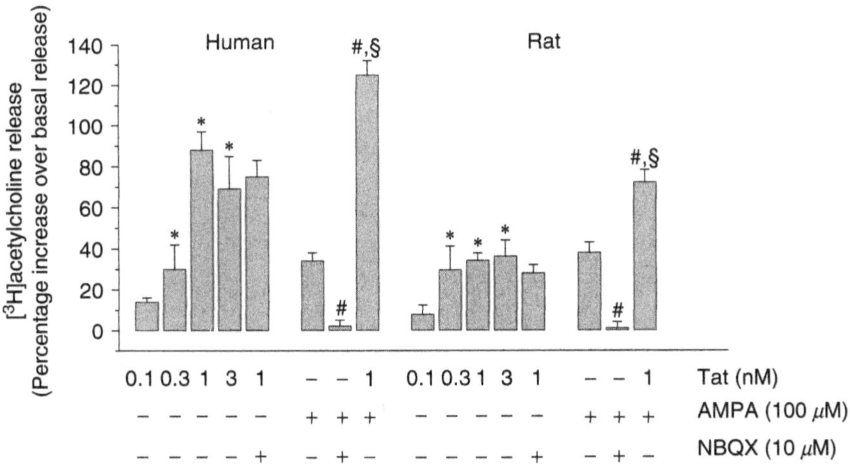

FIG. 1. Effects of Tat on the spontaneous and the AMPA-evoked release of ACh from human and rat cortical synaptosomes: antagonism by NBQX. Human cortical synaptosomes were prepared from samples of human cerebral cortex obtained from informed and consentient HIV-1 negative patients undergoing neurosurgery to reach deeply seated tumors. Synaptosomes were prepared within 90 min after tissue removal. Rat synaptosomes were prepared from cortices of adult male rats (Sprague-Dawley; 200–250 g) killed by decapitation. Animal care and experimental procedures were approved by the Ethical Committee in accordance with the European legislation (European Communities Council Directive of November 24, 1986, 86/609/EEC). Human and rat cortical synaptosomes were prelabeled with [^{3}H]Ch (final concentration 80 nM) and then stratified at the bottom of superfusion chambers. Synaptosomes were superfused at the flow rate of 0.5 ml/min for 36 min to equilibrate the system and then six consecutive 3-min fractions were collected. Human and rat cortical synaptosomes were exposed in superfusion to Tat, AMPA, or Tat plus AMPA (concentrations as indicated) a $t = 39$ min of superfusion. When indicated, NBQX was present starting from 8 min before agonist to the end of the superfusion. Samples collected were analyzed for radioactive content. Results are expressed as percentage increase over basal (no drug added) release. Data are mean ± SEM of at least three experiments run in triplicate. *$p < 0.05$ at least versus control; # $p < 0.05$ at least versus AMPA; § $p < 0.05$ at least versus Tat.

subunits. NR1 subunits exist in nine different splice variants, varying for H^+ and PKC sensitivities, while NR2 subunits consist of four different types, namely, the NR2A to D subunits, and NR3 consists of two (NR3A and B) subunits. Receptor activation is caused by glutamate acting at the recognition-binding site located on NR2 subunits, but it is permitted and regulated by the coagonist glycine. Glycine binds to the site located on NR1 subunits, then enabling glutamate-mediated receptor activation (Dingledine *et al.*, 1999). In CNS, NMDARs exhibit both pre- and postsynaptic localization. Under resting conditions, Mg^{2+} blocks the NMDAR-associated ion channel and the activation of the receptor only occurs when Mg^{2+} is removed from the channel, that is, by omitting Mg^{2+} ions in the external mileau or by contemporary application of a mild-depolarizing (i.e., K^+ or AMPA) stimulus (Mela *et al.*, 2006; Raiteri *et al.*, 1992).

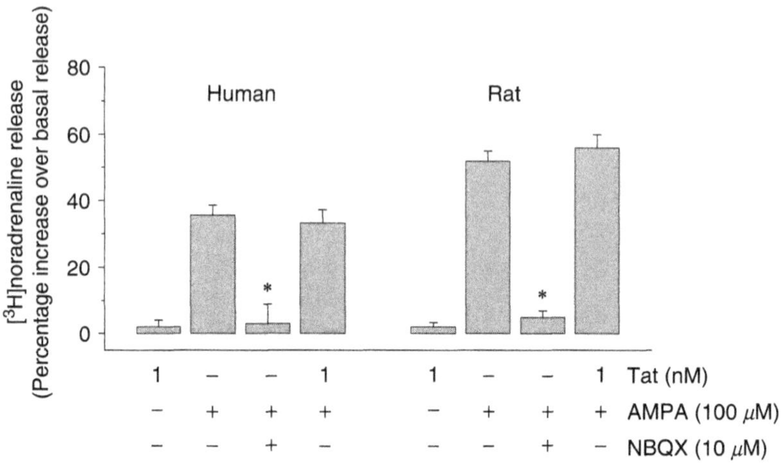

FIG. 2. Effects of Tat on the spontaneous and the AMPA-evoked release of NA from human and rat cortical synaptosomes: antagonism by NBQX. Human and rat cortical synaptosomes were prelabeled with [^{3}H]NA (final concentration 50 nM) in the presence of the transporter blockers 6-nitroquipazine (0.1 μM) and GBR 12909 (0.1 μM) to avoid false labeling with [^{3}H]NA of serotonergic and dopaminergic nerve terminals, respectively, and then superfused as described in the legend to Fig. 1. Results are expressed as percentage increase over basal (no drug added) release. Data are means $\pm$ SEM of at least three experiments run in triplicate. *$p < 0.05$ at least versus control.

Tat has been reported to modulate positively NMDARs in neuronal and organotypic slices (Chandra *et al.*, 2005; Haughey *et al.*, 2001; Prendergast *et al.*, 2002; Self *et al.*, 2003; Song *et al.*, 2003), although the site of action, whether on NMDARs or on receptive sites that can cross talk with NMDARs, or on both, is not well understood. Trying to elucidate the mode of interaction of Tat with NMDARs, we investigate whether and to what extent Tat can modulate the functions of NMDARs presynaptically located on human and rat cortical nerve endings.

Presynaptic NMDARs have been reported to be present on rat cholinergic cortical nerve terminals (Mela *et al.*, 2006), but they do not participate to the Tat-mediated releasing effect in cortical cholinergic nerve endings since the release of [^{3}H]ACh caused by 1-nM Tat was unaffected, in both human and rat synaptosomal preparations, by contemporary administration of CGS19755, a competitive antagonist at the glutamate-binding site located on NMDAR complex (Feligioni *et al.*, 2003).

Presynaptic NMDARs also exist on human and rat cortical noradrenergic nerve endings: these receptors are physiologically blocked by Mg^{2+}, but they became active when external Mg^{2+} is removed or when vicinal AMPARs, colocalized on the same synaptosomal membrane, are contemporary activated (Pittaluga and Raiteri, 1990, 1994; Raiteri *et al.*, 1992).

Results in Fig. 3 indicate that Tat does not mimic glutamate at the NMDARs controlling the release of noradrenaline from human and rat cortical terminals. Tat did not evoke the release of amine, also when superfusion of synaptosomes was performed in absence of Mg^{2+} ions and in presence of saturating glycine, conditions known to permit almost maximal activation of NMDARs by glutamate-like ligand (Longordo *et al.*, 2006; Pittaluga and Raiteri, 1990, 1994).

A striking result was obtained when studying the effect of the protein on the NMDA/glycine-evoked release of noradrenaline. Tat was found to strongly reinforce the release of [^{3}H]NA caused by NMDA/glycine from both human and rat cortical synaptosomal preparations (i.e., higher concentration of NMDA

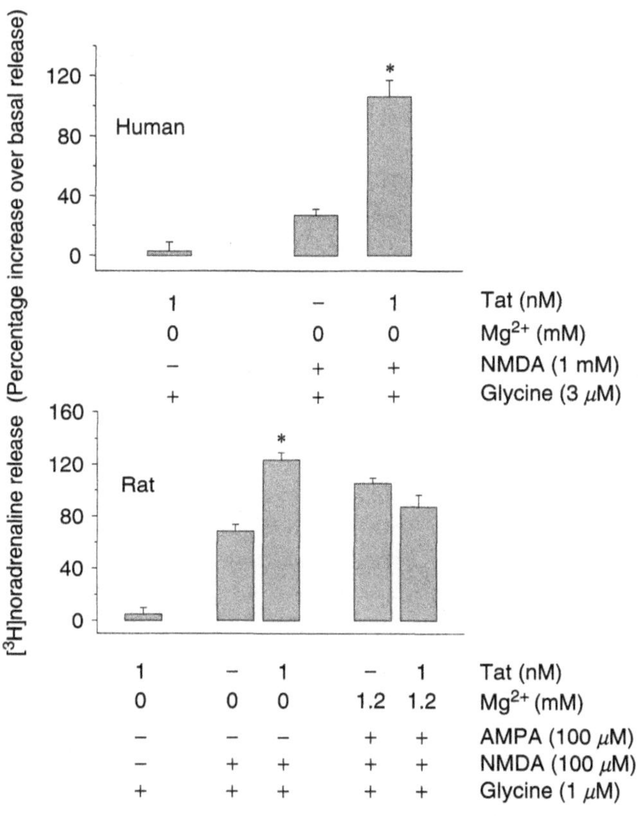

FIG. 3. Effects of Tat on the spontaneous and the NMDA/glycine-evoked release of NA from human and rat cortical synaptosomes. Human (upper panel) and rat (lower panel) cortical synaptosomes were exposed to protein and agonists as indicated starting at $t = 39$ min of superfusion. Results are expressed as percentage increase over basal (no drug added) release. Data are means ± SEM of at least three experiments run in triplicate. $*p < 0.05$ at least versus control.

and glycine are needed to disclose a releasing effect from human synaptosomes when compared to the rodent ones, possibly because of the better viability of rodent preparation, see also Longordo *et al.*, 2006). Curiously, however, the potentiation by Tat of the NMDA/glycine-mediated releasing effect was unveiled only when NMDAR activation was induced by removing Mg^{2+} from superfusion medium. When NMDAR functions were disclosed by contemporary activation of colocalized AMPARs, Tat was unable to potentiate the AMPA/NMDA/glycine-evoked release of tritium (Fig. 3, lower panel). All together these results indicate that (1) the viral protein affects, although indirectly, the function of NMDAR presynaptically located on noradrenergic terminals and (2) the occurrence of the protein-induced potentiation of the NMDAR-mediated releasing effect depends on the mode of activation of the ionotropic receptor.

It is worth reminding that studies with neuronal and organotypic slice cultures of rat hippocampus aimed at understanding the mechanisms of NMDAR upregulation by Tat proposed that the protein activate the polyamine-sensitive site located on NMDARs (Prendergast *et al.*, 2002; Self *et al.*, 2003). Actually, the neurotoxic effects produced by Tat in the CA1 and CA3 pyramidal cell layers (Prendergast *et al.*, 2002) as well as the elevation in intracellular Ca^{2+} caused by the protein in the same regions (Self *et al.*, 2003) were significantly attenuated by the presence of polyamine antagonist arcaine. In our experimental conditions, the potentiation by Tat of the NMDA/glycine-evoked release of [^{3}H]NA in human and rat synaptosomes was insensitive to arcaine (not shown, see Longordo *et al.*, 2006), suggesting that the effect of Tat on [^{3}H]NA release does not involve the binding of the protein to the polyamine allosteric binding site.

Finally, the possibility that Tat act at a Zn^{2+}-sensitive binding site on NMDAR as proposed by Song *et al.* (2003) or that the effect of the protein is due to its ability to chelate endogenous Zn^{2+} (Chandra *et al.*, 2005) is indirectly excluded by the observation that the NMDARs located on noradrenergic terminals are constituted by coassembly of $NR1_{00x}$ with NR2B, but not NR2A, subunits (Pittaluga *et al.*, 2001), a condition known to make the receptor insensitive to the presence of Zn^{2+}.

VII. Effects of Tat on Presynaptic Metabotropic Glutamate Receptors

Haughey *et al.* (1999) demonstrated that Tat activates pertussis toxin–sensitive G-protein–coupled receptors (GPCRs), then eliciting the release of Ca^{2+} from IP_3-sensitive intraterminal stores. The authors also hypothesized that the GPCR-mediated phosphorylative processes triggered by Tat can reverberate on NMDARs colocalized with the GPCRs, reinforcing the function of the ionotropic glutamate receptors (Haughey *et al.*, 2001). Given the hydrophobic nature of the protein, Tat was proposed to penetrate neuronal membranes then interacting *directly* with G-protein in the inner side of neuronal membranes (Haughey *et al.*, 2001).

Well in keeping with these findings, Tat-mediated releasing effect in human, but not in rat, cortical cholinergic synaptosomes was found to be dependent on IP_3-mediated mechanisms. The releasing effect caused by 1-nM Tat was significantly prevented by xestospongin C, a selective antagonist at IP_3 receptors located on the endoplasmic reticulum (Feligioni *et al.*, 2003). Differently, the effect of Tat on rat cortical cholinergic terminals was found to depend on activation of an intraterminal pathway involving (1) overproduction of cyclic adenosine diphosphate ribose (cADP-ribose, cADPR); (2) activation of ryanodine receptors (RYRs); and (3) the release of Ca^{2+} ions from intraterminal stores (Feligioni *et al.*, 2003). So far, the nature of the cellular target involved in this chain of events is unknown. Interestingly, however, data are present in literature demonstrating that in Aplysia buccal ganglion neurons (Mothet *et al.*, 1998), the injection of cADPR facilitates acetylcholine release in a Ca^{2+}-dependent fashion. A schematic representation of the two intraterminal pathways is reported in Fig. 4.

Since the Tat-mediated releasing effect in human cortical cholinergic terminals was independent of either AMPA or NMDARs and taking into account that

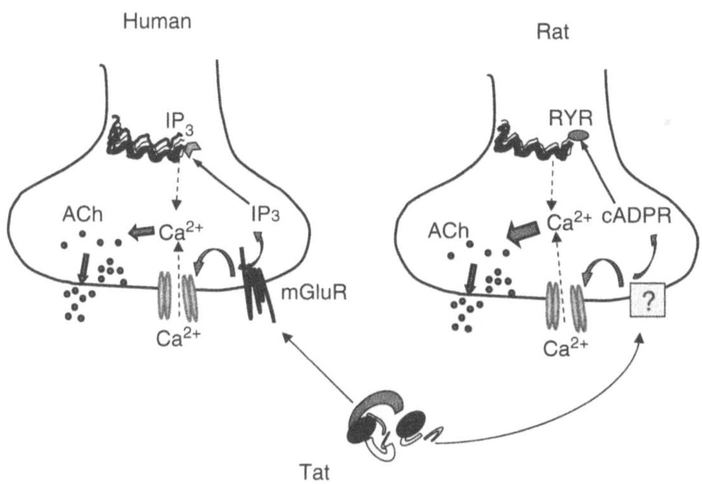

FIG. 4. Effects of Tat on the release of ACh from human and rat cortical cholinergic synaptosomes. In human cortex, Tat activates glutamate I group metabotropic receptors on ACh terminals leading to inositol triphosphate (IP_3) production, IP_3 receptor (IP_3R) activation, mobilization of Ca^{2+} from the endoplasmic reticulum (ER), and vesicular ACh release. In rat cortex cholinergic terminals, Tat binds to an unidentified receptor whose activation leads to ACh release. This release also is dependent on intraterminal Ca^{2+}, but this is mobilized by ryanodine receptor (RYR) activation via the endogenous cADPR.

the involvement of IP_3-mediated signaling may favor a role of GPCRs, we sought to evaluate if metabotropic receptors could be involved. Again, due to the excitotoxic nature of Tat-mediated detrimental effects, we focused on metabotropic glutamate receptors (mGluRs) and in particular on the receptors belonging to group I since these receptors are known to mobilize Ca^{2+} from the intraterminal IP_3-sensitive stores (Conn and Pin, 1997).

Human cortical synaptosomes prelabeled with [^{3}H]choline were then exposed to Tat in the presence of CPCCOEt, a highly selective noncompetitive negative allosteric modulator of mGluR1, or in presence of MPEP, a highly selective noncompetitive negative allosteric modulator of mGluR5. Either CPCCOEt (5 μM) or MPEP (1 μM) significantly inhibited the Tat-evoked release of [^{3}H]ACh; the inhibitory effects of the two compounds were not additive (Feligioni *et al.*, 2003). The Tat-evoked release of [^{3}H]ACh was also totally prevented by the compound LY367385, a competitive antagonist at the glutamatergic binding site on mGluR1. These results favored the involvement of glutamate metabotropic receptors belonging to the group I, a thesis further supported by the finding that the broad spectrum group I mGluR agonist 3,5-DHPG, concentration-dependently released [^{3}H]ACh from human cortical nerve endings. Also in this case, the releasing effect induced by 3,5-DHPG (100 μM) was significantly antagonized by CPCCOEt or MPEP and the effects of the two antagonists were not additive (Feligioni *et al.*, 2003). All together these results suggest that (1) human cortical cholinergic terminals are endowed with group I mGluRs, and (2) these receptors can be selective targeted by the viral protein Tat to elicit the release of [^{3}H]ACh in an IP_3-sensitive manner.

This finding prompted us to verify whether the Tat-mediated modulation of NMDARs site on noradrenergic terminals also involve mGluRs. Human synaptosomes were exposed to NMDA/glycine/Tat in the presence of CPCCOEt or MPEP, respectively. CPCCOEt totally prevented the Tat-mediated component of the NMDA/glycine/Tat-evoked release, whereas MPEP exerted only a slight, nonsignificant inhibition. To confirm the involvement of an mGlu1 receptor subtype, LY367385 was found able to impede the Tat-mediated component of the NMDA/glycine/Tat-evoked [^{3}H]NA release from human nerve terminals.

Almost superimposable results were obtained when rat synaptosomes were exposed to NMDA/glycine/Tat in the presence of the two noncompetitive mGluR antagonists. CPCCOEt totally prevented the Tat-mediated component of the evoked release, while MPEP was almost inactive (Longordo *et al.*, 2006). Moreover, LY367385 concentration-dependently (10 nM–1 μM) inhibited the Tat-mediated potentiation of the NMDA-induced responses, causing complete inhibition when added at 0.1 μM. Since LY367385 is a competitive antagonist of mGluR1, we investigated if Tat could indeed compete with this drug. As already stated, 0.1-μM LY367385 inhibited the Tat-mediated component of the NMDA/glycine/Tat-evoked [^{3}H]NA release; increasing Tat concentration up to 10 nM did not further potentiate the NMDA/glycine-evoked release, but it reversed completely the antagonism by LY367385 (Longordo *et al.*, 2006). This result in our opinion strongly supports the thesis that Tat acts as an mGluR1 preferential agonist that binds to the outer side of mGluR1 receptor.

The effect of Tat on NA release thus requires mGluR1-NMDAR cross talks. Experiments aimed to investigate the transducing mechanism involved in the mGluR1-NMDAR-receptor interaction revealed that receptor cross talk occurs through intraterminal pathways involving PLC activation and downstream signaling transduced by a PKC/Src-mediated pathway (summarized in Fig. 5), while other enzymes, like PKA or CaMKII, appear not to be involved (Longordo *et al.*, 2006). Notably, this cascade has been already proposed to mediate upregulation of NMDA currents by various GPCRs, including group I mGluRs (Salter and Kalia, 2004).

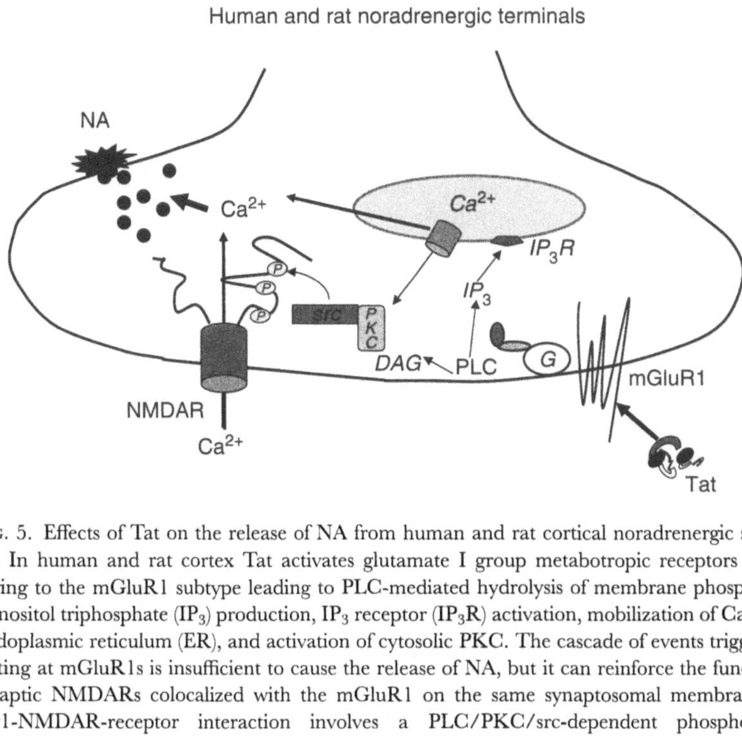

FIG. 5. Effects of Tat on the release of NA from human and rat cortical noradrenergic synaptosomes. In human and rat cortex Tat activates glutamate I group metabotropic receptors possibly belonging to the mGluR1 subtype leading to PLC-mediated hydrolysis of membrane phosphoinositides, inositol triphosphate (IP_3) production, IP_3 receptor (IP_3R) activation, mobilization of Ca^{2+} from the endoplasmic reticulum (ER), and activation of cytosolic PKC. The cascade of events triggered by Tat acting at mGluR1s is insufficient to cause the release of NA, but it can reinforce the functions of presynaptic NMDARs colocalized with the mGluR1 on the same synaptosomal membrane. The mGlur1-NMDAR-receptor interaction involves a PLC/PKC/src-dependent phosphorylative pathway.

VIII. Specie Specificity of Tat-Mediated Effects and Amino Acid Sequences Involved

The effect of the full-length protein in human cholinergic and in human and rat noradrenergic synaptosomes was found to be retained by the sequence 32–62 and 41–60, but not by 49–85, suggesting that the Tat-recognition binding site to mGluR1 resides within this sequence. To note, the two peptides contain the "arginine-rich basic region"(region 49–57), a sequence known to be critical for efficient transactivation (Ensoli *et al.*, 1993). This sequence was found to be responsive of the binding of the protein to synaptosomal membranes (Sabatier *et al.*, 1991) and to induce depolarization in hippocampal neurons and human fetal neurons (Kruman *et al.*, 1998; Nath *et al.*, 1996).

On the contrary, Tat 49–86, but not Tat 32–62, affected the [^{3}H]ACh release from rat cortical synaptosomes. The potentiating effect was retained by the peptide 61–80, while "flanking" regions 51–70 and 71–80 were inactive. The sequence 61–80 contains the RGD motif (aa 78–80), a high conservative domain that is common to extracellular matrix proteins (ECM, Kolson *et al.*, 1993). ECM molecules recognize and bind to integrin receptors through the sequence RGD and exogenous proteins expressing this sequence in an epitopic region, like Tat, are likely candidate for binding to integrin receptors (Jones, 1996).

All together these results suggest that the rodent model should be exploited with caution. Actually, while the NMDAR-mediated NA release appears appropriate to study effects of Tat and fragments on glutamate receptors, the mechanisms involved in the Tat-mediated releasing effects in human terminals strikingly differ from those in rodent terminals.

IX. Conclusions

It is concluded that human cortical cholinergic and noradrenergic terminals as well as rat noradrenergic terminals are endowed with group I mGluRs, in particular with mGluR1s. These receptors can be selectively targeted by the HIV-1 viral protein Tat. In human cortical cholinergic terminals, Tat elicits the release of [^{3}H]ACh in a IP_3-induced, Ca^{2+}-dependent fashion (Feligioni *et al.*, 2003). In human and rat noradrenergic terminals Tat is ineffective on its own, but acts in cooperation with coexisting NMDARs. The mGluR1–NMDAR interaction, triggered by concomitant exposure to Tat and NMDA, leads to enhancement of NE release (Longordo *et al.*, 2006).

The protein recognizes with picomolar affinity the group I mGluRs so that these receptors may be considered preferential targets of the protein. The amount of protein needed to activate group I mGluRs is significantly lower (about two orders of magnitude) than the one known to cause neuronal death, indicating that impairments of central neurotransmission and neurotoxicity are distinct. The synthesis of orally available, blood–brain barrier-permeant mGluR1 antagonists may be fundamental to propose new approaches for the cure of HAD, complementary to antiretroviral therapy.

Acknowledgments

The authors thank the Centralised Facility for AIDS Reagents and the UK Medical Research Council. We are grateful to Mrs. Maura Agate for careful editorial assistance. This work was supported by grants from Istituto Superiore di Sanità (Programma nazionale di Ricerca sull'AIDS—Progetto "Patologia, Clinica e Terapia dell'AIDS") and from Italian MIUR.

References

Brake, D. A., Debouck, C., and Biesecker, G. (1990). Identification of an Arg-Gly-Asp (RDG) cell adhesion site in human immunodeficiency virus type 1 transactivation protein Tat. *J. Cell Biol.* **111,** 1275–1281.

Chandra, T., Mier, W., Konig, H. G., Hirzel, K., Kogel, D., Schuler, T., Chandra, A., Demirhan, I., and Laube, B. (2005). Molecular interactions of the type 1 human immunodeficiency virus transregulatory protein Tat with N-methyl-D-aspartate receptor subunits. *Neuroscience* **134,** 145–153.

Cheng, J., Nath, A., Knudsen, B., Hochman, S., Geiger, J. D., Ma, M., and Magnuson, D. S. K. (1998). Neuronal excitatory properties of human immunodeficiency virus type 1 Tat protein. *Neuroscience* **82,** 97–106.

Conn, P. J., and Pin, J. P. (1997). Pharmacology and functions of metabotropic glutamate receptors. *Annu. Rev. Pharmacol. Toxicol.* **37,** 205–237.

Diesing, T. S., Swindells, S., Gelbard, H., and Gendelman, H. E. (2002). HIV-1-associated dementia: A basic science and clinical perspective. *AIDS Read.* **12,** 358–368.

Dingledine, R., Borges, K., Bowie, D., and Traynelis, S. F. (1999). The glutamate receptor ion channels. *Pharmacol. Rev.* **51,** 7–61.

Ensoli, B., Buonaguro, L., Barillari, G., Fiorelli, V., Gendelman, R., Morgan, R. A., Wingfield, P., and Gallo, C. (1993). Release, uptake and effects of extracellular HIV-1 tat protein on cell growth in viral transactivation. *J. Virol.* **67,** 277–287.

Everall, I. P., Heaton, R. K., Marcotte, T. D., Ellis, R. J., McCutchan, J. A., Atkinson, J. H., Grant, I., Mallory, M., and Masliah, E. (1999). Cortical synaptic density is reduced in mild to moderate human immunodeficiency virus neurocognitive disorder. *Brain Pathol.* **9,** 209–217.

Feligioni, M., Raiteri, L., Pattarini, R., Grilli, M., Bruzzone, S., Cavazzani, P., Raiteri, M., and Pittaluga, A. (2003). The human immunodeficiency virus-1 protein Tat and its discrete fragments evoke selective release of acetylcholine from human and rat cerebrocortical terminals through species-specific mechanisms. *J. Neurosci.* **23,** 6810–6818.

Ghersi, C., Bonfanti, A., Manzari, B., Feligioni, M., Raiteri, M., and Pittaluga, A. (2003). Pharmacological heterogeneity of release-regulating presynaptic AMPA/kainate receptors in the rat brain: Study with receptor antagonists. *Neurochem. Int.* **42,** 283–292.

Green, J. D., and Lowenstein, P. M. (1988). Autonomous functional domains of chemically synthesized human immunodeficiency virus tat trans-activator protein. *Cell* **55,** 1179–1188.

Haughey, N. J., Holden, C. P., Nath, A., and Geiger, J. D. (1999). Involvement of inositol 1,4,5-trisphospahte-regulated stores of intracellular calcium in calcium dysregulation and neuron cell death caused by HIV-1 protein Tat. *J. Neurochem.* **73,** 1363–1374.

Haughey, N. J., Nath, A., Mattson, M. P., Slevin, J. T., and Geiger, J. D. (2001). HIV-1 Tat through phosphorylation of NMDA receptors potentiates glutamate excitotoxicity. *J. Neurochem.* **78,** 457–467.

Jones, S. L. (1996). Integrins: Possible functions in the adult CNS. *Trends Neurosci.* **19,** 68–72.

Kelly, G. D., Ensoli, B., Gunthel, C. J., and Offermann, M. K. (1998). Purified Tat induces inflammatory response genes in Kaposi's sarcoma cells. *AIDS* **12,** 1753–1761.

Kolson, D. L., Buchhalter, J. U., Collman, R., Hellmig, B., Farell, C. F., Debourek, C., and Gonzales-Scarano, F. (1993). HIV-1 Tat alters normal organization of neurones and astrocytes in primary rodent brain cell cultures: RDG sequence dependence. *AIDS Res. Hum. Retroviruses* **9,** 677–685.

Kruman, I. I., Nath, A., and Mattson, M. P. (1998). HIV-1 protein Tat induces apoptosis of hippocampal neurons by a mechanism involving caspase activation, calcium overload, and oxidative stress. *Exp. Neurol.* **154,** 276–288.

Lawrence, D. M., and Major, E. O. (2002). HIV-1 and the brain: Connections between HIV-1-associated dementia, neuropathology and neuroimmunology. *Microbes Infect.* **4,** 301–308.

Longordo, F., Feligioni, M., Chiaramonte, G., Sbaffi, P., Raiteri, M., and Pittaluga, A. (2006). The human immunodeficiency virus-1 protein Tat upregulates NMDA receptor function by acting at mGluR1 receptors coexisting on human and rat brain noradrenergic neurons. *J. Pharmacol. Exp. Ther.* **317,** 1097–1105.

Magnuson, D. S. K., Knudsen, B. E., Geiger, J. D., Brownstone, R. M., and Nath, A. (1995). Human immunodeficiency virus type 1 Tat activates non-N-methyl-D-aspartate excitatory amino acid receptors and causes neurotoxicity. *Ann. Neurol.* **37,** 373–380.

Masliah, E., Achim, C. L., Ge, N., DeTeresa, R., Terry, R. D., and Wiley, C. A. (1992). Spectrum of human immunodeficiency virus-associated neocortical damage. *Ann. Neurol.* **32,** 321–329.

Masliah, E., Ge, N., and Mucke, L. (1996). Pathogenesis of HIV-1 associated neurodegeneration. *Crit. Rev. Neurobiol.* **10,** 57–67.

Masliah, E., Heaton, R. K., Marcotte, T. D., Ellis, R. J., Wiley, C. A., Mallory, M., Achim, C. L., McCutchan, J. A., Nelson, J. A., Atkinson, J. H., and Grant, I. (1997). Dendritic injury is a pathological substrate for human immunodeficiency virus-related cognitive disorders. *Ann. Neurol.* **42,** 963–972.

McArthur, J. C. (2004). HIV-1 dementia: An evolving disease. *J. Neuroimmunol.* **157,** 3–10.

McArthur, J. C., Hoover, D. R., Bacellar, H., Miller, E. N., Cohen, B. A., Becker, J. T., Graham, N. M., McArthur, J. H., Selnes, O. A., Jacobson, L. P., Visscher, B. R., Concha, M., *et al.* (1993). Dementia in AIDS patients: Incidence and risk factors. Multicenter AIDS Cohort Study. *Neurology* **43,** 2245–2252.

McArthur, J. C., Brew, B. J., and Nath, A. (2005). Neurological complications of HIV infection. *Lancet Neurol.* **4,** 543–555.

Mela, F., Marti, M., Fiorentini, C., Missale, C., and Morari, M. (2006). Group-II metabotropic glutamate receptors negatively modulate NMDA transmission at striatal cholinergic terminals: Role of P/Q-type high voltage activated Ca^{2+} channels and endogenous dopamine. *Mol. Cell. Neurosci.* **31,** 284–292.

Minghetti, L., Visentin, S., Patrizio, M., Franchini, L., Ajmone-Cat, M. A., and Levi, G. (2004). Multiple actions of the human immunodeficiency virus type-1 Tat protein on microglial cell functions. *Neurochem. Res.* **29,** 965–978.

Mothet, J. P., Fossier, P., Meunier, F. M., Stinnakre, J., Tauc, L., and Baux, G. (1998). Cyclic ADP-ribose and calcium-induced calcium release regulate neurotransmitter release at a cholinergic synapse of Aplysia. *J. Physiol.* **507,** 405–414.

Nath, A., and Geiger, J. D. (1998). Neurobiological aspects of HIV infections: Neurotoxic mechanisms. *Prog. Neurobiol.* **54,** 19–33.

Nath, A., Psooy, K., Martin, C., Knudsen, B., Magnuson, D. S. K., Haughey, N., and Geiger, J. D. (1996). Identification of a human immunodeficiency virus type 1 Tat epitope that is neuroexcitatory and neurotoxic. *J. Virol.* **70,** 1475–1480.

Navia, B. A., Jordan, B. D., and Price, R. W. (1986). The AIDS dementia complex: I. Clinical features. *Ann. Neurol.* **19,** 517–524.

Peruzzi, F., Bergonzini, V., Aprea, S., Reiss, C., Sawaya, B. E., Rappaport, J., Amini, S., and Khalili, K. (2005). Cross talk between growth factors and viral and cellular factors alters neuronal signalling pathways: Implication for HIV-associated dementia. *Brain Res. Brain Res. Rev.* **50,** 114–125.

Pittaluga, A., and Raiteri, M. (1990). Release-enhancing glycine-dependent presynaptic NMDA receptors exist on noradrenergic terminals of hippocampus. *Eur. J. Pharmacol.* **191,** 231–234.

Pittaluga, A., and Raiteri, M. (1994). HIV-1 envelope protein gp120 potentiates NMDA-evoked noradrenaline release by a direct action at rat hippocampal and cortical noradrenergic nerve endings. *Eur. J. Neurosci.* **6,** 1743–1749.

Pittaluga, A., Bonfanti, A., and Raiteri, M. (1997). Differential desensitization of ionotropic non-NMDA receptors having distinct neuronal location and function. *Naunyn Schmiedeberg's Arch. Pharmacol.* **356,** 29–38.

Pittaluga, A., Pattarini, R., Feligioni, M., and Raiteri, M. (2001). N-methyl-D-aspartate receptors mediating hippocampal noradrenaline and striatal dopamine release display differential sensitivity to quinolinic acid, the HIV-1 envelope protein gp120, external pH and protein kinase C inhibition. *J. Neurochem.* **76,** 139–148.

Pittaluga, A., Segantini, D., Feligioni, M., and Raiteri, M. (2005). Extracellular protons differentially potentiate the responses of native AMPA receptors subtypes regulating neurotransmitter release. *Br. J. Pharmacol.* **144,** 293–299.

Prendergast, M. A., Trent Rogers, D., Mulholland, P. J., Littleton, J .M., Wilkins, L. H., Jr., Self, R. L., and Nath, A. (2002). Neurotoxic effects of the human immunodeficiency virus type-1 transcription factor Tat require function of a polyamine sensitive-site on the N-methyl-D-aspartate receptor. *Brain Res.* **954,** 300–307.

Raiteri, M., Angelini, F., and Levi, G. (1974). A simple apparatus for studying the release of neurotransmitters from synaptosomes. *Eur. J. Pharmacol.* **25,** 411–414.

Raiteri, M., Garrone, B., and Pittaluga, A. (1992). N-Methyl-D-aspartic acid (NMDA) and non-NMDA receptors regulating hippocampal norepinephrine release. II. Evidence for functional cooperation and for coexistence on the same axon terminal. *J. Pharmacol. Exp. Ther.* **260,** 238–242.

Rosenmund, C., Stern-Bach, Y., and Stevens, C. F. (1998). The tetrameric structure of a glutamate receptor channel. *Science* **280,** 1596–1599.

Sabatier, J. M., Vives, E., Mabrouk, K., Benjouad, A., Rochat, H., Duval, A., Hue, B., and Bhraoui, E. (1991). Evidence for neurotoxic activity of tat from human immunodeficiency virus type 1. *J. Virol.* **65,** 961–967.

Salter, M. W., and Kalia, L. V. (2004). Src kinases: A hub for NMDA receptor regulation. *Nat. Rev. Neurosci.* **5,** 317–328.

Self, R. L., Mulholland, P. J., Nath, A., Harris, B. R., and Prendergast, M. A. (2003). The human immunodeficiency virus type-1 transcription factor Tat produces elevation in intracellular Ca^{2+} that requires function of an N-methyl-D-aspartate receptor polyamine-sensitive site. *Brain Res.* **995,** 39–45.

Song, L., Nath, A., Geiger, J. D., Moore, A., and Hochman, S. (2003). Human immunodeficiency virus type 1 Tat protein directly activates neuronal N-methyl-D-aspartate receptors at an allosteric zinc-sensitive site. *J. Neurovirol.* **9,** 399–403.

Ungvarski, P. J., and Trzcianowska, H. (2000). Neurocognitive disorders seen in HIV-1 disease. *Issues Ment. Health Nurs.* **21,** 51–70.

Wang, P., Barks, J. D. E., and Silverstein, F. S. (1999). Tat, a human immunodeficiency virus-1-derived protein, augments excitotoxic hippocampal injury in neonatal rats. *Neuroscience* **88,** 585–597.

Wessenligh, S. L., Power, C., Glass, J. D., Tyor, W. R., Mc Arthur, J. C., Farber, J. M., Griffin, J. W., and Griffin, D. E., (1993). Intracerebral cytokine messenger RNA expression in acquired immunodeficiency syndrome dementia. *Ann. Neurol.* **33,** 576–582.

Westendorp, M. O., Frank, R., Ochsenbauer, C., Stricker, K., Dhein, J., Walczak, H., Debatin, K. M., and Krammer, P. H. (1995). Sensitization of T cells to CD95-mediated apoptosis by HIV-1 Tat and gp120. *Nature* **375,** 497–500.

Wiley, C. A., Baldwin, M., and Achim, C. L. (1996). Expression of regulatory and structural mRNA in the central nervous system. *AIDS* **10,** 943–947.

EVIDENCE TO IMPLICATE EARLY MODULATION OF INTERLEUKIN-1β EXPRESSION IN THE NEUROPROTECTION AFFORDED BY 17β-ESTRADIOL IN MALE RATS UNDERGONE TRANSIENT MIDDLE CEREBRAL ARTERY OCCLUSION

Olga Chiappetta,* Micaela Gliozzi,‡ Elisa Siviglia,‡ Diana Amantea,* Luigi A. Morrone,*,† Laura Berliocchi,*,† G. Bagetta,*,† and M. Tiziana Corasaniti‡,§

*Department of Pharmacobiology, University of Calabria, Via P. Bucci
87036 Arcavacata di Rende (CS), Italy
†UCHAD Center of Neuropharmacology for Normal and Pathological Neuronal Plasticity
University of Calabria, Via P. Bucci, 87036 Arcavacata di Rende (CS), Italy
‡Department of Pharmacobiological Sciences
University "Magna Graecia" of Catanzaro, 88100 Catanzaro, Italy
§Center for Experimental Neuropharmacology, Mondino-Tor Vergata
University of Rome Tor Vergata, 00133 Rome, Italy

Neuroprotection exerted by 17β-estradiol (17β-E_2) has been widely investigated in animal models of acute cerebral ischemia. Estrogens interact with intracellular receptors (ERα and ERβ) to modulate the transcription of target genes, including those implicated in neuronal survival. Neuroprotection may also occur via interaction with ER-like membrane receptors mediating rapid, non-genomic, actions or via receptor-independent mechanisms. There is also evidence that blockade of inflammatory factors may represent an important mechanism involved in estrogenic neuroprotection. Here we investigate whether reduced brain damage by acute pharmacological treatment with 17β-E_2 in male rats subjected to transient (2 h) middle cerebral artery occlusion (tMCAo) involves modulation of interleukin-1β (IL-1β), a proinflammatory cytokine strongly implicated in the pathophysiology of ischemic stroke.

INTERNATIONAL REVIEW OF
NEUROBIOLOGY, VOL. 82
DOI: 10.1016/S0074-7742(07)82019-8

Administration of 17β-E_2 (0.2 mg/kg, i.p., 1 h before tMCAo) results in significant reduction of brain infarct volume, and this is reverted by the ER antagonist ICI 182,780 (0.25 mg/kg, i.p.) administered 1 h before 17β-E_2. Two hours MCAo followed by 2-h reperfusion results in a significant, threefold increase of IL-1β levels in the cortical tissue ipsilateral to the ischemic damage. Interestingly, a pretreatment with a neuroprotective dose of 17β-E_2 attenuates the cytokine elevation and this appears to occur through ER activation. In addition, neuroprotection by 17β-E_2 is accompanied by reduced cytochrome *c* translocation both in the striatum and in the cortex as revealed by Western blotting 3 h after reperfusion.

In conclusion, we report the original observation that neuroprotection exerted by 17β-E_2 in a rat model of transient focal brain ischemia is accompanied by reduced cytochrome *c* translocation to the cytosol and involves early modulation of IL-1β production.

I. Introduction

It is widely recognized that stroke represents a major cause of death and long-term disability in the Western world. However, current therapeutic approaches, including antiplatelet and thrombolytic drugs, only partially ameliorate the clinical outcome of stroke patients, since they are aimed at preserving or restoring cerebral blood flow rather than targeting molecular mechanisms implicated in neuronal cell death (Gladstone *et al.*, 2002).

Several studies have highlighted the ability of estrogens to enhance recovery from ischemic brain injury resulting from cardiovascular disease or cerebrovascular stroke. 17β-estradiol (17β-E_2) has been reported to reduce mortality and cerebral damage in a variety of animal models of acute cerebral ischemia, including transient and permanent middle cerebral artery occlusion (MCAo) (Alkayed *et al.*, 1998; Dubal *et al.*, 1998; Simpkins *et al.*, 1997), photothrombotic focal ischemia (Fukuda *et al.*, 2000), and global forebrain ischemia (Bagetta *et al.*, 2004; Sudo *et al.*, 1997). Administration of either pharmacological or physiological doses of 17β-E_2 determines neuroprotection in ovariectomized female rodents subjected to focal brain ischemia (Alkayed *et al.*, 1998; Dubal *et al.*, 1998; Rusa *et al.*, 1999; Simpkins *et al.*, 1997). Although less investigated, neuroprotection by female sex hormones has also been demonstrated in male rats since either acute or chronic 17β-E_2 administration significantly reduces brain damage following tMCAo (Toung *et al.*, 1998).

In humans, there is evidence that estrogen enhances recovery from traumatic brain injury following cerebral ischemia (Paganini-Hill, 1995; Schmidt *et al.*, 1996), and continued use of estrogens has been shown to significantly reduce the risk of stroke (Falkeborn *et al.*, 1993; Longstreth *et al.*, 1994; Paganini-Hill *et al.*, 1988). This is also confirmed by epidemiological evidence, indicating that women are more protected than men against stroke until the menopause (Paganini-Hill, 1995). Yet, large, randomized, clinical trials have questioned the effectiveness of female sex hormones in the prevention of coronary heart disease and stroke (Hulley *et al.*, 1998; Simon *et al.*, 2001; Viscoli *et al.*, 2001; see also Hurn and Brass, 2003). However, these studies focused on the effects of chronic treatments with physiological levels of estrogen whereas little is known about the clinical effectiveness of acute administration of pharmacological doses of the hormone.

The mechanisms underlying neuroprotection by 17β-E_2 are not completely understood. A number of mechanisms have been proposed, including modulation of synaptogenesis, protection against apoptosis, anti-inflammatory activity, and increased cerebral blood flow. Estrogens exert their activity through the interaction with intracellular receptors (ER), ERα and ERβ, that results in the modulation of the transcription of estrogen target genes, including those implicated in neuronal survival. Experimental evidence supports the notion that neuroprotection may also occur via interaction with ER-like membrane receptors, mediating rapid nongenomic actions, or receptor-independent mechanisms, mainly due to the antioxidant free radical scavenging properties of the steroidal molecules (Amantea *et al.*, 2005).

There is strong evidence suggesting that the beneficial effects of estrogens may be dependent on their anti-inflammatory activity (Maggi *et al.*, 2004). Data suggest that estrogens regulate the expression of proinflammatory cytokines and interfere with their signaling pathways in a range of tissues and cell types (Corasaniti *et al.*, 2005; Evans *et al.*, 2002; Jilka *et al.*, 1992; Nordell *et al.*, 2003; Polan *et al.*, 1989). Since neurodegeneration induced by experimental cerebral ischemia in rodents has been associated with increased brain levels of the proinflammatory cytokine interleukin (IL)-1β (Rothwell, 2003), we have investigated whether neuroprotection afforded by 17β-E_2 involves modulation of the cytokine in the ischemic rat brain. Thus, we have observed that early induction of IL-1β expression in the cortex of rats subjected to tMCAo is attenuated by pretreatment with a dose of 17β-E_2 that reduces cytochrome *c* translocation and results in significant neuroprotection. Moreover, IL-1β modulation appears to occur via an ER-dependent mechanism, although further studies are needed to elucidate the molecular (genomic vs nongenomic) pathways implicated in the observed effect.

II. Methods

A. Focal Cerebral Ischemia and Drug Treatments

Adult male Wistar rats (Charles River, Calco, Como, Italy) were housed under controlled environmental conditions with ambient temperature of 22 °C, relative humidity of 65%, and 12-h light:12-h dark cycle, with free access to food and water.

Brain ischemia was induced by occlusion of the middle cerebral artery (MCAo) in rats weighing 280–320 g by intraluminal filament, using the relatively noninvasive technique previously described by Longa *et al.* (1989). Briefly, rats were anesthetized with 5% isoflurane in air and were maintained with the lowest acceptable concentration of the anesthetic (1.5–2%). Body temperature was measured with a rectal probe and was kept at 37 °C during the surgical procedure with a heating pad. Under an operating microscope, the external and internal right carotid arteries were exposed through a neck incision. The external carotid artery was cut approximately 3 mm above the common carotid artery (CCA) bifurcation and a silk suture was tied loosely around the external carotid stump. A silicone-coated nylon filament (diameter, 0.28 mm) was then inserted into the external carotid artery and gently advanced into the internal carotid artery, approximately 18 mm from the carotid bifurcation, until mild resistance was felt, thereby indicating occlusion of the origin of the middle cerebral artery in the Willis circle. The silk suture was tightened around the intraluminal filament to prevent bleeding. The wound was then sutured and anesthesia discontinued. Sham rats were exposed to the same surgical procedure without occlusion of MCA.

One hour after surgery, the animals were grossly assessed for neurological deficit as follows: 0 = no deficit, 1 = failure to extend left forelimb, 2 = decreased resistance to lateral push, 3 = circling to contralateral side, 4 = walks only when stimulated, and 5 = no spontaneous motor activity. Only rats with clear neurological deficits ($\geq$3) were included in the study.

To allow reperfusion, rats were briefly reanesthetized with isoflurane, and the nylon filament was withdrawn 2 h after MCAo. After the discontinuation of isoflurane and wound closure, the animals were allowed to awake and were kept in their cages with free access to food and water.

17β-E_2 (Sigma-Aldrich, Milan, Italy) was dissolved in vegetable oil and administered i.p., 1 h prior to MCAo, at a dose of 0.20 mg/kg. The ER antagonist ICI 182,780 (0.25 mg/kg i.p.; Tocris Bioscience, Avonmouth, United Kingdom) was dissolved in 4% DMSO in vegetable oil and administered 1 h before 17β-E_2.

All the experimental procedures were in accordance to the guidelines of the European Community Council Directive 86/609, included in the D.M. 116/1992 of the Italian Ministry of Health.

B. Neuropathology and Quantification of Ischemic Damage

Cerebral infarct volume was evaluated 22 h after reperfusion in rats subjected to 2-h MCAo. Rats were sacrificed by decapitation and the brains were rapidly removed. Eight serial sections from each brain were cut at 2-mm intervals from the frontal pole using a rat brain matrix (Harvard Apparatus, Massachusetts). To measure ischemic damage, brain slices were stained in a solution containing 2% 2,3,5-triphenyltetrazolium chloride (TTC) in saline, at 37 °C. After 10 min incubation, the slices were transferred to 10% neutral buffered formaldehyde and stored at 4 °C prior to analysis. Images of TTC-stained sections were captured using a digital scanner and analyzed using an image analysis software (ImageJ, version 1.30). The infarct volume (mm^3) was calculated by summing the infarcted area (unstained) of the eight sections and multiplying by the interval thickness between sections as previously described (Li *et al.*, 2000).

C. Subcellular Fractionation

Three hours after reperfusion, rats were sacrificed and their cortices and striata were dissected out and processed for subcellular fractionation as described by Corasaniti *et al.* (2001). Individual brain tissue samples were homogenized in 1:6 volumes (w/v) of ice-cold homogenization buffer [320-mM sucrose, 10-mM Tris–HCl, pH 7.4, containing a cocktail of protease inhibitors (Sigma, Milan, Italy)] and the homogenates were centrifuged twice at 1300 × *g* for 3 min to pellet nuclei. The supernatants were centrifuged at 17,000 × *g* for 10 min to pellet mitochondria; the postmitochondrial supernatants were centrifuged at 100,000 × *g* for 1 h and the high-speed supernatant retained as cytosolic fraction.

Protein concentration was determined by using the Bio-Rad DC protein assay (Bio-Rad Laboratories, Milan, Italy). All procedures were performed at 4 °C.

D. Western Blotting

For Western blot analysis of cytochrome *c*, proteins (20 μg) from cytosolic fraction were resolved by 15% sodium dodecyl sulphate-polyacrylamide gel electrophoresis (SDS-PAGE) and transferred to nitrocellulose membranes (Optitran BA-S 83, Schleicher & Schuell Biosciences, Dassel, Germany); these were then cut to obtain two strips: the first, from the origin to the 30-kDa position, to be immunoprobed overnight at 4 °C with a mouse monoclonal anti-actin (clone AC-40, 1:2000 dilution; Sigma) antibody, and the second, from the 30-kDa position to the front, to be probed overnight at 4 °C with a mouse anti-cytochrome c monoclonal antibody (clone 7H8.2C12, 1:2000 dilution; PharMingen, CA). The membranes were then incubated with horseradish peroxidase-conjugated anti-mouse IgG (1:8000 dilution; Pierce, Rockford, IL) for 1 h at room temperature and the immunoreactivity visualized by chemiluminescent detection (Amersham Biosciences, GE Healthcare, Milan, Italy).

E. IL-1β ELISA

Immunoreactive IL-1β levels were analyzed in individual brain cortical tissue homogenates by an established, rat specific, sandwich ELISA previously described (Corasaniti *et al.*, 2001; Hagan *et al.*, 1996), using an immunoaffinity-purified polyclonal sheep anti-rat IL-1β coating antibody (1 μg/ml) and a biotinylated, immunoaffinity-purified polyclonal sheep anti-rat IL-1β detecting antibody (1:1000 dilution) kindly provided by Dr. Stephen Poole (National Institute of Biological Standards and Controls, NIBSC, Hertfordshire, United Kingdom). Poly-horseradish peroxidase-conjugated streptavidin (CLB, Amsterdam, the Netherlands) was used at 1:5000 dilution and the color was developed by using the chromogen *o*-phenylenediamine. Optical densities (OD) were read at 492 nm by using an automated plate reader (Multiscan MS, Labsystems, Helsinki, Finland) and cytokine levels were calculated by interpolation from a standard curve obtained from recombinant rat IL-1β (0.0–1000 pg/ml). Data were corrected for protein concentration and the results expressed as picograms of IL-1β per milligram of protein.

F. Statistical Analysis

Data are expressed means ± SEM. Statistical analysis was performed by ANOVA followed by Dunnett's or Tukey-Kramer's post hoc tests using the Prism 3 program (GraphPAD Software for Science, San Diego, CA). Differences were considered statistically significant when $p < 0.05$.

III. Results

Focal brain ischemia induced by 2-h MCAo resulted in a significant damage to the striatum and the frontoparietal cortex as assessed by TTC staining 22 h after reperfusion (Fig. 1A). Brain infarct volume detected in vehicle-treated rats did not differ significantly from that measured in noninjected animals subjected to tMCAo (data not shown). In comparison to vehicle-injected rats, acute treatment with a pharmacological dose of 17β-E_2 (0.20 mg/kg, given i.p. 1 h before the ischemic insult), resulted in a significant reduction of brain infarct area and volume sensitive to

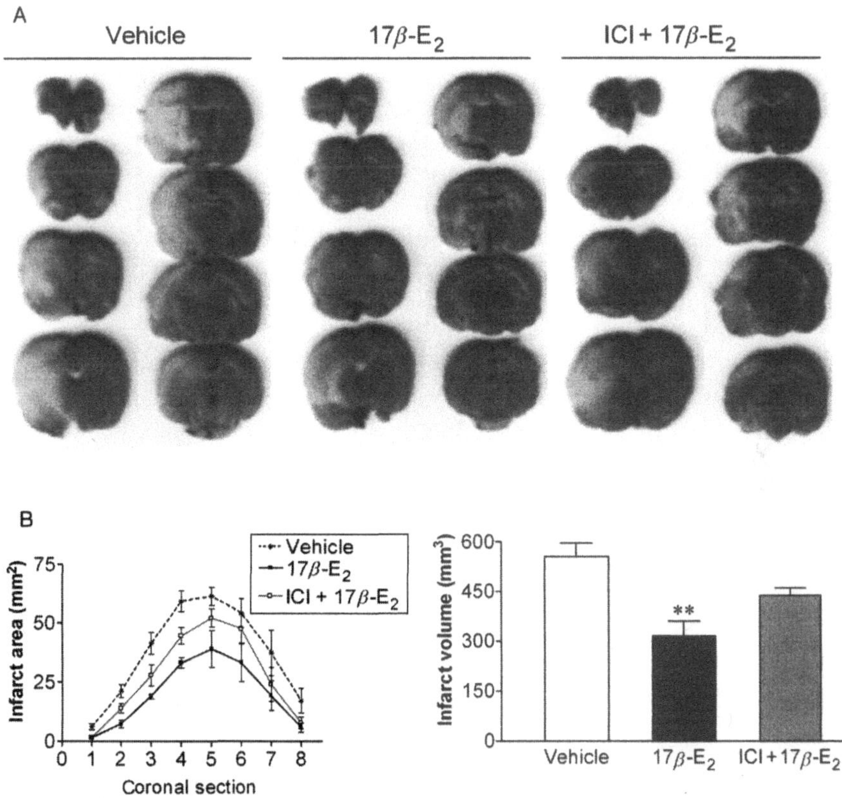

FIG. 1. Neuroprotection afforded by 17β-E_2 against brain damage produced by MCAo is reverted by ICI 182,780, an ER antagonist. The right MCA was occluded for 2 h with a nylon suture, as described in Section II, and cerebral infarct area and volume were evaluated 22 h after reperfusion in rats pretreated with the vehicle used to dissolve 17β-E_2 (Vehicle; see Section II), with 17β-E_2 (0.20 mg/kg, given i.p. 1 h before MCAo), or with ICI 182,780 (ICI; 0.25 mg/kg, given i.p. 1 h prior to 17β-E_2) plus 17β-E_2. Panel (A) shows three series of eight serial coronal sections obtained from the brain of three individual animals cut at 2-mm intervals from the frontal pole and stained with TTC to measure infarct area (unstained); quantitative assessment of the infarct area is reported in the left panel in (B). The infarct volume shown in the right panel in (B) results from the sum of the infarct area of the eight tissue sections multiplied by the interval thickness among sections. Values are expressed as mean ± SEM ($n = 5$ per experimental group), and analyzed by ANOVA followed by Dunnett's post hoc test. ** denote $p < 0.01$ versus vehicle. (See Color Insert.)

the reversal operated by systemic administration of the ER antagonist ICI 182,780 (0.25 mg/kg, given i.p. 1 h prior to 17β-E_2) (Fig. 1A and B).

In order to assess the involvement of IL-1β in the mechanisms of neuroprotection exerted by estrogen, we measured cortical levels of the cytokine during the early stages of reperfusion in the absence or presence of hormone treatment. Elevation of IL-1β content in the cortex ipsilateral to the ischemic insult was observed as early as 1 h after reperfusion (data not shown). Cytokine levels further increased following 2 h of reperfusion both in the ipsilateral and contralateral cortex though statistical significance was reached in the ischemic, but not contralateral, side of the brain as compared to cortical samples from sham-operated animals (Fig. 2). Pretreatment with a neuroprotective dose of 17β-E_2 (0.2 mg/kg, given i.p. 1 h prior to MCAo) attenuated the enhanced levels of the cytokine induced in the ipsilateral cortex by the ischemic insult (Fig. 2). This effect was reverted by administration of the ER antagonist ICI 182,780 (0.25 mg/kg, given i.p. 1 h before E_2) (Fig. 2).

Focal brain ischemia produced by MCAo was associated with an early translocation of cytochrome *c* to the cytosol both in the striatal and in the cortical tissue, as revealed by Western blotting 3 h after reperfusion (data not shown).

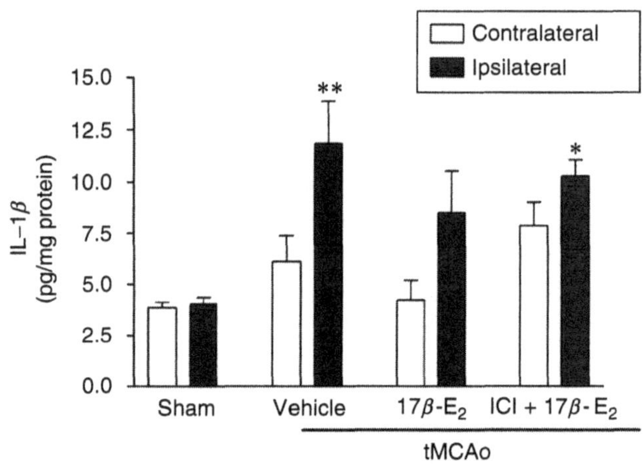

FIG. 2. Systemic administration of 17β-E_2 in rats undergone tMCAo minimizes the increase of IL-1β in the ischemic cortex: reversal by the ER antagonist ICI 182,780. Immunoreactive IL-1β levels are increased in the ipsilateral, ischemic cortex of rats ($n = 4$) subjected to 2-h MCAo followed by 2-h reperfusion as compared to sham-operated animals ($n = 3$). This effect is prevented by administration of 17β-E_2 (0.20 mg/kg, given i.p. 1 h prior MCAo, $n = 3$). Pretreatment with ICI 182,780 (ICI, 0.25 mg/kg, given i.p. 1 h before 17β-E_2, $n = 3$) abolished the attenuation of IL-1β release induced by 17β-E_2 in the ipsilateral, ischemic cortex. Immunoreactive IL-1β levels were assayed in individual brain cortical tissue samples by an established, rat specific, sandwich ELISA. IL-1β levels were corrected for protein concentration and the results expressed as picograms of IL-1β per milligram of protein. Data are expressed as mean ± SEM and the resulting means evaluated statistically for differences using ANOVA followed by Dunnett's test. **$p < 0.01$ and *$p < 0.05$ versus sham ipsilateral cortex.

Administration of a neuroprotective dose of 17β-E_2 (0.20 mg/kg, given i.p. 1 h before the ischemic insult) resulted in reduced cytosolic cytochrome *c* levels, as detected both in cortical (Fig. 3) and in striatal (not shown) tissue samples isolated from hormone-treated rats, after 2-h MCAo followed by 3-h reperfusion and this was reverted by administration of the ER antagonist ICI 182,780 (0.25 mg/kg, given i.p. 1 h before 17β-E_2) (Fig. 4).

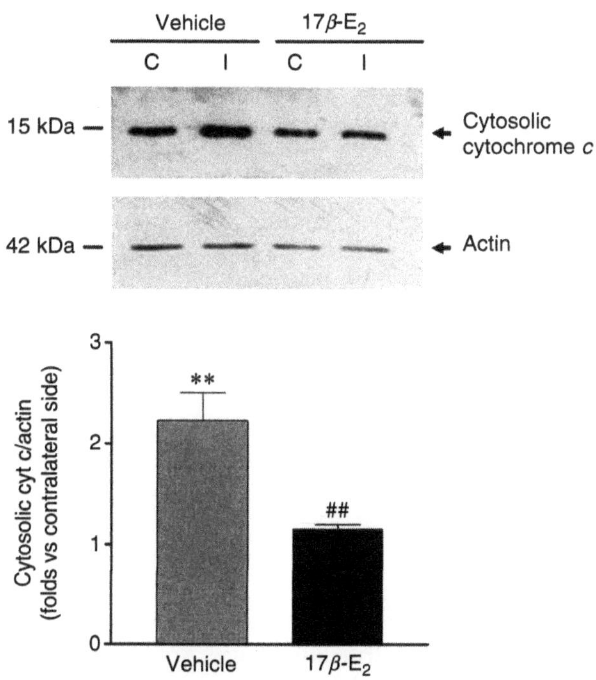

FIG. 3. Neuroprotection by 17β-E_2 is associated with reduced cytosolic cytochrome *c* levels in the ischemic cortex. Western blot analysis of cytochrome *c* in the cytosolic fraction of individual cortices ipsilateral and contralateral to brain ischemia obtained from vehicle- and 17β-E_2-treated rats. Two hours transient MCAo, followed by 3-h reperfusion, enhances cytochrome *c*–like immunoreactivity in the cytosol, and this is minimized by 17β-E_2 (0.2 mg/kg i.p.) given 1 h before ischemia. Actin immunoreactivity from the same gel is shown as internal control. Histograms in lower panel show results of computer-assisted (Quantiscan, Biosoft, United Kingdom) densitometric analysis of autoradiographic bands corresponding to cytosolic cytochrome *c* (cyt *c*); data have been normalized to the values yielded by actin and reported as folds versus contralateral, nonischemic side of the brain (mean ′ SEM; $n = 3$ per group). ** Denote $p < 0.01$ versus contralateral side; ## denote $p < 0.01$ versus vehicle and not significant versus contralateral side (ANOVA followed by Tukey-Kramer multiple comparisons test).

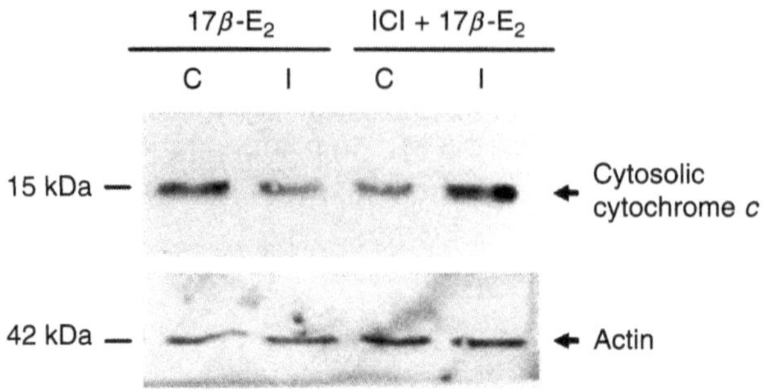

FIG. 4. The ER antagonist ICI 182,780 prevents the inhibition of cytochrome *c* translocation induced by 17β-E$_2$ in the cerebral cortex of rats undergone tMCAo. Pretreatment with ICI 182,780 (ICI; 0.25 mg/kg, $n = 3$), given i.p. 1 h before 17β-E$_2$ (0.20 mg/kg, given i.p. 1 h prior MCAo, $n = 3$) abolishes the inhibition of cytochrome *c* translocation induced by 17β-E$_2$ in rats subjected to 2 h transient MCAo, followed by 3-h reperfusion. Representative immunoblot from three independent experiments is shown.

IV. Discussion

The present study shows that acute administration of a pharmacological dose of 17β-E$_2$ results in significant neuroprotection against tMCAo-induced brain damage in male rats. This is coincident with reduced translocation of cytosolic cytochrome *c* both in the cortex and in the striatum. More interestingly, hormone treatment attenuates the early increase of IL-1β levels detected in the cortex 2 h after reperfusion and this occurs via an ER-dependent mechanism. Thus, inhibition of proinflammatory cytokine production appears to underlie the neuroprotective action of 17β-E$_2$ against a focal ischemic insult.

Deleterious mitochondrial responses have been detected during ischemia and reperfusion thus suggesting a contribution of these organelles to cerebral ischemic damage. These responses include impairment of ATP production, induction of the mitochondrial permeability transition pore during early reperfusion, and release of factors promoting apoptotic cell death (Sims and Anderson, 2002). Release of cytochrome *c* into the cytoplasm of cells has been observed with both permanent and transient ischemia and this is known to trigger apoptosis, a death process that contributes to the expansion of the ischemic lesion. Accordingly, increased levels of cytosolic cytochrome *c* have been detected within few hours after permanent MCAo (Zhao *et al.*, 2003) and during the early stages of reperfusion following transient global ischemia (Sugawara *et al.*, 1999; Zhang *et al.*, 2002) and tMCAo (Babu *et al.*, 2000; Li *et al.*, 2002) in rats.

We have previously reported that neuroprotection exerted by 17β-E$_2$ against hippocampal damage produced by transient global ischemia is associated with reduced translocation of cytochrome *c* to the cytosol in rat (Bagetta *et al.*, 2004). Accordingly, here we demonstrate that hormone treatment does revert the increase of cytosolic cytochrome *c* levels induced in the cortex and striatum of rats subjected to a transient focal ischemic insult. The latter observation is of importance in view of the role played by translocated cytochrome *c* in the control of cell death (Kroemer and Reed, 2000). In the presence of deoxy-ATP (dATP), cytochrome *c* binds to Apaf-1, and this, in turn, is responsible of the activation of caspase-9, which represents an initiator of the cytochrome *c*–dependent caspase cascade (Li *et al.*, 1997). Activated caspase-9 directly cleaves procaspase-3, and active caspase-3 triggers activation of additional caspases and leads to apoptosis (Liu *et al.*, 1996; Slee *et al.*, 1999). Interestingly, it has been shown that Bcl-2 expression is significantly reduced in the cortex of rats following focal cerebral ischemia, and this effect can be prevented by previous administration of 17β-E_2 (Dubal *et al.*, 1999). Similar results have been obtained by Alkayed *et al.* (2001) who demonstrated that the neuroprotection exerted by 17β-E_2 against brain damage induced by tMCAo in rat is associated with increased Bcl-2 mRNA and protein levels within the periinfarct regions. Mitochondrial Bcl-2 promotes the release of survival mediators that stabilize mitochondrial membrane potential (Dispersyn *et al.*, 1999), ATP depletion (Kim *et al.*, 2001), and lipid peroxidation induced by cytotoxic stimuli such as oxidative stress (Howard *et al.*, 2002), and, besides its antiapoptotic activity, it promotes axonal growth and regeneration (Chen *et al.*, 1997; Holm and Isacson, 1999) and attenuates the generation of reactive oxygen species (Bogdanov *et al.*, 1999). More importantly, Bcl-2 prevents the release of cytochrome *c* from mitochondria, thus blocking the activation of the caspase cascade (Desagher and Martinou, 2000; Reed, 2002). Thus, modulation of this antiapoptotic protein might represent a likely mechanism underlying the observed inhibition of cytochrome *c* translocation by estrogen during ischemic stroke, and may represent a pivotal mechanism involved in hormone-mediated neuroprotection.

Although the regulation of mitochondrial apoptotic pathway has been strongly implicated in the neuroprotective effects of estrogens (Garcia-Segura *et al.*, 1998, 2001; Kajta and Beyer, 2003), these hormones also appear to affect cytokine-mediated apoptosis. In fact, estradiol has been shown to inhibit the activation of caspase-3 induced by IL-1β in hippocampal cultures at early stages of development (Kajta *et al.*, 2006). Moreover, modulation of cytokine-mediated neuroinflammatory responses has been implicated in the beneficial effects of estrogens against a variety of brain insults (Maggi *et al.*, 2004). In an animal model of multiple sclerosis, estradiol has been shown to inhibit cytokine, chemokine, and chemokine receptor mRNA expression in the central nervous system of female rats (Matejuk *et al.*, 2001), thus leading to the hypothesis that blockade of inflammatory factors may represent an important mechanism involved in estrogenic neuroprotection. By contrast, in an animal model of neuroAIDS we have previously shown that inhibition of apoptotic cell death by estrogen is associated with increased IL-1β levels in the rat neocortex, further emphasizing the dualistic role (neurotoxic vs neurotrophic) of this cytokine (Corasaniti *et al.*, 2005). 17β-E_2 replacement has also been reported to differentially modulate NMDA-induced IL-1β increase in the female rat forebrain depending on animal age (Nordell *et al.*, 2003). These observations suggest that estrogens may positively or negatively modulate cytokine expression depending on the reproductive age of the organism, but also on the type of insult (Johnson *et al.*, 2006). Thus, we aimed at investigating whether neuroprotection exerted by estradiol against ischemic stroke involves modulation of IL-1β endogenous levels in adult male rats.

Here we report that 17β-E_2 attenuates the early elevation of IL-1β in the ischemic cortex produced by tMCAo via an ER-dependent mechanism and this might represent a pivotal mechanism involved in the neuroprotection exerted by the hormone.

Neurodegeneration induced by experimental cerebral ischemia in rodents has been associated with increased brain levels of the proinflammatory cytokine IL-1β (Rothwell, 2003). In fact, induction of IL-1β mRNA and increased IL-1β protein levels have been reported to occur very early after either permanent (Buttini *et al.*, 1994; Davies *et al.*, 1999; Legos *et al.*, 2000; Liu *et al.*, 1993) or tMCAo (Hara *et al.*, 1997; Wang *et al.*, 1994; Zhang *et al.*, 1998), reaching maximum levels within hours of reperfusion. Accordingly, we observe that cytokine levels increase significantly 2 h after the beginning of reperfusion in the cortex ipsilateral to the ischemic damage. This increase may contribute to deleterious effects as demonstrated by the evidence that intracerebral injection of neutralizing anti-IL-1β antibody to rats reduces ischemic brain damage (Yamasaki *et al.*, 1995) and both intracerebroventricular, and systemic administration of IL-1 receptor antagonist (IL-1ra) markedly reduces brain damage induced by focal stroke (Garcia *et al.*, 1995; Mulcahy *et al.*, 2003; Relton and Rothwell, 1992; Relton *et al.*, 1996).

In conclusion, our data demonstrate that acute administration of a pharmacological dose of 17β-E_2 results in neuroprotection in male Wistar rats subjected to tMCAo. Reduced ischemic damage by the hormone is associated with inhibition of cytosolic cytochrome *c* translocation in the ischemic cortex. More interestingly, 17β-E_2 reverts the early increase of IL-1β detected in the cortex 2 h after reperfusion and this occurs via an ER-dependent mechanism.

Acknowledgments

Financial support from MIUR (COFIN2004 prot. n. 2004053099_004) and from University of Calabria (Ex quota 60%) to G.B. is gratefully acknowledged. C.O., G.M., and S.E. are listed in alphabetical order to confirm that they have contributed equally.

References

Alkayed, N. J., Harukuni, I., Kimes, A. S., London, E. D., Trayatman, R. J., and Hurn, P. D. (1998). Gender-linked brain injury in experimental stroke. *Stroke* **29,** 159–165.

Alkayed, N. J., Goto, S., Sugo, N., Joh, H.-D., Klaus, J., Crain, B. J., Bernard, O., Traystman, R. J., and Hurn, P. D. (2001). Estrogen and Bcl-2: Gene induction and effect of transgene in experimental stroke. *J. Neurosci.* **21,** 7543–7550.

Amantea, D., Russo, R., Bagetta, G., and Corasaniti, M. T. (2005). From clinical evidence to molecular mechanisms underlying neuroprotection afforded by estrogens. *Pharmacol. Res.* **52,** 119–132.

Babu, P. P., Yoshida, Y., Su, M., Segura, M., Kawamura, S., and Yasui, N. (2000). Immunohistochemical expression of Bcl-2, Bax and cytochrome c following focal cerebral ischemia and effect of hypotermia in rat. *Neurosci. Lett.* **291,** 196–200.

Bagetta, G., Chiappetta, O., Amantea, D., Iannone, M., Rotiroti, D., Costa, A., Nappi, G., and Corasaniti, M. T. (2004). Estradiol reduces cytochrome c translocation and minimizes hippocampal damage caused by transient global ischemia in rat. *Neurosci. Lett.* **368,** 87–91.

Bogdanov, M. B., Ferrante, R. J., Mueller, G., Ramos, L. E., Martinou, J. C., and Beal, M. F. (1999). Oxidative stress is attenuated in mice overexpressing Bcl-2. *Neurosci. Lett.* **262,** 33–36.

Buttini, M., Sauter, A., and Boddeke, H. (1994). Induction of interleukin-1beta mRNA after focal ischemia in the rat. *Mol. Brain Res.* **23,** 126–134.

Chen, D. F., Schneider, G. E., Martinou, J. C., and Tonegawa, S. (1997). Bcl-2 promotes regeneration of severed axons in mammalian CNS. *Nature* **385,** 434–439.

Corasaniti, M. T., Turano, P., Bilotta, A., Malorni, W., Stingaro, A. R., Nisticò, R., Finazzi-Agrò, A., and Bagetta, G. (2001). Evidence that increases of mitochondrial immunoreactive IL-1β by HIV-1 gp120 implicate in situ cleavage of pro-IL-1β in the neocortex of rat. *J. Neurochem.* **78,** 611–618.

Corasaniti, M. T., Amantea, D., Russo, R., Piccirilli, S., Leta, A., Corazzari, M., Nappi, G., and Bagetta, G. (2005). 17β-Estradiol reduces neuronal apotosis induced by HIV-1 gp120 in the neocortex of rat. *NeuroToxicol.* **26,** 893–903.

Davies, C. A., Loddick, S. A., Toulmond, S., Stroemer, R. P., Hunt, J., and Rothwell, N. J. (1999). The progression and topographic distribution of interleukin-1 beta expression after permanent middle cerebral artery occlusion in the rat. *J. Cereb. Blood Flow Metab.* **19,** 87–98.

Desagher, S., and Martinou, J. C. (2000). Mitochondria as the central control point of apoptosis. *Trends Cell Biol.* **10,** 369–377.

Dispersyn, G., Nuydens, R., Connors, R., Borgers, M., and Geerts, H. (1999). Bcl-2 protects against FCCP-induced apoptosis and mitochondrial membrane potential depolarization in PC12 cells. *Biochim. Biophys. Acta* **1428,** 357–371.

Dubal, D. B., Kashon, M. L., Pettigrew, L. C., Ren, J. M., Finklestein, S. P., Ran, S. W., and Wise, P. M. (1998). Estradiol protects against ischemic injury. *J. Cereb. Blood Flow Metab.* **18,** 1253–1258.

Dubal, D. B., Shughrue, P. J., Wilson, M. E., Merchenthaler, I., and Wise, P. M. (1999). Estradiol modulates Bcl-2 in cerebral ischemia: A potential role for estrogen receptor. *J. Neurosci.* **19,** 6385–6396.

Evans, M. J., Lai, K., Shaw, L. J., Harnish, D. C., and Chadwick, C. C. (2002). Estrogen receptor α inhibits IL-1β induction of gene expression in the mouse liver. *Endocrinology* **143,** 2559–2570.

Falkeborn, M., Persson, I., Terent, A., Adami, H. O., Lithell, H., and Bergstrom, R. (1993). Hormone replacement therapy and the risk of stoke. Follow-up of a population-based cohort in Sweden. *Arch. Intern. Med.* **153,** 1201–1209.

Fukuda, K., Yao, H., Ibayashi, S., Nakahara, T., Uchimura, H., Fujishima, M., and Hall, E. D. (2000). Ovariectomy exacerbates and estrogen replacement attenuates photothrombotic focal ischemic brain injury in rats. *Stroke* **31,** 155–160.

Garcia, J. H., Liu, K. F., and Relton, J. K. (1995). Interleukin-1 receptor antagonist decreases the number of necrotic neurons in rats with middle cerebral artery occlusion. *Am. J. Pathol.* **147,** 1477–1486.

Garcia-Segura, L. M., Cardona-Gomez, P., Naftolin, F., and Chowen, J. A. (1998). Estradiol upregulates Bcl-2 expression in adult brain neurons. *NeuroReport* **9,** 593–597.

Garcia-Segura, L. M., Azcoitia, I., and Don Carlos, L. L. (2001). Neuroprotection by estradiol. *Prog. Neurobiol.* **63,** 29–60.

Gladstone, D. J., Black, S. E., Hakim, A. M., and Heart and Stroke Foundation of Ontario Centre of Excellence in Stroke Recovery (2002). Toward wisdom from failure. Lessons from neuroprotective stroke trials and new therapeutic directions. *Stroke* **33,** 2123–2136.

Hagan, P., Poole, S., Bristow, A. F., Tilders, F, and Silverstein, F. S. (1996). Intracerebral NMDA injection stimulates production of interleukin-1β in perinatal rat brain. *J. Neurochem.* **67,** 2215–2218.

Hara, H., Friedlander, R. M., Gagliardini, V., Ayata, C., Fink, K., Huang, Z., Shimizu-Sasamata, M., Yuan, J., and Moskowitz, M. A. (1997). Inhibition of interleukin 1β converting enzyme family proteases reduces ischemic and excitotoxic neuronal damage. *Proc. Natl. Acad. Sci. USA* **94,** 2007–2012.

Holm, K., and Isacson, O. (1999). Factors intrinsic to the neuron can induce and maintain its ability to promote axonal outgrowth: A role for Bcl-2? *Trends Neurosci.* **22,** 269–273.

Howard, S., Bottino, C., Brooke, S., Cheng, E., Giffard, R. G., and Sapolsky, R. (2002). Neuroprotective effects of bcl-2 overexpression in hippocampal cultures: Interactions with pathways of oxidative damage. *J. Neurochem.* **83,** 914–923.

Hulley, S., Grady, D., Bush, T., Furberg, C., Herrington, D., Riggs, B., and Vittinghoff, E. (1998). Randomized trial of estrogen plus progestin for secondary prevention of coronary heart disease in postmenopausal women. Heart and Estrogen/progestin Replacement Study (HERS) Research Group. *JAMA* **280,** 605–613.

Hurn, P. D., and Brass, L. M. (2003). Estrogen and stroke: A balanced analysis. *Stroke* **34,** 338–341.

Jilka, R. L., Hangoc, G., Girasole, G., Passeri, G., Williams, D. C., Abrams, J. S., Boyce, B., Broxmeyer, H., and Manolagas, S. C. (1992). Increased osteoclast development after estrogen loss: Mediation by interleukin-6. *Science* **257,** 88–91.

Johnson, A. B., Bake, S., Lewis, D. K., and Sohrabji, F. (2006). Temporal expression of IL-1b protein and mRNA in the brain after systemic LPS injection is affected by age and estrogen. *J. Neuroimmunol.* **174,** 82–91.

Kajta, M., and Beyer, C. (2003). Cellular strategies of estrogen-mediated neuroprotection during brain development. *Endocrine* **21,** 3–9.

Kajta, M., Trotter, A., Lason, W., and Beyer, C. (2006). Impact of 17β-estradiol on cytokine-mediated apoptotic effects in primary hippocampal and neocortical cell cultures. *Brain Res.* **1116,** 64–74.

Kim, H. E., Yoon, S. Y., Lee, J. E., Choi, W. S., Jin, B. K., Oh, T. H., Markelonis, G. J., Chun, S. Y., and Oh, Y. J. (2001). MPP(+) downregulates mitochondrially encoded gene transcripts and their activities in dopaminergic neuronal cells: Protective role of Bcl-2. *Biochem. Biophys. Res. Commun.* **286,** 659–665.

Kroemer, G., and Reed, J. C. (2000). Mitochondrial control of cell death. *Nat. Med.* **6,** 513–519.

Legos, J. J., Whitmore, R. G., Erhardt, J. A., Parsons, A. A., Tuma, R. F., and Barone, F. C. (2000). Quantitative changes in interleukin proteins following focal stroke in the rat. *Neurosci. Lett.* **282,** 189–192.

Li, F., Omori, N., Sato, K., Jin, G., Nagano, I., Manabe, Y., Shoji, M., and Abe, K. (2002). Coordinate expression of survival p-ERK and proapoptotic cytochrome c signals in rat brain neurons after transient MCAO. *Brain Res.* **958,** 83–88.

Li, H., Colbourne, F., Sun, P., Zhao, Z., Buchan, A. M., and Iadecola, C. (2000). Caspase inhibitors reduce neuronal injury after focal but not global cerebral ischemia in rats. *Stroke* **31,** 176–182.

Li, P., Nijhawan, D., Budihardjo, I., Srinvasula, S. M., Ahmad, M., Alnemri, E. S., and Wang, X. (1997). Cytochrome x and dATP-dependent formation of Apaf-1/caspase-9 complex initiates an apoptotic protease cascade. *Cell* **91,** 479–489.

Liu, T., McDonnell, P. C., Young, P. R., White, R. F., Siren, A. L., Hallenbeck, J. M., Barone, F. C., and Feurerstein, G. Z. (1993). Interleukin-1-beta messenger RNA expression in ischemic rat cortex. *Stroke* **24,** 1746–1751.

Liu, X., Kim, C. N., Yang, J., Jemmerson, R., and Wang, X. (1996). Induction of apoptotic program in cell-free extracts: Requirement for dATP and cytochrome *c*. *Cell* **86,** 147–157.

Longa, E. Z., Weinstein, P. R., Carlson, S., and Cummins, R. (1989). Reversible middle cerebral artery occlusion without craniotomy in rats. *Stroke* **20,** 84–91.

Longstreth, W. T., Nelson, L. M., Koepsell, T. D., and van Belle, G. (1994). Subarachnoid hemorrhage and hormonal factors in women. A population-based case-control study. *Ann. Intern. Med.* **121,** 168–173.

Maggi, A., Ciana, P., Belcredito, S., and Vegeto, E. (2004). Estrogens in the nervous system: Mechanisms and nonreproductive functions. *Annu. Rev. Physiol.* **66,** 291–313.

Matejuk, A., Adlard, K., Zamora, A., Silverman, M., Vandenbark, A. A., and Offner, H. (2001). 17 beta-estradiol inhibits cytokine, chemokine, and chemokine receptor mRNA expression in the central nervous system of female mice with experimental autoimmune encephalomyelitis. *J. Neurosci. Res.* **65,** 529–542.

Mulcahy, N., Ross, J., Rothwell, N. J., and Loddick, S. A. (2003). Delayed administration of interleukin-1 receptor antagonist protects against transient cerebral ischemia in the rat. *Br. J. Pharmacol.* **140,** 471–476.

Nordell, V. L., Scarborough, M. M., Buchanan, A. K., and Sohrabji, F. (2003). Differential effects of estrogen in the injured forebrain of young adult and reproductive senescent animals. *Neurobiol. Aging* **24,** 733–743.

Paganini-Hill, A. (1995). Estrogen replacement therapy and stroke. *Prog. Cardiovasc. Dis.* **38,** 223–242.

Paganini-Hill, A., Ross, R. K., and Henderson, B. E. (1988). Postmenopausal oestrogen treatment and stroke: A prospective study. *BMJ* **297,** 519–522.

Polan, M. L., Loukides, J., Nelson, P., Carding, S., Diamond, M., Walsh, A., and Bottomly, K. (1989). Progesterone and estradiol modulate interleukin-1 beta messenger ribonucleic acid levels in cultured human peripheral monocytes. *J. Clin. Endocrinol. Metab.* **69,** 12200–12206.

Reed, J. C. (2002). Apoptosis-based therapy. *Nat. Rev. Drug Discov.* **1,** 111–121.

Relton, J. K., and Rothwell, N. J. (1992). Interleukin-1 receptor antagonist inhibits ischaemic and excitotoxic neuronal damage in the rat. *Brain Res. Bull.* **29,** 242–246.

Relton, J. K., Martin, D., Thompson, R. C., and Russell, D. A. (1996). Peripheral administration of interleukin-1 receptor antagonist inhibits brain damage after focal cerebral ischemia in the rat. *Exp. Neurol.* **138,** 206–213.

Rothwell, N. (2003). Interleukin-1 and neuronal injury: Mechanisms, modification, and therapeutic potential. *Brain Behav. Immun.* **17,** 152–157.

Rusa, R., Alkayed, N. J., Crain, B. J., Traystman, R. J., Kimes, A. S., London, E. D., Klaus, J. A., and Hurn, P. D. (1999). 17β-estradiol reduces stroke injury in estrogen-deficient female animals. *Stroke* **30,** 1665–1670.

Schmidt, R., Fazekas, F., Reinhart, B., Kapeller, P., Fazekas, G., Orfenbacher, S., Eber, B., Schumacher, M., and Freidl, W. (1996). Estrogen replacement therapy in older women: A neuropsychological and brain MRI study. *J. Am. Geriatr. Soc.* **44,** 1307–1313.

Simon, J. A., Hsia, J., Cauley, J. A., Richard, C., Harris, F., Fong, J., Barrett-Connor, E., and Hulley, S. B., for the HERS Research Group (2001). Postmenopausal hormone therapy and risk of stroke: The Heart and Estrogen-Progestin Replacement Study (HERS). *Circulation* **103,** 638–642.

Simpkins, J. W., Rajekumar, G., Zhang, Y. Q., Simpkins, C. E., Greenwald, D., Yu, C. J., Bodor, N., and Day, A. L. (1997). Estrogens may reduce mortality and ischemic damage caused by middle cerebral artery occlusion in the female rat. *J. Neurosurg.* **87,** 724–730.

Sims, N. R., and Anderson, M. F. (2002). Mitochondrial contribution to tissue damage in stroke. *Neurochem. Int.* **40,** 511–526.

Slee, E. A., Harte, M. T., Kluck, R. M., Wolf, B. B., Casiano, C. A., Newmeyer, D. D., Wang, H. G., Reed, J. C., Nicholson, D. W., Alnemri, E. S., Green, D. R., and Martin, S. J. (1999). Ordering the cytochrome c-initiates caspase cascade: Hierarchical activation of caspases-2, -3, -6, -7, -8, and -10 in a caspase-9-dependent manner. *J. Cell Biol.* **144,** 281–292.

Sudo, S., Wen, T. C., Desaki, J., Matsuda, S., Tanaka, J., Arai, T., Maeda, N., and Sakanaka, M. (1997). β-estradiol protects hippocampal CA1 neurons against transient forebrain ischemia in gerbil. *Neurosci. Res.* **29,** 345–354.

Sugawara, T., Fujimura, M., Morita-Fujimura, Y., Kawase, M., and Chan, P. H. (1999). Mitochondrial release of cytochrome c corresponds to the selective vulnerability of hippocampal CA1 neurons in rats after transient global cerebral ischemia. *J. Neurosci.* **19,** 1–6.

Toung, T. J., Traystman, R. J., and Hurn, P. D. (1998). Estrogen-mediated neuroprotection after experimental stroke in male rats. *Stroke* **29,** 1666–1670.

Viscoli, C. M., Brass, L. M., Kernan, W. N., Sarrel, P. M., Suissa, S., and Horwitz, R. I. (2001). A clinical trial of estrogen-replacement therapy after ischemic stroke. *N. Engl. J. Med.* **345,** 1243–1249.

Wang, X. K., Yue, T. L., Barone, F. C., White, R. F., Gagnon, R. C., and Feuerstein, G. Z. (1994). Concomitant cortical expression of TNF-α and IL-1β mRNAs follows early response gene expression in transient focal ischemia. *Mol. Chem. Neuropathol.* **23,** 103–114.

Yamasaki, Y., Matsuura, N., Shizuhara, H., Onodera, H., and Itoyama, Y. K. K. (1995). Interleukin-1 as a pathogenetic mediator of ischemic brain damage in the rats. *Stroke* **26,** 676–681.

Zhang, C., Shen, W., and Zhang, G. (2002). N-methyl-D-aspartate receptor and L-type voltage-gated Ca^{2+} channel antagonists suppress the release of cytochrome c and the expression of procaspase-3 in rat hippocampus after global brain ischemia. *Neurosci. Lett.* **328,** 265–268.

Zhang, Z., Chopp, M., Goussev, A., and Powers, C. (1998). Cerebral vessels express interleukin 1 beta after focal cerebral ischemia. *Brain Res.* **784,** 210–217.

Zhao, H., Yenari, M. A., Cheng, D., Sapolsky, R. M., and Steinberg, G. K. (2003). Bcl-2 overexpression protects against neuron loss within the ischemic margin following experimental stroke and inhibits cytochrome c translocation and caspase-3 activity. *J. Neurochem.* **85,** 1026–1036.

A ROLE FOR BRAIN CYCLOOXYGENASE-2 AND PROSTAGLANDIN-E_2 IN MIGRAINE: EFFECTS OF NITROGLYCERIN

Cristina Tassorelli,[*,†] Rosaria Greco,[*] Marie Therèse Armentero,[*] Fabio Blandini,[*] Giorgio Sandrini,[*,†] and Giuseppe Nappi[*,†,‡]

[*]IRCCS "C. Mondino Institute of Neurology" Foundation, Pavia, Italy
[†]University Centre for the Study of Adaptive Disorder and Headache (UCADH), Pavia, Italy
[‡]Department of Neurology and Otorhinolaryngology
University of Rome "La Sapienza," Rome, Italy

Cyclooxygenase-2 (COX-2) may increase prostaglandin E_2 (PGE_2) production in central nervous system (CNS) and contribute to the severity of pain responses in inflammatory pain. In this chapter, we sought to evaluate the possible role of COX-2 induction and prostaglandins (PGs) synthesis within neuronal areas proposed to be involved in migraine genesis in the animal model of migraine based on the administration of systemic nitroglycerin (NTG). Male Sprague-Dawley rats were injected with NTG (10 mg/kg, i.p.) or vehicle and sacrificed 2 and 4 h later. The hypothalamus and the lower brain stem were dissected out and utilized for the evaluation of COX-2 expression by means of Western blotting and for the determination of PGE_2 levels by means of ELISA immunoassay. COX-2 expression increased in the hypothalamus at 2 h and in the lower brain stem at 4 h. PGE_2 levels showed an opposite pattern of change with a decrease in PGE_2 levels at 2 h in the hypothalamus and an increase at 4 h in the lower brain stem.

These data support the hypothesis that NTG administration is capable of activating the COX-2 pathway within cerebral areas. This activity may explain the pronociceptive effect of NTG described in animal and human models of pain.

INTERNATIONAL REVIEW OF
NEUROBIOLOGY, VOL. 82
DOI: 10.1016/S0074-7742(07)82020-4

Most importantly, these findings point to mediators and areas that may be relevant for migraine pathogenesis and treatment.

I. Introduction

Several reports have shown a constant cross talk between nitric oxide (NO) and prostaglandins (PGs) release in different biological systems, particularly in pain modulation (see Mollace *et al.*, 2005 for a review). The relations between NO- and PG-generating machinery may occur at various levels and diverse mechanisms have been hypothesized. Under inflammatory states, NO and PGs, released simultaneously in large amounts (Heitsch, 2000; Mollace *et al.*, 2005), may modulate neuronal activity and mediate the processing of nociceptive information in the central nervous system (CNS). NO may interfere directly with cyclooxygenase (COX) expression and therefore with PGs biosynthesis (Mollace *et al.*, 2005). It is also true that COX-2 may increase prostaglandin E_2 (PGE_2) production in the CNS and contribute to the severity of pain responses in inflammatory pain (Guay *et al.*, 2004).

It is known that inflammatory mediators may activate dural trigeminal afferents and brain stem neurons with meningeal input (Ebersberger *et al.*, 1997; Strassman *et al.*, 1996). PGE_2 synthesized in the brain is probably involved in modulating trigeminal nociception (Tuca *et al.*, 1989). NO and PGE_2 hyperproduction is likely involved in the neurovascular modifications leading to migraine attacks (Sarchielli *et al.*, 2000). Clinical studies, indeed, have reported an upregulation of NO synthase (NOS) and COX in the monocytes of migraineurs (Stirparo *et al.*, 2000), whereas, during the attacks, elevated levels of PGE_2 have been detected in blood and saliva (Tuca *et al.*, 1989).

Nitroglycerin (NTG), an NO donor, is capable of inducing spontaneous-like migraine attacks in migraineurs (Olesen *et al.*, 1993). Increasing evidence suggests that NTG-induced neuronal activation in rat brain is mediated by multiple mechanisms that include direct/indirect neuronal and vascular activities of NTG-derived and/or endogenously synthesized NO (Bergerot *et al.*, 2006; Pardutz *et al.*, 2004; Reuter *et al.*, 2001; Tassorelli *et al.*, 1995, 2005). Systemic administration of NTG activates cerebral nuclei via the intervention of selected neurotransmitters and neuromediators, with a specific time pattern in different brain areas (Tassorelli *et al.*, 2003b, 2005), which are likely to represent the neuroanatomical substrate of migraine (Tassorelli *et al.*, 1999), and include the paraventricular nucleus of the hypothalamus and the nucleus trigeminalis caudalis.

In this chapter, we sought to investigate the role of COX-2 and PGE_2 synthesis within neuronal areas involved in migraine genesis following NTG administration in rats.

II. Materials and Methods

Male Sprague-Dawley rats ($n = 4$–6 per group) were injected i.p. with NTG (10 mg/kg) or vehicle (PEG) and sacrificed 2 and 4 h after the injection. The lower brain stem and the hypothalamus were dissected out and used for the evaluation of COX-2 expression in Western blotting analysis and for the determination of PGE_2 levels.

A. COX-2 Western Blotting

Rats in the treated and control groups were perfused transcardially with 250-ml cold saline, 2 and 4 h after NTG or vehicle administration. Brains were immediately removed and lower brain stem and hypothalamus were dissected out and used for the preparation of total extracts. The samples were homogenized on ice with an Ultra-Turrax homogenizer (Janke & Kunkel GMBH & Co. KG, Staufen, Germany) in at least five volumes of modified RIPA buffer (Tris 50 mM, pH 7.4, NaCl 150 mM, EDTA 1 mM, SDS 0.2%) supplemented with cocktail inhibitors protease and then incubated on ice for 20 min. The tissue lysate was centrifuged at 10,000 × g for 45 min at 4°C and supernatants stored at −80°C.

Protein assay was performed by the bicinchoninic acid (BCA) method. Twenty micrograms of protein were submitted to SDS-polyacrylamide gels 10% and transferred onto a PVDF membrane (Amersham Biosciences). After blocking with 5% dry milk, the blots were probed overnight at 4°C with rabbit polyclonal anti-COX-2 serum (1:1000; Cayman Chemical) and then probed for 1 h with an anti-rabbit horseradish peroxidase–coupled secondary antibody (1:10,000; Amersham Biosciences). An enhanced chemoluminescence system (ECL Advance; Amersham Biosciences) was used for visualization. Membranes were also probed with a rabbit polyclonal anti-β-actin antibody (1:1000; Santa Cruz Biotechnology) as a housekeeping protein. COX-2 expression was evaluated as the ratio between the optical density of the COX-2 band and the β-actin band. The specificity of the antibody was confirmed by immunoprecipitation with a specific blocking peptide (Cayman Chemical, Ann Arbor, MI).

B. PGE_2 Assay

PGE_2 levels were quantified using a commercially available ELISA kit (Cayman Chemical, Ann Arbor, MI), according to the manufacturer's instructions.

C. Statistics

Data are expressed as mean values ± standard deviations. Comparisons between groups (NTG and Control) were performed using the Student's *t*-test and statistical significance was assumed if $p < 0.05$.

III. Results

A. Western Blotting Analysis

In the brain stem, COX-2 protein was visualized as a single band of 70–72 kDa, while in the hypothalamus our antibody recognized an additional band at 65 kDa, which, on subsequent evaluation with the blocking peptide, did not prove specific for COX-2; therefore, for statistical purposes, we considered exclusively the band at 70–72 kDa.

COX-2 protein was constitutively expressed in the cerebral areas evaluated, in both the control animals. NTG caused a massive (+92%) increase of the enzyme expression 2 h after drug administration in the hypothalamus (Fig. 1); a significant increase of COX-2 expression, although of a lesser magnitude

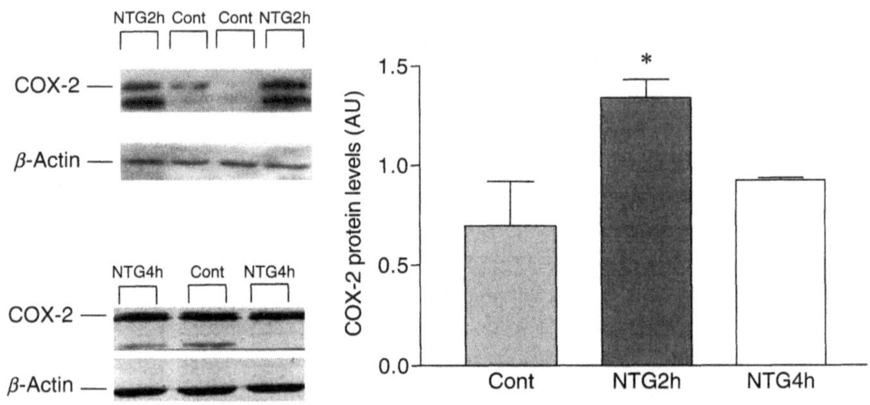

Fig. 1. COX-2 expression in the hypothalamus of rats treated with NTG or vehicle. Left panel: representative Western blots of COX-2 protein (70–72 kDa) in homogenates of hypothalamus of animals evaluated 2 h (NTG2 h) or 4 h (NTG4 h) after NTG administration or of animals treated with vehicle (Cont). β-Actin (39 kDa) was used as a housekeeping protein in the same blot. Right panel: densitometric measurement of COX-2 protein levels in the three groups described above. Values are expressed as mean ± SD. Student's *t*-test $*p < 0.01$ versus control group.

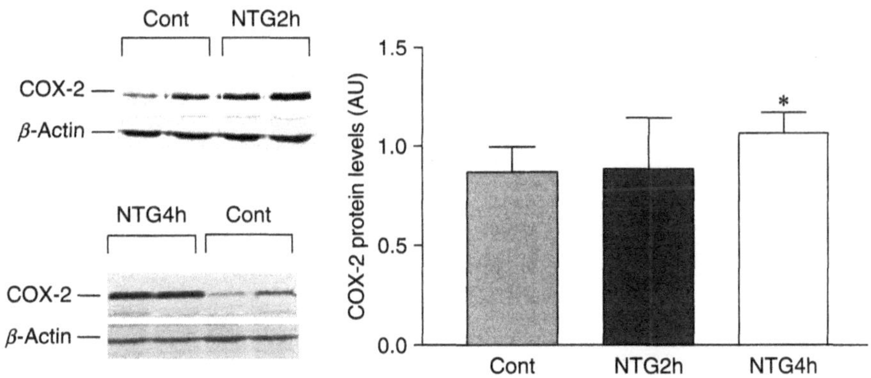

FIG. 2. COX-2 expression in the lower brain stem of rats treated with NTG or vehicle. Left panel: representative Western blots of COX-2 protein (70–72 kDa) in homogenates of lower brain stem of rats evaluated 2 h (NTG2 h) or 4 h (NTG4 h) after NTG administration or of animals treated with vehicle (Cont). β-Actin (39 kDa) was used as a housekeeping protein in the same blot. Right panel: densitometric measurement of COX-2 protein levels in the three groups described above. Values are expressed as mean ± SD. Student's *t*-test *$p < 0.05$ versus control group.

(+23%), was also observed in the lower brain stem of NTG-treated rats, at 4 h postadministration (Fig. 2).

B. ELISA Assays

In the hypothalamus of NTG-treated rats, PGE$_2$ levels were significantly decreased 2 h after the drug administration (Fig. 3). In the lower brain stem, a slight, nonsignificant increase in PGE$_2$ levels was observed at 2 h and reached a statistically significant magnitude at 4 h (Fig. 4).

IV. Discussion

PGs—generally considered as peripheral mediators of inflammation—are also known to be involved in other regulatory processes in the CNS, nociception in particular. Pain-evoked PGE$_2$ release in the spinal cord contributes considerably to the development of hyperalgesia and allodynia (Hofacker *et al.*, 2005). COX-2 is the most important contributor to elevation of spinal PGE$_2$; PGE$_2$, in turn, increases the processing of pain stimuli following inflammation (Vanegas and Schaible, 2001). Therefore, selective COX-2 inhibitors, but not COX-1

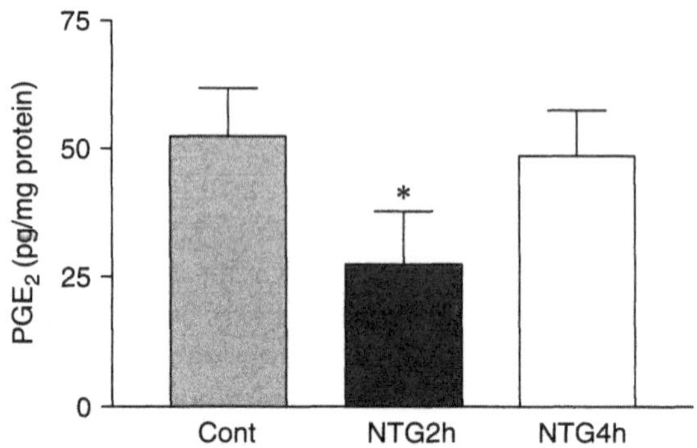

FIG. 3. PGE_2 levels in homogenates of hypothalamus of rats treated with NTG or vehicle. The graph illustrates the data obtained with animals injected with vehicle (Cont) or with NTG 2 h (NTG2 h) and 4 h (NTG4 h) before evaluation. Values are expressed as mean ± SD. Student's *t*-test $*p < 0.05$ versus control group.

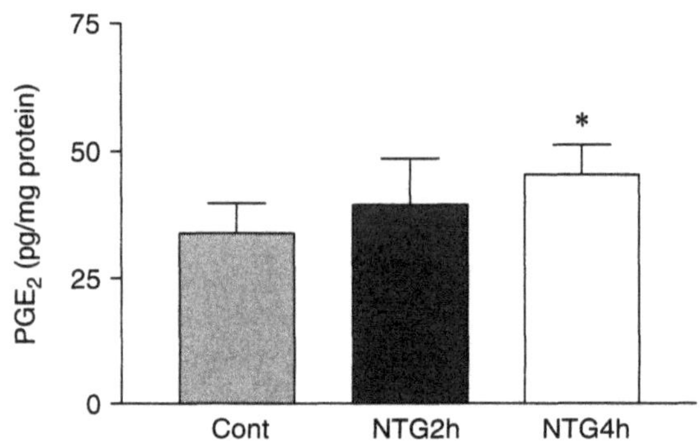

FIG. 4. PGE_2 levels in homogenates of lower brain stem of rats treated with NTG or vehicle. The graph illustrates the data obtained with animals injected with vehicle (Cont) or with NTG 2 h (NTG2 h) and 4 h (NTG4 h) before evaluation. Values are expressed as mean ± SD. Student's *t*-test $*p < 0.05$ versus control group.

inhibitors, reduce hyperalgesia and the levels of PGE_2 in the cerebrospinal fluid (CSF) (Riendeau *et al.*, 1997; Smith *et al.*, 1998). In the rat brain, COX-2 is constitutively expressed in areas involved in the processing of pain and in the integration of autonomic and endocrine functions (Breder *et al.*, 1995). Some studies have demonstrated that PGs are also released from intracranial tissue in experimental models of headache and in the blood of migraineurs during the ictal phase (Ebersberger *et al.*, 1999; Sarchielli *et al.*, 2000). Clinically, administration of specific COX-2 inhibitors is effective in treating acute migraine (Kudrow *et al.*, 2005;

Silberstein *et al.*, 2004), although there is evidence that PGE_2 synthesis is lower at the end of migraine attacks compared with the levels measured at the onset (Sarchielli *et al.*, 2000).

The animal model of migraine based on administration of NTG has provided interesting insights into the neuroanatomical circuitry and neuropharmacological mechanisms involved in the initiation and recurrence of migraine attacks (Tassorelli *et al.*, 1995, 2005, 2006). In this chapter, we report a significant increase in COX-2 expression in the lower brain stem and hypothalamus—two key structures for migraine genesis—after NTG administration. In this regard, it is also noteworthy that an *in vitro* study has demonstrated that NO released by NTG and sodium nitroprusside may directly cause an increase in COX-2 activity (Salvemini *et al.*, 1993). In this chapter, COX-2 activation in cerebral structures was associated to important variations in PGE_2 levels in the same areas. In the lower brain stem, an increase in COX-2 expression was paralleled by an increase in PGE_2 levels 4 h after drug administration, which clearly points to the activation of the COX-2-PGs pathway by NTG. At variance, in the hypothalamus, the increase of density for COX-2 band, 2 h after the drug administration, coincided with a significant reduction in PGE_2 levels in the same area. This finding, which was quite surprising, appears difficult to reconcile with the increased expression of hypothalamic COX-2; however, prominent activation of the biosynthetic pathway for PGE_2, leading to local synthesis, may have been associated with rapid clearance from the hypothalamic tissue, across the blood–brain barrier (BBB); such mechanisms may have played a role in this seemingly paradoxical response of PGE_2 in the hypothalamus (Jones *et al.*, 1993). In fact, as previously reported by Bito *et al.* (1976), rapid PGs transport across the BBB may be a primary mechanism for the termination of the action of autacoids endogenously produced in the course of neuronal activity. Alternatively, it is possible that the massive expression of COX-2 in the hypothalamus may have activated a protective negative feedback, to shut down the CNS production of PGE_2 (Turrin and Rivest, 2004). Further studies are needed for a more evidence-based interpretation of this point.

Theoretically, the changes in COX-2 and PGE_2 levels observed in the two brain areas under investigation could also be the result of the activation of COX-2 by cells of microvasculature penetrating these regions. Numerous papers have indeed provided clear evidence that COX-2 protein is expressed within the endothelium of cerebral capillaries (Engblom *et al.*, 2003; Rivest, 1999). However, the long latencies associated with the changes detected in our experimental paradigm (2 and 4 h) rather suggest that systemically administered NTG is likely to involve also the activation of COX-2 neuronal elements. This is in agreement with a previous report by Guay *et al.* (2004), where a significant increase in COX-2 expression and PGE_2 levels was observed in the rat brain 3–6 h after carrageenan injection.

Previous reports demonstrated that systemic NTG induces a condition of hyperalgesia (Tassorelli *et al.*, 2003a,b). The local formation of PGE_2 in the lower brain stem and their accessibility to the second order nociceptive trigeminal neurons may be relevant for this phenomenon (Pardutz *et al.*, 2000, 2004; Sandrini *et al.*, 2002; Tassorelli *et al.*, 2006). Indeed, central production of PGE_2 by COX-2 is known to induce hyperalgesia and vasodilatation (Hosoi *et al.*, 1997, 1998; Parantainen *et al.*, 1985).

The findings are in agreement with previous studies, where we have demonstrated that nimesulide, a preferential COX-2 inhibitor, inhibits NTG-induced neuronal activation in areas involved in the control of nociception, autonomic, and neuroendocrine functions (Tassorelli *et al.*, 2003a). In addition, nimesulide proved able to abolish NTG-induced hyperalgesia at the formalin test (Tassorelli *et al.*, 2003a,b, 2006). Several studies have demonstrated that NO and NO donors modulate PGs biosynthesis (Salvemini *et al.*, 1993). Furthermore, Pardutz *et al.* (2004) have demonstrated that lysine acetylsalicylate inhibits the formation of neuronal NOS following NTG administration in the nucleus trigeminalis caudalis of rat, confirming the interaction between NO and COX.

Some authors have reported that COX-2 upregulation in inflammatory and neuropathic pain may be mediated by NF-Kappa B (NF-κB) activation (Lee *et al.*, 2004; Sakaue *et al.*, 2001). In this regard, we have previously seen that NTG is capable of promoting NF-κB activity in the lower brain stem (Greco *et al.*, 2005), and more specifically in the nucleus trigeminalis caudalis. In addition, pretreatment with parthenolide, an inhibitor of NF-κB, reduced NTG-induced fos expression in the very same area (Tassorelli *et al.*, 2005).

Taken together, these findings suggest that NTG may induce a very complex series of events that are relevant for pain modulation in the brain.

Acknowledgments

Supported by the EU Euro head (LSHM-CT-2004-504837).

References

Bergerot, A., Holland, P. R., Akerman, S., Bartsch, T., Ahn, A. H., Maassenvandenbrink, A., Reuter, U., Tassorelli, C., Schoenen, J., Mitsikostas, D. D., van den Maagdenberg, A. M., and Goadsby, P. J. (2006). Animal models of migraine: Looking at the component parts of a complex disorder. *Eur. J. Neurosci.* **24,** 1517–1534.

Bito, L. Z., Davson, H., and Hollingsworth, J. R. (1976). Facilitated transport of prostaglandins across the across the blood-cerebrospinal fluid and blood-brain barriers. *J. Physiol.* **256,** 273–285.

Breder, C. D., Dewitt, D., and Kraig, R. P. (1995). Characterization of inducible cyclooxygenase in rat brain. *J. Comp. Neurol.* **355,** 296–315.

Ebersberger, A., Ringkamp, M., Reeh, P. W., and Handwerker, H. O. (1997). Recordings from brain stem neurons responding to chemical stimulation of subarachnoid space. *J. Neurophysiol.* **77,** 3122–3133.

Ebersberger, A., Averbeck, B., Messlinger, K., and Reeh, P. W. (1999). Release of substance P, calcitonin gene- related peptide and prostaglandin E2 from dura mater encephali following electrical chemical stimulation *in vitro*. *Neuroscience* **89,** 901–907.

Engblom, D., Satha, S., Engstrom, L., Westman, M., Audloy, L. P., Jakobsson, P. J., and Blomqvist, A. (2003). Microsomial prostaglandin E synthase-1 is central switch during immune-induced pyresis. *Nat. Neurosci.* **6,** 1137–1138.

Greco, R., Tassorelli, C., Cappelletti, D., Sandrini, G., and Nappi, G. (2005). Activation of the transcription factor NF-kappaB in the nucleus trigeminalis caudalis in an animal model of migraine. *Neurotoxicology* **26,** 795–800.

Guay, J., Bateman, K., Gordon, R., Mancini, J., and Riendeau, D. (2004). Carrageenan-induced paw edema in rat elicits a predominant prostaglandin E2 (PGE2) response in central nervous system associated with the induction of microsomial PGE2 synthase-1. *J. Biol. Chem.* **279,** 24866–24872.

Heitsch, H. (2000). Bradykinin B2 receptor as a potential therapeutic target. *Drug News Perspect.* **13,** 213–225.

Hofacker, A., Coste, O., Nguyen, H. V., Marian, C., Scholich, K., and Geisslinger, G. (2005). Downregulation of cytosolic prostaglandin E2 synthase results in decreased nociceptive behavior in rats. *J. Neurosci.* **25,** 9005–9009.

Hosoi, M., Oka, T., and Hori, T. (1997). Prostaglandin E receptor EP3 subtype is involved in thermal hyperalgesia through its actions in the preoptic hypothalamus and the diagonal band of Broca in rats. *Pain* **71,** 303–311.

Hosoi, M., Oka, T., and Aou, S. (1998). Pain modulatory actions of cytokines and prostaglandin E2 in the brain. *Ann. N. Y. Acad. Sci.* **840,** 269–281.

Jones, S. A., Adamson, S. L., Engelberts, D., Bishai, I., Norton, J., and Coceani, F. (1993). PGE2 in the perinatal brain: Local synthesis and transfer across the blood brain barrier. *J. Lipid Mediat.* **6,** 487–492.

Kudrow, D., Thomas, H. M., Ruoff, G., Ishkanian, G., Sands, G., Le, V. H., and Brown, M. T. (2005). Valdecoxib for treatment af single, acute, moderate to severe migraine headache. *Headache* **45,** 1151–1162.

Lee, K. M., Kang, B. S., Lee, H. L., Son, S. J., Hwang, S. H., Kim, D. S., Park, J. S., and Cho, H. J. (2004). Spinal NF-kB activation induces COX-2 upregulation and contribute to inflammatory pain hypersensitivity. *Eur. J. Neurosci.* **19,** 3375–3381.

Mollace, V., Muscoli, C., Masine, E., Cuzzocrea, S., and Salvemini, D. (2005). Modulation of prostaglandin biosynthesis by nitric oxide and nitric oxide donors. *Pharmacol. Rev.* **57,** 217–252.

Olesen, J., Iversen, H. K., and Thomsen, L. (1993). Nitric oxide supersensitivity: A possible molecular mechanism of migraine pain. *Neuroreport* **4,** 1027–1030.

Parantainen, J., Vapaatalo, H., and Hokkanen, E. (1985). Relevance of prostaglandins in migraine. *Cephalalgia* **2,** 93–97.

Pardutz, A., Krizbai, I., Multon, S., Vecsei, L., and Schoenen, J. (2000). Systemic nitroglycerin increases nNOS levels in rat trigeminal nucleus caudalis. *Neuroreport* **11,** 3071–3075.

Pardutz, A., Szatmari, E., Vecsei, L., and Schoenen, J. (2004). Nitroglycerin-induced nNOS increase in rat trigeminal nucleus caudalis is inhibited by systemic administration of lysine acetylsalicylate but not of sumatriptan. *Cephalalgia* **24,** 439–445.

Reuter, U., Bolay, H., Jansen-Olesen, I., Chiarugi, A., Sanchez del Rio, M., Letourneau, R., Theoharides, T. C., Waeber, C., and Moskowitz, M. A. (2001). Delayed inflammation in rat meninges: Implications for migraine pathophysiology. *Brain* **12,** 2490–2502.

Riendeau, D., Percival, M. D., Boyce, S., Brideau, C., Charleston, S., Cromlish, W., Hethier, D., Evans, J., Falqueyret, J. P., Ford-Hutchinson, A. W., Gordon, R., Greig, G., *et al.* (1997).

Biochemical and pharmacological profile of a tetrasubstituted furanone as a highly selective COX-2 inhibitor. *Br. J. Pharmacol.* **121,** 105–117.

Rivest, S. (1999). What is cellular source of prostaglandins in the brain in response to systemic inflammation? Fact and controversies. *Mol. Psychiatry* **4,** 501–507.

Sakaue, G., Shimaoka, M., Fukuoka, T., Hiroi, T., Inoue, T., Hashimoto, N., Sakaguchi, T., Sawa, Y., Morishita, R., Kiyono, H., Noguchi, K., and Mahimo, M. (2001). NF-kappa B decoy suppress cytokine expression and thermal hyperalgesia in a rat neuropathic pain model. *Neuroreport* **12,** 2079–2084.

Salvemini, D., Misko, T. P., Masferrer, J. L., Seibert, K., Currie, M. G., and Needleman, P. (1993). Nitric oxide activates cyclooxygenase enzymes. *Proc. Natl. Acad. Sci. USA* **90,** 7240–7244.

Sandrini, G., Tassorelli, C., Proietti Cecchini, A., Alfonsi, E., and Nappi, G. (2002). Effects of nimesulide on nitric oxide-induced hyperalgesia in humans – A neurophysiological study. *Eur. J. Pharmacol.* **450,** 259–262.

Sarchielli, P., Alberti, A., Codini, M., Floridi, A., and Gallai, V. (2000). Nitric oxide metabolites, prostaglandins and trigeminal vasoactive peptides in internal jugular vein blood during spontaneous migraine attacks. *Cephalalgia* **20,** 907–918.

Silberstein, S., Tepper, S., Brandes, J., Diamond, M., Goldstein, J., Winner, P., Venkatraman, S., Vrijens, F., Malbecq, W., Lines, C., Visser, W. H., Reines, S., *et al.* (2004). Randomized, placebo-controlled trial of rofecoxib in the acute treatment of migraine. *Neurology* **62,** 1552–1557.

Smith, C. J., Zhang, Y., Koboldt, C. M., Muhammad, J., Zweifel, B. S., Shaffer, A., Talley, J. J., MasferreR, J. L., Seibert, K., and Isakson, P. C. (1998). Pharmacological analysis of cyclooxygenase-1 in inflammation. *Proc. Natl. Acad. Sci. USA* **95,** 13313–13318.

Stirparo, G., Zicari, A., Favilla, M., Lipari, M., and Martelletti, P. (2000). Linked activation of nitric oxide synthase and cyclooxygenase in peripheral monocytes of asymptomatic migraine without aura patients. *Cephalalgia* **20,** 100–106.

Strassman, A. M., Raymond, S. A., and Burstein, R. (1996). Sensitization of meingeal sensory neurons and origin of headaches. *Nature* **384,** 60–564.

Tassorelli, C., and Joseph, S. A. (1995). Systemic nitroglycerin induces Fos-ir in brainstem and forebrain structures of the rat. *Brain Res.* **682,** 167–178.

Tassorelli, C., Joseph, S. A., Buzzi, G., and Nappi, G. (1999). The effect on the central nervous system of nitroglycerin—Putative mechanisms and mediators. *Progr. Neurobiol* **57,** 607–624.

Tassorelli, C., Greco, R., Sandrini, G., and Nappi, G. (2003a). Central components of the analgesic/antihyperalgesic effect of nimesulide: Studies in animal models of pain and hyperalgesia. *Drugs* **63,** 9–22.

Tassorelli, C., Greco, R., Wang, D., Morelli, G., and Nappi, G. (2003b). Nitroglycerin induces Hyperalgesia in Rats: A Time-Course Study. *Eur. J. Pharmacol.* **464,** 159–162.

Tassorelli, C., Greco, R., Morazzoni, P., Riva, A., Sandrini, G., and Nappi, G. (2005). Parthenolide is the component of tanacetum parthenium that inhibits nitroglycerin-induced Fos activation: Studies in an animal model of migraine. *Cephalalgia* **25,** 612–621.

Tassorelli, C., Greco, R., Wang, D., Sandrini, G., and Nappi, G. (2006). Prostaglandins, glutamate and nitric oxide synthase mediate nitroglycerin-induced hyperalgesia in the formalin test. *Eur. J. Pharmacol.* **534,** 103–107.

Tuca, J. O., Planas, J. M., and Parellada, P. P. (1989). Increase in PGE2 and TXA2 in the saliva of common migraine patients Action of calcium channel blockers. *Headache* **29,** 498–501.

Turrin, N. P., and Rivest, S. (2004). Unraveling the Molecular Details Involved in the Intimate Link between the Immune and Neuroendocrine Systems. *Exp. Biol. Med.* **229,** 996–1006.

Vanegas, H., and Schaible, H. G. (2001). Prostaglandin and cyclooxygenase in the spinal cord. *Prog. Neurobiol.* **64,** 327–363.

THE BLOCKADE OF K^+-ATP CHANNELS HAS NEUROPROTECTIVE EFFECTS IN AN *IN VITRO* MODEL OF BRAIN ISCHEMIA

Robert Nisticò,[*,†] Silvia Piccirilli,[*] L. Sebastianelli,[*] Giuseppe Nisticò,[§] G. Bernardi,[*,‡] and N. B. Mercuri[*,‡]

[*]Department of Experimental Neurology, S. Lucia Foundation IRCCS, Rome, Italy
[†]Department of Pharmacobiology and University Centre for Adaptive Disorders and Headache (UCHAD), Section of Neuropharmacology of Normal and Pathological Neuronal Plasticity University of Calabria, Arcavacata di Rende, Italy
[‡]Clinica Neurologica, University of Rome 'Tor Vergata', Rome, Italy
[§]Centre of Pharmaceutical Biotechnology, University of Rome 'Tor Vergata', Rome, Italy

There is a common belief that the opening of K^+-ATP channels during an ischemic episode has protective effects on neuronal functions by inducing a reduction in energy consumption. However, recent studies have also proposed that activation of these channels might have deleterious effects on cell's survival possibly after a stroke or during long-lasting neurodegenerative processes. Considering these contrasting results, we have used a hippocampal *in vitro* slice preparation in order to investigate the possible effects of K^+-ATP channel blockers on the electrophysiological and morphological changes induced by a transient episode of ischemia (oxygen and glucose deprivation) on CA1 pyramidal neurons. Therefore, we found that tolbutamide and glibenclamide, both nonselective

INTERNATIONAL REVIEW OF NEUROBIOLOGY, VOL. 82
DOI: 10.1016/S0074-7742(07)82021-6

K^+-ATP channel blockers, produce neuroprotective effects against *in vitro* ischemia. Interestingly, the mitochondrial K^+-ATP channel blocker 5-hydroxydecanoate and various K^+ channel blockers did not exert neuroprotection. Our results are consistent with the concept that a decreased activity of the plasmalemmal K^+-ATP conductances may have a protective effect during episodes of transient cerebral ischemia.

I. Introduction

It is well known that the occurrence of an ischemic stroke in the mammalian brain, that is mainly characterized by oxygen and glucose deprivation (OGD), leads to the development of severe tissue damage. This is caused by the unwanted consequences of energy failure that usually occurs either during or even after a stroke (Calabresi *et al.*, 2003a; Lo *et al.*, 2003). However, several studies have also reported that neurons, during ischemia, activate endogenous protective mechanisms, which increase their survival and limit the extent of injury (Lo *et al.*, 2003). Among the protective responses, the opening of K^+-ATP channels has been thought to have a prominent role. These channels are readily activated when the intracellular level of ATP drops, and the consequences of K^+-ATP channel opening will depend on their location: activation of plasmalemmal K^+-ATP channels leads to hyperpolarization of cells, while opening the mitochondrial K^+-ATP channels causes depolarization of mitochondria. It is assumed that the plasmalemmal K^+-ATP channel-induced hyperpolarization is an endogenous defensive mechanism against the early pathological events of cerebral hypoxia/ischemia by reducing neuronal energy demand and preventing, at least for an initial period, the occurrence of irreversible and deleterious membrane depolarization (Guatteo *et al.*, 1998; Mercuri *et al.*, 1994a,b). Thus, several K^+-ATP channel openers have received attention as possible therapeutic strategies in animal models of stroke (Ballanyi, 2004; Busija *et al.*, 2004). In spite of this, there are also infrequent experimental data proposing that the opening of K^+ channels might have pejorative effects either in stroke or in degenerative diseases (Liss and Roeper, 2001; Liss *et al.*, 2005). Based on these contradictive results, the aim of this chapter was to further investigate whether the blockade of K^+-ATP channels is neurotoxic or neuroprotective in an *in vitro* model of oxygen and glucose deprivation. In addition, we wanted to establish if the blockade of other K^+ channels might exert neuroprotection. To achieve this goal, we have used field potential recordings and histological analysis of the hippocampal CA1 pyramidal cells.

II. Materials and Methods

A. Brain Slice Preparation and Electrophysiology

Male Wistar rats (30–35 days old) were anesthetized with halothane and killed by decapitation, following international guidelines on the ethical use of animals from the European Communities Council Directive of November 24, 1986 (86/609/EEC). The brain was rapidly removed from the skull and placed in ice-cold oxygenated (95% O_2–5% CO_2) artificial cerebral spinal fluid (aCSF) of the following composition (mM): NaCl 126, KCl 2.5, $MgCl_2$ 1.2, $CaCl_2$ 2.4, NaH_2PO_4 1.2, $NaHCO_3$ 19, glucose 10. Parasagittal slices (400-μm thick) were cut using a vibratome and kept in oxygenated aCSF for at least 1 h at room temperature. A single slice was then placed on a nylon mesh, completely submerged in a small chamber (0.5 ml) and superfused with oxygenated aCSF (30–31 °C) at a constant flow rate of 3 ml/min. The treated solutions reached the preparation in 90 s and this delay was taken into account in our calculations.

B. Extracellular Recordings

Test pulses (80 ms, 0.06 Hz) were delivered through a bipolar nichrome electrode positioned in the stratum radiatum. Evoked extracellular potentials were recorded with glass microelectrodes (2–10 MΩ), (Clark Electromedical Instruments, Panghourne, United Kingdom) filled with 3-M NaCl, placed in the CA1 region of the stratum radiatum. Stimulus–response curves were obtained by gradual increases in stimulus strength at the beginning of each experiment. The test stimulus pulse was then adjusted to produce a field excitatory postsynaptic potential (fEPSP) whose slope was 40–50% of the maximum and was kept constant throughout the experiment. The amplitude of fEPSP was routinely measured and expressed as the percentage of the average amplitude of the potentials measured during 10 min after establishing a stable baseline of evoked response under normoxic conditions. In some experiments, both the amplitude and the initial slope of fEPSP were quantified, but since no appreciable differences were observed in the effect of drugs and of *in vitro* ischemia, only the measure of the amplitude was expressed in figures. *In vitro* OGD was obtained by perfusing the slice with aCSF without glucose and gassed with nitrogen (95% N_2, 5% CO_2). At the end of the ischemic period, the slice was again superfused with normal (glucose-containing) oxygenated aCSF. Data were collected and analyzed online using the LTP program (Anderson and Collingridge, 2001).

C. Morphological Studies

After the recording session, the slices were stained with Cresyl Fast Violet solution to evaluate the number of cells irreversibly damaged either by the handling procedures alone (control) or by the putative ischemic insult (ischemic untreated). Once the cellular damage induced by ischemia was established by histology, the group of slices pretreated with glibenclamide (ischemic treated) was also examined.

Briefly, after dissection, slices were transferred in a holding chamber and left to recover for 30 min (30–31 °C) in standard aCSF medium saturated with 95% O_2, 5% CO_2. They were then divided into three groups: the first group (control) consisted of slices maintained for 30 min in standard aCSF saturated with the O_2–CO_2 mixture; the second group (ischemia) was placed in a chamber containing aCSF saturated with 95% N_2–5% CO_2 gas mixture for 14 min and no glucose; the third group (ischemia + glibenclamide) was first placed in a chamber containing glibenclamide (10 μM) in aCSF saturated with O_2–CO_2 for 30 min and then transferred in a chamber containing glibenclamide (10 μM) in aCSF saturated with N_2–CO_2 gas mixture for 14 min and no glucose. After each treatment, slices from all three groups were left to recover for 1 h in standard oxygenated aCSF, then fixed overnight in 4% paraformaldehyde in 0.1-M phosphate buffer (PB) and after washings in PB, they were transferred to 30% sucrose/PB at 4 °C until they sank. Finally, the slices were frozen in liquid nitrogen and cut into 20-μm-thick horizontal sections using a freezing microtome. The sections were mounted on slides, stained with 0.1% Cresyl Fast Violet, and analyzed under light microscopy (Zeiss-Axioplan 2). All neurons with intact morphological appearance in each of the two 25-μm^2 fields of the CA1 hippocampal area of each brain section ($n = 6$ per brain; 3–8 rats per group) were counted at 40×.

D. Drug Application

All compounds were obtained from Sigma-Aldrich (Milan, Italy). Drugs were applied by switching the standard aCSF to one containing a known concentration of the drug(s). Full exchange of the solution in the recording chamber occurred over 1 min and this was taken into account for the analysis.

E. Statistical Analysis

All numerical data are expressed as the mean ± SEM. Data were tested for statistical significance with paired two-tailed Student's *t*-test or by ANOVA test followed by Tukey-Kramer multiple comparisons test as appropriate. Significance was set at $p < 0.001$. Each slice was used only for a single experiment.

III. Results

A. The Effect of Different Time Courses OGD Application on fEPSP

We first wanted to assess the threshold ischemic period that induces an irreversible loss of the field potentials of the CA1 pyramidal neurons. As shown in Fig. 1A, an OGD of 9 min caused a transient fEPSP depression that was always reversible after returning to normal oxygenated aCSF ($101 \pm 1\%$, $n = 6$, paired Student's t-test $p < 0.001$).

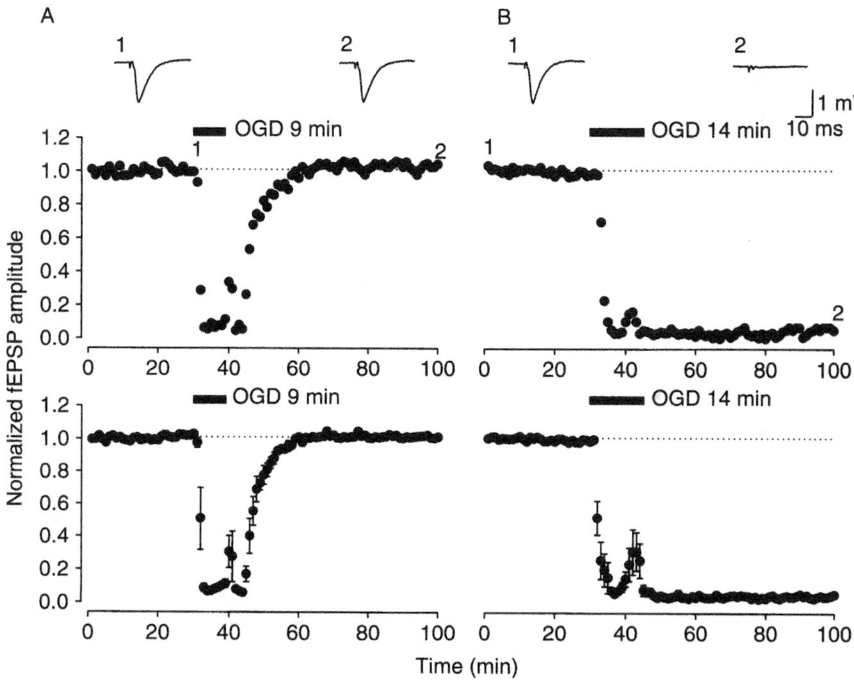

FIG. 1. Effects of ischemia on fEPSPs evoked by electrical stimulation of the stratum radiatum in the CA1 hippocampal region. The upper graph shows a single example of the time course of fEPSPs amplitude, before, during 9-min OGD (A) or 14-min OGD (B) and after reperfusion in normal oxygenated aCSF. An OGD of 9 min caused a transient fEPSP depression that was always reversible after returning to normal oxygenated aCSF, whereas the application of 14-min OGD irreversibly abolished synaptic neurotransmission. Scale bar indicates the time duration of OGD. The fEPSP traces were taken at the times indicated by numbers in the corresponding graph. The lower graph indicates pooled data from 6 (A) and 15 (B) equivalent experiments. Data, expressed as percentage of baseline values, are mean ± SEM. Each point represents fEPSP amplitude expressed as percentage of the mean baseline responses recorded before OGD application.

Prolonging the exposure of the slices to 11-min OGD caused a loss of the field potentials in 50% of the experiments recorded (6/12, data not shown). In the remaining cases, the field recovered to 100% on reperfusion with control aCSF. Therefore, we have chosen a slightly more prolonged ischemic period (14 min) that determined an irreversible loss of the field potentials in 100% of the cases, to test the effects of various K^+ channel blockers on neuronal survival. In fact, the application of a longer, 14-min OGD irreversibly abolished synaptic neurotransmission, and the mean recovery of fEPSP amplitude was 2 ± 0.09% ($n = 15$, paired Student's *t*-test $p < 0.001$) (Fig. 1B). The amplitude of the field potential rapidly decreased and in 5 min OGD was almost abolished. In fact, no recovery of the field was observed during reoxygenation even after 60-min perfusion in normal aCSF.

B. Different Effects of Plasmalemmal Versus Mitochondrial K^+-ATP Channel Blockers on the Irreversible Loss of Synaptic Transmission Induced by *In Vitro* Ischemia

In subsequent experiments, we have shown that the K^+-ATP channel blockers tolbutamide (1 mM) and the more selective blocker glibenclamide (10 μM), superfused for 30 min before and during the ischemic period, were able to rescue the irreversible loss of field potentials typically obtained with 14-min OGD. Figure 2A shows that the suppression of the field potential caused by *in vitro* ischemia only recovered in slices incubated with 1-mM tolbutamide (103 ± 1%,

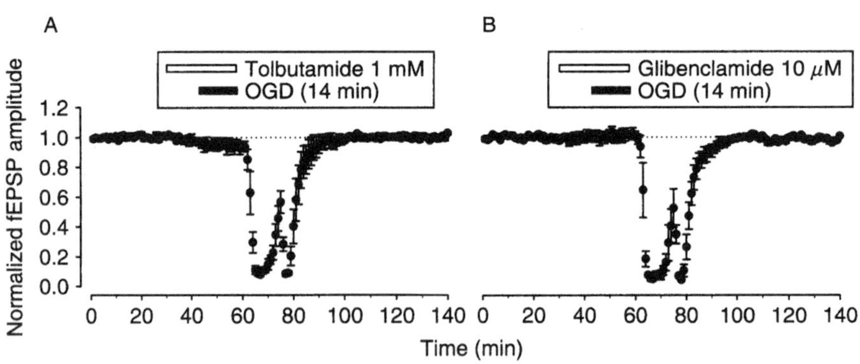

FIG. 2. Blockade of the K^+-ATP channels results in neuroprotection. (A) Pooled data from six experiments showing the protective effect of a pretreatment with tolbutamide (1 mM) on the irreversible fEPSP depression induced by 14-min OGD. (B) Similar effects obtained when glibenclamide (10 μM) was applied ($n = 6$).

paired Student's *t*-test $p < 0.001$; $n = 6$). As expected, neuroprotection also resulted from a treatment of the slices with glibenclamide (101 ± 1%, paired Student's *t*-test $p < 0.001$, $n = 6$) (Fig. 2B). On the contrary, no recovery of the field potential's amplitude was detected in the presence of either 300-μM barium chloride, or 5-mM tetraethylammonium (TEA), or 1-mM 4-aminopyridine (4-AP). These blockers were also applied 30 min before OGD (data not shown). To assess whether tolbutamide and glibenclamide exert their protective effects acting either on the plasmalemmal, or the mitochondrial, or both K^+-ATP channels, we carried out further experiments by applying the selective mitochondrial K^+-ATP channel blocker 5-hydroxydecanoate. Interestingly, we observed no protection when the ischemic insult was induced in slices treated with the mitochondrial K^+-ATP channel blocker 5-hydroxydecanoate (100 μM) (2 ± 1% recovery, paired Student's *t*-test $p < 0.001$, $n = 6$) (Fig. 3).

C. Actions of the K^+-ATP Channel Blockers on the Morphological Modifications Induced by *In Vitro* Ischemia in the Hippocampal Neurons

The neuroprotection afforded by glibenclamide (10 μm) was further confirmed by morphological studies. As expected, in histological sections, the CA1 region of 14-min OGD treated slices exhibited a marked decline in viable neurons when compared to control animals (OGD vs ctrl: 48.41 ± 2.17 vs 79.92 ± 1.49; $p < 0.001$, $n = 6$) (Figs. 4A–D and 5). However, histological examinations revealed that

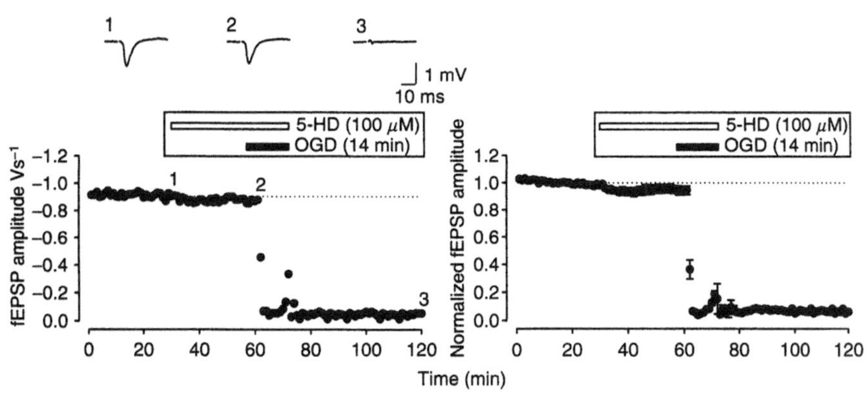

FIG. 3. Selective pharmacological blockade of mitochondrial K^+-ATP channel does not prevent the irreversible loss of synaptic field potentials exposed to 14-min OGD. A single example (left graph) and pooled data from six experiments (right graph) showing no recovery of synaptic transmission after 14-min OGD when applying the selective mitochondrial K^+-ATP channel blocker 5-hydroxydecanoate (100 μM), both in pretreatment and during OGD.

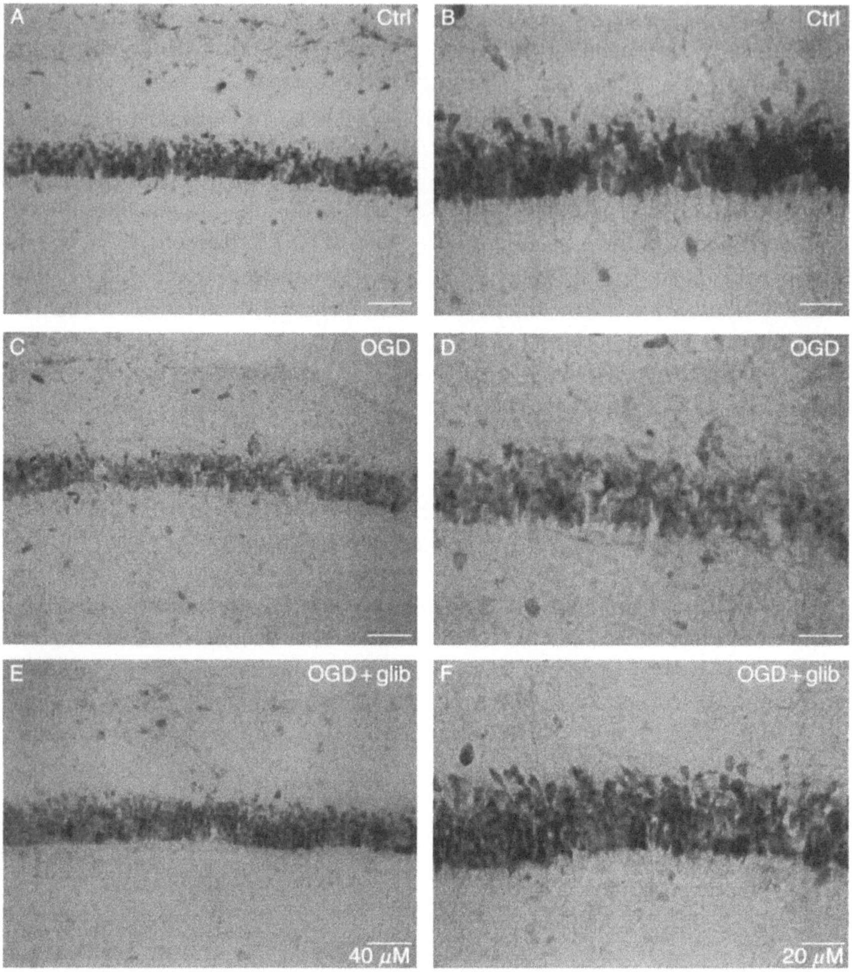

FIG. 4. Representative photomicrographs showing the protective effects of glibenclamide (10 μM) against 14-min OGD. (A–F) Photomicrographs of rat hippocampal sections stained with cresyl violet; (A, C, and E) low magnification (20×); (B, D, and (F) high magnification (40×). Panels (A) and (B) refer to slices taken from a control animal; panels (C) and (D) to slices treated with 14-min OGD; finally panels (E) and (F) show the effects of a pretreatment with glibenclamide (10 μM). (See Color Insert.)

glibenclamide (10 μM added to the aCSF 30 min before and during the 14-min OGD) was able to rescue neurons that otherwise were destined to deteriorate during OGD (glibenclamide vs OGD: 67.79 ± 2.18 vs 48.42 ± 2.17; $p < 0.001$, $n = 6$) (Figs. 4C–F and 5).

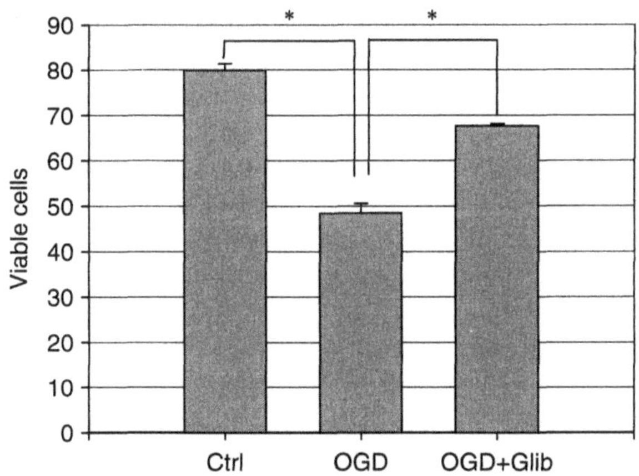

FIG. 5. Density of cresyl violet viable cells per unit test area (25 μm^2) in the CA1 hippocampal area. Slices ($n = 6$) exposed for 14-min ischemia *in vitro* contained more damaged cells than control slices (OGD vs ctrl: 48.41 ± 2.17 vs 79.92 ± 1.49; $p < 0.001$). A treatment of the slices ($n = 6$) with glibenclamide (10 μM) significantly reduced the severity of cell damage after ischemia compared with the untreated ischemic slices (glibenclamide vs OGD: 67.79 ± 2.18 vs 48.42 ± 2.17; $p < 0.001$).

IV. Discussion

The emerging evidence of this study is that K^+-ATP channel blockers have neuroprotective actions in an *in vitro* model of brain ischemia. Interestingly, neuroprotection was confirmed either by using electrophysiological or histological methods. We have utilized the CA1 hippocampal pyramidal neurons because they are highly vulnerable to ischemia. It is believed that the ischemic neuronal damage is certainly due to a sequence of early and delayed events that are triggered by an immediate energy failure. Therefore, OGD causes an increased membrane permeability to sodium and calcium ions throughout voltage-dependent channels and ionotropic glutamate receptors. These phenomena have a critical role in neuronal vulnerability (Calabresi *et al.*, 1998; Centonze *et al.*, 2001; Lo *et al.*, 2003; Zou *et al.*, 2005).

On the contrary, increased membrane permeability to potassium ions caused by the opening of K^+-ATP channels consequent to energy failure is believed to play a protective role (Wu *et al.*, 2006). This neuroprotection might derive from a reduced opening of voltage-activated sodium and calcium conductances causing a limitation in cellular excitability (Calabresi *et al.*, 2003b; Davis *et al.*, 1994; Hanon and Klitgaard, 2001; Kobayashi and Mori, 1998; Mora *et al.*, 1999; Pisani *et al.*, 1998; Ren *et al.*, 2004; Siniscalchi *et al.*, 1997).

Alternatively, it might be also argued that an excessive activation of the K^+-ATP channels during OGD might contribute to the ischemic damage by increasing the efflux of potassium, thus reducing the intracellular content of this ion. Therefore, to compensate for the ionic disequilibrium caused by OGD, an increased operation of the Na^+/K^+-pump (extruding sodium and intruding potassium; Fig. 6) certainly activates an initial but inadequate mechanism, which, at last, could lead to a further dissipation of the remaining cellular fuel, that is ATP. In accordance with the above theory, our electrophysiological, pharmacological, and histological experiments show that a reduced activation of K^+-ATP channels is of importance in limiting the OGD-induced neuronal damage. Therefore, the K^+-ATP channel blockers tolbutamide and glibenclamide exert neuroprotective effects, allowing a full recovery of the hippocampal field potentials recorded from hippocampal slices toward control levels. Conversely, the specific blocker of the mitochondrial K^+-ATP channel 5-hydroxydecanoate (Liu *et al.*, 1999) did not show neuroprotective effects. This suggests that only a reduced activity of plasmalemmal K^+-ATP channels

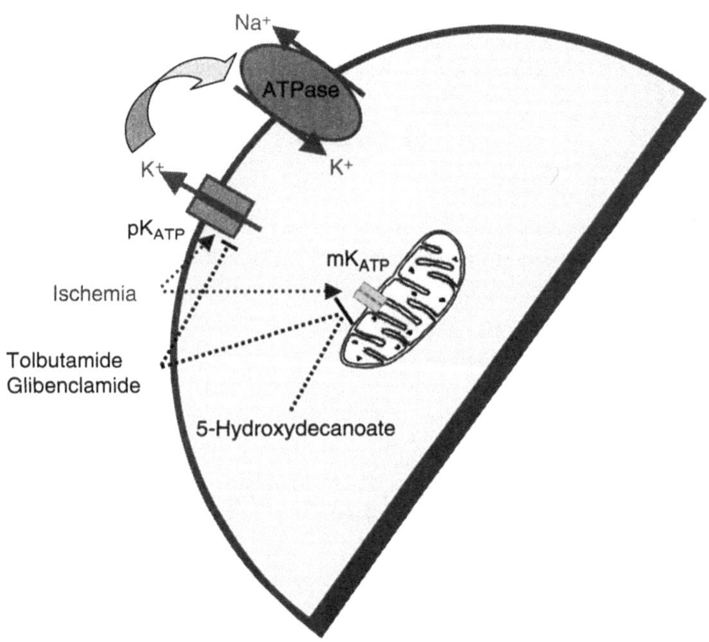

Fig. 6. A scheme representing the differential pharmacological regulation of the plasmalemmal and mitochondrial ATP-sensitive potassium channels. The targets of tolbutamide, glibenclamide, and 5-hydroxydecanoate are indicated. The diagram takes into account the possibility that a decreased efflux of K^+, direct consequence of blocking the plasmalemmal K^+-ATP conductances, might contribute to preserving the Na^+K^+-ATPase pump activity. This in turn may lead to a reduced consumption of ATP that may be vital in pathological conditions of energy failure such as ischemia. (See Color Insert.)

may result in a protective effect. On the other hand, it is still possible that, contrary to what is happening for the plasmalemmal channels, an increased activation of the mitochondrial K^+-ATP channel might exert neuroprotection (Wu *et al.*, 2006).

Since the broad spectrum K^+ channel blocker barium, the blocker of the delayed rectifier K^+ channels TEA and the IA K^+ channel blocker 4-AP, did not result in neuroprotection, the effect of the K^+-ATP channel blockers are rather specific (Zou *et al.*, 2005).

Therefore, a sparing of ATP during ischemia could be obtained by reducing the opening of K^+-ATP channels using pharmacological procedures. Consequently, in spite of a declining ATP production during ischemia, the reduction of ATP consumption could permit the essential functioning of the cells, therefore prolonging survival.

Noteworthy, the present data are partially in-line with the experimental evidence (Huang *et al.*, 2001; Zou *et al.*, 2005), which have proposed that blockers of the delayed rectifier K^+ channels have neuroprotective effects in models of brain ischemia.

Moreover, it has been proposed that expression of the nonselective cation channel NC_{Ca}-ATP is upregulated after middle cerebral artery occlusion (MCAo) and that their block by low-dose glibenclamide results in major improvements in stroke outcome (Simard *et al.*, 2006). Interestingly, it has been demonstrated that a reduced expression of K^+-ATP channels might preserve dopaminergic neurons from degeneration in Parkinson's disease and promote the differential degeneration of dopaminergic midbrain neurons (Liss and Roeper, 2001; Liss *et al.*, 2005). Taken together, our experimental observations may have profound clinical implications since the selective and rapid targeting of plasmalemmal ATP-sensitive potassium channels might represent a new therapeutic approach for the development of potential protective compounds in the early phases of cerebral ischemia.

Acknowledgments

This work was supported by grants from Ministero dell'Istruzione e Ministero dell'Università e della Ricerca, e dal Consiglio Nazionale delle Ricerche (CNR).

References

Anderson, B., and Collingridge, G. L. (2001). The LTP Program: A data acquisition program for on-line analysis of long-term potentiation and other synaptic events. *J. Neurosci. Methods* **108,** 71–83.

Ballanyi, K. (2004). Protective role of neuronal KATP channels in brain hypoxia. *J. Exp. Biol.* **207,** 3201–3212.

Busija, D. W., Lacza, Z., Rajapakse, N., Shimizu, K., Béla, K., Bari, F., Domoki, F., and Horiguchi, T. (2004). Targeting mitochondrial ATP-sensitive potassium channels—a novel approach to neuroprotection. *Brain Res. Rev.* **46,** 282–294.

Calabresi, P., Centonze, D., Pisani, A., Sancesario, G., Gubellini, P., Marfia, G. A., and Bernardi, G. (1998). Striatal spiny neurons and cholinergic interneurons express differential ionotropic glutamatergic responses and vulnerability: Implications for ischemia and Huntington's disease. *Ann. Neurol.* **43,** 586–597.

Calabresi, P., Centonze, D., Pisani, A., Cupini, L., and Bernardi, G. (2003a). Synaptic plasticity in the ischaemic brain. *Lancet Neurol.* **2,** 622–629.

Calabresi, P., Cupini, L. M., Centonze, D., Pisani, A., and Bernardi, G. (2003b). Antiepileptic drugs as a possible neuroprotective strategy in brain ischemia. *Ann. Neurol.* **53,** 693–702.

Centonze, D., Marfia, G. A., Pisani, A., Picconi, B., Giacobini, P., Bernardi, G., and Calabresi, P. (2001). Ionic mechanisms underlying differential vulnerability to ischemia in striatal neurons. *Prog. Neurobiol.* **63,** 687–696.

Davis, R., Peters, D. H., and McTavish, D. (1994). Valproic acid. A reappraisal of its pharmacological properties and clinical efficacy in epilepsy. *Drugs* **47,** 332–372.

Guatteo, E., Federici, M., Siniscalchi, A., Knopfel, T., Mercuri, N. B., and Bernardi, G. (1998). Whole cell patch-clamp recordings of rat midbrain dopaminergic neurons isolate a sulphonylurea and ATP-sensitive component of potassium currents activated by hypoxia. *J. Neurophysiol.* **79,** 1239–1245.

Hanon, E., and Klitgaard, H. (2001). Neuroprotective properties of the novel antiepileptic drug levetiracetam in the rat middle cerebral artery occlusion model of focal cerebral ischemia. *Seizure* **10,** 287–293.

Huang, H., Gao, T. M., Gong, L., Zhuang, Z., and Li, X. (2001). Potassium channel blocker TEA prevents CA1 hippocampal injury following transient forebrain ischemia in adults rats. *Neurosci. Lett.* **305,** 83–86.

Kobayashi, T., and Mori, Y. (1998). Ca^{2+} channel antagonists and neuroprotection from cerebral ischemia. *Eur. J. Pharmacol.* **363,** 1–15.

Liss, B., and Roeper, J. (2001). ATP-sensitive potassium channels in dopaminergic neurons: Transducers of mitochondrial dysfunction. *News Physiol. Sci.* **16,** 214–217.

Liss, B., Haeckel, O., Wildmann, J., Miki, T., Seino, S., and Roeper, J. (2005). K-ATP channels promote the differential degeneration of dopaminergic midbrain neurons. *Nat. Neurosci.* **12,** 1742–1751.

Liu, Y., Sato, T., Seharaseyon, J., Szewczyk, A., O'Rourke, B., and Marban, E. (1999). Mitochondrial ATP-dependent potassium channels. Viable candidate effectors of ischemic preconditioning. *Ann. N. Y. Acad. Sci.* **874,** 27–34.

Lo, E. H., Dalkara, T., and Moskowitz, M. A. (2003). Mechanisms, challenges and opportunities in stroke. *Nat. Rev. Neurosci.* **4,** 399–415.

Mercuri, N. B., Bonci, A., Calabresi, P., Stratta, F., and Bernardi, G. (1994a). Responses of rat mesencephalic dopaminergic neurons to a prolonged period of oxygen deprivation. *Neuroscience* **63,** 757–764.

Mercuri, N. B., Bonci, A., Johnson, S. W., Stratta, F., Calabresi, P., and Bernardi, G. (1994b). Effects of anoxia on rat midbrain dopamine neurons. *J. Neurophysiol.* **71,** 1165–1173.

Mora, A., Gonzalez-Polo, R. A., Fuentes, J. M., Soler, G., and Centeno, F. (1999). Different mechanisms of protection against apoptosis by valproate and Li+. *Eur. J. Biochem.* **266,** 886–891.

Pisani, A., Calabresi, P., Tozzi, A., D'Angelo, V., and Bernardi, G. (1998). L-type Ca^{2+} channel blockers attenuate electrical changes and Ca^{2+} rise induced by oxygen/glucose deprivation in cortical neurons. *Stroke* **29,** 196–201.

Ren, M., Leng, Y., Jeong, M., Leeds, P. R., and Chuang, D. M. (2004). Valproic acid reduces brain damage induced by transient focal cerebral ischemia in rats: Potential roles of histone deacetylase inhibition and heat shock protein induction. *J. Neurochem.* **89,** 1358–1367.

Simard, J. M., Chen, M., Tarasov, K. V., Bhatta, S., Ivanova, S., Melnitchenko, L., Tsymbalyuk, N., West, G. A., and Gerzanich, V. (2006). Newly expressed SUR1-regulated NC (Ca-ATP) channel mediates cerebral edema after ischemic stroke. *Nat. Med.* **12,** 433–440.

Siniscalchi, A., Bonci, A., Mercuri, N. B., and Bernardi, G. (1997). Effects of riluzole on rat cortical neurones: An *in vitro* electrophysiological study. *Br. J. Pharmacol.* **120**(2), 225–230.

Wu, L., Shen, F., Lin, L., Zhang, X., Bruce, I. C., and Xia, Q. (2006). The neuroprotection conferred by activating the mitochondrial ATP-sensitive K(+) channel is mediated by inhibiting the mitochondrial permeability transition pore. *Neurosci. Lett.* **402,** 184–189.

Zou, B., Li, Y., Deng, P., and Xu, Z. C. (2005). Alterations of potassium currents in ischemia-vulnerable and ischemia-resistant neurons in the hippocampus after ischemia. *Brain Res.* **1033,** 78–89.

RETINAL DAMAGE CAUSED BY HIGH INTRAOCULAR PRESSURE–INDUCED TRANSIENT ISCHEMIA IS PREVENTED BY COENZYME Q10 IN RAT

Carlo Nucci,[*,†] Rosanna Tartaglione,[‡] Angelica Cerulli,[*] R. Mancino,[*] A. Spanò,[*] Federica Cavaliere,[‡] Laura Rombolà,[‡] G. Bagetta,[‡,§] M. Tiziana Corasaniti,[†,¶] and Luigi A. Morrone[‡,§]

[*]Physiopathological Optics, Department of Biopathology, University of Rome "Tor Vergata" 00133 Rome, Italy

[†]IRCCS (Istituti di Ricovero e Cura a Carattere Scientifico) Neurological Institute C. Mondino Foundation, "Mondino-Tor Vergata," Center for Experimental Neuropharmacology, Laboratori of Neurochemistry, 00133 Rome, Italy

[‡]Department of Pharmacobiology, University of Calabria, 87036 Arcavacata di Rende, Italy

[§]University Center for Adaptive Disorders and Headache (UCHAD) Section of Neuropharmacology of Normal and Pathological Neuronal Plasticity University of Calabria, 87036 Arcavacata di Rende, Italy

[¶]Department of Pharmacobiological Sciences, University "Magna Graecia" of Catanzaro 88100 Catanzaro, Italy

Recent studies support a role for excitotoxicity in the development of retinal ganglion cell (RGC) damage in subjects suffering from glaucoma. Coenzyme Q10 (CoQ10), an essential cofactor of the electron transport chain, has been reported to afford neuroprotection, preventing the formation of the mitochondrial permeability transition pore. Using an established animal model of retinal ischemia/reperfusion here, we show that synaptic glutamate increases at 130 min from beginning of reperfusion and delayed apoptosis in the RGC layer is seen at 24 h. Intraocular administration of CoQ10 minimizes glutamate increase and affords

INTERNATIONAL REVIEW OF NEUROBIOLOGY, VOL. 82
DOI: 10.1016/S0074-7742(07)82022-8

0074-7742/07 $35.00

neuroprotection, suggesting that oxidative stress and energy failure might be implicated in the mechanisms of RGC death.

I. Introduction

Glaucoma, a leading cause of blindness in the world (Quigley, 1996), is a progressive optic neuropathy often associated with increased intraocular pressure and characterized by progressive death of retinal ganglion cells (RGCs) (Osborne *et al.*, 2004). Increasing evidence supports an important, causative, role for oxidative stress in the reported loss of RGC (Tezel, 2006). Several studies have shown that free radical species can cause neuronal cell death by inhibition of key enzymes of the tricarboxylic acid cycle, the mitochondrial electron transport chain, and mitochondrial calcium homeostasis, leading to defective energy metabolism (Duchen, 2000; Patel *et al.*, 1996). Growing evidence also supports that oxidative stress is the leading mechanism of excitotoxic, glutamate-induced, RGC loss *in vitro* (Luo *et al.*, 2001) and *in vivo* (Nucci *et al.*, 2005) experimental conditions. It is well established that free radical scavengers are useful pharmacological tools to improve mitochondrial function and prevent neuronal cell death under excitotoxic conditions (Lipton and Rosenberg, 1994). Coenzyme Q10 (CoQ10) is an important component of the mitochondrial electron transport chain endowed with potent antioxidant properties that have been shown to mediate neuroprotection afforded by this agent (Beal, 1999).

Accordingly, here we now report that topical administration of CoQ10 in the eye prevents glutamate-induced apoptosis of RGC under experimental conditions of high intraocular pressure (IOP)–induced transient ischemia in rat (Nucci *et al.*, 2005).

II. Materials and Methods

A. Ischemia Model

Male Wistar rats (250–300 g) (Charles River, Lecco, Italy) were maintained on a 12-h light–dark cycle. Before ischemia was induced, animals were anesthetized with chloral hydrate (400 mg/kg, i.p.). Corneal analgesia was achieved using topical drops of oxibuprocaine 0.4% (Novesina, Novartis Farma, Italy). Pupillary dilation was maintained using 0.5% tropicamide (Visumidriatic 0.5%, Visufarma, Italy). The anterior chamber of the right eye was cannulated with a 27-gauge infusion needle connected to a 500-ml plastic container of sterile saline, the IOP was raised to

120 mmHg for 45 min by elevating the saline reservoir as previously reported (Osborne *et al.*, 2004). Retinal ischemia was confirmed by observing whitening of the iris and loss of the red reflex of the retina. Sham procedure was performed without the elevation of the bottle in control eyes.

B. Microdialysis

Retinal extracellular glutamate was monitored in urethane (1500 mg/kg, i.p.) anesthetized rats during and after pressure-induced ischemia using a microdialysis technique. For implantation, a microdialysis probe (concentric design, 2-mm regenerated cellulose membrane, molecular weight cutoff 5 kDa) was advanced through the lumen of a guide cannula placed in the vitreous body. The surface of the dialysis membrane was secured perpendicularly to the retina for stable sampling during the experiment. Superfusion medium was continuously delivered via the probe at a rate of 2 μl/min. The composition of the medium (in mM) was: NaCl, 125; KCl, 2.5; $MgCl_2$, 1.18; $CaCl_2$, 1.26; NaH_2PO_4, 0.2; pH adjusted to 7.0. After 1-h stabilization period, dialysate samples (20 μl) were collected at 10-min intervals after reperfusion. For analysis, the dialysate samples were derivatized with *o*-phthalaldehyde (OPA) and the concentration of glutamate determined as previously reported (Richards *et al.*, 2000) by means of a high-performance liquid chromatography (HPLC) equipped with a fluorescence detector. Briefly, separation was achieved with a Hypersil ODS column (5 μm, 150 mm × 3 mm, Chrompack, Milan, Italy) using a short methanol gradient (7–14% over 15 min) in 50-mM sodium acetate buffer, pH 6.95, followed by elution of remaining peaks with 95% methanol. Total run time was 17 min. The baseline concentration of glutamate was the mean concentration obtained by averaging the three samples collected consecutively at 10-min intervals immediately prior to the onset of ischemia and was used as control. Glutamate concentration in each dialysate sample was calculated as percentage of control.

C. Morphometric Analysis

Twenty-four hours after reperfusion rats were anesthetized as described above and perfused through the left ventricle of the heart with 100 ml of heparinized phosphate buffered (pH 7.4) saline followed by 200 ml of 4% paraformaldehyde in phosphate buffered saline. Two hours after the perfusion procedure had been completed, the eyes were enucleated and postfixed in 4% paraformaldehyde for 72 h. Serial coronal sections (5-μm thick), cut along the vertical meridian of the eye passing through the optic nerve head, were stained with hematoxylin and eosin (H&E). The number of cells in the RGC layer was counted in six areas (25 × 25 μm each) of each section ($n = 5$ per eye) at 300 μm from the optic nerve head on the superior and inferior hemisphere, using light microscopy (40× magnification). Morphological characteristics of adjacent tissue sections were assessed under light microscopy (100× magnification) and DNA fragmentation was detected by terminal transferase (TdT) dUDP nick end-labeling (TUNEL) technique (Gavrieli *et al.*, 1992) using a commercial kit (Promega, Madison, WI). For analysis, nuclei were counterstained with 40–60-diamino-2-phenylindole (DAPI; 1.5 mg/ml)-supplemented mounting medium (Vectashield, Vector Laboratories, Inc.).

D. Drug Application

For neurochemical studies, animals received (30 min before ischemia) intravitreal administration (via the microdialysis probe; 2-μl/min rate, 5-min duration of treatment in all instance) of (1) CoQ10 (solution of CoQ10 0.1% + Vit-E 0.5%), (2) Vit-E (0.5%), or (3) vehicle (saline + EDTA 0.1%). For neuropathologic studies, animals have been treated topically with CoQ10, Vit-E, or vehicle; in all instance, rats received eye application of 50 μl of solution every 15 min during 1 h before and after ischemia had been induced. CoQ10 and its vehicle were kindly donated by Visufarma (Italy).

E. Statistical Analysis

All data are expressed as the mean ± SEM and evaluated statistically for difference by one-way analysis of variance (ANOVA) followed by Tukey-Kramer test for multiple comparisons as appropriate.

III. Results

A. CoQ10 Minimizes Glutamate Increase Induced by Ischemia/Reperfusion

Extracellular glutamate was monitored in the retina of anesthetized rats during and after IOP-induced ischemia. Control levels of glutamate (set to 100%), measured during the pre-ischemia period, correspond to 0.307 ± 0.044 μM (n = 6 rats), and these showed a trend toward increase during the reperfusion phase to peak (199.3% ± 43.7% increase vs control) at 130 min (Fig. 1). Focal administration (30 min before ischemia) via the microdialysis probe of CoQ10 (n = 3 rats) reduces

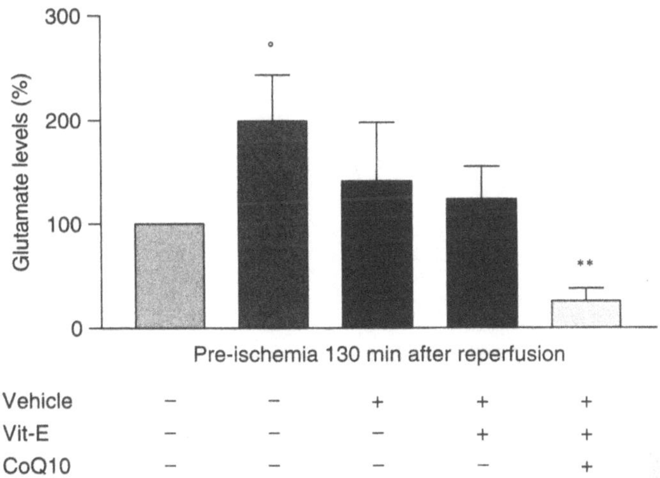

FIG. 1. Neurochemical data obtained by intraocular microdialysis experiments in anesthetized rats demonstrate that ischemia/reperfusion insult increases intraretinal glutamate sensitive to the reversal of CoQ10. The observed increase in glutamate (199.3 ± 43.7% vs pre-ischemia levels set to 100%; pre-ischemia glutamate levels = 0.307 ± 0.044 μM, n = 6) peaks at 130 min after the beginning of reperfusion. Retinal administration, via the probe, of CoQ10 (n = 3) (30 min before ischemia), significantly, prevents glutamate increase (data not shown) and prevents the peak increase seen after 130 min of reperfusion (26.06 ± 12.1% vs 130-min reperfusion levels), whereas vehicle or Vit-E are without significant effect [141.4 ± 56.5 (n = 3) and 123.7 ± 31.6 (n = 5), respectively]. Glutamate values are expressed as mean percentage changes ± SEM. Statistical significance was assessed by one-way ANOVA with post hoc comparisons using Tukey test. $p < 0.05$ versus pre-ischemia, **$p < 0.001$ versus 130 min after reperfusion levels.

glutamate peak increase to 26.06% ± 12.1% ($p < 0.001$), whereas vehicle or Vit-E are without significant effect [141.4 ± 56.5 (n = 3) and 123.7 ± 31.6 (n = 5), respectively].

B. CoQ10 Affords Neuroprotection Against Cell Loss Yielded by Ischemia/Reperfusion in the RGC Layer

Retinal ischemia for 45 min followed by 24-h reperfusion causes retinal damage (Fig. 2) and cell death that seems to occur via an apoptotic mechanism (Fig. 3). In no instance, retinal damage (Fig. 2) or apoptotic, TUNEL-positive, cells (Fig. 3) are observed in sections obtained form sham-operated eyes. Under these experimental conditions, reduction in the number of cells in the RGC layer by 25.75% is also observed (Fig. 4). Topical treatment with Vit-E or CoQ10 minimizes retinal damage (Fig. 2) and cell death for apoptosis (Fig. 3).

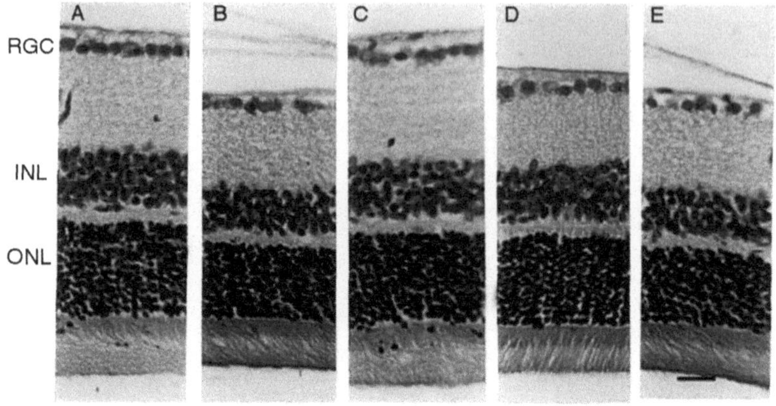

FIG. 2. Retinal ischemia for 45 min followed by 24-h reperfusion reduces the number of cells in the retinal ganglion cell layer (B) as compared to sham-operated eyes (A). Topical treatment with CoQ10 (C), or with Vit-E (D), prevents the tissue damage observed in (B). Treatment with vehicle does not prevent RGC layer loss (E). H&E staining. RGC, retinal ganglion cell layer; INL, inner nuclear layer; and ONL, outer nuclear layer. Scale bar, 50 μm. (See Color Insert.)

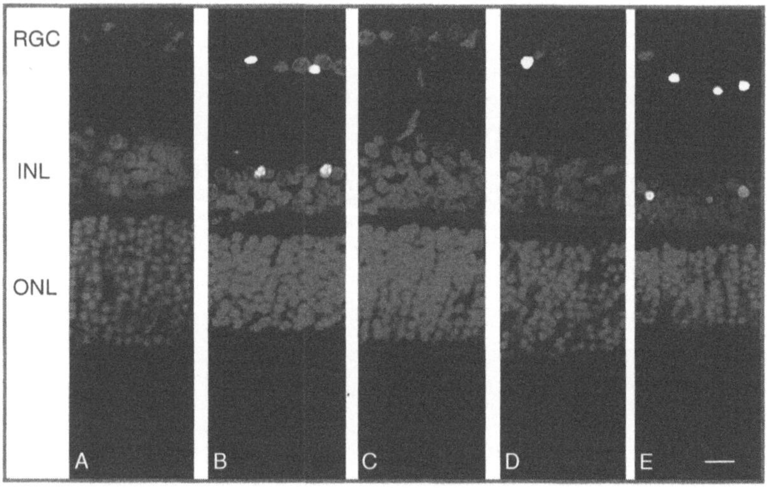

FIG. 3. (A) Elevated IOP-induced ischemia for 45 min followed by an interval of 24-h reperfusion induces DNA fragmentation (TUNEL-positive cells) in a tissue section from rat retina (B). Topical treatment with CoQ10 (C) and Vit-E (D), but not with vehicle (E), prevents retinal cell apoptosis induced by acute raise of IOP in rats. Note the presence of TUNEL-positive cells both in the retinal ganglion cell layer and in the inner nuclear layer. No TUNEL-positive cells are detected in retinal tissue sections obtained from sham-operated eyes (A). RGC, retinal ganglion cell layer; INL, inner nuclear layer; and ONL, outer nuclear layer. Scale bars, 50 μm. (See Color Insert.)

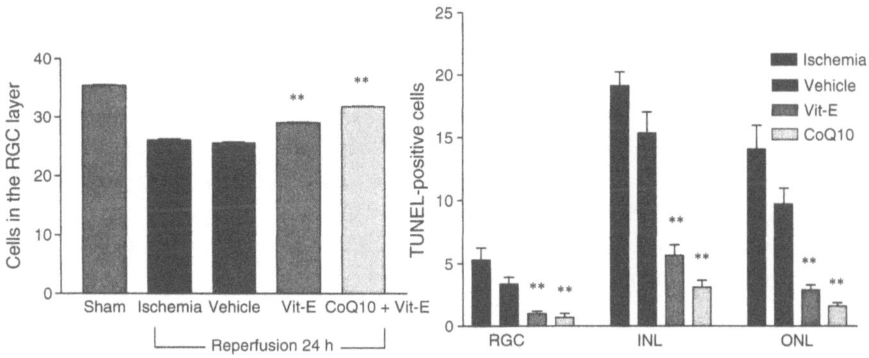

FIG. 4. Elevated IOP-induced ischemia for 45 min followed by an interval of 24-h reperfusion induces cell loss and DNA fragmentation sensitive to the reversal of CoQ10 and Vit-E in the retinal ganglion cell layer. Ischemia causes a reduction in the number of cells in the RGC layer (stained with H&E) per counted area compared to controls (-25.7%, $n = 11$). Topical treatment with Vit-E and CoQ10 significantly prevents RGC loss (-17.8%, $n = 6$, and -10.3%, $n = 7$, respectively). DNA fragmentation is detected by TdT-dUDP TUNEL technique in RGC layer (RGC 5.3 ± 0.9, $n = 3$), inner nuclear layer (INL, 19.1 ± 1.1, $n = 3$); outer nuclear layer (ONL, 14.0 ± 1.9, $n = 3$). No TUNEL-positive cells are detected in retinal tissue sections obtained from sham-operated eyes (data not shown). Vit-E ($n = 6$) and CoQ10 ($n = 5$) prevent DNA fragmentation in RGC layer (1.0 ± 0.1 and 0.7 ± 0.3, respectively), INL (5.6 ± 0.8 and 3.1 ± 0.5), and ONL (2.9 ± 0.4 and 1.6 ± 0.2). Data ere expressed as mean ± SEM per area and are evaluated statistically for differences using ANOVA followed by Tukey-Kramer's test. **$p > 0.001$ versus ischemia.

Accordingly, significant ($p < 0.001$) reduction in the percentage loss of cells in the RGC layer to 17.8% ($n = 6$ rats) and 10.3% ($n = 7$ rats), respectively, is also observed (Fig. 4).

IV. Discussion

High IOP-induced ischemia is an established animal model to study the mechanisms underlying RGC death observed under clinical conditions of central retinal artery or ophthalmic artery occlusion that also recapitulate features of acute angle closure glaucoma (POAG) (Osborne *et al.*, 2004). We have reported that, under these experimental conditions, a delayed and progressive loss of viable cells in the RGC layer is observed starting from 6 h after the beginning of the reperfusion to peak at 7 days (Nucci *et al.*, 2005). The mechanism underlying cell loss implicates overactivation of NMDA and non-NMDA subtypes of glutamate receptors and consequent accumulation of nitric oxide (NO) being the loss minimized by systemic pretreatment with MK801 and GYKI 52466, respectively, and by l- NAME, a nonselective NO synthase (NOS) inhibitor (Nucci *et al.*, 2005). The excitotoxic, glutamate-mediated, nature of the underlying mechanism of RGC death has also been confirmed by microdialysis experiments demonstrating that during the early phase of reperfusion extracellular glutamate increases significantly in the retina of the ischemic eye and this is sensitive to the prevention afforded by systemic MK801 (Nucci *et al.*, 2005).

Here we have used this experimental model in combination with a neurochemical and neuropathologic approach to gain more insight in the neuroprotective profile of CoQ10, an essential cofactor of the electron transport chain that has been reported to afford neuroprotection under several experimental circumstances (Beal, 2004). Thus, we confirmed that during the reperfusion phase, the ischemic insult elevates significantly retinal glutamate leading to cell death in the RGC layer via an apoptotic mechanism. More importantly, here we have observed that administration of CoQ10 prevents glutamate increase and this is accompanied by minimization of cell death in rat. The mechanism underlying neuroprotection afforded by CoQ10 is not known, though it is conceivable that a free radical scavenging mechanism may play a minor role. In fact, neuroprotection afforded by CoQ10 was far greater than that provided by treatment with Vit-E that, however, failed to affect significantly elevation of retinal glutamate. Increasing experimental evidence implicates failure of mitochondrial energy metabolism in the pathogenesis of diseases such as Alzheimer's and Parkinson's diseases, neurodegenerative disorders where CoQ10 has been reported to afford neuroprotection (Beal, 1999, 2004; Beal and Matthews, 1997; Ferrante *et al.*, 2002; Matthews *et al.*, 1998; Shults *et al.*, 2002, 2004). Accordingly, an alternative hypothesis would be that CoQ10 reduces the detrimental action of ischemia/reperfusion on mitochondrial energy metabolism and, consequently, on the function of glutamate transporters, thus limiting accumulation of extracellular glutamate and preventing apoptotic death of RGC in the rat. In fact, it is well documented that excessive activation of glutamate receptors via the excitotoxic cascade leads to the permeability transition pore (PTP) formation and cytochrome *c*, a member of the mitochondrial electron transport chain, to be released from the mitochondrial intermembrane space into the cytosol, where it functions as a proapoptotic factor committing the RGC to death (see Kroemer and Reed, 2000). Incidentally, CoQ10 has been shown to inhibit apoptosis by maintaining PTP in the closed conformation via a mechanism independent from free radical scavenging (Papucci *et al.*, 2003). Preliminary cytofluorimetric experiments show that exposure to CoQ10 increases by 30% the viability of RGC cultures stained with Annexin V and 7-AAD and maintained under hypoxia conditions (data not shown) supporting the antiapoptotic, neuroprotective, action of this agent.

In conclusion, the present data support the usefulness of the model of retinal ischemia provided by high IOP in rat to study the mechanisms underlying RGC death with the goal to discover novel molecular targets for neuroprotection.

Acknowledgments

Financial support from the Ministry of Health (RF 2005) and Visufarma (Rome, Italy) is gratefully acknowledged.

References

Beal, M. F. (1999). Coenzyme Q10 administration and its potential for treatment of neurodegenerative diseases. *Biofactors* **9,** 261–266.

Beal, M. F. (2004). Therapeutic effects of coenzyme Q10 in diseases. *Methods Enzymol.* **382,** 473–487.

Beal, M. F., and Matthews, R. T. (1997). Coenzyme Q10 in the central nervous system and its potential usefulness in the treatment of neurodegenerative diseases. *Mol. Aspects Med.* **18**(Suppl.), 169–179.

Duchen, M. R. (2000). Mitochondria and calcium: From cell signaling to cell death. *J. Physiol.* **529,** 57–68.

Ferrante, R. J., Andreassen, O. A., Dedeoglu, A., Ferrante, K. L., Jenkins, B. G., Hersch, S. M., and Beal, M. F. (2002). Therapeutic effects of coenzyme Q10 and remacemide in transgenic mouse models of Huntington's disease. *J. Neurosci.* **22,** 592–599.

Gavrieli, Y., Sherman, Y., and Ben-Sasson, S. A. (1992). Identification of programmed cell death *in situ* via specific labeling of nuclear DNA fragmentation. *J. Cell Biol.* **119,** 493–501.

Kroemer, G., and Reed, J. C. (2000). Mitochondrial control of cell death. *Nat. Med.* **6,** 513–519.

Lipton, S. A., and Rosenberg, P. A. (1994). Excitatory amino acids as a final common pathway for neurologic disorders. *N. Engl. J. Med.* **330,** 613–622.

Luo, X., Heidinger, V., Picaud, S., Lambrou, G., Dreyfus, H., Sahel, J., and Hicks, D. (2001). Selective excitotoxic degeneration of adult pig retinal ganglion cells *in vitro*. *Invest. Ophthalmol. Vis. Sci.* **42,** 1096–1106.

Matthews, R. T., Yang, L., Browne, S., Baik, M., and Beal, M. F. (1998). Coenzyme Q10 administration increases brain mitochondrial concentrations and exerts neuroprotective effects. *Proc. Natl. Acad. Sci. USA* **95,** 8892–8897.

Nucci, C., Tartaglione, R., Rombolà, L., Morrone, L. A., Fazzi, E., and Bagetta, G. (2005). Neurochemical evidence to implicate elevated glutamate in the mechanisms of high intraocular pressure (IOP)-induced retinal ganglion cell death in rat. *NeuroToxicology* **26,** 935–941.

Osborne, N. N., Casson, R. J., Wood, J. P. M., Childlow, G., Graham, M., and Melena, J. (2004). Retinal ischemia: Mechanisms of damage and potential therapeutic strategies. *Prog. Ret. Eye Res.* **23,** 91–147.

Papucci, L., Schiavone, N., Witort, E., Donnini, M., Lapucci, A., Tempestini, A., Formigli, L., Zecchi-Orlandini, S., Orlandini, G., Carella, G., Brancato, R., and Capaccioli, S. (2003). Coenzyme Q10 prevents apoptosis by inhibiting mitochondrial depolarization independently of its free radical scavenging property. *J. Biol. Chem.* **278,** 28220–28228.

Patel, M., Day, B. J., Crapo, J. D., Fridovich, I., and McNamara, J. O. (1996). Requirement for superoxide in excitotoxic cell death. *Neuron* **16,** 345–355.

Quigley, H. A. (1996). Number of people with glaucoma worldwide. *Br. J. Ophthalmol.* **80,** 389–393.

Richards, D. A., Morrone, L. A., Bagetta, G., and Bowery, N. G. (2000). Effects of α-dendrotoxin and dendrotoxin k on extracellular excitatory amino acids and on electroencephalograph spectral power in the hippocampus of anaesthetised rats. *Neurosci. Lett.* **293,** 183–186.

Shults, C. W., Oakes, D., Kieburtz, K., Beal, M. F., Haas, R., Plumb, S., Juncos, J. L., Nutt, J., Shoulson, I., Carter, J., Kompoliti, K., Perlmutter, J. S., *et al.* (2002). Effects of coenzyme Q10 in early Parkinson disease: Evidence of slowing of the functional decline. *Arch. Neurol.* **59,** 1541–1550.

Shults, C. W., Beal, M. F., Song, D., and Fontaine, D. (2004). Pilot trial of high dosages of coenzyme Q10 in patients with Parkinson's disease. *Exp. Neurol.* **188,** 491–494.

Tezel, G. (2006). Oxidative stress in glaucomatous neurodegeneration: Mechanisms and consequences. *Prog. Ret. Eye Res.* **25,** 490–513.

EVIDENCE IMPLICATING MATRIX METALLOPROTEINASES IN THE MECHANISM UNDERLYING ACCUMULATION OF IL-1β AND NEURONAL APOPTOSIS IN THE NEOCORTEX OF HIV/gp120-EXPOSED RATS

Rossella Russo,[*,†,‡] Elisa Siviglia,[§] Micaela Gliozzi,[§] Diana Amantea,[*,†] Annamaria Paoletti,[¶] Laura Berliocchi,[*,†,||] G. Bagetta,[*,†] and M. Tiziana Corasaniti,[§,||]

*Department of Pharmacobiology, University of Calabria, Via P. Bucci 87036 Arcavacata di Rende (CS), Italy

†UCHAD Center of Neuropharmacology for Normal and Pathological Neuronal Plasticity University of Calabria, Via P. Bucci, 87036 Arcavacata di Rende (CS), Italy

‡Burnham Institute for Medical Research, Del E. Webb Center for Neurosciences and Aging, La Jolla, California 92037, USA

§Department of Pharmacobiological Sciences, University "Magna Graecia" of Catanzaro 88100 Catanzaro, Italy

¶CNR Institute of Neurological Science, Section of Pharmacology Roccelletta di Borgia, 88100 Catanzaro, Italy

||Center for Experimental Neuropharmacology, Mondino-Tor Vergata University of Rome Tor Vergata, 00133 Rome, Italy

Neuroinflammation is often associated with neurodegenerative diseases, including multiple sclerosis (MS), stroke, Alzheimer's disease, and HIV-1-associated dementia (HAD). The proinflammatory cytokine interleukin-1β (IL-1β) is one of the main mediators of inflammation, and IL-1β expression in the brain is rapidly upregulated in response to acute and chronic insults. IL-1β is synthesized as biologically inactive precursor (pro-IL-1β), which is classically processed by caspase-1 [also known as interleukin-converting enzyme (ICE)] into the active,

INTERNATIONAL REVIEW OF NEUROBIOLOGY, VOL. 82
DOI: 10.1016/S0074-7742(07)82023-X

0074-7742/07 $35.00

mature cytokine. However, caspase-1/ICE-independent mechanisms of IL-1β processing have recently been suggested. Here we report that matrix metalloproteinases (MMPs) participate in the maturation process (cleavage and activation) of IL-1β in an *in vivo* model of HIV-associated neurodegeneration based on the intracerebroventricular injection of the HIV-1 envelope glycoprotein gp120. We show that, following gp120 exposure, MMP-9 and MMP-2, but not caspase-1/ICE, are rapidly induced. Pharmacological manipulation of MMPs activity, using the broad spectrum MMPs inhibitor GM6001, reduces the increase in IL-1β immunoreactivity and the neuronal apoptosis induced by gp120. Taken together, these findings point to a critical role for MMPs in IL-1β increase and consequent neurotoxicity triggered by gp120 in the neocortex of rat and suggest new links between IL-1β processing and MMP activation during the neuroinflammatory process.

I. Introduction

Interleukin-1β (IL-1β) is a pleiotropic cytokine playing a key role in immunologic response and inflammation. Under physiological conditions in the brain, IL-1β is expressed at low level, whereas it is rapidly induced in response to acute and chronic insults (Allan *et al.*, 2005). IL-1β is synthesized as an inactive 33-kDa precursor (pro-IL-1β), which requires proteolytic cleavage to generate the bioactive, 17-kDa, mature form (Dinarello, 1998). This process is classically mediated by caspase-1 [also known as interleukin-converting enzyme (ICE)] (Cerretti *et al.*, 1992; Thornberry *et al.*, 1992). However, alternative mechanisms of IL-1β processing have been recently suggested. In ICE-deficient mice, inflammation induced by turpentine results in the production of a biological active form of IL-1β, indicating that the IL-1β precursor can be processed by proteases other than ICE (Fantuzzi *et al.*, 1997). Similarly, human keratinocytes, which do not produce active ICE, have been shown to produce mature IL-1β (Nylander-Lundqvist *et al.*, 1996). Moreover, it has been reported that several matrix metalloproteinases (MMPs), in particular MMP-2, MMP-3, and MMP-9, can process IL-1β precursor into bioactive forms *in vitro* (Schonbeck *et al.*, 1998).

MMPs constitute a family of proteolytic enzymes that are mainly responsible for the remodeling of the extracellular matrix (ECM) (Visse and Nagase, 2003). However, their upregulation has been implicated in the pathogenesis of several inflammatory and neurodegenerative disorders, including ischemia (Gasche *et al.*, 2006), Alzheimer's disease (Adair *et al.*, 2004; Deb and Gottschall, 1996), multiple sclerosis (Gijbels *et al.*, 1992; Rosenberg *et al.*, 1996), and HIV-associated dementia (HAD) (Conant *et al.*, 1999). Interestingly, active MMPs, as well as IL-1β, are involved in the inflammatory responses and are present at the site of inflammation (Dinarello, 1998; Rosenberg, 2002; Rothwell and Luheshi, 2000).

HAD is a neurological syndrome characterized by cognitive and motor dysfunction, which is manifest in a substantial proportion of patients infected with *Human immunodeficiency virus 1* (HIV-1) (Gendelman *et al.*, 2005; Price and Brew, 1988). Neuronal damage and cell loss have been described postmortem in the brain of patients suffering from HAD in the absence of marked neuronal infection (Kaul *et al.*, 2001; Masliah *et al.*, 1992).

The HIV-1 envelope glycoprotein gp120 seems to play a crucial role in the development of the neurodegenerative processes associated with HIV infection. Transgenic mice expressing gp120 showed neuropathological changes similar to those observed in HAD patients (Toggas *et al.*, 1994). Also, gp120 induced degeneration and death of neurons *in vitro* (Lipton and Gendelman, 1995) and *in vivo* (Bagetta *et al.*, 1995; Toggas *et al.*, 1994). In the rat brain, gp120 has been shown to induce apoptotic cell death (Bagetta *et al.*, 1995). This latter event was preceded by microglial cell activation (Bagetta *et al.*, 1996) and enhanced expression of IL-1β (Bagetta *et al.*, 1999). Although IL-1β antagonists and ICE inhibitors prevented gp120-induced apoptosis (Bagetta *et al.*, 1999; Corasaniti *et al.*, 2005), the mechanism underlying gp120-induced IL-1β accumulation and cell death is still unclear.

Here we report evidence to implicate proteases, other than ICE, in the processing of the pro-form and accumulation of the active cytokine, IL-1β, and consequent apoptosis in the brain neocortex following intracerebroventricular (i.c.v.) injection of HIV-1/gp120 in rat.

II. Materials and Methods

A. Drugs

Lyophilized, full-length glycosylated recombinant HIV-1 gp120$_{\text{IIIB}}$ (>90% pure by SDS-PAGE and tested randomly for endotoxin contamination), dissolved in PBS, was from INTRACEL (catalog no. 120011). Acetyl-Tyr-Val-Ala-Asp-chloromethylketone (Ac-YVAD-cmk) was from Bachem (Bubendorf, Switzerland); *N*-[(2R)-2-(Hydroxamidocarbonylmethyl)-4-methylpenthanoyl]-L-tryptophan methylamide (GM6001, also known as Galardin) and *N*-*t*-butoxycarbonyl-L-leucyl-L-tryptophan methylamide (GM6001 negative control) were from Calbiochem (La Jolla, CA). Stock solutions of Ac-YVAD-cmk (25 mg/ml) and GM6001 (5 mg/ml) were made in dimethyl sulphoxide (DMSO) and further diluted in PBS prior to administration. Final solutions never contained more than 0.2% DMSO.

B. Subjects

Male Wistar rats (250–280 g), housed in a temperature (22 °C) and humidity (65%) controlled colony room, were anesthetized with chloral hydrate (400 mg/Kg intraperitoneally) for a chronic implantation of a guide cannula (25 gauge) into one lateral cerebral ventricle (i.c.v.) under stereotaxic guidance (Paxinos and Watson, 1998) as previously described (Bagetta *et al.*, 1995). Drug solutions and gp120 were administered with a 5-μl Hamilton syringe (1–2 volume; 1-μl/min rate) connected via a Teflon tube to an injector exceeding the length of the guide cannula by 2 mm.

For ELISA, gel zymography and caspase-1 activity studies, gp120 (100 ng/rat) or PBS, the vehicle used to dilute gp120, were administered immediately after the implantation of the guide cannula. In the antagonism study, the caspase-1 preferring inhibitor Ac-YVAD-cmk (100 pmol/rat, corresponding to 54 ng/rat) or, alternatively, the broad spectrum MMP inhibitor GM6001 (10 ng/rat) and the GM6001 negative control (10 ng/rat) were administered i.c.v 1 h before the injection of gp120. Because stock solutions of these drugs were made up in DMSO and these were further diluted in PBS, preliminary experiments were carried out by injecting control animals by i.c.v. route with PBS containing an equal proportion of DMSO, 1 h before gp120 administration; in no instance this experimental protocol affected IL-1β levels modified by gp120. For the neuropathological evaluation of brain damage, gp120 (100 ng/rat) or BSA (60 ng/rat) was administered once daily for 5 consecutive days; GM6001 (1–10 ng/rat) and the GM6001 negative control (10 ng/rat) were administered i.c.v 1 h before the daily injection of gp120 or BSA.

The experimental protocol was in accordance to the guidelines of the Italian Ministry of Health for animal care (D.M. 116/1992).

C. Neuropathology

Twenty-four hours after the last injection, rats were anesthetized and perfused-fixed; then coronal sections (14 μm) of brain tissue were prepared as detailed elsewhere (Bagetta *et al.*, 1999). Brain sections were processed for *in situ* analysis of DNA fragmentation according to the terminal deoxynucleotidyl transferase-mediated dUTP-biotin nick end-labeling (TUNEL) method (Gavrieli *et al.*, 1992), with minor modifications (Bagetta *et al.*, 1999).

D. IL-1β ELISA

Individual brain cortical tissue samples ($n = 3$–5 per experimental group) were homogenized in ice-cold phosphate buffer saline (PBS) containing a protease inhibitor cocktail (Sigma, Milan, Italy) and centrifuged at 14,000 × *g* for 10 min at 4 °C.

After centrifugation, the supernatant was collected and the protein concentration determined by using the Bio-Rad DC protein assay (Bio-Rad Laboratories, Milan, Italy).

Immunoreactive IL-1β levels were analyzed by an established, rat-specific, sandwich ELISA, as previously described (Corasaniti *et al.*, 2001a; Hagan *et al.*, 1996). The immunoaffinity-purified polyclonal sheep anti-rat IL-1β-coating antibody (1 μg/ml), the biotinylated, immunoaffinity-purified, polyclonal sheep anti-rat IL-1β-detecting antibody (1:1000 dilution), as well as the recombinant rat IL-1β standard were kindly provided by Dr. Stephen Poole (National Institute of Biological standards and Controls, NIBSC, Potters Bar, Hertfordshire, United Kingdom). Poly-horseradish peroxidase-conjugated streptavidin (CLB, Amsterdam) was used at 1:5000 dilution and the color was developed using the chromogen *o*-phenylenediamine (Sigma, Milan). Optical densities (OD) were read at 492 nm by using an automated plate reader (Multiscan MS, Labsystems, Helsinki, Finland) and cytokine levels calculated by interpolation from the standard curve (0.0–1000 pg/ml). Data were corrected for protein concentration and the results expressed as pg of IL-1β per mg of protein.

E. Gel Zymography

MMP-2 (gelatinase A) and MMP-9 (gelatinase B) gelatinolytic activities were detected by gelatin gel zymography (Gu *et al.*, 2002; Planas *et al.*, 2001; Zhang and Gottschall, 1997). Individual brain cortical tissue samples ($n = 3$ per experimental group) were homogenized in ice-cold Tris-buffered saline (TBS), containing 150-mM NaCl, 5-mM $CaCl_2$, 0.05% Brij35, pH 7.6, 0.02% NaN_3, 1% Triton X-100, 100-μM PMSF, and a protease inhibitor cocktail (Sigma, Milan), and centrifuged at 14,000 $\times$ *g* for 20 min at 4°C. Supernatants were subjected to affinity precipitation with gelatin-conjugated Sepharose beads (Gelatine-Sepharose 4B, Amersham Biosciences, GE Healthcare, Milan, Italy), overnight at 4°C. The bound proteins were eluted from the beads in TBS containing 10% DMSO by shaking for 1 h at 4°C. Ten microliters of the samples were diluted 1:1 in a nonreducing SDS loading buffer (0.0625-M Tris–HCl, pH 6.8, 2% SDS, 10% glycerol, 0.25% bromophenol blue) and run in 10% polyacrylamide gels containing 0.1% porcine gelatin as substrate. After electrophoretic separation, the proteins were renaturated by washing out the SDS with two washes (30 min each) in 2.5% Triton X-100 and then incubated with developing buffer (50-mM Tris–HCl, pH 7.78, 10-mM $CaCl_2$, 0.02% NaN_3) at 37°C for 40 h. Gels were stained with 0.25% Comassie Brilliant Blue R-250 for 1 h and then appropriately destained in a solution of acetic acid:methanol:water (1:3:6). Gel images were transferred onto Polaroid films.

F. Fluorimetric Caspase-1 Activity Assay

Individual brain cortical tissue samples ($n = 3$ per experimental group) were rapidly dissected out and homogenized in ice-cold lysis buffer (50-mM HEPES, pH 7.4, 150-mM NaCl, 5-mM $MgCl_2$, 5-mM EDTA, 0.1% CHAPS, 5-mM DTT, 10-μg/ml pepstatin A, 10-μg/ml leupeptin, 10-μg/ml aprotinin); following 10-min incubation on ice, samples were centrifuged at 12,000 × *g* for 10 min at 4°C and protein concentration in supernatants was determined by the DC protein assay (Bio-Rad Laboratories). Brain cortical supernatants were diluted in assay buffer (100-mM HEPES, pH 7.4, 5-mM EDTA, 0.1% CHAPS, 5-mM dithiothreitol, 10% glycerol) to a final concentration of 1.2-μg protein/μl and incubated in triplicate in a 96-well clear-bottom plate with the fluorogenic substrate acetyl-Trp-Glu-His-Asp-7-amino-4-methylcoumarin (Ac-WEHD-AMC; 10 μM; Bachem) (Rano *et al.*, 1997). Production of fluorescent-free AMC, released by caspase-1 activity, was monitored over 60 min at 37°C using a microplate fluorometer (Victor^2 multilabel counter, Perkin-Elmer Life Sciences; excitation: 355 nm, emission: 460 nm). Specific contribution of caspase-1 activity in each brain extract was determined by preincubating parallel sample aliquots with the caspase-1 preferring inhibitor acetyl-Trp-Glu-His-Asp-aldehyde (Ac-WEHD-CHO; 10 μM; Bachem) (Garcia-Calvo *et al.*, 1998) for 10 min at 37°C prior to the addition of the caspase substrate; the difference between the substrate cleavage activity in the absence and presence of Ac-WEHD-CHO was regarded as specific caspase-1 activity. The increase in fluorescence was linear for 40 min after addition of the fluorogenic substrate. Data were analyzed by linear regression within the linear range of the enzymatic reaction and the results expressed as relative fluorescence units (RFU)/min/mg of protein.

G. Statistical Analysis

Data are expressed as the mean ± SEM of the indicated number of independent experiments and evaluated statistically for difference by Student's *t*-test or by one-way analysis of variance (ANOVA) followed by Tukey-Kramer or by Dunnett tests for multiple comparisons. $p < 0.05$ was considered to be significant.

III. Results

Measurement of IL-1β levels in the neocortex of rats at 1 and 6 h after the implantation of the guide cannula and the injection of vehicle (PBS) used to dilute the viral protein revealed that cytokine levels were elevated in the injected side of

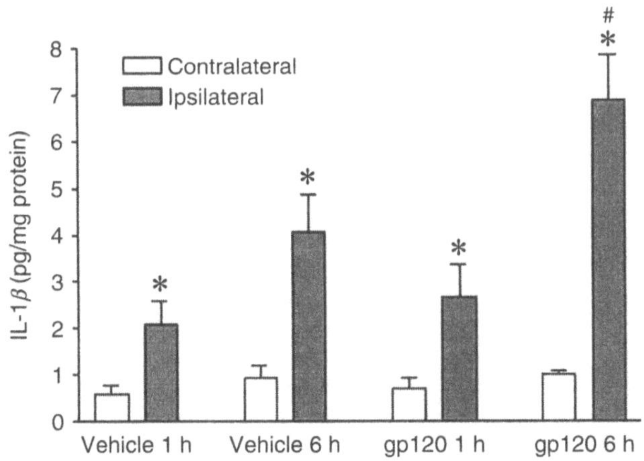

FIG. 1. Immunoreactive IL-1β levels are increased in the ipsilateral neocortex of rats treated 1 and 6 h beforehand with vehicle or HIV-1 gp120 (100 ng) given into one lateral cerebral ventricle (i.c.v.) and this as compared to contralateral, noninjected side of the brain. Immunoreactive IL-1β levels were assayed in individual brain tissue samples by an established, rat-specific, sandwich ELISA. IL-1β levels were corrected for protein concentration and the results expressed as picograms of IL-1β per milligram of proteins. Data are expressed as mean ± SEM (bars) values ($n = 3$–5 per group). * denotes $p < 0.05$ versus respective contralateral (Student's t-test); # denotes $p < 0.001$ versus respective contralateral (ANOVA followed by Tukey-Kramer test for multiple comparisons).

the brain compared to contralateral, noninjected side ($p < 0.05$ by Student's t-test) (Fig. 1). This observation suggests an ongoing inflammatory reaction triggered by the surgical procedure. Compared to contralateral side, a significant ($p < 0.05$) elevation of IL-1β levels also occurred in the ipsilateral neocortex of gp120-injected animals at both 1 and 6 h following treatment; however, statistical analysis of the difference between means using an ANOVA followed by a Tukey-Kramer test for multiple comparison indicates that a significant ($p < 0.001$) increase in IL-1β levels occurred only in the injected side 6 h after gp120 exposure (Fig. 1).

Quite surprisingly, enhancement of IL-1β levels in the brain neocortex ipsilateral to the i.c.v. injection of both vehicle and HIV-1 gp120 was not associated with a significant elevation in caspase-1 activity (Fig. 2) in the same brain region, although a trend toward an increase was observed in the neocortex of rats i.c.v. injected with vehicle 1 h but not 6 h beforehand. These observations suggest that enzymes other than caspase-1 may be responsible for elevation of IL-1β levels. To test this hypothesis, rats were treated with the caspase-1 preferring inhibitor Ac-YVAD-cmk (100 pmol, given i.c.v. 1 h before vehicle or gp120), and cytokine levels were measured 6 h after injection of vehicle or gp120. Interestingly, Ac-YVAD-cmk failed to prevent the observed increase of IL-1β levels in the neocortex of vehicle-treated rats (data not shown) or rats exposed to gp120 (Fig. 3).

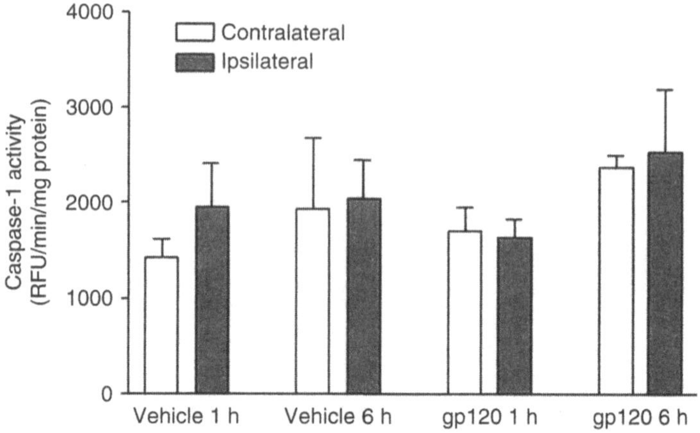

FIG. 2. Increase in brain cortical IL-1β induced by HIV-gp120 is not associated to elevation of caspase-1 activity. Caspase-1 activity was assayed by measuring the cleavage of the fluorogenic substrate Ac-WEHD-AMC in cortical homogenates from individual noninjected (contralateral) and injected (ipsilateral) side of the brain ($n = 3$ animals per group). Results are expressed as relative fluorescence units (RFU)/min/mg of protein and reported as mean ± SEM. Statistical analysis shows no significant variations among means (ANOVA followed by Tukey-Kramer for multiple comparisons test).

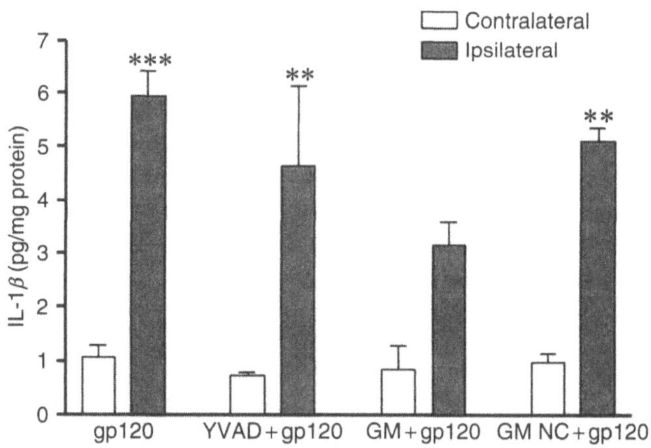

FIG. 3. Increase in rat brain cortical IL-1β levels produced by a single injection of HIV-gp120 is not minimized by the caspase-1 inhibitor Ac-YVAD-cmk, but it is reduced by the MMPs inhibitor GM6001. Individual brain cortical homogenates from rats treated with gp120 (100 ng, given i.c.v. 6 h beforehand) and Ac-YVAD-cmk (100 pmol, given 1 h before gp120), GM6001 (10 ng, given 1 h before gp120) or the GM6001 negative control (10 ng; GM NC) were analyzed for IL-1β levels by a rat-specific ELISA assay. Data are expressed as mean ± SEM (bars) values ($n = 3$ per group). ** and ***denote $p < 0.01$ and $p < 0.001$, respectively, versus respective contralateral (ANOVA followed by Tukey-Kramer test for multiple comparisons).

Incidentally, compared to vehicle-treated rats, administration of the HIV-1 coat protein gp120 (100 ng, i.c.v.) enhanced the expression of pro-MMP-9, active MMP-9, and MMP-2 as early as 1 h after administration (Fig. 4). Expression of pro-MMP-9, active MMP-9, and MMP-9 dimers all increased at a later stage, that is 6 h after viral protein administration, a time at which only an increase in active MMP-9 but not in pro-MMP-9 was observed in the neocortex of vehicle-treated rats. This result suggested that this latter effect might be a consequence of the surgical procedure (Fig. 4).

To implicate the observed MMP-9/MMP-2 induction in the mechanism underlying accumulation of IL-1β, we tested the effect of GM6001, a broad spectrum MMP inhibitor (Levy *et al.*, 1998); as negative control, we used an analogue of GM6001 with an inactive *t*-butylcarbamido group substituted for the idroxamic reactive group (GM6001 negative control, GM NC). It is worth noting that GM6001 (10 ng/rat, given i.c.v. 1 h before gp120) partially prevented gp120-induced increase of IL-1β levels in the brain cortex (Fig. 3). Administration of the negative control (GM6001 negative control, 10 ng/rat, i.c.v.) yielded no significant effect (Fig. 3). GM6001 or the negative control given alone did not affect the basal level of IL-1β in the rat cortex (data not shown).

In agreement with a previously described role for accumulation of IL-1β in the mechanism underlying gp120-induced neuronal apoptosis in the rat neocortex (Corasaniti *et al.*, 2001b), the apoptosis was minimized by administration of

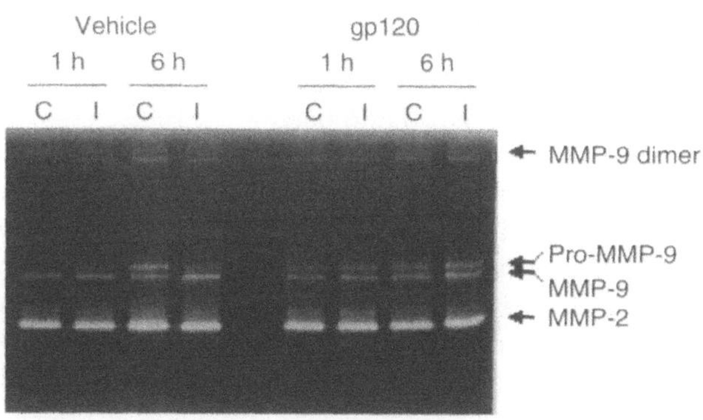

FIG. 4. HIV-1 gp120 enhances brain cortical expression of pro-MMP-9, active MMP-9, and MMP-2 as early as 1 h after administration. Contralateral (C) and ipsilateral (I) cortex from rats ($n = 3$ per experimental group) treated 1 and 6 h beforehand with vehicle or gp120 (100 ng) given into one lateral cerebral ventricle were homogenized, subjected to MMPs precipitation by using gelatin-Sepharose 4B and analyzed by gel zymography to detect MMP-2 and MMP-9 gelatinolytic activities. MMP-9 dimer ($\geq$250 kDa), pro-MMP-9 (90–95 kDa), MMP-9 (87 kDa), and MMP-2 (65–67 kDa) were identified on the zymogram by their molecular weight. Representative gelatin gel zymography from three independent experiments is shown.

TABLE I

HIV-gp120 Administered Once Daily into One Lateral Cerebral Ventricle (i.c.v.) for 5 Consecutive Days Increases the Number of Apoptotic Cells in the Neocortex of Rat: Neuroprotection by GM6001, a Specific, Broad Spectrum, Inhibitor of MMPs

	Apoptotic cells per microscopic field (mean ± SEM)	
Treatment	Noninjected side	Injected side
GM6001 10 ng + BSA	1.15 ± 0.35	2.46 ± 0.29
GM6001 NC 10 ng + gp120	2.30 ± 0.30	8.20 ± 0.20*
GM6001 1 ng + gp120	2.14 ± 0.15	4.85 ± 0.35*
GM6001 10 ng + gp120	0.77 ± 0.25	1.68 ± 0.22

*denotes $p < 0.001$ versus GM6001 10 ng + BSA and GM6001 10 ng + gp120 (ANOVA followed by Tukey-Kramer multiple comparison test; statistical analysis relative to the injected side of the brain).

GM6001 (10 ng/rat, given i.c.v. for 5 consecutive days, 1 h before gp120) (Table I); a lower dose of GM6001 (1 ng/rat) yielded partial protection (Table I).

IV. Discussion

The classical paradigm for the cleavage of the 33-kDa pro-IL-1β into the 17-kDa biologically active form involves the proteolytic removal of the N-terminal portion from the precursor and this process is known to be mediated by the ICE (or caspase-1) (Cerretti *et al.*, 1992; Thornberry *et al.*, 1992).

Here, using an experimental model of HAD, characterized by microglia activation, IL-1β production, and death of neocortical neurons consequent to i.c.v. administration of the HIV-1 glycoprotein gp120 in rat (Bagetta *et al.*, 1999; see Corasaniti *et al.*, 2001b), we show that the observed increase in IL-1β is not paralleled by enhanced caspase-1 activity. It is conceivable, therefore, that mechanisms other than caspase-1 may participate in the maturation process leading to accumulation of mature IL-1β. This hypothesis is strengthened by our evidence that pretreatment with a neuroprotective dose of Ac-YVAD-cmk (Bagetta *et al.*, 1999), a caspase-1-preferring inhibitor (Milligan *et al.*, 1995), did not prevent IL-1β increase induced by gp120. Interestingly, under similar experimental conditions, we have reported that Ac-YVAD-(aciloxy)mk, another caspase-1 preferring inhibitor, prevented gp120-induced translocation of cytochrome *c* without affecting accumulation of IL-1β in the rat neocortex (Corasaniti *et al.*, 2005). Interestingly, it has been reported that administration of the caspase-1 inhibitor Ac-YVAD-cmk to rats can reduce LPS-induced lethality without reducing IL-1β production (Mathiak *et al.*, 2000).

Collectively, these findings suggest that caspase-1 preferring inhibitors mediate neuroprotection through an alternative, caspase-1-independent mechanism.

Quite importantly, active IL-1β has been detected in ICE-deficient mice following tissue inflammation by turpentine (Fantuzzi *et al.*, 1997) and in human keratinocytes which do not express caspase-1/ICE (Nylander-Lundqvist *et al.*, 1996), supporting the notion that mechanism(s) in addition to caspase-1/ICE-operated cleavage of pro-IL-1β might be implicated. In fact, besides caspase-1 other proteases are able to cleave pro-IL-1β, producing peptides with biological activity. For instance, a role for a member of the Kallikrein family (mK13) in pro-IL-1β cleavage and activation has been suggested by Yao *et al.* (2006). Also, various MMPs, especially MMP-3, MMP-2 (gelatinase A), and MMP-9 (gelatinase B), have been reported as alternative enzymes able to process the IL-1β precursor into bioactive form, though this has been shown *in vitro* only (Schonbeck *et al.*, 1998).

Under physiological conditions, MMPs are expressed in the brain at low levels, although these can be upregulated under pathological circumstances, including acute and chronic neurodegenerative disorders such as stroke, Alzheimer's disease, multiple sclerosis, and HAD (Lukes *et al.*, 1999; Yong, 2005). Indeed, upregulation of MMPs, including MMP-2 (gelatinase A) and MMP-9 (gelatinase B), has been shown in the cerebrospinal fluid of patients with HAD (Conant *et al.*, 1999; Liuzzi *et al.*, 2000) as well as in the brain of gp120 transgenic mice (Marshall *et al.*, 1998).

Here we report evidence to support the hypothesis that MMPs participate in the processing of IL-1β in the neocortex of rats bearing neuropathological features recapitulating those of HIV-associated neurodegeneration. In fact, following gp120 treatment, MMP-9/MMP-2 expression is rapidly increased in the brain neocortex, and pharmacological manipulation aimed at inhibiting their enzymatic activity support the conclusion that MMPs are involved in IL-1β accumulation in this same region of the rat brain. Thus, gp120-induced accumulation of IL-1β was minimized by specific, broad spectrum MMP inhibitor, GM6001, which also prevented neuronal apoptosis typically induced by the viral protein in the neocortex (Bagetta *et al.*, 1995, 1999; Corasaniti *et al.*, 2001c).

As components of the inflammatory response, both MMPs and IL-1β are expressed at the site of inflammation. We have previously reported that the main sources for gp120-enhanced IL-1β immunoreactivity in the rat neocortex are neurons and microglia (Bagetta *et al.*, 1999). Furthermore, it has been shown that in response to an inflammatory stimulus, microglial cells not only produce active IL-1β, but preferentially release IL-1β precursor (pro-IL-1β) (Chauvet *et al.*, 2001).

Following a neuroinflammatory insult, neurons and microglia are the resident cellular sources of MMP-9 and MMP-2 (Rosenberg, 2002). This suggests a cell-type colocalization of pro-IL-1β and MMPs that may favor the interaction of the two proteins, once they have been released into the extracellular space, and further supports possible MMP-mediated cleavage of the proinflammatory cytokine.

IL-1β is one of the key inducer of biosynthesis and secretion of pro-MMPs. In particular, IL-1β stimulates expression and activity of MMP-2 and MMP-9 (Gottschall and Yu, 1995; Vecil *et al.*, 2000) through the activation of the transcription factor NF-κB (Wu *et al.*, 2004; Xie *et al.*, 2004). Here we suggest a model which implies an amplifying connection between IL-1β and MMPs at the site of inflammation. Following acute insult caused by gp120, an initial (within 1 h), although not pronounced, increase in the active form of IL-1β may be sufficient to trigger the observed induction of MMP-9 and MMP-2 that, in turn, may be responsible for the larger increase in mature IL-1β observed in the brain neocortex 6 h after injection of the coat protein.

In conclusion, MMP-2 and MMP-9, but not caspase-1/ICE, are involved in the accumulation of IL-1β induced by gp120 in the rat neocortex. Accordingly, systemic administration of a specific, broad spectrum MMP inhibitor, GM6001, minimized accumulation of this proinflammatory cytokine and prevented the consequent neuronal cell death typically caused by gp120.

Acknowledgments

This work has been carried out in collaboration with and with the financial support of the Istituto Superiore di Sanità (ISS, Rome, Italy) in the frame of "Progetti di ricerca su neoplasie, patologie cardiovascolari, malattie respiratorie, salute della donna, neuroscienze, riabilitazione, malattie infettive, tabagismo." Partial financial support from the V National Programme on AIDS, Istituto Superiore di Sanità, Rome (Italy) and from MIUR (FIRB 2001, RBNE01E7YX_005) to G.B. is gratefully acknowledged. We thank Dr. Stuart Lipton (La Jolla, California) for comments on the chapter.

References

Adair, J. C., Charlie, J., Dencoff, J. E., Kaye, J. A., Quinn, J. F., Camicioli, R. M., Stetler-Stevenson, W. G., and Rosenberg, G. A. (2004). Measurement of gelatinase B (MMP-9) in the cerebrospinal fluid of patients with vascular dementia and Alzheimer disease. *Stroke* **35,** 159–162.

Allan, S. M., Tyrrell, P. J., and Rothwell, N. J. (2005). Interleukin-1 and neuronal injury. *Nat. Rev. Immunol.* **5,** 629–640.

Bagetta, G., Corasaniti, M. T., Berliocchi, L., Navarra, M., Finazzi-Agro, A., and Nisticó, G. (1995). HIV-1 gp120 produces DNA fragmentation in the cerebral cortex of rat. *Biochem. Biophys. Res. Commun.* **211,** 130–136.

Bagetta, G., Corasaniti, M. T., Malorni, W., Rainaldi, G., Berliocchi, L., Finazzi-Agro, A., and Nisticó, G. (1996). The HIV-1 gp120 causes ultrastructural changes typical of apoptosis in the rat cerebral cortex. *Neuroreport* **7,** 1722–1724.

Bagetta, G., Corasaniti, M. T., Berliocchi, L., Nistico, R., Giammarioli, A. M., Malorni, W., Aloe, L., and Finazzi-Agro, A. (1999). Involvement of interleukin-1beta in the mechanism of human immunodeficiency virus type 1 (HIV-1) recombinant protein gp120-induced apoptosis in the neocortex of rat. *Neuroscience* **89,** 1051–1066.

Cerretti, D. P., Kozlosky, C. J., Mosley, B., Nelson, N., Van Ness, K., Greenstreet, T. A., March, C. J., Kronheim, S. R., Druck, T., Cannizzaro, L. A., Huebner, K., and Black, R. A. (1992). Molecular cloning of the interleukin-1 beta converting enzyme. *Science* **256,** 97–100.

Chauvet, N., Palin, K., Verrier, D., Poole, S., Dantzer, R., and Lestage, J. (2001). Rat microglial cells secrete predominantly the precursor of interleukin-1beta in response to lipopolysaccharide. *Eur. J. Neurosci.* **14,** 609–617.

Conant, K., McArthur, J. C., Griffin, D. E., Sjulson, L., Wahl, L. M., and Irani, D. N. (1999). Cerebrospinal fluid levels of MMP-2, 7, and 9 are elevated in association with human immunodeficiency virus dementia. *Ann. Neurol.* **46,** 391–398.

Corasaniti, M. T., Turano, P., Bilotta, A., Malorni, W., Stringaro, A. R., Nisticó, R., Finazzi-Agro, A., and Bagetta, G. (2001a). Evidence that increases of mitochondrial immunoreactive IL-1beta by HIV-1 gp120 implicate *in situ* cleavage of pro-IL-1beta in the neocortex of rat. *J. Neurochem.* **78,** 611–618.

Corasaniti, M. T., Maccarrone, M., Nisticò, R., Malorni, W., Rotiroti, D., and Bagetta, G. (2001b). Exploitation of the HIV-1 coat protein, gp120, in neurodegenerative studies *in vivo*. *J. Neurochem.* **79,** 1–8.

Corasaniti, M. T., Piccirilli, S., Paoletti, A., Nisticó, R., Stringaro, A., Malorni, W., Finazzi-Agró, A., and Bagetta, G. (2001c). Evidence that the HIV-1 coat protein gp120 causes neuronal apoptosis in the neocortex of rat via a mechanism involving CXCR4 chemokine receptor. *Neurosci. Lett.* **312,** 67–70.

Corasaniti, M. T., Russo, R., Amantea, D., Gliozzi, M., Siviglia, E., Stringaro, A. R., Malorni, W., Melino, G., and Bagetta, G. (2005). Neuroprotection by the caspase-1 inhibitor Ac-YVAD-(acyloxy)mk in experimental neuroAIDS is independent from IL-1beta generation. *Cell Death Differ.* **12,** 999–1001.

Deb, S., and Gottschall, P. E. (1996). Increased production of matrix metalloproteinases in enriched astrocyte and mixed hippocampal cultures treated with beta-amyloid peptides. *J. Neurochem.* **66,** 1641–1647.

Dinarello, C. A. (1998). Interleukin-1 beta, interleukin-18, and the interleukin-1 beta converting enzyme. *Ann. N. Y. Acad. Sci.* **856,** 1–11.

Fantuzzi, G., Ku, G., Harding, M. W., Livingston, D. J., Sipe, J. D., Kuida, K., Flavell, R. A., and Dinarello, C. A. (1997). Response to local inflammation of IL-1 beta-converting enzyme- deficient mice. *J. Immunol.* **158,** 1818–1824.

Garcia-Calvo, M., Peterson, E. P., Leiting, B., Ruel, R., Nicholson, D. W., and Thornberry, N. A. (1998). Inhibition of human caspases by peptide-based and macromolecular inhibitors. *J. Biol. Chem.* **273,** 32608–32613.

Gasche, Y., Soccal, P. M., Kanemitsu, M., and Copin, J. C. (2006). Matrix metalloproteinases and diseases of the central nervous system with a special emphasis on ischemic brain. *Front. Biosci.* **11,** 1289–1301.

Gavrieli, Y., Sherman, Y., and Ben-Sasson, S. A. (1992). Identification of programmed cell death *in situ* via specific labeling of nuclear DNA fragmentation. *J. Cell Biol.* **119,** 493–501.

Gendelman, H. E., Grant, I., Lipton, S. A., Everall, I., and Swindells, S. (2005). "The Neurology of AIDS," 2nd Oxford University Press, London.

Gijbels, K., Masure, S., Carton, H., and Opdenakker, G. (1992). Gelatinase in the cerebrospinal fluid of patients with multiple sclerosis and other inflammatory neurological disorders. *J. Neuroimmunol.* **41,** 29–34.

Gottschall, P. E., and Yu, X. (1995). Cytokines regulate gelatinase A and B (Matrix Metalloproteinases 2 and 9) activity in cultured rat astrocytes. *J. Neurochem.* **64,** 1513–1520.

Gu, Z., Kaul, M., Yan, B., Kridel, S. J., Cui, J., Strongin, A., Smith, J. W., Liddington, R. C., and Lipton, S. A. (2002). S-nitrosylation of matrix metalloproteinases: Signaling pathway to neuronal cell death. *Science* **297,** 1186–1190.

Hagan, P., Poole, S., Bristow, A. F., Tilders, F., and Silverstein, F. S. (1996). Intracerebral NMDA injection stimulates production of interleukin-1β in perinatal rat brain. *J. Neurochem.* **67,** 2215–2218.

Kaul, M., Garden, G. A., and Lipton, S. A. (2001). Pathways to neuronal injury and apoptosis in HIV associated dementia. *Nature* **410,** 988–994.

Levy, D. E., Lapierre, F., Liang, W., Ye, W., Lange, C. W., Li, X., Grobelny, D., Casabonne, M., Tyrrell, D., Holme, K., Nadzan, A., and Galardy, R. E. (1998). Matrix metalloproteinase inhibitors: A structure-activity study. *J. Med. Chem.* **41,** 199–223.

Lipton, S. A., and Gendelman, H. E. (1995). Dementia associated with the acquired immunodeficiency syndrome. *New Engl. J. Med.* **332,** 934–940.

Liuzzi, G. M., Mastroianni, C. M., Santacroce, M. P., Fanelli, M., D'Agostino, C., Vullo, V., and Riccio, P. (2000). Increased activity of matrix metalloproteinases in the cerebrospinal fluid of patients with HIV-associated neurological diseases. *J. Neurovirol.* **6,** 156–163.

Lukes, A., Mun-Bryce, S., Lukes, M., and Rosenberg, G. A. (1999). Extracellular matrix degradation by metalloproteinases and central nervous system diseases. *Mol. Neurobiol.* **19,** 267–284.

Marshall, D. C., Wyss-Coray, T., and Abraham, C. R. (1998). Induction of matrix metalloproteinase-2 in human immunodeficiency virus-1 glycoprotein 120 transgenic mouse brains. *Neurosci. Lett.* **254,** 97–100.

Masliah, E., Ge, N., Achim, C. L., Hansen, L. A., and Wiley, C. A. (1992). Selective neuronal vulnerability in HIV encephalitis. *J. Neuropathol. Exp. Neurol.* **51,** 585–593.

Mathiak, G., Grass, G., Herzmann, T., Luebke, T., Zetina, C. C., Bohlen, S. A., Neville, L. F., and Hoelscher, A. H. (2000). Caspase-1-inhibitor ac-YVAD-cmk reduces LPS-lethality in rats without affecting haematology or cytokine responses. *Br. J. Pharmacol.* **131,** 383–386.

Milligan, C. E., Prevette, D., Yaginuma, H., Homma, S., Cardwell, C., Fritz, L. C., Tomaselli, K. J., Oppenheim, R. W., and Schwartz, L. M. (1995). Peptide inhibitors of the ICE protease family arrest programmed cell death of motoneurons *in vivo* and *in vitro*. *Neuron* **15,** 385–393.

Nylander-Lundqvist, E., Back, O., and Egelrud, T. (1996). IL-1 beta activation in human epidermis. *J. Immunol.* **157,** 1699–1704.

Paxinos, G., and Watson, C. (1998). "The Rat Brain in Stereotaxic Coordinates." Academic Press, San Diego, CA.

Planas, A. M., Sole, S., and Justicia, C. (2001). Expression and activation of matrix metalloproteinase-2 and -9 in rat brain after transient focal cerebral ischemia. *Neurobiol. Dis.* **8,** 834–846.

Price, R. W., and Brew, B. J. (1988). The AIDS dementia complex. *J. Infect. Dis.* **158,** 1079–1083.

Rano, T. A., Timkey, T., Peterson, E. P., Rotonda, J., Nicholson, D. W., Becker, J. W., Chapman, K. T., and Thornberry, N. A. (1997). A combinatorial approach for determining protease specificities: Application to interleukin-1beta converting enzyme (ICE). *Chem. Biol.* **4,** 149–155.

Rosenberg, G. A. (2002). Matrix metalloproteinases in neuroinflammation. *Glia* **39,** 279–291.

Rosenberg, G. A., Dencoff, J. E., Correa, N., Jr., Reiners, M., and Ford, C. C. (1996). Effect of steroids on CSF matrix metalloproteinases in multiple sclerosis: Relation to blood-brain barrier injury. *Neurology* **46,** 1626–1632.

Rothwell, N. J., and Luheshi, G. N. (2000). Interleukin 1 in the brain: Biology, pathology and therapeutic target. *Trends Neurosci.* **23,** 618–625.

Schonbeck, U., Mach, F., and Libby, P. (1998). Generation of biologically active IL-1 beta by matrix metalloproteinases: A novel caspase-1-independent pathway of IL-1 beta processing. *J. Immunol.* **161,** 3340–3346.

Thornberry, N. A., Bull, H. G., Calaycay, J. R., Chapman, K. T., Howard, A. D., Kostura, M. J., Miller, D. K., Molineaux, S. M., Weidner, J. R., Aunins, J., Elliston, K. O., Ayala, J. M., *et al.* (1992). A novel heterodimeric cysteine protease is required for interleukin-1 beta processing in monocytes. *Nature* **356,** 768–774.

Toggas, S. M., Masliah, E., Rockenstein, E. M., Rall, G. F., Abraham, C. R., and Mucke, L. (1994). Central nervous system damage produced by expression of the HIV-1 coat protein gp120 in transgenic mice. *Nature* **367,** 188–193.

Vecil, G. G., Larsen, P. H., Corley, S. M., Herx, L. M., Besson, A., Goodyer, C. G., and Yong, V. W. (2000). Interleukin-1 is a key regulator of matrix metalloproteinase-9 expression in human neurons in culture and following mouse brain trauma *in vivo*. *J. Neurosci. Res.* **61,** 212–224.

Visse, R., and Nagase, H. (2003). Matrix metalloproteinases and tissue inhibitors of metalloproteinases: Structure, function, and biochemistry. *Circ. Res.* **92,** 827–839.

Wu, C. Y., Hsieh, H. L., Jou, M. J., and Yang, C. M. (2004). Involvement of p42/p44 MAPK, p38 MAPK, JNK and nuclear factor-kappa B in interleukin-1beta-induced matrix metalloproteinase-9 expression in rat brain astrocytes. *J. Neurochem.* **90,** 1477–1488.

Xie, Z., Singh, M., and Singh, K. (2004). Differential regulation of matrix metalloproteinase-2 and -9 expression and activity in adult rat cardiac fibroblasts in response to interleukin-1beta. *J. Biol. Chem.* **279,** 39513–39519.

Yao, C., Karabasil, M. R., Purwanti, N., Li, X., Akamatsu, T., Kanamori, N., and Hosoi, K. (2006). Tissue Kallikrein mK13 is a candidate processing enzyme for the precursor of interleukin-1beta in the submandibular gland of mice. *J. Biol. Chem.* **281,** 7968–7976.

Yong, V. W. (2005). Metalloproteinases: Mediators of pathology and regeneration in the CNS. *Nat. Rev. Neurosci.* **6,** 931–944.

Zhang, J. W., and Gottschall, P. E. (1997). Zymographic measurement of gelatinase activity in brain tissue after detergent extraction and affinity-support purification. *J. Neurosci. Methods* **76,** 15–20.

NEUROPROTECTIVE EFFECT OF NITROGLYCERIN IN A RODENT MODEL OF ISCHEMIC STROKE: EVALUATION OF Bcl-2 EXPRESSION

Rosaria Greco,* Diana Amantea,†,‡ Fabio Blandini,* Giuseppe Nappi,*,§,¶ Giacinto Bagetta,†,‡ M. Tiziana Corasaniti,||,** and Cristina Tassorelli*,§

*IRCCS "C. Mondino Institute of Neurology" Foundation, 27100 Pavia, Italy
†Department of Pharmacobiology, University of Calabria, 87036 Arcavacata di Rende (Cosenza), Italy
‡University Centre for Adaptive Disorders and Headache (UCADH), Section of Neuropharmacology of Normal and Pathological Neuronal Plasticity, University of Calabria 87036 Arcavacata di Rende (Cosenza), Italy
§University Centre for the Study of Adaptive Disorder and Headache (UCADH), Section of Pavia, Italy
¶Department of Neurology and Otorhinolaryngology, University of Rome "La Sapienza", Rome, Italy
||Department of Pharmacobiological Sciences, University Magna Graecia of Catanzaro, Italy
**Center for Experimental Neuropharmacology, "Mondino-Tor Vergata", University of Rome Tor Vergata, Rome, Italy

Transient focal ischemia caused by middle cerebral artery occlusion (MCAo) produces apoptotic cell death in the penumbra area. Bcl-2 is a protooncogene that plays a major antiapoptotic role, at the cellular level, by counteracting the activation of apoptosis effectors, that is, caspases. It has been suggested that nitroglycerin (NTG), a nitric oxide donor, reduces ischemia/reperfusion-induced brain damage via the inhibition of caspase activity and NMDA receptor. In this chapter, we evaluated the protective effects of NTG against cerebral damage caused by

INTERNATIONAL REVIEW OF NEUROBIOLOGY, VOL. 82
DOI: 10.1016/S0074-7742(07)82024-1

0074-7742/07 $35.00

transient (2 h) MCAo (tMCAo) focusing our interest on the potential effects on Bcl-2 expression. Male Wistar rats were administered intraperitoneally (i.p.) with NTG (10 mg/kg) or vehicle (PEG, 1 ml/kg) 20 min before the induction of MCAo by intraluminal silicon-coated filament (0.37-mm diameter). Cerebral infarct volume was measured 22 h after reperfusion, while cortical Bcl-2 expression was evaluated at the end of 2-h MCAo (without reperfusion) and at 5 h of reperfusion. The results show significant reduction of the infarct volume in rats preinjected with NTG, as compared to the vehicle group. After 2 h of occlusion, no significant difference was seen in Bcl-2 expression in the ipsilateral and contralateral cortex of either experimental groups (NTG and vehicle). However, 5 h after reperfusion, a significant increase of Bcl-2 expression was detected in the damaged cortex of control rats, probably reflecting a compensatory response aiming at counteracting the cell death process; this increase was absent in the NTG-treated rats. These data, while confirming the neuroprotective effect of NTG in an *in vivo* ischemia/reperfusion model, seem to suggest that the drug may act by downsizing the complex chain of events underlying apoptosis activation and consequent activation of antiapoptotic responses.

I. Introduction

Increasing evidence has shown that transient brain ischemia initiates a cascade of injurious events, with accumulation of intracellular calcium, generation of free radicals, and release of cytokines, which lead to disruption of cellular homeostasis and structural damage of cerebral tissue (Feuerstein *et al.*, 1994; Kochanek and Hallenbeck, 1992). Ischemic neuronal death has been ascribed in part to programmed cell death (Chopp and Li, 1996; Linnik *et al.*, 1995a). During apoptosis, free radicals are known to induce lipid peroxidation, DNA damage (Lo *et al.*, 2003), release of cytochrome *c* from mitochondria (Chen *et al.*, 1996; Fujimura *et al.*, 2005; Kluck *et al.*, 1997), and activation of caspases (Bredesen, 1995). In the central nervous system (CNS), antiapoptotic protein Bcl-2 counteracts apoptotic neuronal death induced by various stimuli, including ischemia (Allsopp *et al.*, 1993; Bidmon *et al.*, 2001; Ferrer *et al.*, 1997; Garcia *et al.*, 1992; Kane *et al.*, 1993; Zhong *et al.*, 1993). Furthermore, Bcl-2 and related proteins are expressed in surviving neurons in models of epilepsy and in regions selectively vulnerable to apoptotic cell death after experimental traumatic brain injury (TBI) (Chen *et al.*, 1995; Clark *et al.*, 1997; Graham *et al.*, 1996; Krajewski *et al.*, 1995; Shimazaki *et al.*, 1994). An overexpression of Bcl-2 produced by viral transfection reduces infarct area after focal ischemia (Lawrence *et al.*, 1996; Linnik *et al.*, 1995b), and transgenic mice overexpressing Bcl-2 have reduced infarction size after cerebral ischemia compared with wild-type mice (Martinou *et al.*, 1994).

Nitric oxide (NO), an unstable gas, has a dichotomous regulatory role in the brain. It is an intra- and intercellular transmitter under physiological conditions, whereas it may exert detrimental effects under pathological conditions. It is known that, as a neurotoxin, NO may mediate the ischemic excitotoxic brain injury induced by glutamate release and overactivation of the *N*-methyl-D-aspartate (NMDA) receptor (see Lipton, 1999 for review). On the other hand, as a signaling molecule, it may induce an increase in blood perfusion of ischemic penumbra in the early phases of cerebral stroke (Iadecola *et al.*, 1996). Thus, the effects of NO are modulated by both direct and indirect interactions that can be dose-dependent and cell-type specific (Razavi *et al.*, 2005). The extraordinary diffusibility of NO in biological systems makes this molecule unique among several regulators of apoptosis (Choi *et al.*, 2002; Kim *et al.*, 2001; Razavi *et al.*, 2005). Numerous researches have demonstrated that protective cerebrovascular activity induced by various NO donors in models of experimental stroke is in part due to their vasodilator activity and hemodynamic effects (Khan *et al.*, 2006; Salom *et al.*, 2000; Zhang and Iadecola, 1994). Organic nitrates, such as nitroglycerin (NTG), are vasodilators that may activate guanylate cyclase synthase (sGC) *in vivo* and increase tissue levels of cGMP important for smooth muscle relaxation (Artz *et al.*, 2001; Tassorelli *et al.*, 2004). However, a previous *in vitro* report (Lipton *et al.*, 1994) suggests that NTG may reduce ischemia/reperfusion-induced brain damage also through the downregulation of caspases and NMDA receptor activity. A proposed theory hypothesizes that NTG inhibits these activities through protein *S*-nitrosylation, via a mechanism that involves the transfer of an NO group to the sulfhydryl moiety of a cysteine residue, to form an RS-NO (*S*-nitroso adduct).

In this chapter, we have examined the potential-protecting effects of NTG against cerebral damage induced by a transient (2 h) middle cerebral artery occlusion (MCAo), focusing our interest on the modulation of expression of Bcl-2 by NO.

II. Materials and Methods

Male Wistar rats (280–300 g) four to six per group were used for the experiments; adopted procedures were in accordance with the European Convention for Care and Use of Laboratory Animals and were approved by the local Animal Care Committee.

A. Focal Cerebral Ischemia

Twenty minutes after systemic administration of NTG (10 mg/kg) or vehicle (PEG, 1 ml/kg), brain ischemia was induced by MCAo, using an intraluminal filament (Longa *et al.*, 1989). Briefly, rats were anesthetized with 5% isoflurane in air and maintained with the lowest acceptable concentration of anesthetic (1.5–2%). Body temperature was measured with a rectal probe and was kept at 37°C during the surgical procedure with a heating pad.

Under an operating microscope, the external and internal right carotid arteries were exposed through a neck incision. The external carotid artery was cut approximately 3 mm above the common carotid artery bifurcation and a silk suture was tied loosely around the external carotid stump. A silicone-coated nylon filament (diameter, 0.38 mm) was then inserted into the external carotid artery and gently advanced into the internal carotid artery, approximately 18 mm from the carotid bifurcation, until mild resistance was felt, thereby indicating occlusion of the origin of the middle cerebral artery in the Willis circle. The silk suture was tightened around the intraluminal filament to prevent bleeding. The wound was then sutured and anesthesia discontinued. To allow reperfusion, rats were briefly reanesthetized with isoflurane, and the nylon filament was withdrawn 2 h after MCAo. After the discontinuation of isoflurane, the animals were kept in their cages with free access to food and water.

One hour after surgery, the animals were grossly assessed for neurological deficit as follows: 0 = no deficit, 1 = failure to extend the left forelimb, 2 = decreased resistance to lateral push, 3 = circling to contralateral side, 4 = walks only when stimulated, and 5 = no spontaneous motor activity. Only rats with clear neurological deficits (≥3) were included in the study.

B. Neuropathology and Quantification of Ischemic Damage

Cerebral infarct volume was evaluated 22 h after reperfusion, in rats subjected to 2-h MCAo, using the triphenyltetrazolium chloride (TTC) vital staining method. Rats were sacrificed by decapitation and the brains were rapidly removed. Eight serial sections from each brain were cut at 2-mm intervals from the frontal pole using a rat brain matrix (RBMA-300C, 2 Biological Instruments, Besozzo, Italy). To measure ischemic damage, brain slices were stained in a solution containing 2% TTC in saline, at 37°C. After a 10-min incubation, the slices were transferred to 10% neutral-buffered formaldehyde and stored at 4°C prior to analysis. Images of TTC-stained sections were captured using a digital scanner and analyzed using an image analysis software (ImageJ, version 1.30). The infarct volume (mm^3) was calculated by summing the infarcted area (unstained) of the eight sections and multiplying by the interval thickness between sections (Li *et al.*, 2000).

C. Western Blotting Analyses

Following 2 h of MCAo (without reperfusion) and 5 h after reperfusion, the samples corresponding to the penumbra zone ipsilateral and contralateral to damage were dissected out and immediately used for the preparation of total extracts. Tissues were lysed in at least 10 volumes of lysis buffer (RIPA buffer) containing Tris 50 mM (pH 7.4), NaCl 150 mM, EDTA 1 mM, SDS 0.2% supplemented with a cocktail of protease inhibitors. Samples were then incubated on ice for 20 min. The tissue lysates were centrifuged at 10,000 × *g* for 45 min at 4°C and supernatants stored at −80°C. Protein assay was performed by BCA method. Forty micrograms of protein were submitted to SDS-polyacrylamide gels 12% and transferred onto a nitrocellulose membrane (Amersham Biosciences, Piscataway, NJ). After blocking with 5% dry milk, the blots were probed overnight at 4°C with rabbit polyclonal anti-Bcl-2 serum (1:200; Santa Cruz Biotechnology, Santa Cruz, CA) and then probed for 1 h with an anti-rabbit horseradish peroxidase–coupled secondary antibody (1:10,000; Amersham Biosciences, Piscataway, NJ). An enhanced chemoluminescence system (ECL Advance; Amersham Biosciences, Piscataway, NJ) was used for visualization. Membranes were also probed with a rabbit polyclonal anti-β-actin antibody (1:1000; Santa Cruz Biotechnology, Santa Cruz, CA) as a housekeeping protein. For semiquantitative analysis, a Bio-Rad GS800 densitometer was used; Bcl-2 expression was quantified as the ratio between the optical densities of Bcl-2 and β-actin bands.

D. Statistical Analysis

Data are expressed as mean ± SEM. Student's *t*-test was used to analyze the effects of drug treatment on infarct size (total volume) and Bcl-2 expression. A value of $p < 0.05$ was considered significant.

III. Results

A. Effect of NTG on Infarct Volume

In control rats, 22 h after reperfusion, the ischemic zone appeared as a distinct unstained area, consistently located in the cortex and striatum of the right cerebral hemisphere (Fig. 1). In the NTG group, we observed a significant reduction of the total infarct volume when compared to the vehicle group (−36%).

B. Effect of NTG on Bcl-2 Protein Expression

Cortical Bcl-2 expression did not change in the injured (right) hemisphere, compared to the intact hemisphere in control animals sacrificed at the end of a 2-h MCAo. Similarly, no significant changes were observed in the NTG-treated group, although a slight reduction in Bcl-2 expression was detected, in the injured

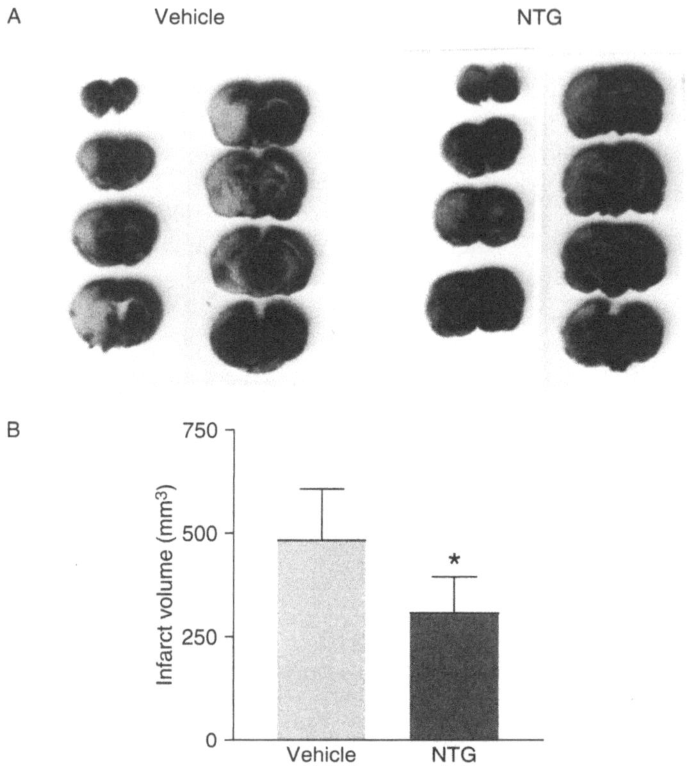

FIG. 1. (A) Representative coronal sections showing infarct area in the cortex and striatum (unstained area) of rats pretreated with vehicle (PEG, 1 mg/kg i.p.) or NTG (10 mg/kg i.p.) and subjected to 2-h ischemia, followed by 22-h reperfusion. (B) Histograms show the infarct volume (mm^3), expressed as mean ± SEM. Student's *t*-test: $*p < 0.05$ versus vehicle group.

side (Fig. 2). At variance with these data, analysis of tissue specimens from the brain of animals sacrificed 5 h after reperfusion showed a significant increase in Bcl-2 protein in the ischemic cortex of control animals (Fig. 3A) but not in the brain cortex of animals pretreated with NTG (Fig. 3B).

IV. Discussion

A large body of evidence suggests that NO may contribute to cellular homeostatic balance by modulating the apoptotic pathway at multiple levels (Choi *et al.*, 2002; Kim *et al.*, 2001). Morphological and biochemical studies have indicated an apoptotic component in neurons as well as the involvement of different factors,

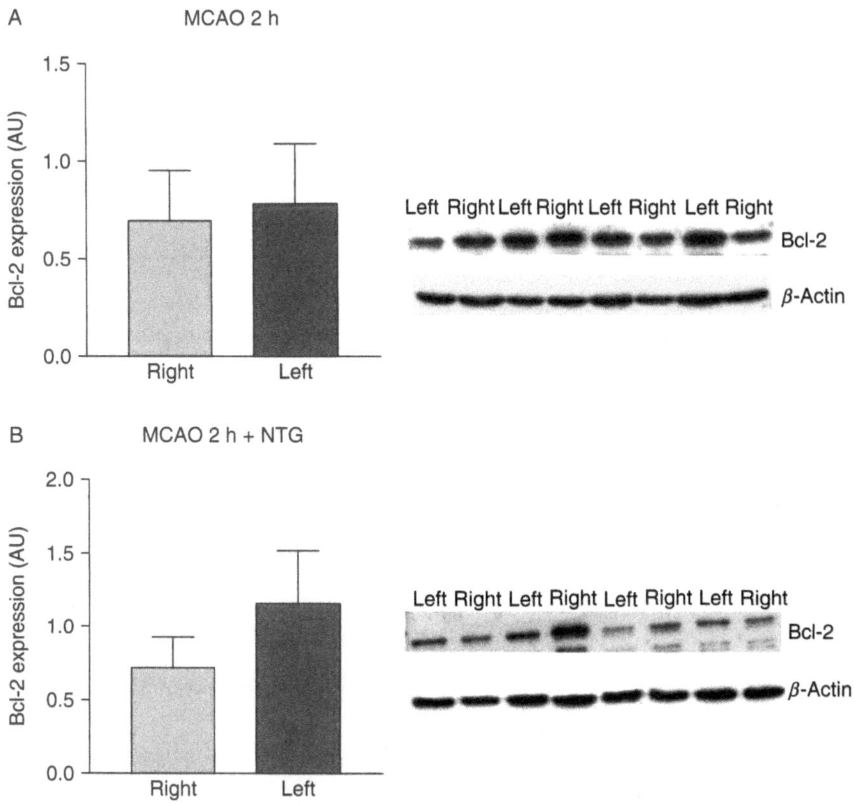

FIG. 2. Expression of Bcl-2 (29 kDa) in homogenates from cortices dissected out at the end of a 2-h occlusion of the right middle cerebral artery (MCAo). Rats were injected with (A) vehicle (PEG, 1 mg/kg i.p.) or (B) NTG (10 mg/kg i.p.) 20 min before the occlusion was started. β-Actin (39 kDa) was used as housekeeping protein in the same blot. Data are expressed mean ± SEM.

including Bcl-2 family, caspases, and trophic factors during both focal and global ischemia (Ferrer, 1999).

In this chapter, we have evaluated the neuroprotective effects of NTG, an NO donor, against cerebral infarct induced by tMCAo in the rat. In addition, in order to test the possible effect of the drug on the apoptotic pathway, we have investigated whether neuroprotection involves modulation of Bcl-2 expression in the ischemic penumbra.

Our findings show that NTG significantly reduces infarct volume (−36%), as compared to the vehicle-treated group, when administered prior to a transient ischemic insult (2-h MCAo), while the drug failed to protect against permanent ischemic damage when injected 1 h after occlusion (data not shown). Lipton's seminal work (Lipton *et al.*, 1993) on the interaction of NO with NMDA receptors

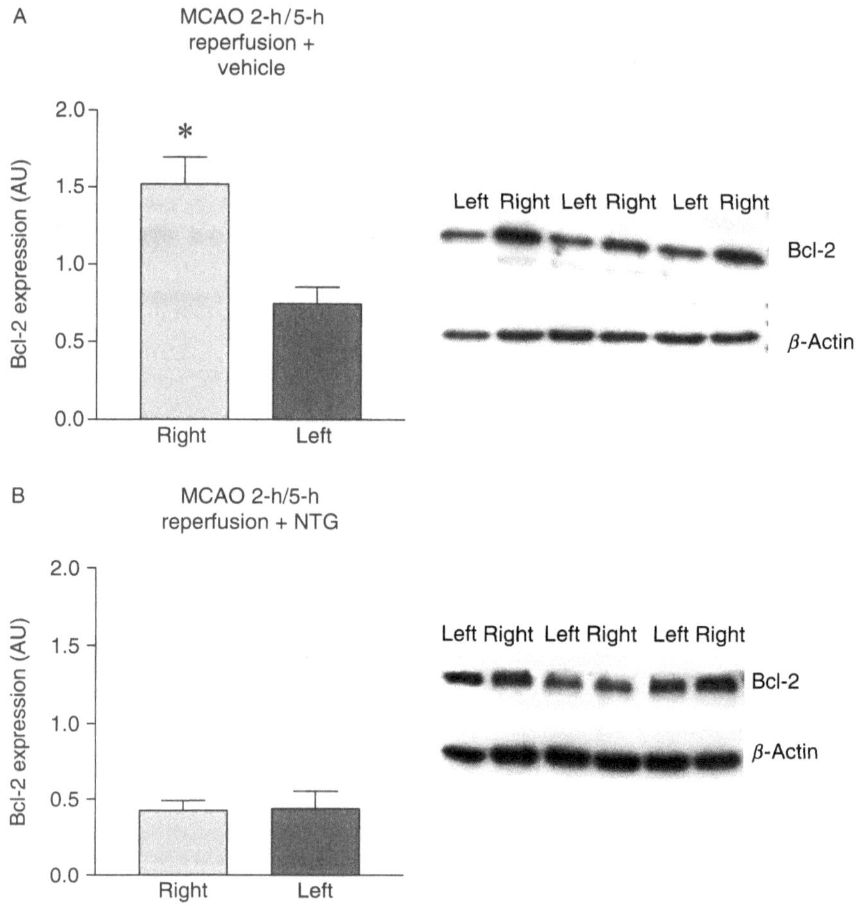

FIG. 3. Expression of Bcl-2 (29 kDa) in homogenates from cortices dissected out 5 h after reperfusion from a 2-h MCAo. Rats were injected with (A) vehicle (PEG, 1 mg/kg i.p.) or (B) NTG (10 mg/kg i.p.) 20 min before the occlusion was started. β-Actin (39 kDa) was used as housekeeping protein in the same blot. Data are expressed mean ± SEM. Student's t-test: $^*p < 0.05$ lesioned versus contralateral side.

showed that NTG is neuroprotective in models of NMDA receptor-mediated excitotoxic neuronal damage. Zhang and Iadecola (1994) have reported that NO donors are unable to protect the brain tissue form infarction and to increase CBF in the penumbra when infused 2 h after occlusion, and some studies have indicated that the vasodilatory action of NO donors is beneficial in the early phases of MCAo (Zhang and Iadecola, 1993, 1994). In contrast, some reports have found that NO donors may be neurotoxic (Dawson *et al.*, 1991). These apparent discrepancies are likely related to differences in the biological chemistry of the NO donors themselves, including rates of NO release, tissue selectivity, and intrinsic biological activity (Beckman *et al.*, 1990; Manzoni *et al.*, 1992). Indeed, under specific circumstances, NO and *S*-nitrosothiols may constitute the biologically active species formed through bioactivation of drugs such as NTG (Sokolowska and Wlodek, 2003).

The findings obtained suggest that systemic NTG administration may exert its neuroprotective effect by interacting with Bcl-2 protein expression in the ischemic cortex. Based on the present data, Bcl-2 expression seems to be associated to events taking place in the early phases of reperfusion since no significant interhemispheric differences were found at the end of the 2 h occlusion, in either experimental group. By contrast, a significant increase in Bcl-2 protein expression was observed 5 h after reperfusion in the ischemic hemisphere of rats pretreated with vehicle (controls) and this is in agreement with previous findings (Alkayed *et al.*, 2001; Ferrer *et al.*, 2003). During the reperfusion phase, upregulation of the antiapoptotic protein Bcl-2 in the ischemic penumbra of vehicle-treated rats reflects a prosurvival mechanism for counteracting the neuronal death process triggered by a detrimental insult (Chen *et al.*, 1995; Clark *et al.*, 1997; Graham *et al.*, 1996; Krajewski *et al.*, 1995; Shimazaki *et al.*, 1994). Accordingly, it has been shown that oxidative stress (Chandra *et al.*, 2000; Hockenbery *et al.*, 1993; Lopez Diazguerrero *et al.*, 2006; Saitoh *et al.*, 2003) and activation of caspases in neurons during reperfusion (MacManus and Linnik, 1997) are likely to increase Bcl-2 expression in the ischemic penumbra. By contrast, a decreased Bcl-2 immunoreactivity was demonstrated after permanent occlusion in the cortical penumbra (Gillardon *et al.*, 1996; Won *et al.*, 2006).

The lack of increase in cortical Bcl-2 in NTG-treated rats may be partly ascribed to enhanced cerebral blood flow in the penumbra after ischemia/reperfusion injury (Zhang *et al.*, 1994). However, it is conceivable that NTG and/or drug-derived NO might interfere with critical steps of the apoptotic processes via different mechanisms, including inhibition of caspase-3-like via *S*-nitrosylation (Kim *et al.*, 2000; Lipton *et al.*, 1998) reaction with thiol groups of NMDA receptor to limit excessive Ca^{2+} influx (Lipton and Wang, 1996), inhibition of cytochrome *c* release, and activation of NF-κB (Greco *et al.*, 2005). As a consequence of an effective blockade of the apoptotic cascade, the need for a compensatory increase in Bcl-2 expression is reduced.

Thus, the pharmacological effects of NTG may be both "vascular," acting on the cerebral blood vessels, and "neuronal," depending on the neuronal effects of NTG-derived NO. A study indeed demonstrates that NO donors are able to reduce oxidative stress both in brain and vasculature (Khan *et al.*, 2006). Experimental data support the idea that NO is an antioxidant agent and inhibits lipid peroxidation of low-density lipoproteins (Jessup *et al.*, 1992). During ischemia/reperfusion, free oxygen radicals are probably produced by several cell types, thus NO released from NTG could have scavenging effects and contribute to stable levels of Bcl-2 in penumbra. It has been reported that NTG has antioxidant capacity (Lo *et al.*, 2003) and it may attenuate the oxidative stress of ischemia/reperfusion injury and improve pulmonary function (Kawashima *et al.*, 2000), while in the heart the drug upregulates antioxidant enzymes (Husain, 2003).

In conclusion, we report that NTG reduces infarct volume in transient, but not in permanent MCAo, demonstrating that the action of the drug might be specific for—or limited to—the events occurring during the reperfusion. In addition, we have shown that the neuroprotective effect of NTG is significant during the early stages of ischemia/reperfusion and that it may partly depend on a direct cytoprotective mechanism involving Bcl-2 modulation.

Acknowledgments

Partial financial support from the Italian Ministry of University and Research (PRIN prot. 2004053099_004 to G.B.) is gratefully acknowledged. Mr. Guido Fico is gratefully acknowledged for skillful technical assistance.

References

Alkayed, N. J., Goto, S., Sugo, N., Joh, H. D., Klaus, J., Crain, B. J., Bernard, O., Traystman, R. J., and Hurn, P. D. (2001). Estrogen and Bcl-2: Gene induction and effect of transgene in experimental stroke. *J. Neurosci.* **21,** 7543–7550.

Allsopp, T. E., Wyatt, S., Paterson, H. F., and Davies, A. M. (1993). The proto-oncogene bcl-2 can selectively rescue neurotrophic factor-dependent neurons from apoptosis. *Cell* **73,** 295–307.

Artz, J. D., Toader, V., Zavorin, S. I., Bennett, B. M., and Thatcher, G. R. (2001). *In vitro* activation of soluble guanylyl cyclase and nitric oxide release: A comparison of NO donors and NO mimetics. *Biochemistry* **40,** 9256–9264.

Beckman, J. S., Beckman, T. W., Chen, J., Marshall, P. A., and Freeman, B. A. (1990). Apparent hydroxyl radical production by peroxynitrite: Implications for endothelial injury from nitric oxide and superoxide. *Proc. Natl. Acad. Sci. USA* **87,** 1620–1624.

Bidmon, H. J., Emde, B., Kowalski, T., Schimitt, M., Mayer, B., Kato, K., Asayama, K., Witte, W., and Zilles, K. (2001). Nitric oxide synthase-I containing cortical interneurons co-express antioxidative enzymes and anti-apoptotic Bcl-2 following focal ischemia: Evidence for direct and indirect mechanisms towards their resistance to neuropathology. *J. Chem. Neuroanat.* **22,** 167–184.

Bredesen, D. E. (1995). Neural apoptosis. *Ann. Neurol.* **38,** 839–851.

Chandra, J., Samali, A., and Orrenius, S. (2000). Triggering and modulation of apoptosis by oxidative stress. *Free Radic. Biol. Med.* **29,** 323–333.

Chen, J., Graham, S. H., Chan, P. H., Lan, J., Zhou, R. L., and Simon, R. P. (1995). Bcl-2 is expressed in neurons that survive focal ischemia in the rat. *Neuroreport* **6,** 394–398.

Chen, J., Zhu, R. L., Nakayama, M., Kawaguchi, K., Jin, K., Stetler, R. A., Simon, R. P., and Graham, S. H. (1996). Expression of the apoptosis-effector gene, bax, is up-regulated in vulnerable hippocampal CA1 neurons following global ischemia. *J. Neurochem.* **67,** 64–71.

Choi, B. M., Pae, H. O., Jang, S., Kim, Y. M., and Chung, H. T. (2002). Nitric oxide as a pro-apoptotic as well as anti-apoptotic modulator. *J. Biochem. Mol. Biol.* **35,** 116–126.

Chopp, M., and Li, Y. (1996). Apoptosis in focal cerebral ischemia. *Acta Neurochir.* **66,** 21–26.

Clark, R. S., Chen, J., Watkins, S. C., Kochanek, P. M., Chen, M., Stetler, R. A., Loeffert, J. E., and Graham, S. H. (1997). Apoptosis-suppressor gene bcl-2 expression after traumatic brain injury in rats. *J. Neurosci.* **17,** 9172–9182.

Dawson, V. L., Dawson, T. M., and London, E. D. (1991). Nitric oxide mediates glutamate neurotoxicity in primary cortical cultures. *Proc. Natl. Acad. Sci. USA* **88,** 6368–6371.

Ferrer, I. (1999). Neuronal death mechanisms in cerebral ischemia. *Rev. Neurol.* **29,** 515–521.

Ferrer, I., Pozas, E., Lopez, E., and Ballabriga, J. (1997). Bcl-2, Bax and Bcl-2 expression following hypoxic ischemia in the infant rat brain. *Acta Neuropathol.* **94,** 583–589.

Ferrer, I., Friguls, B., Dalfo, E., Justicia, C., and Planas, A. M. (2003). Caspase–dependent and Caspase-independent signalling of apoptosis in the penumbra following middle cerebral artery occlusion in the adult rat. *Neuropathol. Appl. Neurobiol.* **29,** 472–481.

Feuerstein, G. Z., Liu, T., and Barone, F. C. (1994). Cytokines, inflammation, and brain injury: Role of tumor necrosis factor-alpha. *Cerebrovasc. Brain Metab. Rev.* **6,** 341–360.

Fujimura, M., Tominaga, T., and Chan, P. H. (2005). Neuroprotective effect of an antioxidant in ischemic brain injury: Involvement of neuronal apoptosis. *Neurocrit. Care* **2,** 59–66.

Garcia, I., Martinou, I., Tsujimoto, Y., and Martinou, J. C. (1992). Prevention of programmed cell death of sympathetic neurons by the bcl-2 proto-oncogene. *Science* **258,** 302–304.

Gillardon, F., Lenz, C., Waschke, K. F., Krajewski, S., Reed, J. C., Zimmermann, M., and Kuschinsky, W. (1996). Altered expression of Bcl-2, Bcl-X, Bax, and c-Fos colocalizes with DNA fragmentation and ischemic cell damage following middle cerebral artery occlusion in rats. *Brain Res. Mol. Brain Res.* **40,** 254–260.

Graham, S. H., Chen, J., Stetler, R. A., Zhu, R. L., Jin, K. L., and Simon, R. P. (1996). Expression of proto-oncogene bcl-2 is increased in the rat brain following kainate-induced seizures. *Restor. Neurol. Neurosci.* **9,** 243–250.

Greco, R., Tassorelli, C., Cappelletti, D., Sandrini, G., and Nappi, G. (2005). Activation of the transcription factor NF-kappaB in the nucleus trigeminalis caudalis in an animal model of migraine. *Neurotoxicology* **26,** 795–800.

Hockenbery, D. M., Oltvai, Z. N., Yin, X. M., Milliman, C. L., and Korsmeyer, S. J. (1993). Bcl-2 functions in an antioxidant pathway to prevent apoptosis. *Cell* **75,** 241–251.

Husain, K. (2003). Interaction of physiacial training and chronic nutroglycerin treatment on blood pressure, nitric oxide, and oxidants/antioxidants in the rat heart. *Pharmacol. Res.* **48,** 253–261.

Iadecola, C., Zhang, F., Casey, R., Clark, H. B., and Ross, M. E. (1996). Inducible nitric oxide synthase gene expression in vascular cells after transient focal cerebral ischemia. *Stroke* **27,** 1373–1380.

Jessup, W., Mohr, D., Gieseg, S. P., Dean, R. T., and Stocker, R. (1992). The participation of nitric oxide in cell free- and its restriction of macrophage-mediated oxidation of low-density lipoprotein. *Biochim. Biophys. Acta* **1180,** 73–82.

Kane, D. J., Sarafian, T. A., Anton, R., Hahn, H., Gralla, E. B., Valentine, J. S., Ord, T., and Bredesen, D. E. (1993). Bcl-2 inhibition of neural death: Decreased generation of reactive oxygen species. *Science* **262,** 1274–1277.

Kawashima, M., Bando, T., Nakamura, T., Isowa, N., Liu, M., Toyokuni, S., Hitomi, S, and Wada, H. (2000). Cytoprotective effects of nitroglycerin in ischemia-reperfusion-induced lung injury. *Am. J. Respir. Crit. Care Med.* **161,** 935–943.

Khan, M., Jatana, M., Elango, C., Paintlia, A. S., Singh, A. K., and Singh, I. (2006). Cerebrovascular protection by various nitric oxide donors in rats after experimental stroke. *Nitric Oxide* **15,** 114–124.

Kim, P. K., Zamora, R., Petrosko, P., and Billiar, T. R. (2001). The regulatory role of nitric oxide in apoptosis. *Int. Immunopharmacol.* **1,** 1421–1441.

Kim, Y. M., Kim, T. H., Chung, H. T., Talanian, R. V., Yin, X. M., and Billiar, T. R. (2000). Nitric oxide prevents tumor necrosis factor alpha-induced rat hepatocyte apoptosis by the interruption of mitochondrial apoptotic signaling through S-nitrosylation of caspase-8. *Hepatology* **32,** 770–778.

Kluck, R. M., Bossy-Wetzel, E., Green, D. R., and Newmeyer, D. D. (1997). The release of cytochrome c from mitochondria: A primary site for Bcl-2 regulation of apoptosis. *Science* **275,** 1132–1136.

Kochanek, P. M., and Hallenbeck, J. M. (1992). Polymorphonuclear leukocytes and monocytes/macrophages in the pathogenesis of cerebral ischemia and stroke. *Stroke* **23,** 1367–1379.

Krajewski, S., Mai, J. K., Krajewska, M., Sikorska, M., Mossakowski, M. J., and Reed, J. C. (1995). Upregulation of Bax protein levels in neurons following cerebral ischemia. *J. Neurosci.* **15,** 6364–6376.

Lawrence, M. S., Ho, D. Y., Sun, G. H., Steinberg, G. K., and Sapolsky, R. M. (1996). Overexpression of bcl-2 with herpes simplex virus vectors protects CNS neurons against neurological insults *in vitro* and *in vivo*. *J. Neurosci.* **16,** 486–496.

Li, H., Colbourne, F., Sun, P., Zhao, Z., Buchan, A. M., and Iadecola, C. (2000). Caspase inhibitors reduce neuronal injury after focal but not global cerebral ischemia in rats. *Stroke* **31,** 176–182.

Linnik, M. D., Miller, J. A., Sprinkle-Cavallo, J., Mason, P. J., Thompson, F. Y., Montgomery, L. R., and Schroeder, K. K. (1995a). Apoptotic DNA fragmentation in the rat cerebral cortex induced by permanent middle cerebral artery occlusion. *Brain Res. Mol. Brain Res.* **32,** 116–124.

Linnik, M. D., Zahos, R., Geschwind, M. D., and Federoff, H. J. (1995b). Expression of Bcl-2 from a defective herpes simplex virus-1 vector limits neuronal death in focal cerebral ischemia. *Stroke* **26,** 1670–1675.

Lipton, S. A. (1999). Neuronal protection and destruction by NO. *Cell Death Differ.* **6,** 943–951.

Lipton, S. A., Choi, Y. B., Pan, Z. H., Lei, S. Z., Chen, H. S., Sucher, N. J., Loscalzo, J., Singel, D. J., and Stamler, J. S. (1993). A redox- based mechanism for the neuroprotective and neurodestructive effect if nitric oxide and related nitroso-compounds. *Nature* **364,** 626–632.

Lipton, S. A., Rayudu, P. V., Choi, Y. B., Sucher, N. J., and Chen, H. S. (1998). Redox modulation of the NMDA receptor by NO-related species. *Progr. Brain Res.* **118,** 73–82.

Lipton, S. A., Singel, D. J., and Stamler, J. S. (1994). Neuroprotective and neurodestructive effects of nitric oxide and redox congeners. *Ann. N. Y. Acad. Sci.* **738,** 382–387.

Lo, E. H., Dalkara, T., and Moskowitz, M. A. (2003). Mechanism, challenges and opportunities in stroke. *Nat. Rev. Neurosci.* **4,** 399–415.

Longa, E. Z., Weinstein, P. R., Carlson, S., and Cummins, R. (1989). Reversible middle cerebral artery occlusion without craniotomy in rats. *Stroke* **20,** 84–91.

Lopez Diazguerrero, N. E., Lopez Araiza, H., CondePerezprina, J. C., Bucio, L., Cardenas Aguayo, M. C., Ventura, J. L., Covarrubias, L., GutierrezRuiz, M. C., Zentella, A., and Konigsberg, M. (2006). Bcl-2 protects against oxidative stress while inducing premature senescence. *Free Radic. Biol. Med.* **40,** 1161–1169.

MacManus, J. P., and Linnik, M. D. (1997). Gene expression induced by cerebral ischemia: An apopototic perspective. *J. Cereb. Blood Flow Metab.* **17,** 815–832.

Manzoni, O., Prezeau, L., Marin, P., Deshager, S., Bockaert, J., and Fagni, L. (1992). Nitric Oxide-induced blockade of NMDA receptors. *Neuron* **8,** 653–662.

Martinou, J. C., Dubois-Dauphin, M., Staple, J. K., Rodriguez, I., Frankowski, H., Missotten, M., Albertini, P., Talabot, D., Catsicas, S., Pietra, C., and Huarte, J. (1994). Overexpression of BLC-2 in transgenic mice protects neurons from naturally occurring cell death and experimental ischemia. *Neuron* **13,** 1017–1030.

Razavi, H. M., Hamilton, J. A., and Feng, Q. (2005). Modulation of apoptosis by nitric oxide: Implications in myocardial ischemia and heart failure. *Pharmacol. Ther.* **106,** 147–162.

Saitoh, Y., Ouchida, R., Kayasuga, A., and Miwa, N. (2003). Anti-apoptotic defense of Bcl-2 gene against hydroperoxide-induced cytotoxicity together with suppressed lipid peroxidation, enhanced ascorbate uptake, and upregulated Bcl-2 protein. *J. Cell. Biochem.* **89,** 321–334.

Salom, J. B., Centino, J. M., Torregrosa, G., and Alborch, E. (2000). Reduction of infarct size by the NO donors sodium nitroprusside and spermine/NO after transient focal cerebral ischemia in rats. *Brain Res.* **865,** 149–156.

Shimazaki, K., Ishida, A., and Kawai, N. (1994). Increase in Bcl-2 oncoprotein and the tolerance to ischemia-induced neuronal death in the gerbil hippocampus. *Neurosci. Res.* **20,** 95–99.

Sokolowska, M., and Wlodek, L. (2003). Influence of N-acetylcysteine on bioactivation of nitroglycerin to nitrox oxide and S-nitrosothiols in the liver and brain of mice. *Polish J. Pharmacol.* **55,** 401–408.

Tassorelli, C., Blandini, F., Greco, R., and Nappi, G. (2004). Nitroglycerin enhances cGMP expression in specific neuronal and cerebrovascular structures of the rat brain. *J. Chem. Neuroanat.* **27,** 23–32.

Won, C. K., Fim, M. O., and Hoh, P. O. (2006). Estrogen modulates Bcl-2 family proteins in ischemic brain injury. *J. Vet. Med. Sci.* **68,** 277–280.

Zhang, F., and Iadecola, C. (1993). Nitroprusside improves blood flow and reduces brain damage after focal ischemia. *Neuroreport* **4,** 559–562.

Zhang, F., and Iadecola, C. (1994). Reduction of focal cerebral ischemia damage by delayed treatment with nitric oxide donors. *J. Cereb. Blood Flow Metab.* **14,** 574–580.

Zhang, F., White, J. G., and Iadecola, C. (1994). Nitric oxide donors increase blood flow and reduce brain damage in focal ischemia: Evidence that nitric oxide is beneficial in the early stages of cerebral ischemia. *J. Cereb. Blood. Flow Metab.* **14,** 217–226.

Zhong, L. T., Sarafian, T., Kane, D. J., Charles, A. C., Mah, S. P., Edwards, R. H., and Bredesen, D. E. (1993). Bcl-2 inhibits death of central neural cells induced by multiple agents. *Proc. Natl. Acad. Sci. USA* **90,** 4533–4537.

INDEX

I

J

K

R

CONTENTS OF RECENT VOLUMES

Volume 39

Volume 40

Volume 41

Volume 42

Volume 43

Volume 44

Volume 45

Volume 46

Volume 47

Volume 48

Volume 49

Volume 50

Volume 51

Volume 52

Volume 53

Volume 54

Volume 55

Volume 56

Volume 57

Volume 58

Volume 59

Volume 60

Volume 61

Volume 62

Volume 63

Volume 67

Volume 68

Volume 69

Volume 70

Volume 71

Volume 72

Volume 73

Volume 74

Volume 75

Volume 76

Volume 77

Volume 78

Volume 79

Volume 81

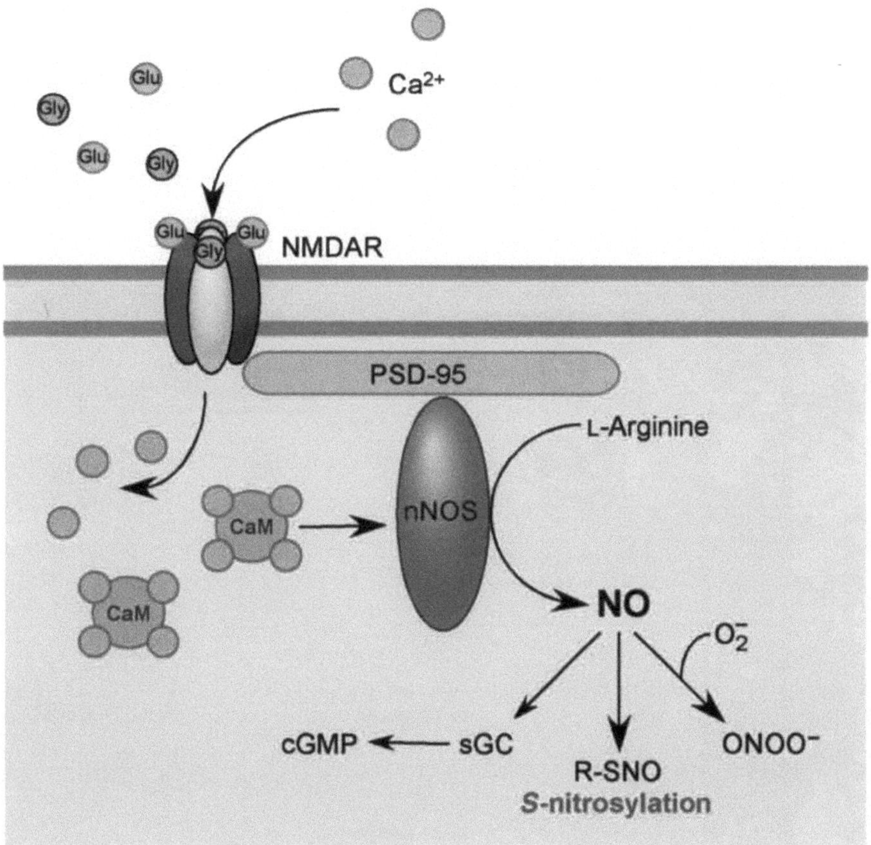

LIPTON *ET AL.*, CHAPTER 1, FIG. 1. (See Legend in Text.)

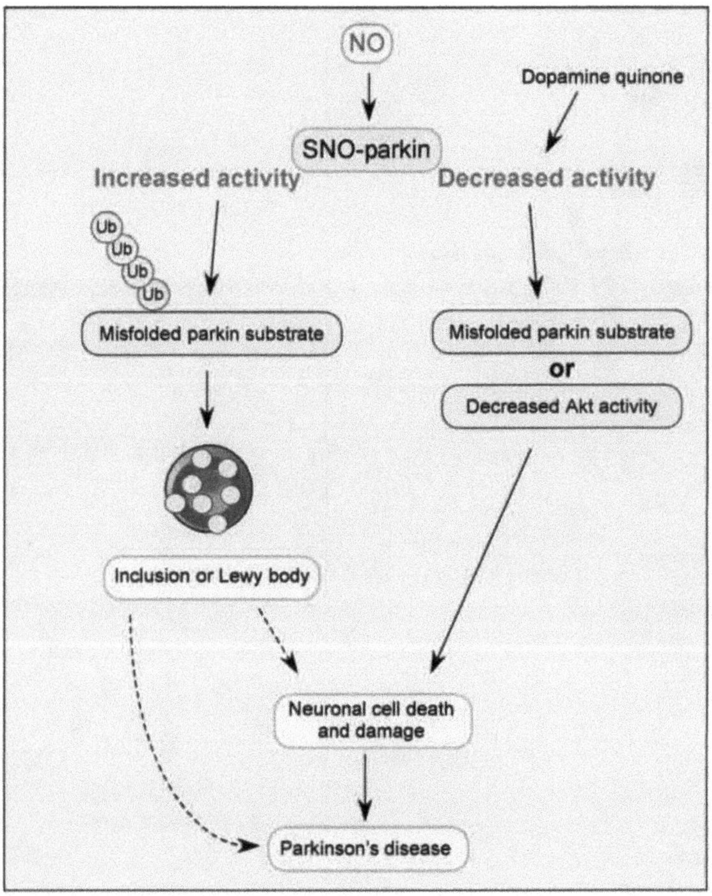

LIPTON *ET AL.*, CHAPTER 1, FIG. 2. (See Legend in Text.)

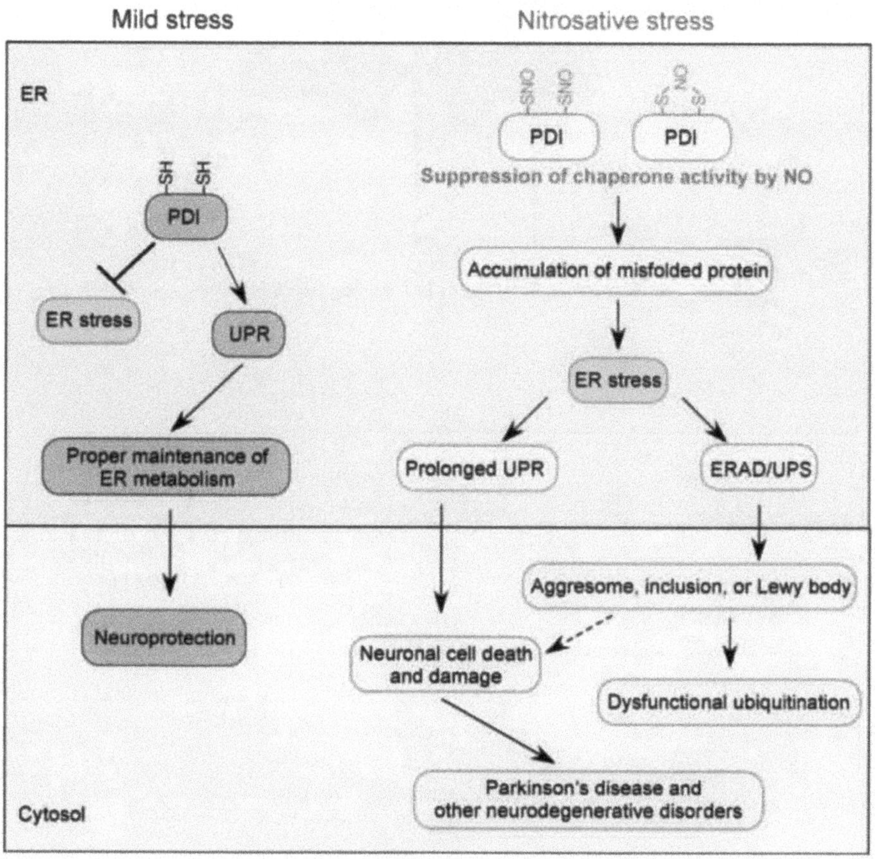

LIPTON *ET AL.*, CHAPTER 1, FIG. 3. (See Legend in Text.)

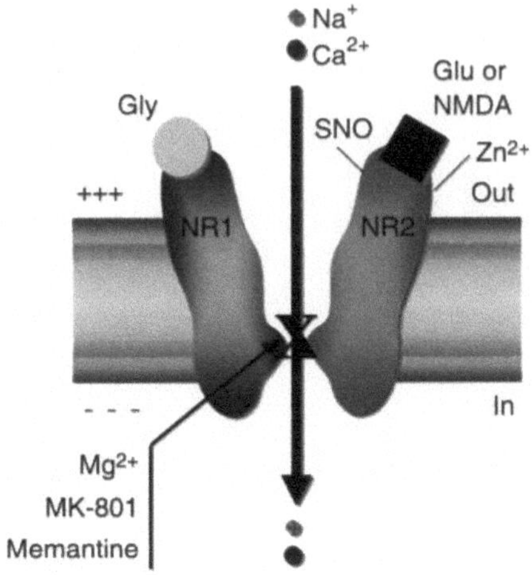

LIPTON *ET AL.*, CHAPTER 1, FIG. 4. (See Legend in Text.)

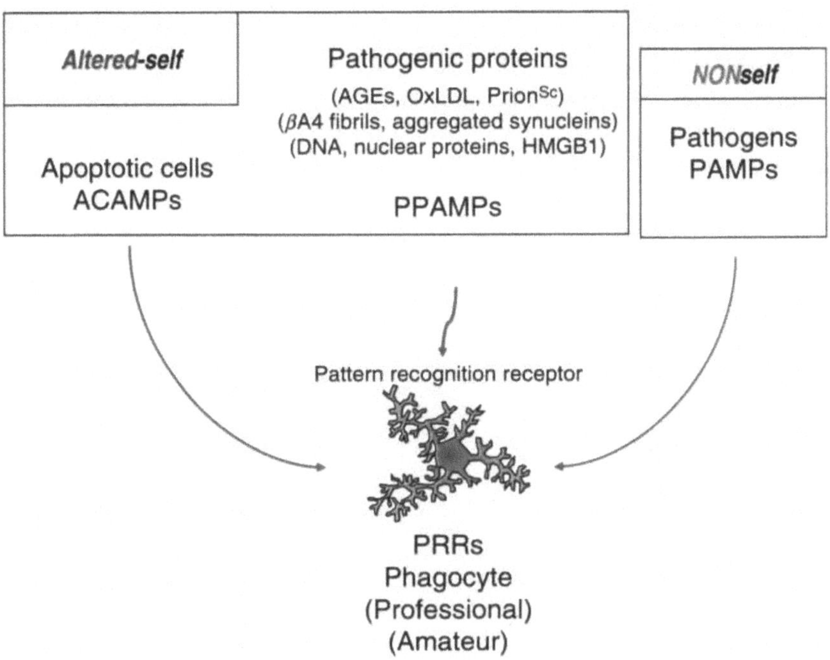

GRIFFITHS *ET AL.*, CHAPTER 2, FIG. 1. (See Legend in Text.)

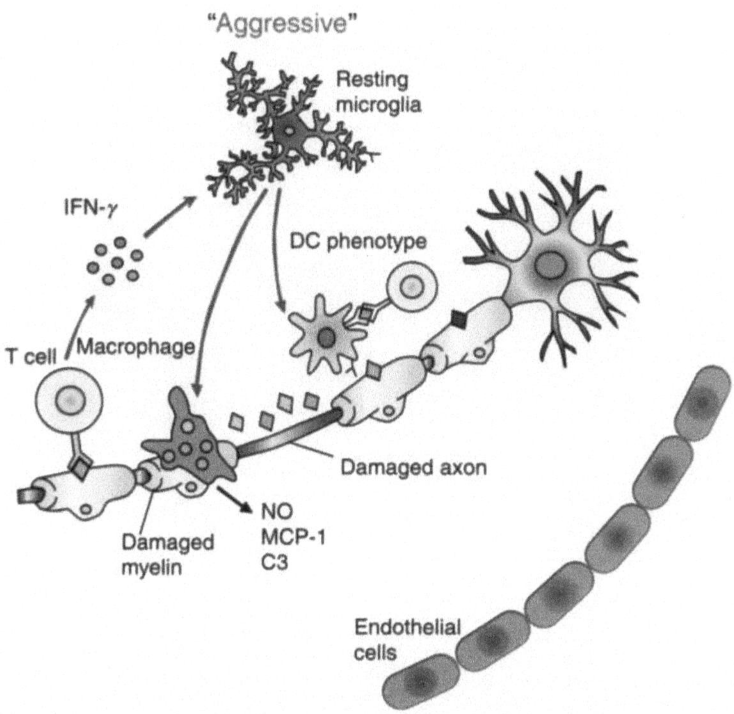

GRIFFITHS *ET AL.*, CHAPTER 2, FIG. 2. (See Legend in Text.)

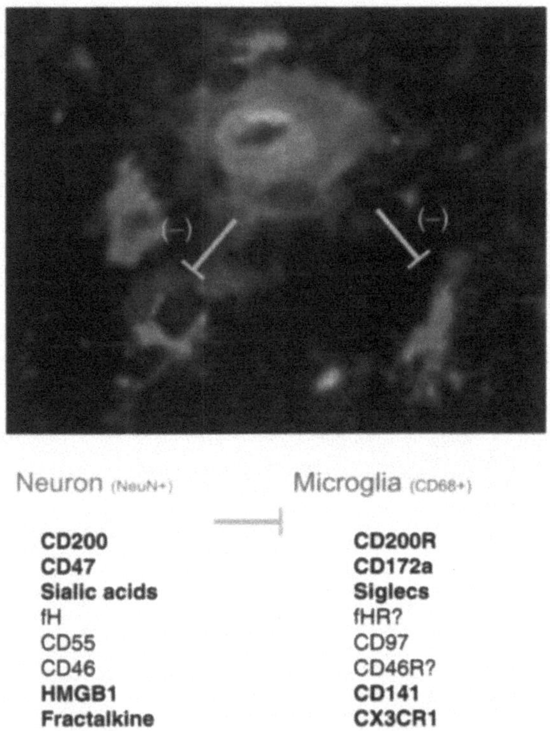

GRIFFITHS *ET AL.*, CHAPTER 2, FIG. 3. (See Legend in Text.)

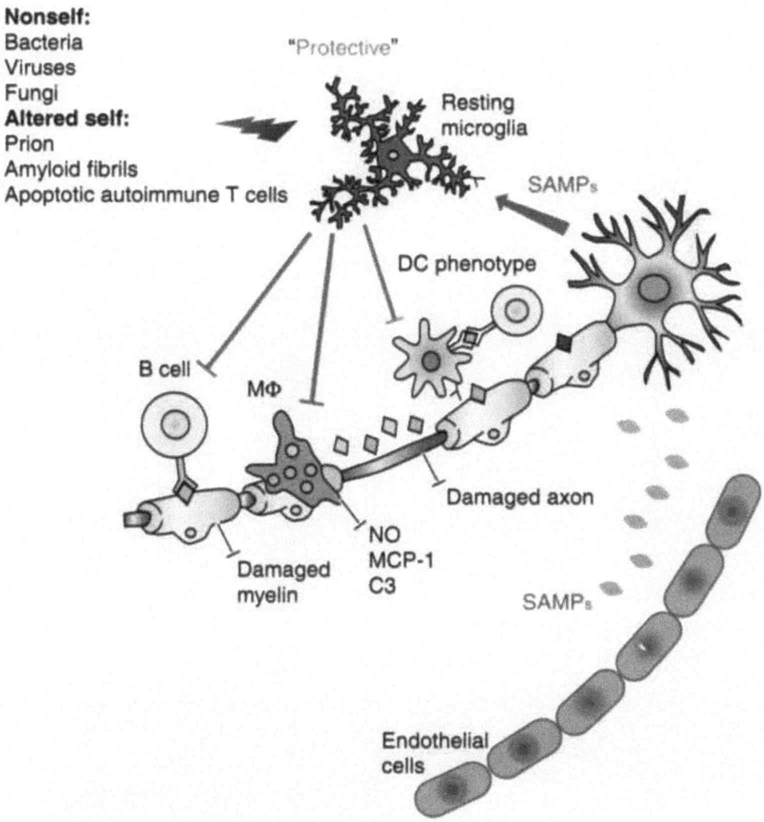

GRIFFITHS *ET AL.*, CHAPTER 2, FIG. 4. (See Legend in Text.)

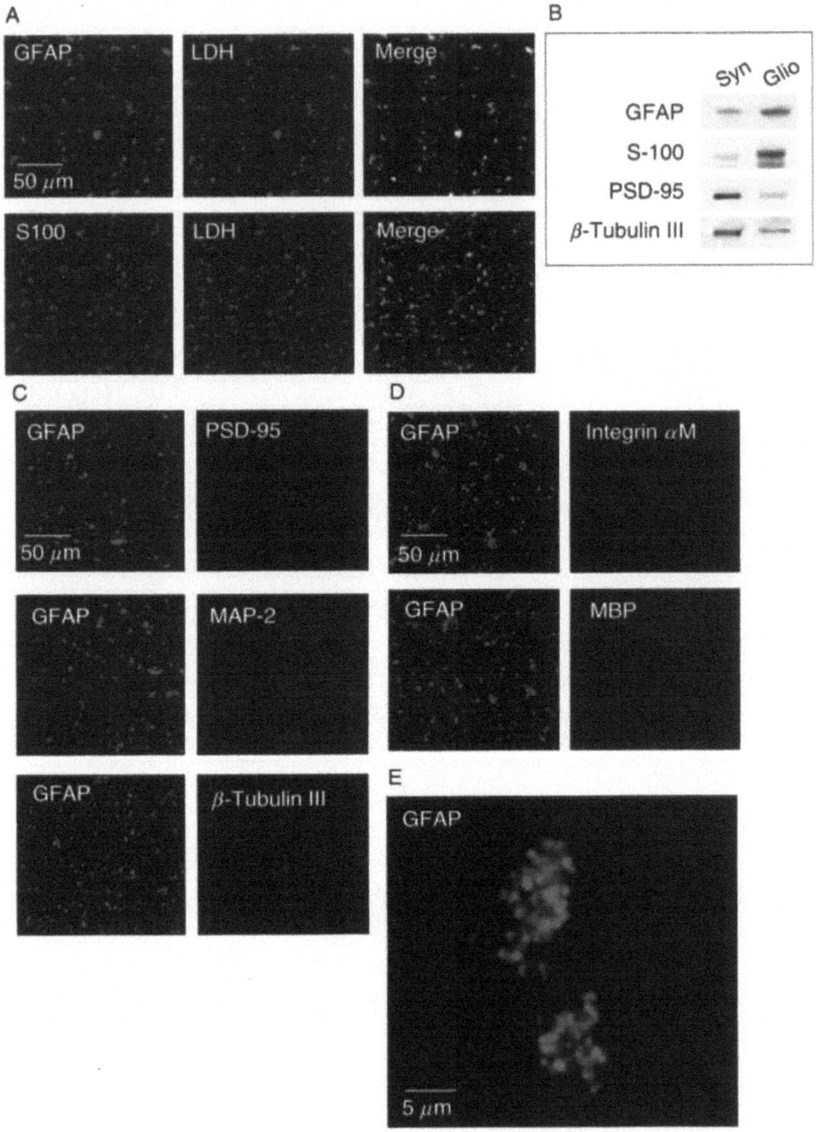

BONANNO *ET AL.*, CHAPTER 4, FIG. 1. (See Legend in Text.)

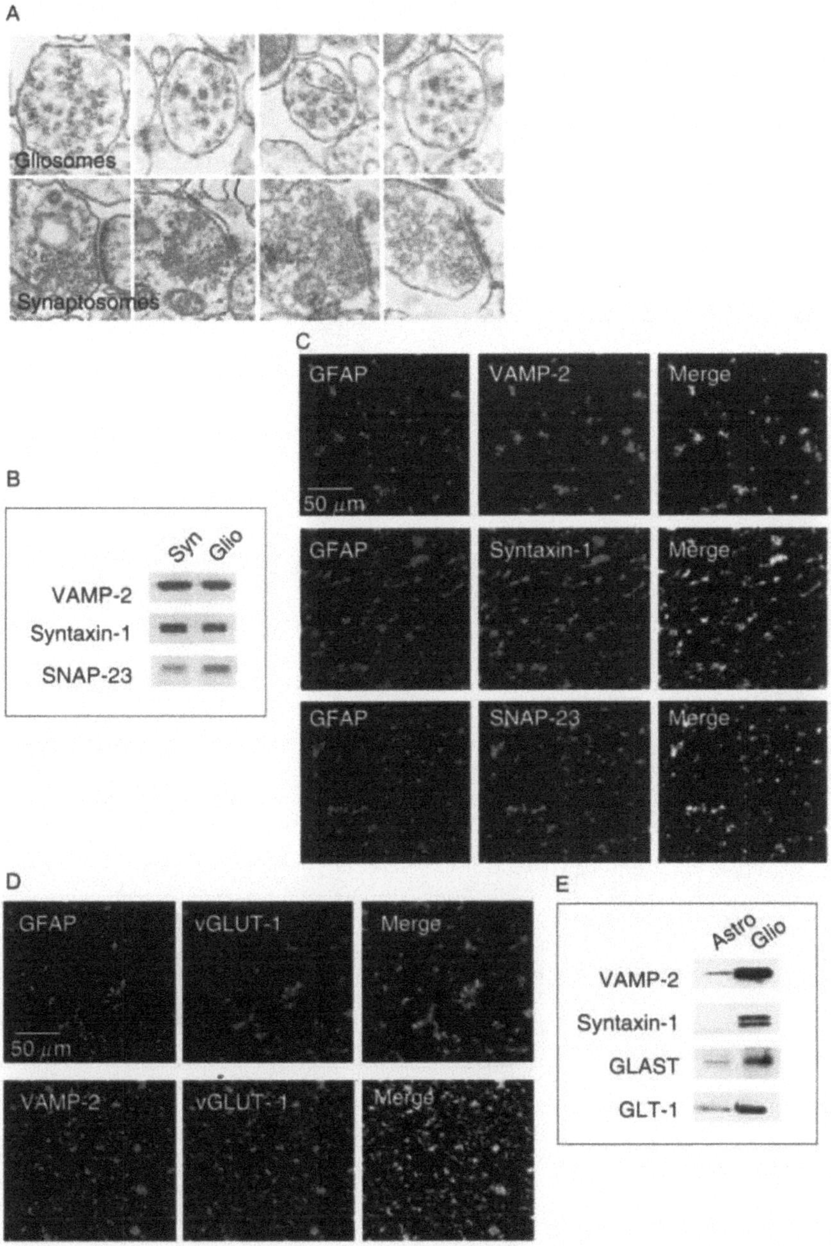

Bonanno *et al.*, Chapter 4, Fig. 2. (See Legend in Text.)

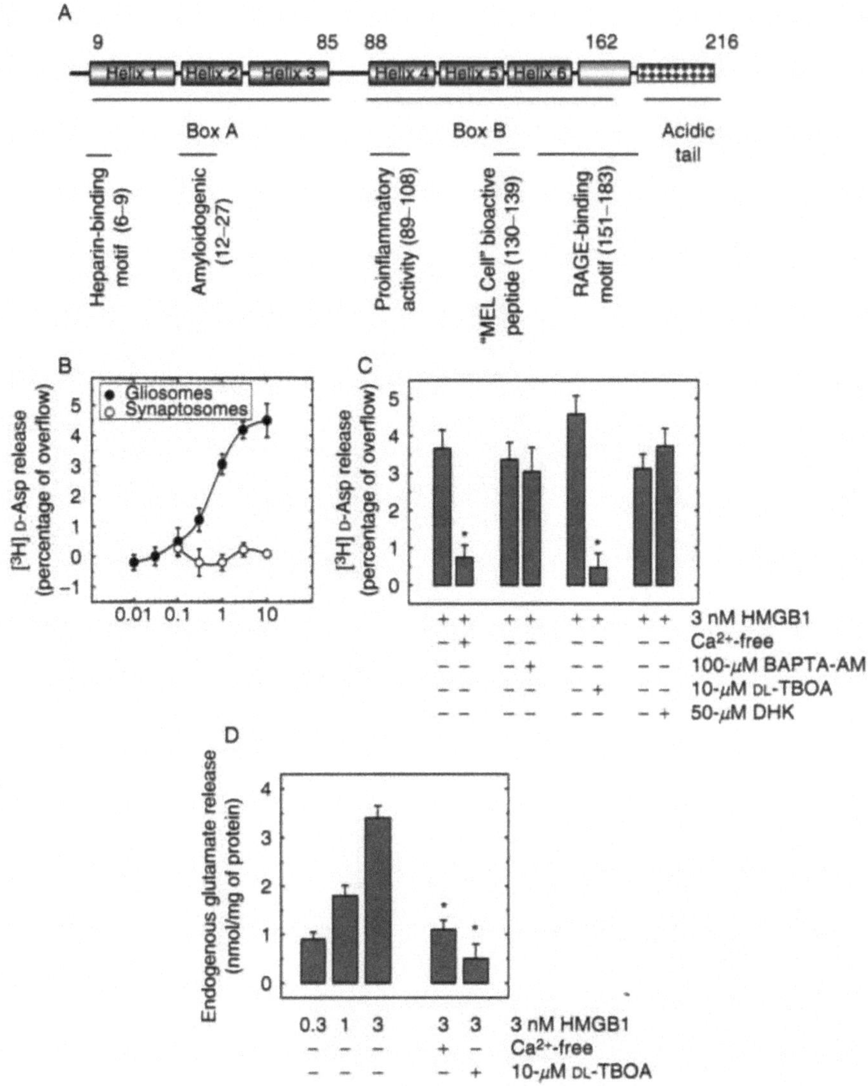

BONANNO *ET AL.*, CHAPTER 4, FIG. 3. (See Legend in Text.)

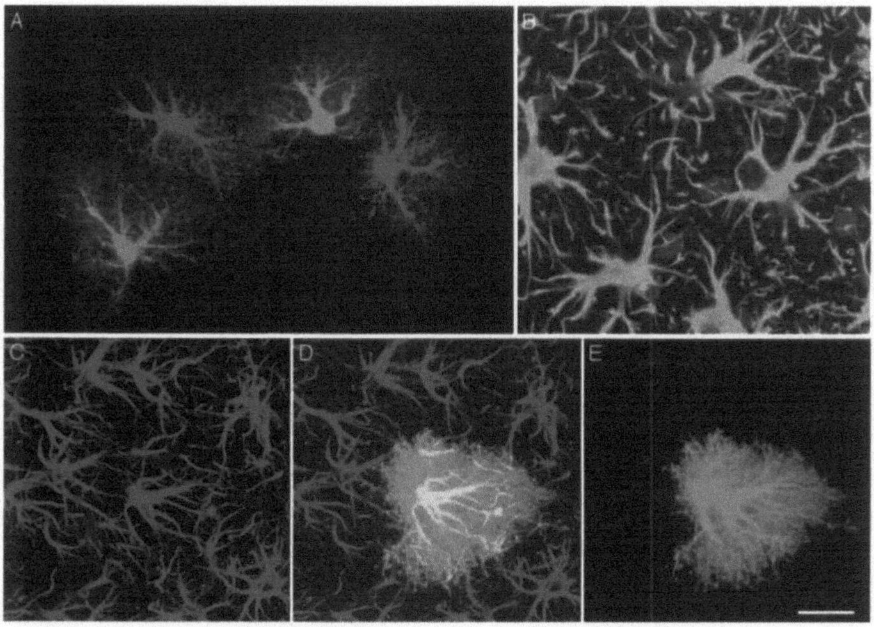

Pekny *et al.*, Chapter 5, Fig. 1. (See Legend in Text.)

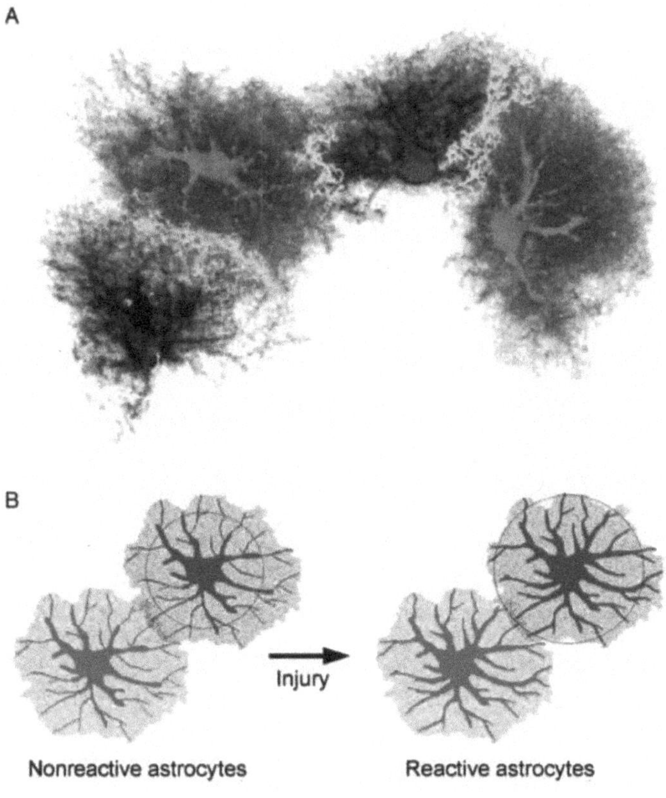

PEKNY *ET AL.*, CHAPTER 5, FIG. 2. (See Legend in Text.)

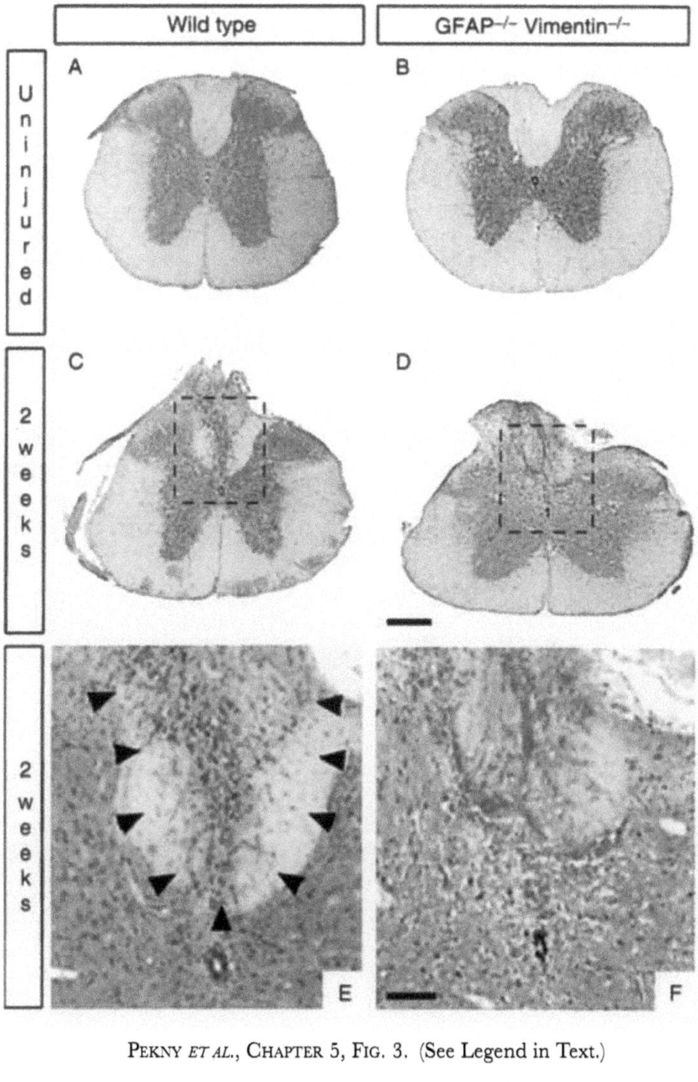

PEKNY *ET AL.*, CHAPTER 5, FIG. 3. (See Legend in Text.)

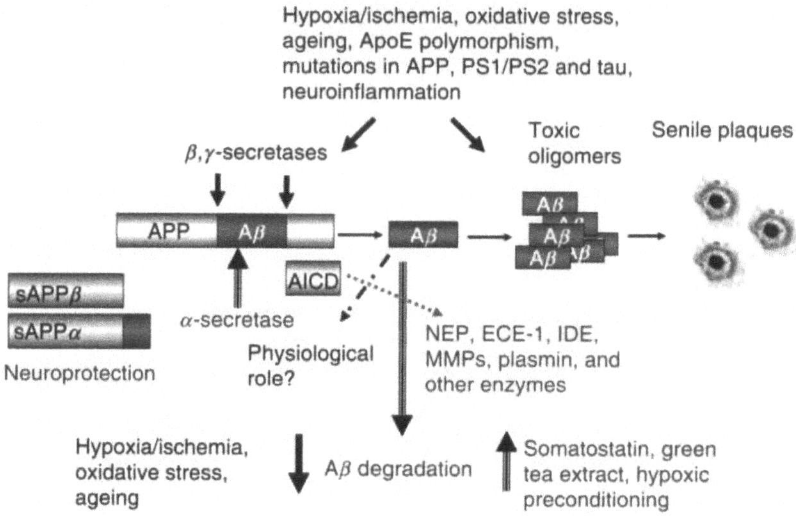

Turner and Nalivaeva, Chapter 6, Fig. 2. (See Legend in Text.)

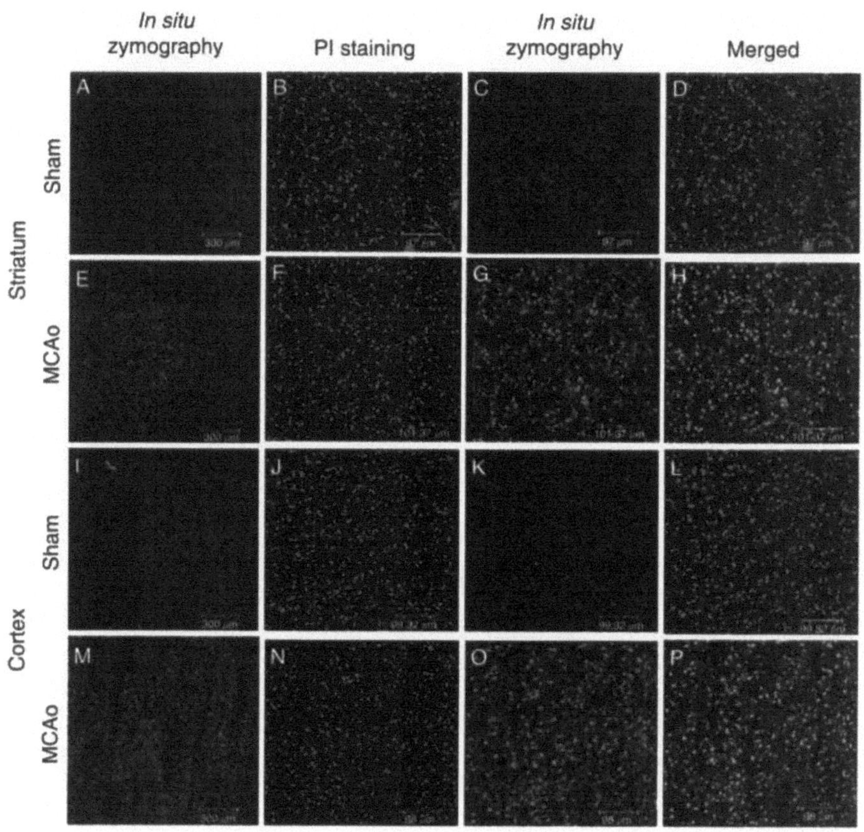

Amantea *et al.*, Chapter 8, Fig. 1. (See Legend in Text.)

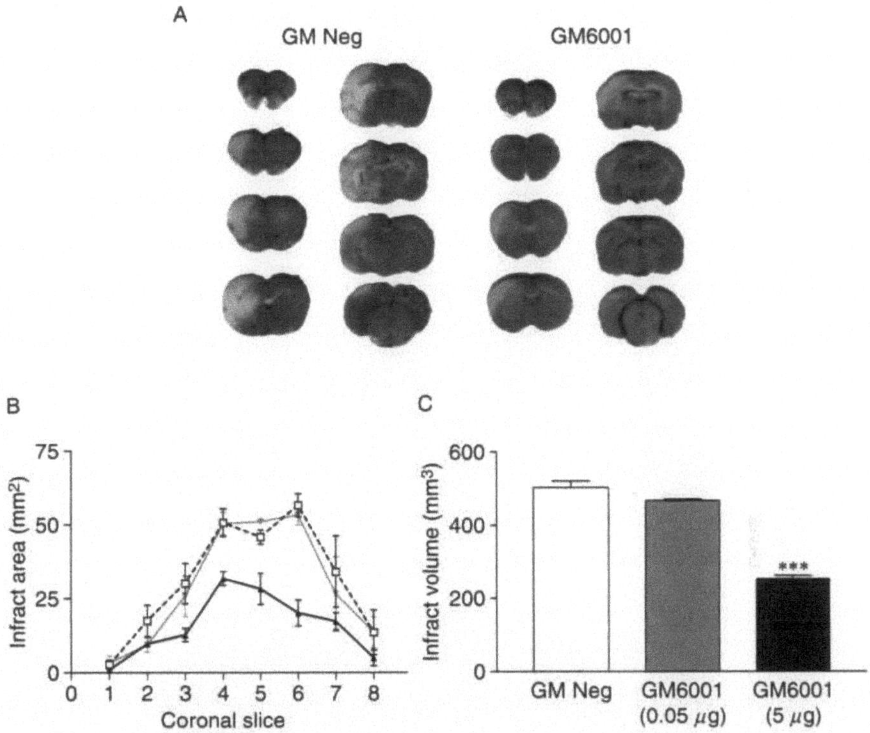

AMANTEA *ET AL.*, CHAPTER 8, FIG. 4. (See Legend in Text.)

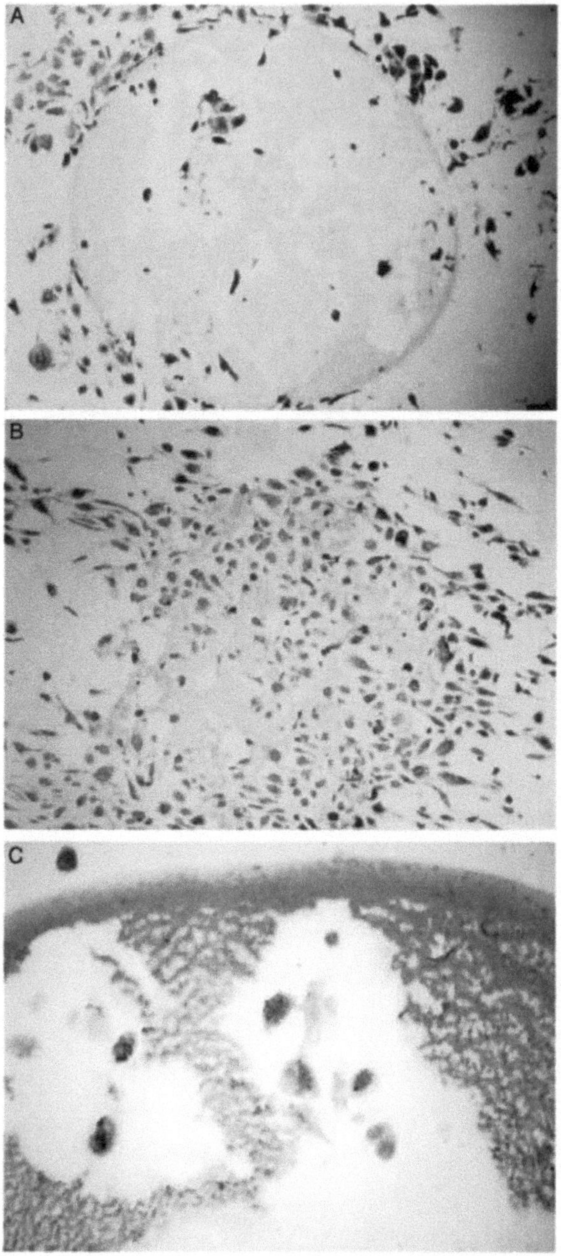

ROGERS *ET AL.*, CHAPTER 12, FIG. 1. (See Legend in Text.)

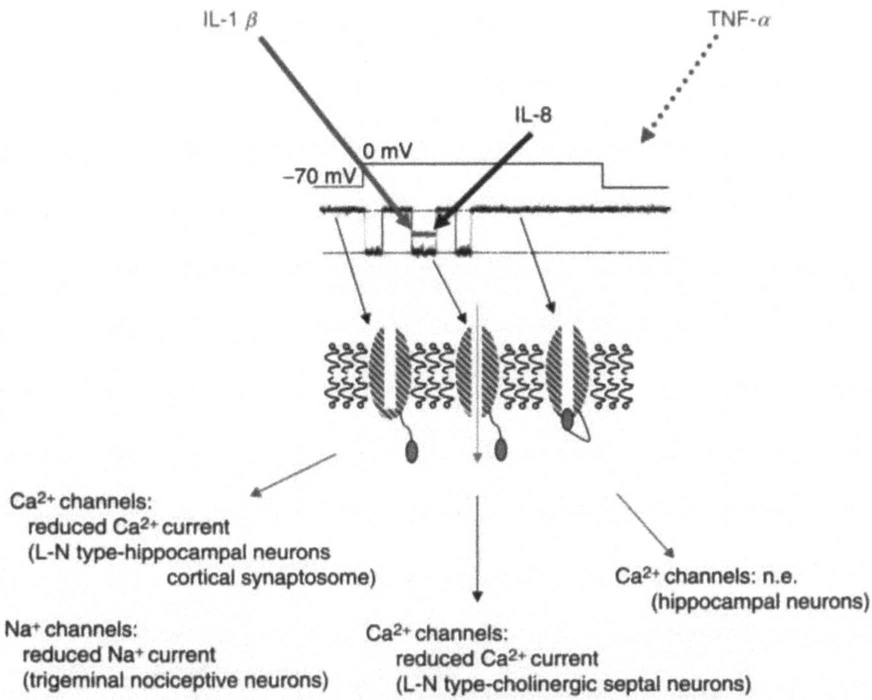

VIVIANI *ET AL.*, CHAPTER 13, FIG. 1. (See Legend in Text.)

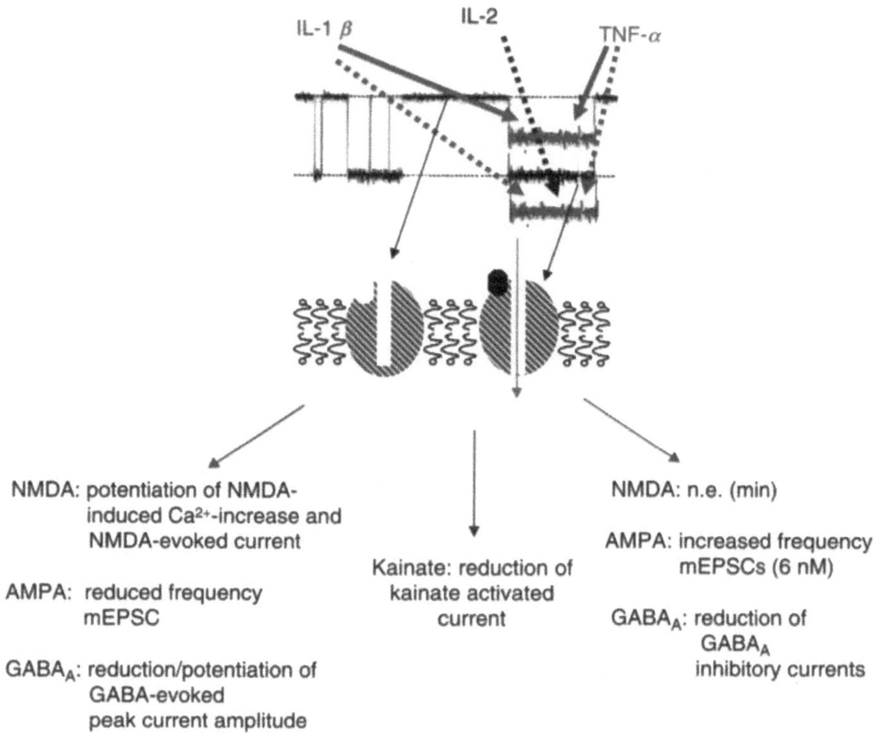

VIVIANI *ET AL.*, CHAPTER 13, FIG. 2. (See Legend in Text.)

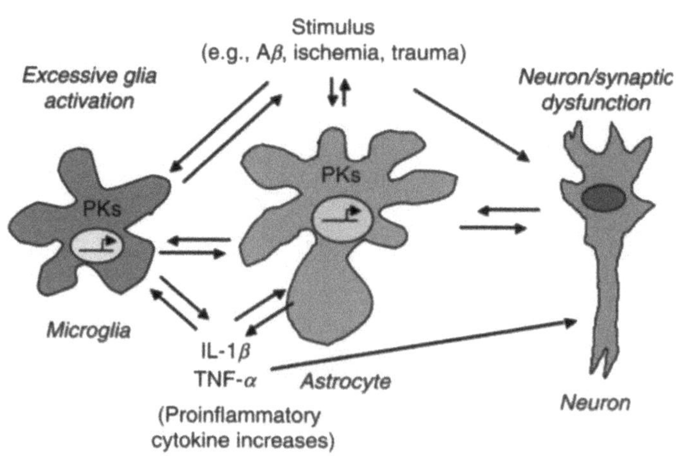

VAN ELDIK *ET AL.*, CHAPTER 15, FIG. 1. (See Legend in Text.)

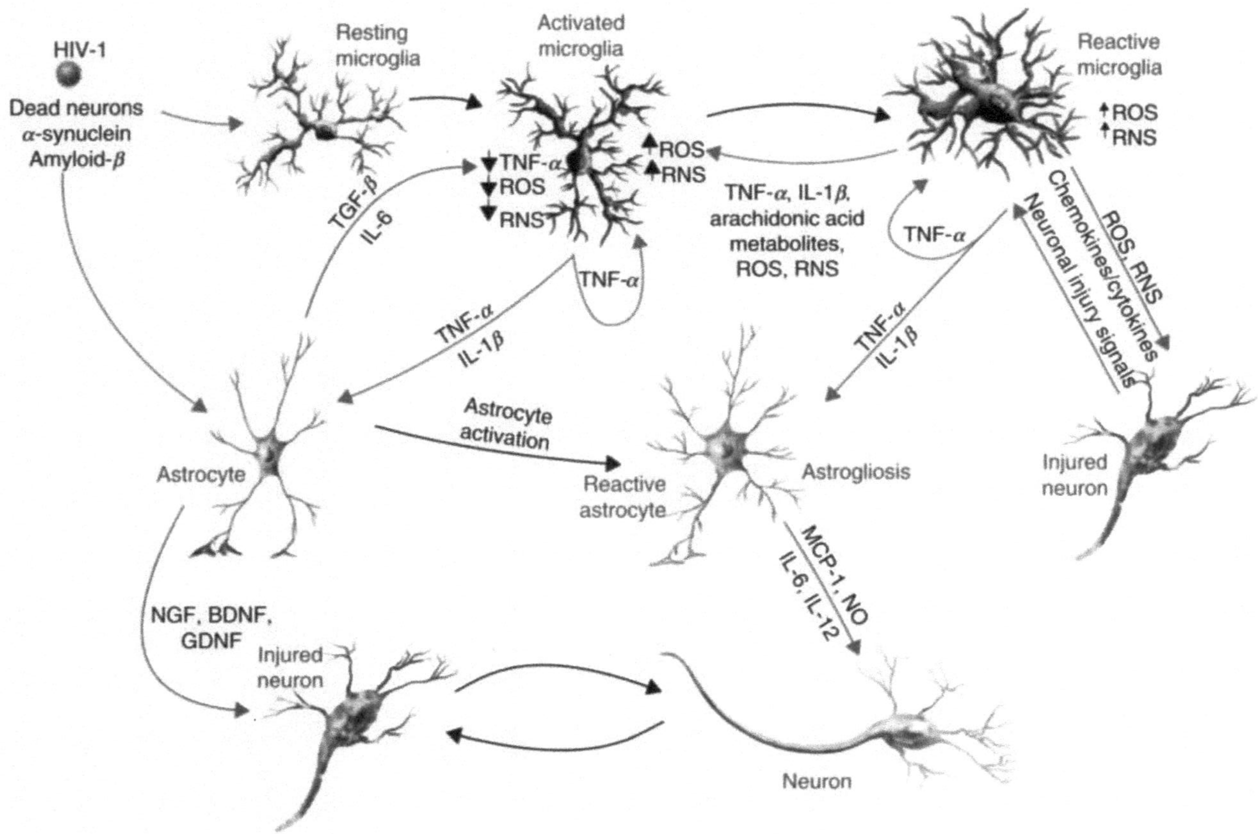

REYNOLDS *ET AL.*, CHAPTER 16, FIG. 1. (See Legend in Text.)

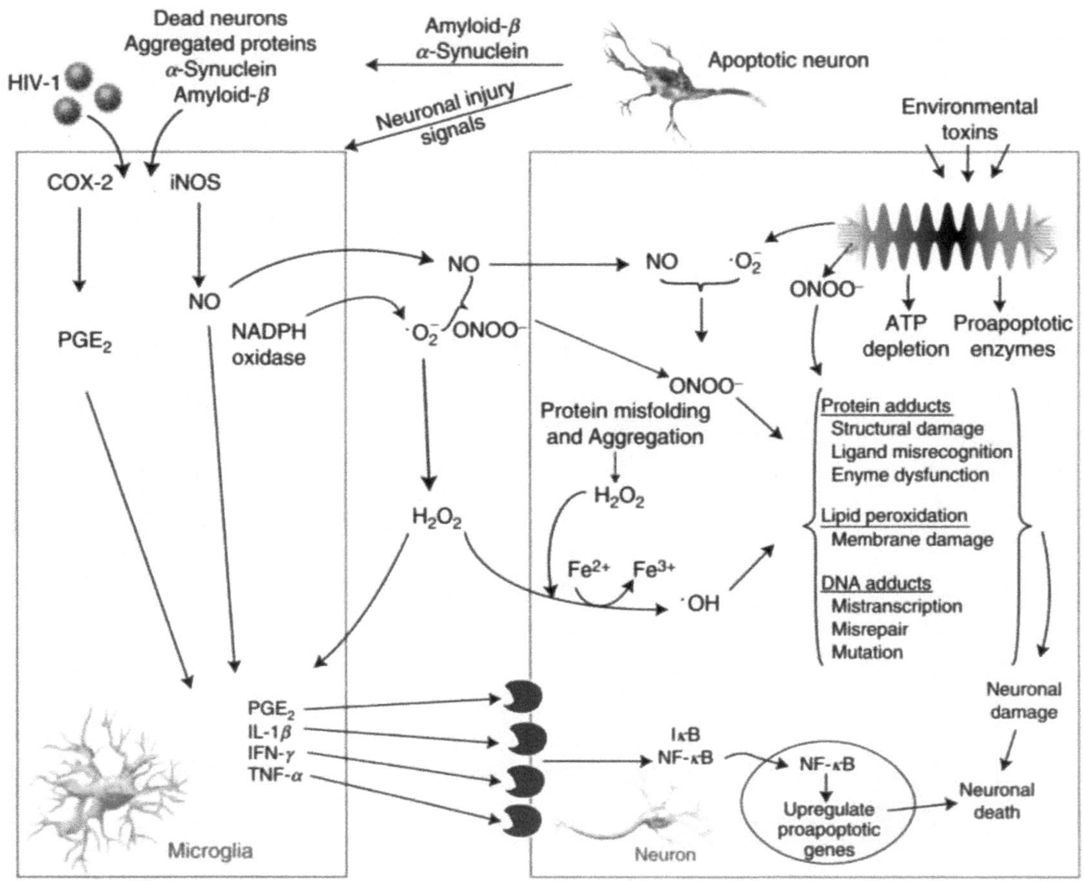

REYNOLDS *ET AL.*, CHAPTER 16, FIG. 2. (See Legend in Text.)

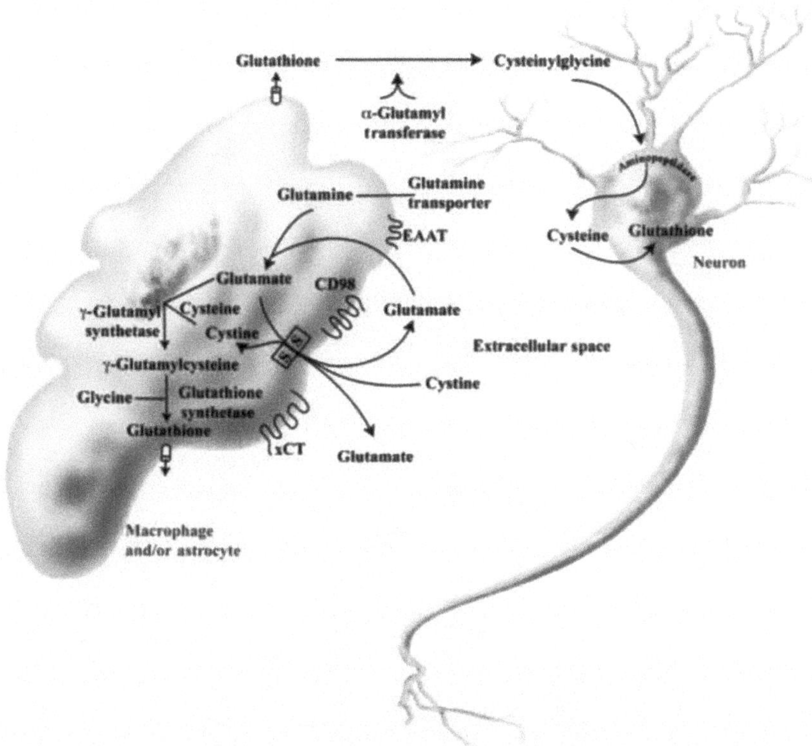

REYNOLDS *ET AL.*, CHAPTER 16, FIG. 3. (See Legend in Text.)

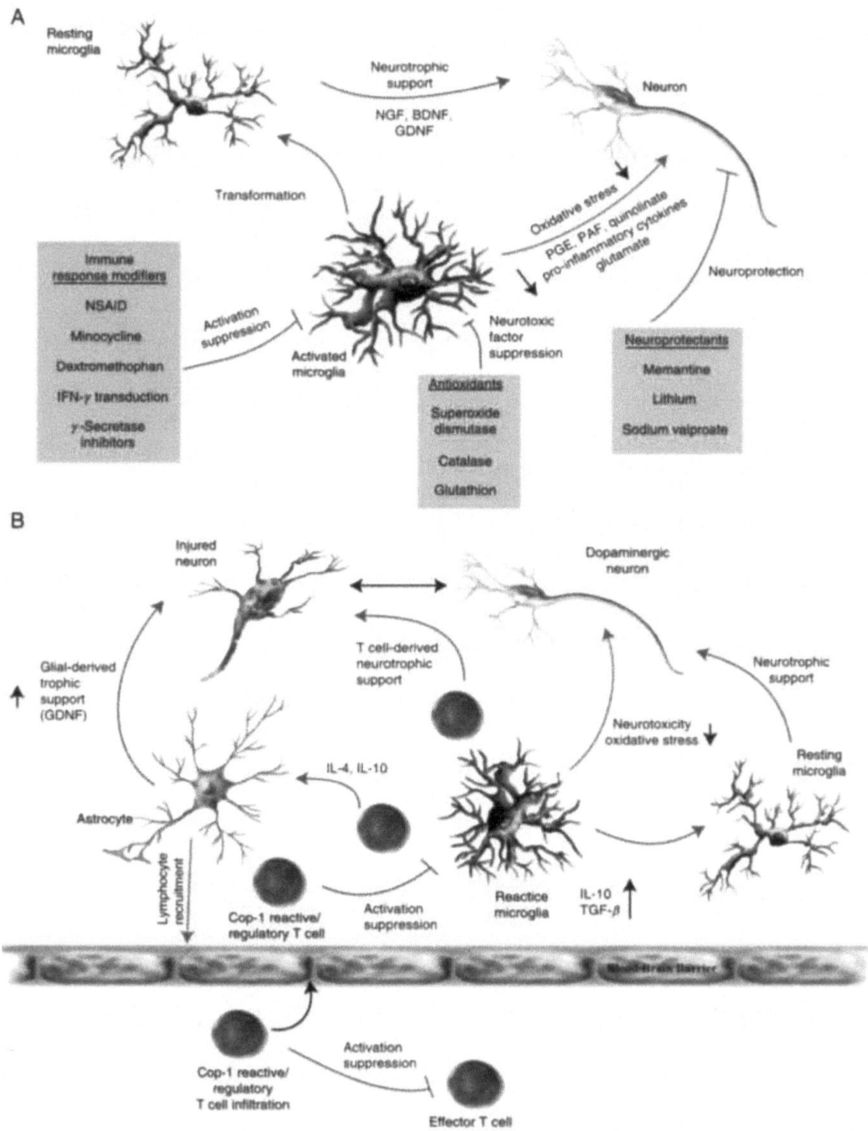

REYNOLDS *ET AL.*, CHAPTER 16, FIG. 4. (See Legend in Text.)

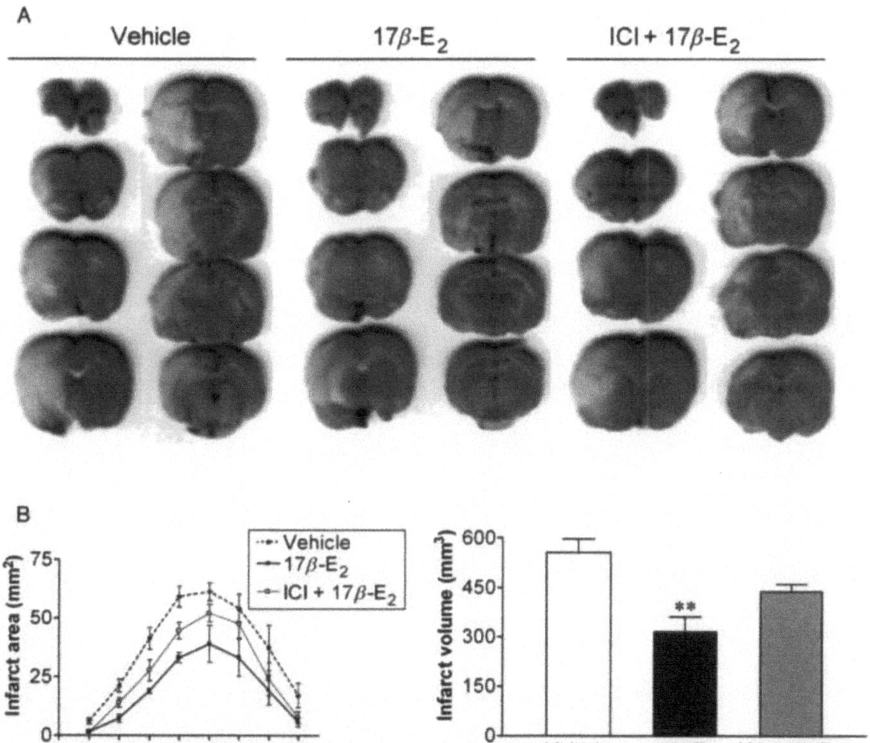

CHIAPPETTA *ET AL.*, CHAPTER 19, FIG. 1. (See Legend in Text.)

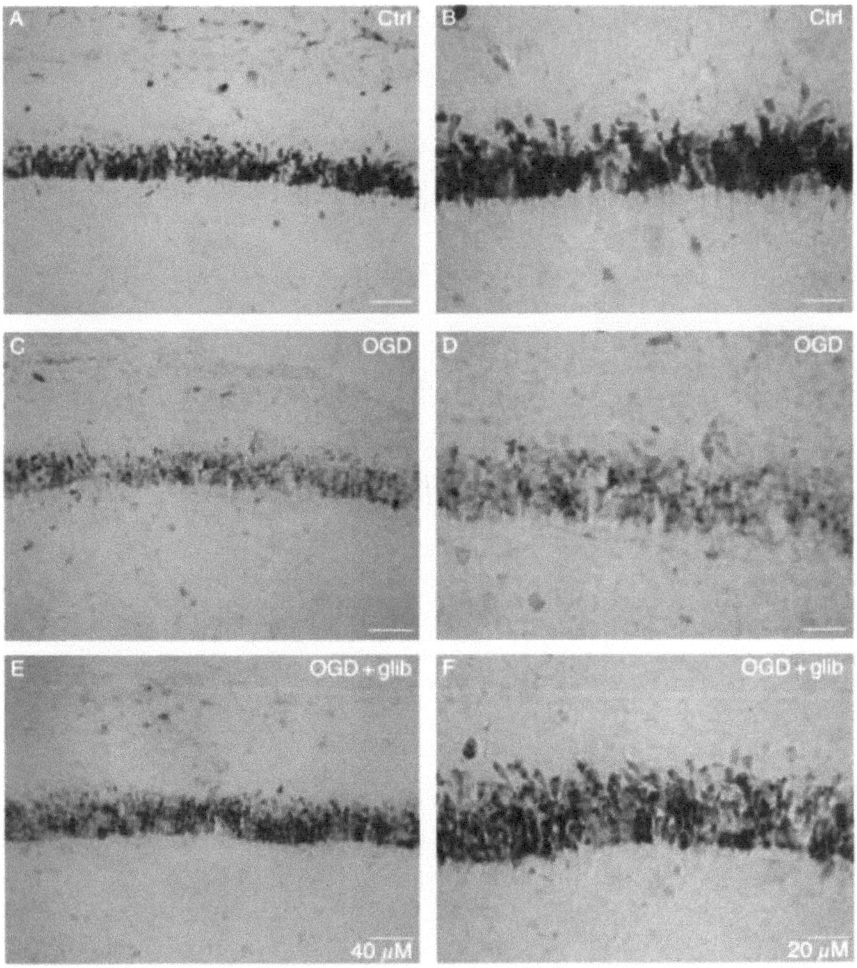

NISTICÒ *ET AL.*, CHAPTER 21, FIG. 4. (See Legend in Text.)